'*Chasing the Sun* is both a grand history of civilization and an irresistible account of an around-the-world odyssey in search of an elusive moving target. Richard Cohen collects fascinating stories with the exuberance and erudition of a Victorian explorer filling a curio cabinet with rare specimens. An amazing tour de force'
Sylvia Nasar, author of *A Beautiful Mind*

'[A] quite extraordinary book, which I absolutely loved. I found it impossible to read in a few sittings, but sort of sat around regularly sun-bathing in its information afterglow. It's a dazzling solar encyclopedia; but also a fabulously provoking history of discoveries, dreams and delusions. I shall bask in its shimmering digressions, crazy cross-references, and dizzy overviews for many moons'
Richard Holmes, author of *The Age of Wonder*

'*Chasing the Sun* is a warming book for short winter days, blending myth with history and science, and guaranteed to please and fascinate almost any reader'
Hilary Mantel, author of *Wolf Hall*

'On every single page there is something utterly fascinating. It's a vast and all-encompassing delight'
Philip Pullman, author of 'His Dark Materials'

'Brave and fascinating . . . Cohen writes with great exuberance'
Boris Johnson, *Mail on Sunday*

'The sun is a big subject, and Richard Cohen does it wonderful justice in this encyclopedic compendium that sees him visit dozens of countries, read hundreds of books and consider our star from every conceivable perspective – and some that are barely conceivable'
Spectator 'Books of the Year'

Selected as one of the *Daily Telegraph*'s top ten books of 2010

'A glorious literary orbit around the sun. This extraordinary book is informed by the most prodigious reading, and the most fabulous detail. His chapters on the sun in literature and art are mesmerising, with perceptive and unexpected observations . . . At every point, the footnotes, asides and interrogations throw up the unexpected and the delightful . . . Some will call it eccentric in its discursiveness: a better word would be concentric. The journey is unpredictable, but always at a proper distance from its subject. It's a glorious literary walk in the sun'
Tim Radford, *Guardian*

'Sprinkled throughout with . . . glittery delights'
Wall Street Journal

'A book of wonders . . . Displaying a brilliant surface sheen of solar materialism, myth, metaphysics and metaphor, the intention and achievement of this book is to irradiate the dazzled reader'
The Times

'A pleasure . . . Cohen writes with charm and lucidity, casting (sun)light on many obscure and unexpected corners. His publishers have done him proud too: the book is handsomely designed and generously illustrated . . . impressive'
Neville Hawcock, *Financial Times*

'Unfailingly riveting'
Boston Globe

'Ambitious but rewarding . . . a rich seam of quirky details'
Lonely Planet Magazine

'A remarkably comprehensive and engrossing synthesis of the sun's influence on science, art, religion, literature, mythology and politics. Former publishing executive Cohen will not be confined in this study of all things that have been touched by the sun. After more than eight years of research and visits to nearly 20 countries, he ranges about in gleeful fascination, marching through the rich, ancient history of stargazing, from the Babylonian and Egyptian astronomers to Pythagoras, Aristotle, Chinese cosmology, Copernicus, Galileo and Newton. The author discusses the story of sunspots, the art of navigation and the art of nudism, the breathtaking presence of Stonehenge, medicine wheels and the Dancing Stones of Namoratunga in Kenya. He traces the evolution of the calendar, the advent of solar power and the daystar's effect on the writing of Euripides, Shakespeare, Nabokov and Styron, and the paintings of Turner and Hockney – though these just touch the surface of Cohen's breadth and depth. "Sometimes it is the direct subject of their creations," he writes, "sometimes a symbol of what they have wanted to convey, infusing their work with an authority, even majesty, that no other force could match." Finally, the author turns to the death of the sun, a brilliant story unto itself. After billions of years of evolution, in its "final stage" it will "simply dwindle away, a dark nuclear waste drifting in the vacuum, its life-giving journey done." Ever enthusiastic, Cohen provides illuminating personal anecdotes, but he includes just the right amount of detail, never allowing the material to sprawl untethered. Apollo, Ra, Inti or Huitzilopochtli – all would rock with delight at Cohen's sweeping endeavor'
Kirkus Reviews (starred review)

'[A] spirited panoramic survey . . . Cohen is cheerfully digressive, covering topics such as nudism, quantum theory, albinos, the rise of khaki uniforms, vampires, seasonal affective disorder, life at the bottom of the sea, and the curiously worded abdication of divinity by the emperor of Japan after the Second World War . . . Embedded in the footnotes are some of the book's most memorable details'
The New Yorker

'Not just grandly informative but peculiarly charming, a book nobody else would have had the chutzpah to write'
David Sexton, *Spectator*

'Cohen is an endurance scholar, unstoppably curious and gifted, with both lucidity and elegance of expression . . . It would be impossible for anyone to emerge unenlightened from this dazzling book. The footnotes alone are worth the price'
Peter Temple, *Melbourne Age*

'Tells non-scientists all they might want to know about the sun, and scientists much that they probably did not know. Richard Cohen supplies plenty of dizzying figures – the sun is 32,000 light years from the centre of its galaxy of a hundred billion stars, and so on – but also considers our star from such perspectives as the mythological, literary and musical. There is an excellent chapter on sunbathing, an activity which was banned in Bournemouth until the early 1930s. Endlessly informative and diverting, *Chasing the Sun* is an encyclopaedic plum pudding of a compendium'
Lewis Jones, *Sydney Morning Herald*

'An unflinching, spellbinding account . . . Cohen's work is prodigious, audacious, and dazzlingly illuminating'
Richard Di Dio, *Philadelphia Enquirer*

'Cohen's charm and erudition keep *Chasing the Sun* consistently lively, and the pages fly by like a dinner with a master raconteur . . . a richly satisfying read . . . Cohen's mission is accomplished. We are left sharing . . . his wonderment at all that modern science has taught us about the sun'
Brian Schofield, *Sunday Times*

'Richard Cohen's exhaustive story of the sun takes us from its beginning to its end, touching on just about everything that has anything to do with the sun in between, from ancient sun worship to modern sunbed worship, the history of astronomy to the future of solar power. Here is the story of the sun in all its splendour'
New Scientist

'To write yet another work about our nearest star is a formidable challenge that Cohen takes up with aplomb. *Chasing the Sun* paints a fascinating and far-reaching scene that incorporates nearly all aspects of solar phenomena . . . The narrative is liberally sprinkled with personal anecdotes, which is largely what makes the book so enjoyable . . . A marvelous read'
Douglas Gough, *Nature*

'Captivating . . . particularly good are the sections that describe how solar science has historically advanced in line with the political climate . . . an enjoyable read'
Sky at Night

'Cohen (*By the Sword*) visited 18 countries to gather information for this ambitious and unusual literary opus, including Peru, where he witnessed the reenactment of an Inca ceremony welcoming the summer solstice, and Japan, where he climbed a snow-covered Mt Fuji. He hunted the mythology embedded in the works of Shakespeare, Nabokov ("I must be the only person to have read *Lolita* for its Sun images"), Dante, Chaucer, and other authors, and personally examined the orientation of the Egyptian Pyramids and European cathedrals ... Cohen was compelled to write "the sort of book I'd like to read", a risky position for a writer seeking a broad readership, but one that more than pays off'
Publishers Weekly (starred review)

'Mighty, lively, thoroughly readable ... Cohen leaves no sunbeam unturned in this dazzling history of our solar system, and epic human efforts to understand it. The science Cohen recounts is mind-boggling'
Metro

'Along the way are fascinating stories and sidetracks, often embedded in footnotes ... Cohen has succeeded in giving back a bit of magic and awe to the star we all take for granted'
Suzi Feay, *The Tablet*

'Everything you could want to know about the scientific and cultural history of the star we orbit'
Daily Express

'Richard Cohen's meaty history of our cultural and scientific relationship with this 4.6 billion year old star is full of fascinating facts and theories without blinding the reader with science. Dazzling'
Hit list, *Metro*

'I love this book! It is so well written and so comprehensive, and I've learned so many things about the Sun that I never knew, or never imagined. The writing is exceptional, as is the author's research, and I simply can't imagine anyone who has read *Chasing the Sun* giving it less than five stars ... It is a true gem, unlike anything previously written about this subject'
Reviewer on Amazon UK (all reviews gave five stars)

'Prepare to be dazzled'
The Week

'Cohen's book is inspirational. It masterfully pulls the mere astronomer away from narrow-mindedness. It should help us share our meagre understandings of our central star with the multitude of other people who appreciate the sun in vastly different ways. Read this book. You will enjoy every page and it will broaden your horizons hugely.'
Astronomy Now

CHASING THE SUN

ALSO BY RICHARD COHEN

By the Sword: A History of Gladiators,
Musketeers, Samurai, Swashbucklers, and
Olympic Champions

CHASING THE SUN

THE EPIC STORY OF THE STAR
THAT GIVES US LIFE

RICHARD COHEN

**SIMON &
SCHUSTER**

London · New York · Sydney · Toronto

A CBS COMPANY

First published in Great Britain by Simon & Schuster UK Ltd, 2010
This paperback edition published by Simon & Schuster UK Ltd, 2011
A CBS COMPANY

1 3 5 7 9 10 8 6 4 2

Simon & Schuster UK Ltd
1st Floor
222 Gray's Inn Road
London WC1X 8HB

www.simonandschuster.co.uk

Simon & Schuster Australia
Sydney

For Toby, Mary and Guy
Stars in their own right

CONTENTS

Preface   XV

Sunrise on Mount Fuji   1

PART ONE THE SUN BEFORE SCIENCE

1   Telling Stories   11
    Solar myths; Max Müller's discoveries; children of the Sun;
    an Inca festival

2   Celebrating the Seasons   24
    Solstices and equinoxes; worshipping through the seasons;
    North American sun dances

3   The Three Thousand Witnesses   40
    The Earth's standing stones; Stonehenge; the Pyramids;
    Newgrange and New York

4   Terrors of the Sky   56
    Auroras and eclipses; their respective histories, from
    2000 BC to Shakespeare and Milton

PART TWO DISCOVERING THE SUN

5   The First Astronomers   75
    The earliest sungazers; the Sumerians; the Babylonians;
    the ancient Egyptians

6   Enter the Greeks   86
From Homer and Hesiod to Pythagoras, Plato and Ptolemy

7   Gifts of the Yellow Emperor   105
*China's separateness; modes of thought: yin and yang,
the 'Way' and the Sun*

8   The Sultan's Turret   121
*Arabia's great astronomers; India and Western Europe,
AD 600–1543*

9   The Earth Moves   139
The great four: Copernicus, Tycho, Kepler and Galileo

10   Strange Seas of Thought   162
*Newton on gravity and colour; the transit of Venus;
Newton's successors to 1850*

11   Eclipses and Enlightenment   181
*From Captain Cook to Einstein's great moment;
Antarctica 2003*

12   The Sun Dethroned   199
*Solar physics 1800–1950; the Copenhagen gang;
making the atomic bomb*

PART THREE   THE SUN ON EARTH

13   Sunspots   221
*Their make-up explained; cycles, flares and eruptions;
their influence on weather*

14   The Qualities of Light   240
Light in war, in sport and in crime; shadows, sunsets and twilight

15   Beneath the Beating Sun   256
*Heat and its effects; skin colour fashions; from Sun avoidance
to sunbathing*

16 Skin Deep   277
    Skin cancers and other solar disorders; how melatonin works;
    sungazing; Seasonal Affective Disorder

17 The Breath of Life   294
    Photosynthesis and its discoverers; plants and the Sun;
    hibernation and migration

18 The Dark Biosphere   312
    Creatures of the deep; currents and tides

PART FOUR HARNESSING THE SUN

19 The Heavenly Guide   329
    Navigation and cartography as aided by the Sun

20 Of Calendars and Dials   344
    Reckoning the years, from before Julius Caesar to after
    Gregory VII

21 How Time Goes By   367
    Calculating the hours: the creation of Standard Time,
    Greenwich Mean Time and Daylight Saving Time

22 The Sun in Our Pocket   385
    Solar energy, from Archimedes' mirrors to solar-powered cars

PART FIVE INSPIRED BY A STAR

23 The Vital Symbol   411
    Flags; the Sun King; the metaphors of gold, blondes and mirrors

24 Drawing on the Sun   426
    The Sun in art, from the Renaissance through Turner,
    Van Gogh, Matisse and Hockney

25 Negative Capabilities   447
    The Sun in photography, film and architecture

26   Talk of the Day   462
     The Sun in classical and pop music, from Mozart to the Beatles

27   Busy Old Fool . . .   475
     Homer and Dante, Chaucer and Donne, Lawrence and Nabokov

28   The Rising Star of Politics   498
     Solar imagery and the Nazis; Japan's divine emperor;
     the Sun as taken up by Mao and Benedict XVI

                    PART SIX THE SUN AND THE FUTURE

29   Over the Horizon   521
     Solar astronomy from 1950 to the present

30   Under the Weather   546
     The ozone layer; the Sun and global warming; will the
     Earth get too hot – or freeze over?

31   The Impossible and Beyond   565
     Science-fiction from ancient Rome, Swift, Verne, H. G. Wells,
     Isaac Asimov and Arthur C. Clarke

32   The Death of the Sun   581
     Solar catastrophes; asteroids and meteors; what will happen
     when the Sun burns out

     Sunset on the Ganges   592
     Afterword to the First Paperback Edition   607
     Acknowledgements   611
     Sources   617
     List of Illustrations and Illustration Credits   665
     Index   671

PREFACE

Once we begin to go back to a closer understanding of nature and man's relationship to the Sun, we're going to start developing whole new concepts of who and what we are, and why, and what our rightful place in the universe really is.

Harold Hay, a pioneer of solar energy, now in his 102nd year[1]

The world is full of obvious things which nobody by any chance ever observes.

Sherlock Holmes, in *The Hound of the Baskervilles*

Volume 4 of the *Encyclopedia of Astronomy and Astrophysics* defines the Sun as 'a main-sequence, yellow dwarf of spectral type G_2 V, mass 1.989×10^{30} kg, diameter 1392000 km, luminosity 3.83×10^{26} W and absolute visual magnitude +4.82'.[2] That is certainly one way of looking at the star – although, as Bertie Wooster put it, not the sort of thing to spring on a lad with a morning head. We describe the Sun in myriad ways and comprehend it so variously too. The Indian sages who wrote the Rig-Veda between the eighth and fourth centuries BC assumed a tone of awed reverence: 'Only when illumining Light shines, /Everything else shines; /The self-revealing Light illumines /the entire universe'; while the philosopher-martyr Giordano Bruno described the Sun with sixteenth-century extravagance: 'Apollo, author of poetry, quiver-bearer, bowman of the powerful arrows, pythian, laurel-crowned, prophetic, shepherd, seer, priest, and physician . . .'

Without the Sun, none of us would exist. Our star, formed out of a vast cloud of hydrogen and stardust, has been active for 4.6 billion years and has enough fuel for a further 100 billion at the present rate, but continuing solar convulsions will shorten its life to a mere 5 billion more.

The Earth famously averages a distance of 93 million miles from it (more accurately, 92, 955, 887.6, but that rounding-up by 44,113 miles seems an irrelevance). The Sun's core burns at a sustained 15 million°C, 27 million°F. To give a sense of perspective: a single photon from the Sun's core takes 150,000 years just to reach the frontier with space. Every second, 5 million metric tonnes of mass are converted into nuclear energy, the equivalent to the detonation of 90 thousand million one-megatonne hydrogen bombs – as scientists express it, 3.8×10^{33} ergs (one erg being roughly the amount of energy a mosquito needs to take off). A constant blast of nuclear reactions pushes energy to the surface, releasing it as light and heat. Yet the Earth receives just 1/2.2 billionth of that output, a figure so startlingly small that 120 years ago it was one of the talking points of science: where was all that power going?

The Sun accounts for 99.8 per cent of the mass of the entire solar system, but as time goes by, while continuing to expand at an accelerating pace, it will inexorably draw in on itself and lose some of that mass (although 1,300,000 Earths could still fit inside it). About 2 billion years from now, a far cooler star, it will widen at great speed into a highly luminous 'red giant', then suddenly collapse into a 'white dwarf' (alas, scientific vocabulary is rarely this poetic), all life on Earth having long since been extinguished. After about a further trillion years, it will have frozen to the ultimate cold.

☼

In the early 1990s, while running a publishing house in London, I decided I wanted to learn more about the Sun (no sudden epiphany; just the realization that I knew so little about what, above all other things, governs our lives), and went looking for an author who would write a wide-ranging account – without success. Five years on, I left England for New York with the idea still in my head, and wondered whether in the intervening period someone had written a book that explained how that great ball of gases exercises its enormous and varied effects, and specifically how it creates and maintains life on Earth in all its diversity. A visit to the New York Public Library revealed 5,836 titles under 'Sun', but none offered both the excitement of scientific discovery – Galileo first detecting blemishes on the Sun, or William Herschel identifying a new planet – and an exploration of the Sun's place in art, religion, literature, mythology, politics. I wanted to know about the great solar temple standing amid the ruins of Machu Picchu; Mozart celebrating the Sun in *The Magic Flute*; Charlie Chaplin mocking Hitler's love of Sun imagery in *The Great Dictator*; and why for 1,500 years the civilized West stuck with the ancient notion of the Sun circling the Earth. But most volumes aimed at a specialized area, and the non-scientists who tackled a particular aspect either looked askance at the science or took minimal interest in it.

Fifty Crows – a total solar eclipse as seen from Chiapas,
the southernmost state of Mexico, in 1991

'There are two worlds,' wrote the essayist and poet Leigh Hunt (1784–1859), 'the world that we can measure with line and rule, and the world that we feel with our hearts and imagination.' Conan Doyle even has Sherlock Holmes assert that he doesn't want to know about the solar system, as that would take up brain space better devoted to more pertinent material.[3] Some people still find such a view sympathetic.

Back in 2003 I recall telling the biographer Richard Holmes, then just setting out on the group portrait of early-nineteenth-century British scientists that would become *The Age of Wonder*, that I intended to follow up my previous work, a 3,000-year history of swordfighting, with one about the Sun. 'Ah,' he said, 'first a study of honour, now one about belief.' Maybe he is right, but I began with no particular agenda. The Sun is wonderful – literally 'full of wonders' – and I wanted to know more about it.

This has meant that I have had to learn nearly all the science – for me, a daunting task, as my secondary school was run by Benedictine monks who had little time for such disciplines, and the main memory I have of any scientific endeavour is of the black-robed, white-haired Father Brendan climbing through a window into our classroom carrying a large copper pan and a Bunsen burner. Proceeding to fry an egg, he told us that science should always have a practical end in mind. Many scientists and science-watchers would disagree – but I enjoyed my egg.

This book has taken eight years to complete, over which time I have visited eighteen countries on six of the seven continents. (During these years, my wife would frequently be asked about my whereabouts. 'Oh,' she'd reply, 'he's out chasing the Sun.') Some readers may wish that I had ventured deeper into solar astronomy, but this book is not a rainbow – it has to end somewhere. What I came to realize was that, in the Western world especially, our relationship with the Sun has gone astray. Having cracked the code of our star's great power, the miracle of nuclear fusion, we have discovered new wonders in the Sun; but something of value has been lost in the process.

Once upon a time we thought that we were the centre of the universe and that even the Sun (as well as the Moon and the planets) revolved around us. We were the centre of everything. Yet, not having mastered any of the primal forces that determined our lives, we were keenly aware of our own powerlessness. Hence our longing for mastery, as expressed so vividly in mythology, in the innumerable stories of men, or gods, or animals, trying to control the Sun: a longing for supremacy to be satisfied only through magical thinking, as if we had no other means at our command.

Thousands of years later, we know for a fact that our earliest, most basic idea about our place in the cosmos was false, and that that cosmos is vastly larger than we ever dreamed. We are mere specks. At the same time, we have created for ourselves powers far beyond any we could have imagined, and, a decade into a new century, solar scientists believe we are on the edge of a golden age of discovery about our star – and poised to answer such vexing questions as what causes sunspots and drives solar winds, how the Sun's rays can bring us usable energy and how its magnetic particles and coronal ejections affect our climates. So, as our understanding of the Sun has increased along with our ability to harness it, we have stripped it of its primitive magical qualities.

The American astronomer John Eddy understood this: 'We have always wanted the Sun to be better than the other stars and better than it really is. We wanted it to be perfect, and when the telescope came along and showed us that it wasn't, we said, "At least it's constant." When we found it inconstant, we said, "At least it's regular." It now seems to be none of these; why we thought it should be when other stars are not says more about us than it does about the Sun.'[4] This book is therefore in part about the desacralization of the universe, using the Sun as the emblem of that ancient but ever-stronger process, and about our sense of loss – sometimes conscious, sometimes not – as the wonder has been stripped away.

But it is more than that. Our star has also inspired artists across the centuries. Sometimes it is the direct subject of their creations, sometimes a symbol of what they have wanted to convey, infusing their work with an authority, even majesty, that no other force could match. Eight years into my subject, I have come to feel that the Sun has powers that are still well beyond our understanding – mythic indeed. We haven't at all managed to reduce it to just a 'scientific little luminary, dwindled to a ball of blazing gas', as the fervent Sun worshipper D. H. Lawrence put it. That ball of gas still has mysteries of a magnitude so cosmic that perhaps only the grandeur of myth can encompass what we now know about the Sun's world-destroying and world-birthing powers, and about its eventual death.

Consider global warming: we have been told that the science is settled, that greenhouse gases threaten the planet. I appreciate that to ask whether that is the prime cause precipitating the changes to our environment is to risk having this book branded a tract against common sense (as my three children repeatedly tell me). There is no doubt that we *are* witnessing extreme droughts, floods, epidemics and food shortages, along with the accelerating disappearance of species and abrupt changes in animal behaviour. Yet many of the world's leading solar physicists have told me that the coming 'golden age' of discoveries about the Sun is likely to bear directly on how we understand what is happening to our planet's climate, and that in fact these revelations may dwarf what we currently know about man-made global warming. Will we ignore the research, as if even major cataclysms have to be *our* doing, not that of our governing star? To be sure, we are not blameless; but neither are we ruinously all-powerful. It is as if it offends us to think that we are still at the Sun's mercy. But we are. And of course the Sun has no mercy.

☼

One lesson I have learned in writing this book is that the Sun gets into everything. One evening in 2004, while sitting in the Carlyle Hotel on Madison Avenue, I found myself in the midst of a half-dozen Masters of Foxhounds, men and women resplendent in hunting pink, there to attend a grand dinner. It transpired that true members of the blood dread days when the Sun is shining, as the heated atmosphere rises, lifting the fox's scent high beyond their packs' noses, and the quarry makes its escape. I cannot think how else I might have pulled such a sunny detail out of the air.

The US Scouting Service's 2009 website promotes a game centred on the Sun (albeit one with scant regard for astronomical realities). 'It is like musical chairs,' its author writes. 'There is one fewer chairs [sic] than boys. The extra boy is the Sun. The other boys are each assigned a planet (Mars, Jupiter, etc.). The Sun orbits around the chairs calling out the names of planets. The planet (boy) called gets up and walks (orbits) around the chairs with the Sun. When all the planets are orbiting, the den leader yells "Blast Off." All the boys scramble for a chair. The one left standing becomes the Sun.'[5]

Well, please take your chairs. It is time for blast-off.

SUNRISE ON MOUNT FUJI

> *You must become an ignorant man again*
> *And see the Sun again with an ignorant eye*
> *And see it clearly in the idea of it.*

Wallace Stevens, 'Notes Toward a Supreme Fiction', 1942[1]

> You run and you run to catch up with the Sun . . .

Pink Floyd, 'Dark Side of the Moon', 1973

My goal was to get to the top of Mount Fuji in time to see the Sun rise on 21 June, the summer solstice of 2005. However, I soon learned that it wasn't to be that simple: the five main trails used by the thousands of pilgrims and tourists who climb the mountain each year are off limits until 1 July, by which time most of the summit snow has melted, the strong winds have weakened and the weather overall is kinder. The official season begins with a day of grand ceremonies, one at each of the several 'entrances' to the mountain, ranging from a mass purification in one of the surrounding lakes to the *owaraji hono*, where a twelve-foot-high rope sandal made from rice stalks is offered to the gods in homage. The sandal, a symbol of safety, health and strong legs, replicates the original peasant footwear that was worn (and worn out) on pilgrimages. I decided to stick with my walking boots.

From my apartment in New York City, I applied for a special police permit to make the climb off season, but the official forms were daunting, demanding my blood type and the contact details of my next of kin, while warning that 'climbing Mount Fuji is not easy, and in foul weather can be a nightmare'. I opted to contact the Japanese Mountain Association instead, where a friendly Mr Nakagawa said that he would raise no objection, so long as I exercised extreme care. Temperature and oxygen levels could drop precipitously and suddenly, he cautioned, and there was real risk of high winds and rockfalls. Altitude sickness was another serious danger, and could bring on nausea and disorientation. 'Only fools climb Fuji-san twice,' runs the local saying, which wags in Tokyo have amended to 'once'. No matter: the chance to see the sunrise – *goraiko* – was incentive enough.

My actual journey started on 20 June, from the small western town of Ise, seat of the two great shrines to Amaterasu, goddess of the Sun and mythical ancestress of the imperial house. The day before, I had visited the shrines, the earliest of which dates back to the third century, although every twenty years each is torn down in an act of purification and renewal, then an identical successor raised. Now, at five o'clock in the morning, I boarded the train for Shin Fuji (literally 'New Everlasting Life'), where I would pick up a bus for the three-hour ride to one of the mid-mountain stops, before beginning the climb to the summit, 12,395 feet above sea level.

Since antiquity, Fuji has been revered as an embodiment of the nation. It rises from the Pacific coast of the great island of Honshu, about seventy miles west-south-west of Tokyo. Some twenty Buddhist shrines scatter across its slopes, to the very peak, Ken-ga-mine. Its majestic balance of beauty and power has for generations been invoked by nationalists as proof that the Japanese are the chosen people of the East – indeed, of the whole world. One sect, the Fujiko, believes the mountain itself to be divine.*

Fuji was formed by three volcanoes, the youngest of which became active some 10,000 years ago. Over millennia, lava and other outpourings from 'New Fuji' boiled over the two older craters, creating a cone nearly 2,000 feet across at their conjoint apex. To this day, geologists classify the mountain as potentially active – Krakatoa, after all, was dormant for two centuries before it blew in 1883 – but Fuji last erupted on 24 November 1707. Eyjafjallajökull was five years away: I crossed this hazard off my list.

As the train pulled into Shin Fuji, the bus I needed was waiting about eighty yards from the station – and, but for the driver, empty. My stop was Station 5 of ten, just short of halfway to the summit and about 7,600 feet

---

*Up until the 1870s, no women were permitted to climb, and male pilgrims made the ascent in white robes. The first woman reached the peak in 1832, disguised as a man.

above sea level. I looked up, out of the window: the mountain was cultivated for the first 1,500 feet, then passed into grassy moorland, soon giving way to thick forest, in the midst of which we would find Station 5. After about twenty minutes, the bus took on a stockily built man in his mid-forties, with cropped black hair and a set, determined expression. My companion turned out to be a San Francisco accountant and computer analyst named Victor, who was also bound for the summit. I suggested that we might go up together. He took his time, scrutinizing me with some care before enquiring what experience I had of this sort of climbing. I admitted I had none, and also acknowledged I was fifty-eight. He was clearly reluctant. 'I guess we can try, but if one of us is holding the other back, we split up. OK?'

At Station 5, several tourists – there to take photos of the mountain and visit the gift shop – were getting ready for home. Wandering off on my own, I bought some clementines, a sturdy rosewood walking stick (complete with a small Japanese flag and bear-discouraging bells – I promptly removed the bells, not having heard of any bears on Fuji recently) and an inexpensive torch. I soon found Victor, prodigiously well turned out in black climbing shorts, black above-the-knee socks (not an inch of skin showing), black cotton gloves and a purpose-designed anorak with see-through canisters (each with its different-coloured liquid and plastic straw) strapped to either shoulder. In his left hand he gripped two shiny steel pitons. 'Use them for skiing?' I ventured. He gave me a pained look before sitting down to put on black leather puttees 'to keep pebbles from getting into the boots'. He checked his altimeter, temperature gauge and compass, and made sure his three maps were ready for easy consultation. In solidarity, I checked my jacket, hitched up my jeans and wiggled my stick with its Japanese flag. And then we were off, quite on our own as we strode up the rough dirt path towards the main route of ascent. Much of the mountain was shrouded in heavy mist, but the outline of its still-snow-capped summit was clearly visible.

Victor and I exchanged life stories as we walked. After about forty minutes he pulled up abruptly. 'Altitude sickness,' he explained, taking a sip from the bottle on his right shoulder. I was surprised, but slowed down. We continued on like this for a while, with Victor lagging several feet behind. What soon became apparent was that he was more than just

stocky – he was seriously out of shape. Were I to make the summit in time
to see daybreak, I would have to go at my own speed, so with Victor's
agreement I set off by myself. It was just after 6 p.m., the Sun already
quite low in the sky – for all that it was solstice eve. Perfectly safe, I told
myself: thousands made this trip every summer. But the going was getting harder, with the trickier sections lined with steel stakes linked by
chains. The signposts were in both Japanese and English. BEWARE FALLING
ROCKS AND DANGEROUS WINDS. OK, I would look out for the rocks, but how
did one avoid winds? KEEP TO THE MAIN PATHS. That seemed simple
enough – only it wasn't, as I was about to find out.

When the Sun finally went down it did so abruptly, sending a chill
through me that was not to do with any loss of heat, but rather that rush of
loneliness one often feels at sunset. I turned on my torch, fixing it around
my head with a plastic band to leave my hands free. The terrain varied
wildly: sometimes I would find myself at the bottom of a great concrete stairway set into the mountain, with maybe thirty steep steps (high for me, even
higher for the average Japanese). At other times there were twenty feet of
shale, which I would have to run up to gain any traction; or a series of grey-
black boulders, of varying shapes and sizes, to clamber around or over. The
angle of ascent changed too, from about twenty degrees to something far
more acute, so I was ascending almost doubled over. The increasing
unevenness of the terrain made it impossible to set up any rhythm, and
keeping my balance was a constant battle – as was breathing. I paused to
catch my breath, turned and looked back down: a great vista of varied
stone, then woodland, then moorland . . . but no Victor, not even the glow
of his torch, or anyone else. I felt an involuntary shiver.

As the ascent became more arduous, the signs became less frequent,
so that more than once I wandered off the main path and had to retrace
my steps. The shale and boulders were now broken by patches of rust-
brown slag and knotted limestone crystals – beautiful, but treacherous
to negotiate. Some of the pieces underfoot were full of airholes, like a
sponge. I put a couple in my knapsack.

I began to feel light-headed, without the energy to take me further.
Altitude sickness in the form of a heavy nausea was hitting me hard: even
without Victor's altimeter, I reckoned I had hiked five or six miles – not
always vertically. I counted the next fifty – seventy – seventy-five steps,

then paused for thirty seconds before attempting the next seventy-five. I passed Station 6 – an empty cabin surrounded by fresh scars of mineral diggings – and a half-hour later slowly climbed the steep concrete steps to Station 7, again empty. Station 8, I knew, just half an hour's further climb in summer conditions, could sleep over a hundred people in season.

An 1857 painting of the sun goddess Amaterasu, with Mount Fuji in the background. Legend holds that she sent her grandson to pacify Japan, and his great-grandson, Jimmu, became the country's first emperor

A few minutes after 8.15 p.m., I stumbled up to its bamboo porch. The door opened, and a dyspeptic Japanese man in his late sixties looked me over, hardly pleased to have a visitor out of season. 'Take off shoes!' he barked. Behind him hovered a woman of similar age, his wife perhaps. She looked the more approachable of the two, and in pidgin English and sparrow Japanese she and I settled that I could stay for a few hours, so long as I carried my boots through to the dormitory and didn't make a noise when I rose to leave.

I followed her to a row of double-decker wooden bunks at the back, to find I wasn't the only visitor. Four reporters for the local television station, two of them women, were there to film the construction of a fast new downslope trail, close to the path I had used: the old route, the most dangerous of the three descents, had claimed too many lives. One of the men spoke English, and together we tiptoed outside, where he could smoke and I could share my clementines. Was I going on to the very top – on my own, in the dark? As it was still June, he reminded me, no part of the trail would be lit. Squinting at the thermometer that hung from the veranda roof, he worked out that it would be below freezing at the summit, and the chill factor from strong winds could bring it even lower. I thanked him for the warning, and the two of us made our way back to our bunks.

Sleep came quickly, all three hours of it, for my TV friend kept his promise and woke me at midnight, so that I could stay on schedule. Sunrise was at 4.37 a.m., and I was determined to greet it at the summit. Altitude sickness had worn off, the path seemed clear and I was making good headway when I looked up and almost lost my footing in alarm. No more than twenty feet ahead of me blazed a bright white light, with two serpentine strips of yellow right below it. 'Hi there,' shouted a familiar voice. The glare, I realized, was no more than Victor's hand torch, while the yellow zigzags were the phosphorescent strips sewn on to his lapels. He was sitting on a rock, his first stop since we had set out together over eight hours before. It was good to see him.

He took out his altimeter and estimated that we had over 2,000 feet to go, so we hurried on, talking only a little. I soon left him behind again, and glanced at my watch: nearly 2.30 a.m. The terrain was still erratic – pumiceous lava one moment, then shale, then great hunks of rock, then man-carved steps. On either side, snow-covered country glistened under my headlamp.

The thin air was draining me again, only this time more acutely, and now I had to stop every thirty steps, with a full minute to recover. I ate some chocolate and my last piece of fruit, and cursed my lack of preparation: why hadn't I brought altitude pills? Well, the best course was to go on at a speed I could manage, but I was racing time – sunrise was in two hours, and it would take every minute to reach the summit. If I ever got there.

So much for the ambition to go 'chasing the Sun', I found myself thinking. The phrase occurs in Samuel Johnson's preface to his famous dictionary. He is complaining about the endlessness of a lexicographer's task, seeing 'that one enquiry only gave occasion to another, that book referred to book, that to search was not always to find, and to find was not always to be informed'. He also speaks of the constricted vision of what could be done, and how that pursuing perfection is like chasing the Sun, which, as soon as one reaches 'the hill where he seemed to rest', seems just as far away as before. Not that I was after perfection, simply an end to the journey. Every step seemed to take an immense act of will, and I found the Pink Floyd song running through my mind:

> You run and you run
> To catch up with the sun
> But it's sinking and racing around
> To come up behind you again.
> The sun is the same in a relative way,
> But you're older,
> Shorter of breath,
> And one day closer to death . . .[2]

Fifteen steps at a time now, then ten, taken slowly and at much shorter length than my earlier confident strides. I tried to beat out a tune with my walking stick, letting it bounce in pleasing rhythm off the stones. More than once I turned an ankle on the uneven surfaces. My torch began to dim before fading away completely, but luckily it was almost four o'clock – enough natural light to climb by. Then at last there rose above me a carved wooden arch, through which passed a stone stairway flanked by two immense statuary lions.[3] I had reached my goal, the summit of the land of sunrise. I was at the *torii*, the temple gate whose curved wooden beam is meant to invite the Sun, the King of Nature, to come bird-like and perch there.

I turned around slowly to take in all that lay below: a commanding view over the coastline, Fuji's southern face sweeping unbroken to the coast, while to the north and west stretched a line of lakes, where the rivers moving down Fuji's slopes are dammed by volcanic ash. It was just after 4.30 a.m. I sank to the ground and waited. The heavens were a mass of different-textured bruises, then all at once the beginning of the sunrise inched above the skyline, shedding its colours, each so vivid, and changing so quickly as it moved upwards that I found I was holding my breath, waiting for the next swathe of indigo, or purple, or ochre. It was not yet so bright that I had to shield my eyes, but only seconds later the full glare struck home. Now I understood why we say something 'dawns' on us, bringing a surge of recognition and inspiration, as if seeing something for the first time. The Sun crested the horizon like an infant newly born, and I savoured the moment – atop Fuji on the longest day of the year, with the volcano, the whole panorama, seemingly mine. I couldn't feel possessive, just lucky.

For a while I wandered along the crater's edge, the whole mile and a half of it. A further half-hour went by until below me I spied a familiar figure. Victor had taken off his jacket and donned a kepi-like construction. Sweating, but beaming widely from beneath its white peak, he too was about to gain the summit of Fuji-san, the Sun goddess's own mountain.

PART ONE

THE SUN BEFORE SCIENCE

The heroic archer Houyi who shot down nine of the ten suns, thus saving the world –
and is said thereafter to have reigned over China from 2077 to 2019 BC

1

TELLING STORIES

I look upon the sunrise and sunset, on the daily return of day
and night, on the battle between light and darkness, on the
whole solar drama in all its details that is acted every day, every
month, every year, in heaven and in earth, as the principal
subject of early mythology.

Max Müller, the nineteenth-century Oxford professor
who transformed the study of solar mythology[1]

Man has weav'd out a net,
And this net throwne
Upon the heavens,
And now they are his owne.

John Donne[2]

Donne's awed yet mocking lines were written in the early years of the
Copernican revolution, but they could apply just as easily to man's
attempt to make sense of the heavens – to make them 'his owne' – by creating stories. Because all societies have myths about the Sun, their sheer
variety is glorious – here it is a magician or trickster, there a ball of fire
some figure must carry, another time a canoe, a mirror or an amazing
menagerie of beasts. In Peru and northern Chile, many tribes knew the
Sun as the god Inti, who descended into the ocean every evening, swam
back to the east, then reappeared, refreshed by his bath.[3] As soon as the
horse became domesticated (early in the second millennium BC) the Sun

was portrayed as guiding a chariot drawn by four flaming steeds. In ancient India, these were termed *arushá*, Sanskrit for 'Sun-bright' (the Greek word 'eros' shares that meaning, having evolved from the same root as 'Sun horse'). Birds are often invoked – a falcon, or an eagle, and of course the phoenix, which dies and is reborn from its own ashes. In Africa and India, the tiger and lion are solar animals, sunrise being represented by a young lion, noon by one in its prime, sunset by one in old age. Where lions are absent, local communities adapt: in the pre-Conquest Americas, the eagle and jaguar are the chosen beasts.

Several cultures described the Sun in more than one way: to the Egyptians, the solar gods numbered not only Ra but Khepri, the 'Self-Transforming One', and Harakhty, 'the Far One'. The Aztecs employed 'Huitzilopochtli' (from *huitzilin*, a humming bird) to mean both the rising Sun and the star at its zenith, and 'Tezcatlipoca', 'smoking (or shining) mirror', twilight or evening. The Sun is continually reborn; so that in all they had a jaguar Sun, a wind Sun, a rain Sun, a rain-of-fire Sun and the god Nanahuatzin ('Full of Sores') who became a fifth solar force, that of the earthquake. The Mayans thought of the Sun as a rattlesnake, and named their towns with an 'x' prefix, to represent the sound of a snake as it sheds its old skin and – like the Sun – takes on a new one. Yet whatever form the Sun takes – an eye, a wing, a boat, a dragon, a fish, a bird – there is a common core, a similarity to these tales that spring up in cultures often hemispheres, and millennia, apart.

Sometimes the Sun is seen as so overwhelming a threat that it must in some way be tamed. In ancient Chinese mythology, for instance, the goddess Xihi gives birth to ten Suns, which rise simultaneously into the heavens, burning the harvests and all plant life – bar one huge mulberry bush, the *fusang*, on which the Suns perch. Every morning the goddess bathes one of them, letting it fly up to her on the back of a crow. One day all the Suns escape, and life on Earth becomes unbearable. A variety of monsters scour the land: the ogre Zuochi with long teeth, Quiying, who kills with water and fire, a giant bird that unleashes the wind, Dafeng, the giant boar Fengxi and the great serpent Xisushe. The wretched people below endlessly beg the Suns to come down, but they refuse. Total destruction impends, until Houyi, a young archer, slays the ogre, the monster and the giant bird, cuts the serpent in two, captures the boar and – his crowning act – shoots down nine of the Suns. Ever since, the story concludes, there has been only the one last Sun.

Aesop's fable 'The Sun Gets Married' has a different plot but the same threat. One hot summer, word comes that the Sun is to marry. All the birds and beasts rejoice, especially the frogs, until a wise old toad calls for order. 'My friends,' he tells them, 'you should temper your enthusiasm. For if the Sun alone dries up the marshes so that we can hardly bear it, what will become of us if he should have half a dozen little Suns in addition?' Two stories, both teaching that one can have too much of a good thing.

Almost all ancient civilizations believed the universe to have existed for unknown ages without benefit of any human intervention. The same did not hold true for the Sun, which in a host of mythologies exists only by virtue of man's nurture. The Hopi of north-east Arizona, for instance, claimed they made the Sun by throwing up a buckskin shield along with a fox's coat and a parrot's tail (to make the colours of sunrise and sunset). But whatever form or character it took, the Sun was rarely cast as fully invulnerable (an old German custom forbade pointing at the star lest one do it harm), and it has been variously depicted as having been freed from a cave, or stolen, or having sprung into life through the self-sacrifice of a god or hero. Among the Inuit of the Bering Strait, all creation is attributed to a Raven Father, who is so annoyed at man's rapacity that he hides the Sun in a bag. The terrified people offer him gifts until he relents, but only to a degree, holding the Sun up in the sky for a time before removing it again.

Every early society personified the cycles of nature, but where the Sun is concerned, cultures differed on its gender. In the Romance languages, the star is male, but in the Germanic and Celtic it is feminine and the Moon masculine: in upper Bavaria the Sun is still spoken of as 'Frau Sonne' and the Moon as 'Herr Mond'. For the Rwala Bedouin of Arabia, the Sun is a mean and destructive old hag, who forces the handsome Moon to sleep with her once a month, and so exhausts him that he needs another month to recover.[4] Other groups, such as the Eskimo, Cherokee and Yuchi, also regard the Sun as female, while in Polish the Sun is neuter, the Moon male. These variations may have arisen from climatic differences: in some areas the day is mild and welcoming, hence the Sun tends to be termed feminine, whereas the Moon, ruling the chill, stern night-time, is male. In equatorial regions, where daytime is searingly forbidding, the night mild and pleasant,

The sun chariot of Surya, whose charioteer is Arum and who is also
personified as the redness that accompanies dawn and dusk

the genders reverse. There are exceptions: on the Malay Peninsula, Sun and
Moon are both regarded as female and the stars as the Moon's children.[5]

Most creation accounts cast the Sun as paramount, both over the
Moon and over the heavens. The Book of Genesis declares: 'God made
the two great lights, the greater light to rule the day, and the lesser light
to rule the night.'[6] The Egyptians referred to the Sun and Moon as 'the
two lights', the right and left eye respectively of Ra – the left being
described as weaker, because damaged. In Central and South America
and among the Mundas of Bengal, the Sun and Moon are man and wife.
The Bengalis charmingly call the Sun *Sing-Bonga*, believing him a gentle
god who does not interfere in human affairs. Another myth of the same
region fashions the star as a man with three eyes and four arms who is
abandoned by his wife because his dazzle wearies her. She installs Chhaya
(Darkness) in his place, but the Sun wins her back by reducing his efful-
gence to seven-eighths of his original brilliance (an interesting example
of the spirit of compromise making companionship possible). Many stories are told about such marital troubles, it being a given that the Sun
and Moon can never live happily together.

It occurred to some of the more sophisticated ancient cultures to wonder why, if the Sun were indeed so powerful, he had to abide by strict laws rather than roam at will. Surely only a slave would perform so repetitively? Numerous legends were devised to explain this thraldom. The Sun was portrayed as erratic, sometimes hurrying too fast, at other times dawdling; coming too close to Earth one moment, the next moving too far away. The Spanish soldier-poet Garcilaso de la Vega, one of the first bicultural Spanish Americans, tells the following story about Huayna Capac, greatest of Inca conquerors:

> One day this ruler stares directly into the rays of the Sun, and his high priest has to remind him that their religion forbids this. Huayna Capac replies that he is his king and pontiff. 'Is there any among you who would dare command me to rise and undertake a long journey?'
>
> The high priest answers that this would be unthinkable.
>
> Huayna Capac continues: 'And would any of my chieftains, no matter what his power or worldly estate, refuse to obey me if I should command him to travel to faraway Chile?'
>
> The high priest acknowledges that no chieftain would.
>
> 'Then,' says the Inca, 'I tell you that this our Father the Sun must have a master greater than he, who thus commands him to journey across the sky day after day with never a respite, for if he were the Supreme Lord he would surely sometime cease travelling and rest.'[7]

The Greeks too put the Sun in a somewhat less than exalted position: Homer does not even grant Helios a place among the Olympians. Nor is the Sun seen as always beneficent: in Mesopotamian myth, the solar god Nergal brings plague and war, his weapons being heat, parching winds and lightning. Throughout history there remains a deep ambivalence: humanity cannot do without the Sun's power, but still wishes to tame or seduce it, to limit its hold over us.

☼

What *is* that hold? In the latter half of the nineteenth century a remarkable scholar would make the Sun the focus of his research: Friedrich Max Müller. He would argue that the Sun lay at the root of language, and thus of all major myths, not just the obviously solar ones.

Müller was born in 1823 in Dessau, then the capital of a small state within the German Confederation, the son of a poet. Initially he studied Sanskrit, which kindled an interest in philology and religion. He embarked on a translation of the Rig-Veda, the sacred hymns of Hinduism, and in 1846 travelled to Britain to research the archives of its Indian empire, supporting himself by writing fiction, his first novel, *German Love*, becoming a best-seller. He stayed on, and in 1854 was appointed Professor of Modern Languages at Oxford. Fourteen years later he was made Professor of Comparative Philology as well, and later yet the university's first Professor of Comparative Theology. The breadth of his knowledge, along with the fact that he spent years preparing a massive fifty-volume English translation of *The Sacred Books of the East*, may well have made him the model for George Eliot's Dr Casaubon – the pedant engrossed in his never-ending lifework, *The Key to All Mythologies* – in her novel *Middlemarch*, published in 1871, when Müller's reputation was at its height.

In his time this German-born Oxford academic was a truly famous figure, his friends and acquaintances spanning two generations of the British intellectual elite: Macaulay, Tennyson, Thackeray, Ruskin, Browning, Matthew Arnold, Gladstone and Curzon, among many others. Queen Victoria twice offered him a knighthood, which he declined as inappropriate. When he died, his widow received condolences from kings and emperors. In all, he wrote over fifty books. His last words were, unsurprisingly: 'I am tired.'[8]

In his masterwork, *On the Philosophy of Mythology* (1871), he set about showing that the same kinds of stories, the same traditions and myths, could be found worldwide, and that the appearance and disappearance of the Sun and its worship as the source of life were the basis of most mythological systems. From the earliest times, man constructed his understanding of the world around the Sun. 'What we call the Morning, the ancient Aryans called the Sun or the Dawn . . . What we call Noon, the Evening, and Night, what we call Spring and Winter, what we call Year, and Time, and Life, and Eternity – all this the ancient Aryans called *Sun*. And yet wise people wonder and say, how curious that the ancient Aryans should have had so many solar myths. Why, every time we say "Good morning" we commit a solar myth . . . Every "Christmas number" of our

Max Müller aged thirty, shortly after arriving in England

newspapers – ringing out the old year and ringing in the new – is brimful of solar myths.'[9]

Over a century on, we tend to take the validity of Müller's main arguments for granted. In his own time, however, he seriously over-reached: his insistence that *every* myth derives from the Sun, as well as his emphasis on the primacy of Aryan mythology and his eagerness to converge *all* languages back towards a single common root, provoked a bitter battle between his camp and those who took different paths. The cause of solar panmythology lost its leading light with his death in 1900. But though Müller's work is now known to only a few, he remains a major figure in our understanding of solar myths.

The Sun's place in the world's mythologies was taken up once again in 1923 by William James Perry (1887–1949), a cultural anthropologist at University College, London, and the anatomist Grafton Elliot Smith (1866–1955) when that year they co-authored *Children of the Sun*, which argued that during mankind's early history there were groups of people on most continents who believed themselves to be the progeny of a Sun

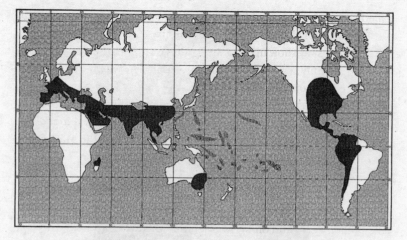

A map of distribution of solar cultures round the world, showing where
'Children of the Sun' were to be found

god. Unrepentant heliocentrists, Perry and Smith contended that 'The
importance of this fact in the history of civilization, and especially in the
study of mythology and tradition, cannot be exaggerated.'[10]

They dated the first appearance of self-proclaimed descendants of the
gods to around 2580 BC. Claiming to be the actual progeny of Ra, the
members of the pharaonic dynasties believed that at some point the Sun
had come down to take the place of the king on Earth, thus making
them his descendants. The subjects of the king were taught never to look
directly at him; rain or sunshine were his to summon; he was master of
magic and giver or withholder of the harvest.[11] The Egyptians took the
divine nature of kingship further than any other society – although it was
the Roman Emperor Vespasian (AD 69–79) who joked on his deathbed:
'Damn, I think I'm turning into a god.'

Perry and Smith discerned similar belief systems among the Asuras of
India, the Timurids of Indonesia, the Abarihu of San Cristobal in the
Solomons, the inhabitants of many parts of Polynesia, New Zealand and
the eastern Pacific, the Inca, the Mayans and several North American
tribes, and concluded: 'Wherever it is possible to examine the ruling
classes of the archaic civilization, it is found that they were what are
termed gods, that they had the attributes of gods, and that they usually

called themselves the Children of the Sun.'[12] Like Müller, they finally pushed a valuable insight too far (the countries where such Children of the Sun held sway show up on a world atlas only in particular areas), yet they did identify a remarkable cultural pattern, which enjoyed pre-eminence for thousands of years.

Perry and Smith then outlined the next cultural shift: as these societies sought to expand, they became more bellicose, and the Children of the Sun were transmogrified into war gods. For instance, as the Maya broke out of Guatemala into southern Mexico, their culture came to institutionalize cruelty and aggressiveness. The same applied to the Aztecs: 'War, at first *defensive*, afterwards *offensive*, became the life of the tribe.'[13] They found the same sequence in Melanesia, Polynesia and North America. Like Müller, Perry and Smith asserted that their insights held true over enormous time spans. Just the major societies so far mentioned stretch over 4,000 years: those in the Nile Valley run from 3200 BC until their conquest by Islam c. AD 700. China's ancient culture can be argued to run from c. 2800 BC until the end of the Tang Dynasty in AD 907, and so on.

In 1925 another man who would play a key role in re-imagining solar mythology, the pioneering psychoanalyst Carl Jung, made two journeys that together shaped his philosophy. On his visit to the Pueblo Indians of Taos, New Mexico, the forty-nine-year-old traveller found himself deep in conversation with one of the tribe's elders. As they sat together on the roof of his house, the old man pointed at the Sun blazing overhead: 'Is not he who moves there our father? How can there be another god? Nothing can be without the Sun.'

Jung in his turn enquired whether the Sun might not instead be a fiery ball shaped by an invisible god. 'My question did not arouse astonishment, let alone anger,' he wrote.[14] 'His only reply was, "The Sun is God. Everyone can see that."

'"We are a people who live on the roof of the world," continued the elder. "We are the sons of Father Sun, and with our religion we daily help our father to go across the sky. We do this not only for ourselves, but for the whole world. If we were to cease practising our religion, in ten years the Sun would no longer rise. Then it would be night for ever."'[15] Jung suddenly grasped the Sun myth's central role in giving the tribe a sense of meaning and purpose.[16]

His own dreams had convinced him that human beings have an unquenchable longing for light,[17] and a trip he took to Africa several months later, to investigate the myths of the mountain peoples dwelling between Mombasa and Nairobi, bore out this theory. There he watched as they waited each day for 'the birth of the Sun in the morning', as they explained: 'The moment when the Sun appears is God.' For Jung, this daily celebration of the Sun in a place so far away from the Pueblo in Taos confirmed the universality of this primal longing (even if the Africans, unlike the Pueblo, saw the Sun as the *means* of creating light, not a supreme being).

For several decades after Jung's journeys, scant organized research was conducted on the religious practices of preliterate peoples, but in 1958 the great Romanian historian of religion, Mircea Eliade (1907–86), took issue with all the heliocentric ideas that had been put forward, arguing that the Sun was 'simply one of the commonplaces of what is in a vague sense religious experience'[18] and that, in the agrarian societies that followed the ancient hunter-gatherer cultures, in which farmers had already become dependent on the changes of the seasons for their crops and especially on warmth and light, the solar deity turned from a god into a 'fecundator' (his special word). For all the innumerable myths, Eliade wrote, only certain cultures (the Egyptian and many of the Indo-European and Meso-American civilizations) had developed true solar religions.

It seems to me, though, that Eliade was building on Jung's insights, not negating them. That so many cultures continued for hundreds, sometimes thousands, of years to hold on to myths of solar supremacy shows that such stories were central to their sense of self – and also tells us something about the Sun's enduring hold over mankind.

Just in what way, I decided to see for myself. In June 2004 I set off for Peru, to witness Inti Raymi, the ancient festival that honours the Inca summer solstice. It takes place in Cuzco, over two miles up in the Andes, and opens in the north-eastern quadrant of the city square. By the time I arrived, the skies had opened, a cloudburst was hammering down and the crowd's red, blue, yellow and green plastic rainwear almost equalled the vividness of the actors – playing the pages of the long-slain Inca's court, the troops of orange-clad 'court virgins', the Inca soldiery (Peruvian regulars, seconded for the occasion) and the royal entourage

Audrey Hepburn and Gregory Peck in *Roman Holiday*, standing in front of the Bocca della Verità ('Mouth of Truth'), the massive marble sun disc (second century BC) in the Church of Santa Maria, Rome. Anyone who tells lies, Peck explains, and inserts their hand in the mouth of the god, will suffer 'the avenging justice of the Sun' and have it bitten off

in all its splendour. After about an hour, we moved on some 300 yards into Qorikancha, the Square of Gold, where messengers sounded conches to placate the 'Apu Inti' (Sun god), and the 'Willaq Uma' (high priest) intoned exhortations in Quechua, the Indian language still spoken by some twenty per cent of Peruvians. The woman next to me explained that local actors had been practising for a month. Her husband, a jeweller, was playing one of the attendant priests; the high priest's acolyte was busily munching gum.

The third and final part of the day's events was held a short bus journey away, at Saqsaywaman – pronounced 'sexy woman' but meaning 'satiated eagle' – where a range of huge joined stones forms the most important Inca monument after Machu Picchu.* Their sheer size is

*In Machu Picchu still stands a stone column called an 'intihuatana', meaning 'hitching post of the Sun' or literally 'for tying the Sun'. The ceremony to tie up the Sun was to prevent it from escaping. The Spanish Conquest never found Machu Picchu, but it destroyed all the other *intihuatana*, so ending the practice.

remarkable – some weigh as much as ninety tons. Between 1533 and 1621 they were rolled on tree trunks greased with vicuna oil from a quarry over half a mile away and arranged into the form of a huge twisting snake: a mighty presence of the old religion, yet somehow permitted by the Spanish. Every five metres, the blocks have been cut either 'male' or 'female', with prominences or sockets accordingly, and fit together so tightly that I couldn't get a sharp blade between them. This great wall formed the backdrop to the main celebration.

Our group's guide, Odilia, a small, well-organized woman in her late forties, mentioned that when she was a child her grandmother (now ninety-four) had told her that behind the stones lay a great earthen hill where once stood a palace and three astronomical observatories. Sure enough, in 1987 all four were discovered. Today excavations are going on all the time, and 328 stones used for solstice celebrations have already been found. Odilia next taught us how to pray to the Sun, leaning forward slightly, not so much out of reverence but to present the softest part of one's head to the great orb above. She removed her shoes and stretched out her arms, fingers splayed, to keep her body in touch with both Sun and Earth, the two great sources of life.

I wandered away from my group, to be swept up by hundreds of locals – women in ancient traditional costume, men rigged out as conquistadors or matadors or draped in nothing but leopard skins, boys (no girls) sounding trumpets, conches or cymbals, strumming guitars or banging drums. Vendors were offering everything from the ubiquitous tapestries of pumas and llamas to hats, coca tea, finger puppets, postcards, sodas, sweets, city flags, ponchos, film equipment, chess sets. There was still a fine drizzle, so the gaudy plastic covers were being bought in numbers. Amid the clamour many were dancing, including an impossibly ancient crone who spun around as energetically as revellers a third her age.

In one corner, just off the main festival precinct, there bustled a travelling fair, complete with slides and merry-go-rounds. Jugglers and street artists performed their acts every few yards, sweaters, necklaces and brooches in every imaginable style were being hawked, and chicken bubbled in huge black stewing pots tended by women in bright shawls, beads and hats. The smells of guinea pig, pork, corn on the cob, vegetable potpourri, and various unnameable concoctions hung thick on the after-noon air. At an ice cream stand a woman vigorously stirred a huge bowl of vanilla. The crowd was raucous, good-natured: this was their day.

The ceremony began at 2 p.m. On this great plateau a large rectangular stage had been constructed, with curtains painted to look like stones. Highland Indians from the four corners of the ancient empire trotted in to join their leader, who was poised to drink the sacred *chicha*, made from fermented corn. Some 500 locals took part, roared on by an audience of 40,000. Fires burst out at each corner of the plateau (the one nearest us sputtered fitfully and kept having to be reignited); the conches were sounded yet again, yet more frenetically. The basic ceremony was much like a Christian Communion service, with girls decked out as Inca princesses offering up baskets of sacred bread (while trying to restrain their giggles). The blazing colours of their costumes took one's breath away. No Inca virgin had her heart ripped out to unveil the future, however, as was essential before the Conquest, and even the llama sacrifice was play-acted – the animal was eventually set free to bleat its tale.

Something of genuine depth seemed to be asserting itself behind the commercial gloss: the pride of the long-conquered in their enduring ways, and even a sense that the star itself was, if no longer divine, then still a very special power. To crown it all, just as the drizzle died away, the Sun came out, and a rainbow magically appeared. I counted its colours, recalling Byron 200 years ago,

> . . . *a heavenly chameleon,*
> *The airy child of vapour and the sun,*
> *Brought forth in purple, cradled in vermilion,*
> *Baptised in molten gold, and swathed in dun* . . . [19]

The Sun god, it seemed, had heard our prayers, and the solstice had been duly honoured. As I walked away from the sacred field I wondered how anyone could doubt the Sun's control over us, in some essential sense still the prime organizing principle of our lives.

2

CELEBRATING THE SEASONS

> 'Tis the year's midnight, and it is the day's,
> Lucy's, who scarce seven hours herself unmasks;
> The Sun is spent, and now his flasks
> Send forth light squibs, no constant rays;
> The world's whole sap is sunk.

John Donne, *A Nocturnal Upon
St Lucy's Day, Being the Shortest Day*, c. 1611[1]

The summer solstice, *the day the Sun is said to pause*, he thought. Pleasing, that idea . . . As though the universe stopped for a moment to reflect, took a day off from work. One could sense it, time slowing down.

Alan Furst, *Dark Star*[2]

At one point in Huckleberry Finn's flight downriver, he comes across a copy of *The Pilgrim's Progress* (a book 'about a man that left his family it didn't say why'), which he leafs through and decides, 'The statements was interesting, but tough'.[3] Studying the Sun may provoke the same reaction, for there is no easy way to explain how it works. We may think of it as rising daily in the east and setting in the west, but it does so at different times, for different durations, in different places. As Barry Lopez observes in his great book on the Arctic: 'In a far northern winter, the Sun surfaces slowly in the south and then disappears at nearly the same spot, like a whale rolling over . . . The thought that a "day" consists of a morning and a forenoon, an afternoon and an evening, is a convention, one so embedded in us that we hardly think about it.'[4] The Sun is not a simple star, particularly in regard to the seasons.

During the course of a year the hours of sunlight vary from the maximum (on the day known as the summer solstice), which augurs the beginning of summer, to the minimum (the winter solstice, which signals winter). In summer, the Sun is brighter and reaches higher into the sky, shortening the shadows that it casts; in winter it rises and sinks closer to the horizon, its light diffuses and its shadows lengthen. As a hemisphere tilts steadily further away from the Sun, daylight becomes shorter and the Sun makes ever lower arcs through the sky. On the first day of winter, the time of briefest day, the Sun rises at the point nearest the equator. On the year's longest day, it rises closest to that hemisphere's pole. On both occasions it seems to stop in its path, before beginning to double back. ('Turnings of the Sun' is an ancient idiom, used by both Hesiod and Homer.) We see that effect at dawn, when, for two or three days, the Sun seems to linger for several minutes – hence the word 'solstice', from *sol* (sun) and *stitium* (from *sistere*, to stand still: 'armistice' translates as 'weapon standstill').

In the northern hemisphere, the first days of spring and autumn fall between 19 and 21 March and 19 and 21 September. These are the equinoxes (from *aequa nox*, equal night), when the hours of dark and light are roughly even all over the planet, and the Sun appears directly overhead at midday, after which it will keep advancing northwards after 21 March and southwards after 21 September. Many astronomers, as well as mariners, believed that the equinoxes brought strong gales – a myth, or at least a misconception, possibly because of the sharp rise in the number of high winds in the second half of September, when the hurricane season does indeed reach its height in the northern hemisphere (but for reasons differently related to the Sun). They were not alone: Herodotus, recording the summer floodings of the Nile, speaks of the Egyptian Sun being 'driven from his old course by storms' during winter, then 'returning to the middle of the sky' in spring.

Societies that were organized around agriculture – such as those in early Egypt and the Americas – intently studied the heavens, carefully noting what happened each year, and ensuring that the solstices and equinoxes were well charted. Despite their best efforts, however, their priests and stargazers came to realize that it was exceptionally hard to pinpoint the moment of solstice by observation alone – even though they could define the successive seasons by the advancing and withdrawal of daylight and darkness (albeit by another twitch of nature the earliest and latest sunrises and sunsets do not fall at the solstices).

The Earth further complicates matters. Our globe tilts on its axis like a spinning top (orbiting the Sun at an angle of 23° 27'), and this determines how much sunlight any part of the planet receives at any given time. Not only does it spin; its shape minutely alters and its axis wobbles, a process called nutation, 'nodding', so that the Earth's orbit fluctuates between greater circularity and greater elongation. If the globe's axis didn't wobble and if its orbit were a true circle, then the equinoxes and solstices would quarter the year into equal parts; but, our orbit being elliptical, the time between the spring and autumn equinoxes in the northern hemisphere is slightly greater than that between autumn and spring, the Earth moving about six per cent faster in January than in July. The Sun takes (in round numbers) ninety-four days to go from the vernal equinox to the summer solstice, ninety-two days from there to the autumnal equinox, eighty-nine days from there to the winter solstice, and ninety days back again to the vernal equinox. At high latitudes, the amount of solar energy received at midsummer can vary by twenty per cent, depending on whether the Earth's various wobbles reinforce each other or cancel each other out. Tough, but interesting.

☼

The apparently supernatural power manifest in solstices and equinoxes has been felt as far back as we know, inducing different reactions in different cultures – fertility rites, fire festivals, offerings to the gods. Many of the wintertime customs in Western Europe descend from the ancient Romans, who believed that their harvest deity, Saturn, had ruled the land during an earlier age of rich crops, and so celebrated the winter solstice, with its promise of a return of summer, with a 'Saturnalia', a high feast of gift-giving, role reversals (slaves berating their masters) and general public holiday each year from 17 to 24 December. They marked planting and harvesting with a ritual that entailed a winning horse from one of the great chariot races being sacrificed to the Sun. A temple stood on the Aventine, one of the city's original Seven Hills, and the populace would watch the races, looking up to see the Sun's image glowing from the roof (it being thought blasphemous to make a spectacle of his greatness indoors). The historian Macrobius explained these celebrations:

> Certainly it is not empty superstition but divine reason that makes them relate almost all the gods – at any rate the celestial gods – to the Sun. For if the Sun, as men of old believed, 'guides and directs the rest of the heavenly lights' and alone presides over the planets in their courses, and if the movements of the planets themselves have power, as some think . . . to foretell the sequence of human destinies, then we have to admit that the Sun, as directing the powers that direct our affairs, is the author of all that goes on around us.[5]

The transition from the Roman Empire and its pagan rituals to Christianity, with its similar rites, took place over several centuries. These were years of great upheaval – between 235 and 284, twenty-six different usurpers seized power in Rome – culminating in Constantine's great triumph at the Milvian Bridge in AD 312, which reunited the empire and ended half a century of civil war. Ascribing his victory to the Christian God, Constantine was later baptized in the new religion and enacted laws to advance Christianity. Many other customs were now appropriated and refashioned, as the Sun and God's Son became inextricably entwined in people's minds. Thus, although the New Testament gives no indication of Christ's actual birthday (early writers preferring a spring date), in 354 Liberius, Bishop of Rome, declared it to have occurred on 25 December. The advantages of 'Christmas Day' being celebrated then were obvious; as the Christian commentator Syrus wrote:

> It was a custom of the pagans to celebrate on the same 25 December the birthday of the Sun, at which they kindled lights in token of festivity. In these solemnities and revelries the Christians also took part. Accordingly, when the Church authorities perceived that the Christians had a leaning to this festival, they took counsel and resolved that the true Nativity should be solemnized on that day.[6]

In Christendom, the Nativity gradually absorbed all other winter solstice rites, and the co-opting of solar imagery was part of the same
process.[7] Thus, the solar discs that had once been depicted behind the
heads of Asian rulers became the haloes of Christian luminaries.*
Then the question arose as to which day of the week should be given
up to celebrating the Mass: the Church father Justin Martyr would
shrewdly explain to the Emperor Marcus Aurelius that Christians
chose the day they did for celebrating the Eucharist because 'It is on
what is called the Sun's day that all who abide in the town or the country come together . . . and we meet on the Sun's day because it is the
first day on which God formed darkness and mere matter into the
world'.

For a long time, the fixing of festivals remained something of a hit-or-
miss affair: over the winter solstice, the Romans were unsure which was
the shortest day of the year. Julius Caesar officially marked it as the 25th.
Pliny, in the first Christian century, put it at 26 December, while his contemporary, Columella, chose the 23rd. By 567 the Council of Tours
declared that the whole period from Christmas to the feast of the
Epiphany on 6 January should form one cycle, and by the eighth century
a system of twelve days of celebration was in place, with the Epiphany –
day 13 – starting a new sequence. Thus was created, in the words of
Ronald Hutton, 'a period between the winter solstice and the traditional
(Roman) time of the New Year, in which politics, education, commerce
were alike suspended, of peace, privacy, domesticity, revelry and charity:
a sacred period in a much broader sense than that associated with any
one religion.'[9]

Despite Christianity's apparent supremacy, however, many of the old

---

*Christian art at first avoided the halo, because of its pagan origins, but from about the
mid-fourth century Christ was portrayed wearing a crown of sun rays, as in effigies of the
Roman emperor, then from about the sixth century on, the Virgin Mary and other saints
were depicted with haloes, and by the ninth the symbol had been completely assimilated.
Again, some Christians were reluctant to use the pagan word 'Sunday'; in Latin-speaking
Christian countries, then in most Romance languages, the first day of the week became
for ever 'the Lord's' – *dominica, dimanche, domingo, domenica*.[8] During Constantine's reign,
the first day of the week was regarded by both pagans and Christians as its high point,
and in 321, when he officially proclaimed Sun Day as the holy day of the week, some
Christians believed it was for their sake, and began calling the day 'Sunday' rather than
the Lord's Day.

customs survived. Not least, feasting during the festive cycle could become an end in itself – so much so that Church elders worried that the veneration of Christ was being lost. St Gregory of Nazianzen (d. 389) urged on his flock 'the celebration of Christ's birth after a heavenly and not after an earthly manner', complaining they were 'feasting to excess, dancing, and crowning the doors [decorating doorways]'. More than a century later, both Augustine of Hippo and Pope Leo the Great again felt compelled to remind their people that Christ, not the Sun, was their proper object of worship; St Patrick, in his *Confessio*, declared outright that any Sun worshipper would be damned eternally.

It should also be remembered that while Roman Christianity was the dominant culture in far the greater part of Western Europe, it was by no means the only one. In the eleventh century the Danes conquered most of England, bringing with them 'Yule', their colloquial term for winter solstice celebrations, probably derived from an earlier term for 'wheel'. For centuries, the most sacred Norse symbol had been the 'wheel of the year', represented by a wheel or by a cross within a wheel, the arms of the cross signifying solar rays. The Norse peoples, many of whom settled in what is now Yorkshire, would construct massive solar wheels and place them on the top of hills, while in medieval times processions bore wheels on chariots or boats; and in some parts of Europe a taboo on using spinning-wheels during solstices lasted well into the twentieth century. The spinning-wheel on which Sleeping Beauty pricks her finger may exemplify this superstition, while the seven dwarfs in *Snow White* may represent the (then-reckoned) seven planets – as medieval lore had it, the seven greatest attendants of Earth.

An early-nineteenth-century miscellany elaborates on another connection between the wheel and the solstice. According to the authors, in much of Western Europe the wheel was an integral part of the summer solstice celebrations: 'In some places they roll a Wheel about, to signify that the Sun, then occupying the highest place in the Zodiac, is beginning to descend . . . we read that this Wheel was taken up to the top of a mountain and rolled down from thence; and that, as it had previously been covered by straw, twisted about it and set on fire, it appeared at a distance as if the Sun had been falling . . . [and] the people imagine that all their ill luck rolls away from them.'[10] Dances celebrating midsummer

Celebrants during summer solstice festivities at St Jean in Alsace

included spinning wheels around, and could go on for a whole month.*

Blazing wheels were often accompanied by festive fires, lit in sacred re-enactment of the creation of the universe, a tradition at both summer and winter solstices. From at least the thirteenth century, the lighting of fires on Midsummer Day became common across the whole of Europe, North-West Africa, Japan and even Brazil. By the mid-nineteenth century, Cornish migrants in South Australia were celebrating the traditional European midsummer with a bonfire on 24 June – midwinter in the southern hemisphere. Up until about the same date, in villages of the north of England, whole communities would meet together at summer solstice 'and be merry over a large fire, which was made for that purpose in the open street'.[11] This, of whatever materials, was called a Bonefire, because such fires generally consisted of bones; there is a nice homily, of unknown date, *De festo sancti Johannis Baptistae*:

In worship of St Johan the people waked at home, and made three maner of fyres: one was clene bones, and noo woode, and that is called

---

*The feast day of St Vitus, the cult of whom took over from the Sun god's in some countries, falls on 15 June, close to Midsummer Day. In 1370 an epidemic of choreia broke out in Germany, especially along the Rhine Valley, its victims falling into involuntary convulsions. Their sad antics so resembled the annual celebrations that the disorder became known as 'St Vitus's Dance'.

a Bone Fyre; another is clene woode, and no bones, and that is called a Wode Fyre, for people to sit and wake therby; the thirde is made of wode and bones, and is called St Johannys fyre.

Midsummer Day was a magical time and, far more than the winter solstice, an occasion for myriad superstitious doings. Shakespeare, inspired by the merrymaking and tomfoolery of the season, set *A Midsummer Night's Dream* at the solstice. Norwegian court records tell of how witches, fortified with home-brewed ale, would fly through the air on Midsummer Eve, often on the backs of cats,[12] while in Russia, Belarus and the Ukraine it was believed that naked she-devils, climbing up through chimneys, would launch themselves on the air. Francis Grose, a late-eighteenth-century original who began as a British officer of dragoons deployed to catch smugglers before switching to a career as an artist and antiquary, wrote: 'Any unmarried woman fasting on Midsummer Eve, and at midnight laying a clean cloth, with bread, cheese and ale, and sitting down as if going to eat, the street-door being left open, the person whom she is afterwards to marry will come into the room and drink to her by bowing; and after filling the glass will leave it on the table, and, making another bow, retire.'[13] He adds that anyone who fasted on solstice eve and sat in a church porch at midnight would see the spirits of those parishioners fated to die that year knocking at the church door.

In some countries, the celebratory fires of Midsummer Eve were also meant to ward off evil spirits, and so were crowned with straw dolls, or witches' brooms and hats, while the encroaching darkness at the time of the winter solstice never failed to arouse fears that the Sun would burn out completely – hence the kindling of the Yule log and the Hanukkah lights (the latter tied to both lunar and solar calendars). The boar's head at Christmas feasts represents the dying Sun of the old year, a suckling pig, with the apple of immortality in its mouth, the new.

Outside Europe, customs varied widely. In Japan, young men known as 'Sun devils', their faces daubed to represent their imagined solar ancestry, would go from farm to farm to ensure the Earth's fertility. In China, the winter solstice honoured the masculine (or *yang*) energies, the summer solstice the feminine (or *yin*); in early times a sacrifice, often human, was offered to the Sun, the smoke joining heaven and Earth. The Aztecs, who believed that the heart held fragments of the Sun's heat, ensured his continual well-being by tearing out this vital organ from hunchbacks, dwarfs or prisoners of war, so releasing the 'divine Sun fragments' entrapped by the body and its desires. Even sporting events were assigned a role: in 2005 I visited a sports field at Chichén Itzá in central Yucatán and was shown a fifteenth-century carving that depicted the bloody aftermath of one game the Mayans had played there. In a contest intended to guarantee the Sun's rebirth at the time of the vernal equinox, two teams, each of seven players and personifying light and darkness, played *ullamaliztli*, in which a solid rubber ball represented the Sun's path on its nightly passage beneath the world, and the players, using upper arms and hips, had to prevent it from hitting the ground or going out of bounds. Alas for the captain of the *winning* team, however, because he became the sacrifice whose blood revitalized Sun and Earth at the culminating ceremony, literally losing his head as a matter of honour.

Above all other rituals, reproducing the Sun by creating fires on

In modern Mexico, this burning sun-ball contest re-enacts games with ritual solar associations that were played for three thousand years in Mesoamerica up until Hispanic times

Earth is the commonest solstice practice, at both midsummer and mid-winter. (In Iran, families still observe the originally Zoroastrian winter solstice festival Zalda with night-long vigils, fires blazing against the darkness, and similar celebrations take place in Tibet and Muslim north-west India.) Thomas Hardy, describing Dorset villagers around a bonfire in *The Return of the Native*, offers an explanation for such a worldwide phenomenon:

> It seemed as if the bonfire-makers were standing in some radiant upper storey of the world, as if these men and boys had suddenly dived into past ages and fetched therefrom an hour and deed which had before been familiar with this spot . . . Indeed, it is pretty well known that such blazes as this the heathmen were now enjoying are rather the lineal descendants from jumbled Druidical rites and Saxon ceremonies than the invention of popular feeling about [the] Gunpowder Plot.*
>
> Moreover to light a fire is the instinctive and resistant act of men when, at the winter ingress, the curfew is sounded throughout Nature. It indicates a spontaneous, Promethean rebelliousness against the fiat that this recurrent season shall bring foul times, cold darkness, misery and death. Black chaos comes, and the fettered gods of the Earth say, Let there be light.[15]

☼

While solstice celebrations have always been the most common, other feasts to mark the changing of the seasons have taken place throughout the year. The French historian Emmanuel Le Roy Ladurie wrote an entire book about one carnival that took place in the town of Romans in 1580, and records the appearance of what has come to be known as the

---

*On 5 November 1605 Guy Fawkes and his fellow conspirators were detected attempting to blow up King, Lords and Commons, an event that became known as the Gunpowder Plot. The Parliament that so narrowly escaped destruction was postponed until the beginning of the following year, when an MP proposed a perpetual annual thanksgiving, whose service remained in the Anglican prayer book until 1859. By the 1620s it had become the most popular state commemoration in the calendar, surpassing by far even the king's or queen's birthday. In the last decades of the eighteenth century, by 'ten o'clock, London was so lit up by bonfires and fireworks, that from the suburbs it looked in one red heat'.[14] No longer witches' brooms but effigies of Fawkes, or sometimes of the Pope (often stuffed with live cats, which would shriek as the flames engulfed them) were paraded atop each fire.

Candlemas bear. Every 2 February the beast would be prodded from its den to peer at the sky.[16] Were it cloudy, superstition held, the bear would be unable to see its shadow and so would behave as if winter were on the wane; but if it saw its shadow (because the weather was bright) it would be frightened back into its hole, and winter would endure for another six weeks. The bear even now is employed at such festivals in the Alps and Pyrenees, while elsewhere other hibernating animals perform a similar role – Ireland has a hedgehog for St Bridget's Day (1 February), while Pennsylvania and other parts of North America celebrate Groundhog Day every 2 February – a folk custom introduced by French and German settlers.

The Romans festival, along with so many others in France, came under attack during the Revolution of 1789. The leaders of that great convulsion were determined to abolish traditional Christian holidays. Sundays were to be workdays; a new calendar would be produced, a new order promulgated. How to begin the year was easy to settle, since the Republic had been declared at the autumn equinox of 1792. On that day (a revolutionary firebrand declared to the Convention in Paris) the flame of liberty illuminated the French nation. The new date – 1 Vendémiaire, the first day of the month of gales – was given an explicitly 'celestial' meaning, as is clear from the countrywide events that were planned to commemorate it. The municipal administration at Aurillac, in the Cantal (one of the eighty-three departments created by the Revolution) in south-central France, described how this glorious pageant should be celebrated:

> Then will appear a superb, shining chariot drawn by twelve horses; this will be the chariot of the Sun; before it will be the signs of the zodiac and the hours according to their new division; beside it will walk young citizenesses, dressed in white, who will represent the hours of the day, while young citizenesses dressed in white and veiled in black will represent the hours of the night; on the chariot will be seen the still-slumbering genius of France covered with a veil decorated with fleurs-de-lys and from time to time making simple movements.[17]

For about a decade, such festivals were enthusiastically celebrated throughout France. But such zeal could not last, and soon the old festivals and calendars returned. The revolutionary system became part of history, with the life-bursting spring month of Germinal (21 March to 19 April) memorialized in the title of Zola's 1885 novel – and Thermidor (19 July to 17 August) in a lobster dish.

☼

Sun celebrations cut across time and culture; but possibly the most dramatic are the Sun dances of the North American Indians, usually held in late June or early July before a buffalo hunt, or to recruit the power of the Sun cycle over crops. Most sun dances centred on a lodge, usually a circle of twenty-eight forked poles that represented the twenty-eight days of the lunar month. These would be joined together by more poles laid horizontally into the upright forks. A witness likened one such to 'a circus tent, the top of which has been ruthlessly torn away by a cyclone'.[18] Entrances to the lodges faced the sunrise. Sometimes a host of lodges – as many as 700, enclosing over 10,000 celebrants – would be arranged in one huge circle as much as six miles in circumference.[19]

For all these clans, the rite often took place under a full Moon (whatever its other benefits, a sun dance was the occasion for energetic couplings). Each tribe had its own characteristics. The Shoshoni and Crow called their ceremony 'the Thirst Lodge', the Cheyenne 'the Medicine [in the sense of big magic] Lodge' and for the Lakota Sioux it was *wiwanyag wachipi*, 'the Dance Staring at the Sun'.[20] In Arizona, the Papago hopped around a solar symbol in alternating directions, extending their hands towards the Sun, then stroking their chests to take in its power.

The spectacle reached its fullest development among the Teton Sioux. Participants congregated in a sweat lodge, a kind of sauna that served as a preliminary purification. The medicine men would pray for fair weather, and a dance leader would bring together an ornamented pipe, a perfectly formed buffalo skull and some buffalo fat; then would follow the setting up of the 'sun pole' (representing a path to the sky), a tobacco offering and a dance of supplication. These preparations would take three to four days, during which the initiates would fast, their mounting hunger sometimes bringing spectacular visions. In more intricate versions, participants would raise tents in a circle representing the aurora borealis around a brush dance enclosure some fifteen feet west of a central pole symbolizing the Sun.

In *The Origin of Table Manners*, Claude Lévi-Strauss explores the ambiguous and equivocal nature of these dances:

> On the one hand, the Indians prayed to the Sun to be favourably disposed towards them . . . On the other hand, they provoked and defied it. One of the final rites consisted in a frenzied dance which was prolonged until after dark, in spite of the exhausted state of the participants. The Arapaho called it 'gambling against the Sun', and the Gros-Ventre 'the dance against the Sun'. The aim was to counteract the opposition of the intense heat of the Sun, who had tried to prevent the ceremony taking place by radiating his warm rays every day during the period preceding the dance. The Indians, then, looked upon the Sun as a dual being: indispensable for human life, yet at the same time representing a threat to mankind.[21]

The dance itself began at sunrise and continued for as long as five days. On the first morning, 'pledgers', perhaps fifty strong, painted their bodies red for sunset, blue for sky, yellow for forked lightning and black for night, often including the form of a sunflower. Each wore an apron of deer or antelope skin, rabbit-fur wristlets and anklets and a downy feather in his hair, carrying in his mouth an eagle-wing bone whistle with porcupine quill or beadwork and eagle-down decoration. The pledgers circled the phallic pole and saluted the Sun to drumbeats and intense singing. With their faces fixed on the great solar orb, they danced on their toes or the balls of their feet. To 'participate in the cosmic renewal of life by shedding their blood for the sake of all life', they would lash their arms and thighs as many as two hundred times and suspend themselves from eight-inch-long wooden skewers inserted by the *kuwa kiyapi* (holy man) into the small of their backs or into their shoulders. The skewers were attached to the central pole by tightly wound rawhide thongs at a height such that only the participants' toes touched the ground; alternately, some participants would suspend themselves upside down, their bodies hanging like corpses – the excruciating ordeal that Richard Harris undergoes in the 1970 Hollywood version of the

The *Okipa*, the dance ritual of the Mandan Indians, as depicted by the artist
George Catlin c. 1835. The last full Mandan died in 1971

ceremony in *A Man Called Horse*. The test of any dancer's honour was not
to scream or cry out.

Urged on by the rest of the tribe, each brave would pull hard on his
wounds until he fell unconscious or tore himself loose, releasing him
from his vow and signifying his willingness to suffer. Once freed, the
liberated dancer would, after a prescribed interval, smoke pipes ('split
the clouds'), enjoy a vapour bath and at last drink and take food (customarily dog soup and buffalo meat) in a concluding feast.

At least ten Plains tribes adopted the main elements of this ritual;
others performed the ceremony without spilling blood. To an outside
eye, such self-mutilation seems monstrous (Jesuits, keen to convert the
heathen Indian, were particularly virulent in their condemnation); but
the historian Joseph Epes Brown, who would collaborate with the
Lakota chief Black Elk in his apologia, *The Sacred Pipe* (1948), compares
the Sun dance with Hindu meditation techniques and the Sun dance
pole with the crucifix, equating the sufferings of the dancers with certain extreme forms of Christian penance. 'When man in awful

ceremony is actually tied to this Tree of the Center by the flesh of his body,' Brown tells us,

> or when women make offerings of pieces cut from their arms, sacrifice through suffering is accomplished that the world and all beings may live, that life be renewed, that man may become who he is. The Sun Dance, thus, is not a celebration by man for man; it is an honoring of all life and the source of all life, that life may go on, and the circle be a cycle . . .[22]

In his moving account of the Kiowa, *One River*, the ethnologist Wade Davis writes that the sun dance was 'by far the most significant religious event of their lives. It was a celebration of war, a time of spiritual renewal, a moment in which the entire tribe partook of the divinity of the Sun.'[23] But its very existence was threatened by the white man. On the night of 13 November 1838, a great meteor shower blazed across the prairie sky, alarming the whole tribe and prompting the elders to predict the end of the world, a forecast that turned out to be accurate for the world of the Kiowa. Their first contact with white soldiers came the following summer. By 1890 East Coast administrators had banned sun dances altogether.*

At the end of his history of the Lakota Sioux, Clyde Holler asks simply: 'Why on earth are they still dancing the Sun Dance?' Among the reasons he offers this: 'To attend the Sun Dance is to feel the power, to approach religious ecstasy, to experience for oneself the essence of a religion based on power. The Sun Dance sweeps away any possible doubts about the existence of sacred power in a searing emotional catharsis.'[25] He might have mentioned physical catharsis too. Dance promotes the uninhibited, the irrational – indeed, even sexual release, orgiastic excess. In *The Birth of Tragedy* Nietzsche set the Dionysian nature of dance against its Apollonian opposites, balance, reason and discipline.[26] The bureaucrats in Washington

---

*Davis (who rejoices in the title of 'Explorer in Residence' at the *National Geographic*) makes the point that modern celebrants, no longer allowed the excesses of old, chew on peyote as a substitute means of achieving enlightenment.[24] A tribesman told Davis that peyote was like the buffalo of ancient times. 'Both are children of the Sun. Peyote is the incarnation of the Sun. As are buffalo.'

were not alone: the Protestants of Tudor and Jacobean England, the inquisitors of the Midi, the Conquistadors who savagely mastered South America, the Gauleiters of National Socialism – all of them tried to regulate and contain solar-related dance rites, if not actually to outlaw them.

In those countries where ceremonial dances came under the sponsorship of the state, attempts were made to fix them to a particular time and place, rather than allowing them to arise spontaneously at some untried location. This made Sun celebrations easier to manage, while still giving the celebrants an outlet. And in truth worshippers, including those who honoured the Sun, yearned for central sites where they could congregate and pay homage. Further, in prehistoric societies the embodiment of their gods in stone was symbolic of the change in religion that had taken place, from nature worship to that of heavenly bodies. Material structures – Stonehenge (known throughout the Middle Ages as *Chorea Giganteum*, 'the Giants' Dance'), the pyramids, those great erections of South America – all such sites came to embody belief, not least about the Sun.

3

THE THREE THOUSAND WITNESSES

To lug those stones was quite an undertaking
technology worth some remark

Pat Wood, 'Dartmoor'

The poet,
Admired for his earnest habit of calling
The Sun the sun, his mind Puzzle, is made uneasy
By these solid statues which so obviously doubt
His antimythological myth

W. H. Auden, 'In Praise of Limestone'[1]

More than three thousand structures, nearly all of stone and scattered worldwide, stand witness to man's obsession with the Sun. Most are built in line with the direction of its rising or setting, while some align with it just once a year, generally at a solstice – although the Great Sphinx at Giza gazes out fixedly towards the vernal equinox. Whether one is looking at the first pillars of Stonehenge* (dating from 2900 BC), the Mayan pyramids in the Yucatán (twelfth century AD), the megaliths of the Great Plains (down to the nineteenth) or the Dancing Stones

---

*A *henge*, while it means 'hanging', is specifically a large Neolithic monument consisting of circular banks of earth paralleled by an internal ditch, and frequently containing a circle of upright stones. Strictly, Stonehenge is not a henge, because the position of its bank and ditch are reversed.[2] A *menhir* is a solitary standing stone, a megalith a 'great stone' (from *megas*, great, and *lithos*, stone).

of Namoratunga, on the grassy plains of Kenya (date unknown), here are the Sun's silent celebrants.

The majority of societies have erected monuments to the Sun with extraordinary care. A large percentage of mid-Neolithic tombs appear

The massive pillars of Stonehenge, still the greatest of solar monuments

to have been built to face the Sun at significant times, and in Britain alone over 900 survive, above all Stonehenge, whose pillars, Samuel Pepys recorded after visiting them in 1668, were 'as prodigious as any tales I ever heard tell of them, and worth going this journey to see'. He then added, with typical honesty: 'God knows what their use was.'

I have visited Stonehenge a dozen times. On occasion, it can seem just a jumble of huge, squat blocks without any special allure. At other times, such as Midsummer Eve, it is so thronged with worshippers that it could be a country fair. But when all but deserted, at either sunrise or sunset, it is magical, and one's imagination soars. Against the lateral sunlight, the stones gain in beauty and majesty. For men to have made such a place at such great effort shows how important the Sun was to him.

Present-day Stonehenge is the outcome of more than five millennia of building and accretion, from ditch banks of earth to chalk, to wood and finally to stone. During its first six centuries, a round enclosure took shape, formed by two banks and a rough 'chalky, dazzling moon-milk white' ditch.[3] Within the circle fifty-six pits were dug, each roughly a metre wide and deep. There was a break in the circle at the north-east, with a pillar 256 feet inward from the entrance, the famous heel stone (a corruption of the Welsh 'hayil' or Norse 'hel', both meaning 'Sun'): 16 feet high (descending another 4 feet below ground), 8 feet in circumference and some 35 tons in weight. On my last visit, one midsummer morning, I stood in the centre of the circle and watched the Sun rise just to the left of the stone: easy to believe then that the site was as much observatory as place of worship.

Between 2300 and 2000 BC at least 82 upright slabs – known as blue-stones from the tint they assume in wet weather – were floated, dragged and hauled for 250 miles from the Preseli Mountains in south-west Wales to form a crescent in the centre of the circle. Over the next four hundred years, a further ring of uprights, consisting of thirty gigantic slabs of stone arranged in pairs that support lintels, was added, in a horseshoe-shaped curve, its opening in line with the Sun's midsummer rising and with the winter solstice sunset. Nearly twenty-three feet high, these trilithons – 'threestones' (two uprights supporting a lintel) – each weighing some forty-five to fifty tons, were cut from sarsen sandstone, a word suggesting pagan origin (from 'Saracen').

Stonehenge first rates a mention in the *Historia Anglorum* (AD 1130). However, that is all there is, a mention; not until 1740 did an English antiquary, William Stukeley, ascribe an astronomical purpose to this mighty engineering feat, arguing that its principal axis pointed towards the dawn at the summer solstice. After Stukeley (who also advanced the notion – not stupidly – that Stonehenge might have been a racetrack), theories came fast and curious, right down to the present day: it was a Bronze Age Lourdes; an ancient Druidic site; a prediction centre for eclipses; a monument to the dead and part of a funerary and processional route (as many as 240 people are known to have been buried there); a tomb for Boadicea, the pagan queen of the Iceni, who led a revolt against the Romans; a mechanism for calculating the calendar; a solar and lunar observatory; a sophisticated forecaster of tides (curious, for a place so far inland); and a monastery or college for a super-elite who were part of a worldwide prehistoric intelligentsia. All these centuries later, it is safe to say that scholars are firmly agreed on scarcely a single aspect of Stonehenge – although the weight of current research leans towards its having been created as a cluster of celestial observatories, some to track the movements of the Moon, others those of the Sun, and possibly to establish a calendar.* Maybe, as the archaeologist Jacquetta Hawkes once famously observed, 'every generation gets the Stonehenge it deserves – and desires'.

With so many different theories, how can we tell the purpose of such buildings – not just Stonehenge, but other monuments too? As Ronald Hutton wryly notes: 'The people who built and used these great

---

*It is easy to take the accuracy of the alignments at Stonehenge for granted, perhaps because transporting and raising the stones was extraordinary achievement enough. But its builders used techniques that would keep predictions reasonably accurate. By contrast, modern map-makers can make huge mistakes: in the early nineteenth century, with American–British Empire relations under extreme stress, the United States erected two fortresses beyond Lake Erie, at the huge cost (for the time) of $113,000. In October 1818, after three years' work, astronomers carrying out a boundary survey discovered that the border was three-quarters of a mile to the south – both forts had been erected in Canada.[4]

structures seem to have taken a positive delight in making exceptions to every rule which we try to discern in their behaviour.'[5] Point up in any direction, he suggests, and you will most likely hit a star or phase of the Sun or Moon. Nor does the notion of monuments laid out according to a sophisticated geometry consort well with evidence on the ground: 'a glance at the plans of Wessex superhenges will reveal that the classic shape is that of a battered car tyre.'[6] Many prehistoric monuments clearly have no connection with the exact occasions of summer or winter. For example, the stone figures, or *moai*, on Easter Island – over nine hundred of them, averaging twenty feet in height – while they probably performed some calendrical function, were based on lunar, not solar cycles.[7] One tomb on the Isle of Arran faces the midsummer sunrise almost exactly, but nineteen similar tombs on the island point in quite other directions. As Bruce Chatwin reports the conversation between an Aborigine and a rough Australian in *The Songlines*,

> The big man leered. 'If all what them says was sacred sites, there'd be 300 bloody billion sacred sites in Australia.'
> 'Not far wrong, mate!' called the thin aboriginal.[8]

However, overall the evidence is irresistible that by far the greater number of structures are consciously connected to the Sun; and while Stonehenge may be the jewel in the crown of such prehistoric monuments, many are the lesser gems spread around the world. A particular favourite of mine is the burial mound at Newgrange – *Liamh Greine* ('the Cave of the Sun') or 'the Irish Stonehenge' – which lies in the Boyne Valley about thirty miles north of Dublin. A megalithic burial place dating back to 3200 BC, it is roughly contemporary with the first Egyptian pyramid. From outside one sees a mass of dull stone atop a small hill; inside it is a place of wonder. At precisely 9.02 a.m. at the winter solstice, sunlight shines through a small window at the end of a nineteen-metre-long gallery, and for the next seventeen minutes creeps along the passage until its eerie glow falls on a circular stone in the opposite wall of an otherwise dark little chamber.

In his 2005 novel *Ireland* Frank Delaney re-creates the lives of those who built the barrow. Inside the central chamber, directly in line with the passageway, the tomb's creator ('the Architect') has placed a 'round, smooth dish made of sand-coloured stone', and now assembles the Elders within the tomb:

> At the farthest end of the passageway, red-gold colours tinged the edges of the rectangular aperture, making it look like a box full of light. Next, a shaft of sunshine slid through the box and lit the floor, just inside the entrance. The Elders watched this glow and became transfixed . . . Steadily, with no jumps, no lurches, the light flowed down the passageway. Its golden-yellow beam, thick as a man's arm, flowed slowly, slowly – it was as though the sunlight was honey from a bowl that someone tipped over and spilled into a long, direct channel . . . When the sunbeam reached the edge of the chamber, its yellow-orange light began to glow upwards into the gathering, gilding the Elders' faces. The ray seemed to hesitate for a moment. And then it surged forward until it splashed into the great stone dish – and filled it exactly. Not a drop of sunshine slipped over an edge here, there or anywhere. The final resting-place of the Sun for that brief moment was precisely within the full circumference of the bowl, leaving no area of the bowl's stone surface dark or cold. It lay there like a golden sphere.[9]

Stonehenge's greatest rival among the prehistoric monuments of Europe is not Newgrange, however, but the forest of alignments, stone circles, dolmens (burial places formed from great graven stones) and tumuli (earthen tombs) stretching beyond the remote seaside town of Carnac on the southern coast of Brittany. Some 10,000 megaliths – notably the Grand Menhir Brisé at Locmariaquer, a massive edifice that once rose sixty feet but now lies shattered into five parts, each of about three hundred and forty tons – huddle together like so many battle-weary giants. Local legend has it that the stones are Roman soldiers literally petrified by Pope Cornelius during the schism of AD 251–3. The site boasts several burial chambers, deep caverns faced towards the rising Sun, a powerful index of the likely interest in things solar among those who dwelt here in the New Stone Age. It is among the great stones of Locmariaquer that Porthos meets his death (in *The Man in the Iron*

Just north of the River Boyne in Ireland: the 19-metre passageway at
Newgrange burial mound, down which the Sun flames each winter solstice

Mask), covering Aramis's escape from a company of the king's guards-
men after being trapped inside one of the great caves.[10]

Most countries in Europe have at least one major site of ancient solar
reverence. The most impressive constructions are hardly limited to the
one continent, however. Halfway around the world, an extraordinary
temple sits within the complex of buildings that makes up Angkor Wat
in Cambodia. Constructed between 1113 and 1150, it forms part of the
largest 'astronomically sited' building in the world. The temple (Wat),
almost a mile across, was designed both as a tomb for its creator,
Suryavarman II, and as an observatory: nearly every measurement
encodes calendrical information, and the bas-reliefs are all oriented west
to catch the setting Sun.

There are five particularly grand solar temples in India, all now in
ruins: at Delhi, Konorak, Mudan, near Ranapur in Rajasthan, and
Modera, in Gujarat. In October 2006 I visited the one at Modera, built
in AD 1026, about 200 years before the next oldest, at Konorak. The
shrine lies about six miles from the state's former capital, Patan, and its

survival is all the more surprising given that it lies in an earthquake zone – badly hit in 1918, again in 1965 and once more on 25 January 2000, when the shock registered 7.9. My guide, a Brahmin who has converted to Buddhism, explained to me that unlike the peoples of Europe or the Americas, Hindus try never to face the Sun directly, which is why statues of Indian gods look eastwards, a hand held up in blessing, so that anyone looking at a god catches the Sun's reflection. Temple domes are designed in the shape of massive lingams (yet one more tribute to the Sun's potency).

Solar-oriented monuments assume an amazing variety of forms. In North America, for instance, the Indians of the Great Plains position 'medicine' (i.e. magical) wheels to register the Sun's course, each wheel consisting of a central circle or rock pile from which radiate spokes of stones. These creations can range up to a hundred metres in diameter, while the hub may be ten metres across and several metres high. They are found along the eastern boundary of the Rocky Mountains from Colorado to Alberta and Saskatchewan. One, dubbed 'the American Stonehenge', sits atop Medicine Mountain, in the Bighorn range of north-central Wyoming. This sandstone edifice was held sacred by several Plains tribes, and a line through two of its cairns (stone memorial mounds) lines up to within a third of a degree with the exact point of sunrise at the summer solstice.[11] In Salem, New Hampshire, stands a rival 'American Stonehenge' – a complex of cairns, chambers, walls and huts between 4,000 and 3,000 years old encircled by notched 'sighting stones', aligned on the sunrise and sunset solstices. In a series of markings, the site also celebrates 'cross-quarter' holidays, which fall halfway between solstice and equinox.[12]

The Americas are particularly rich in such places. North-east of St Louis, Missouri, over a hundred mounds, dating from 800 to 1550 and covering sixteen acres, form the largest prehistoric earthwork in the world, many of its mounds arranged along the sunrise line of the winter solstice. As elsewhere, some are no more than elaborate burial tumuli, while others served as temples or great effigies of gods, or as fortifications.

The Maya, Aztecs and Inca all constructed such works too. At Teuchitlán (literally, 'Where Men Become Gods'), in the present Mexican state of Jalisco, arise some 600 pyramids of various sizes, dominated by a great Pyramid of the Sun, one of three temples covering over an acre and aligned east–west, suggesting they functioned too as solar observatories. They were built between 300 BC and AD 200, one of the three being initially slightly misaligned, which fault so upset Montezuma that he had it razed and rebuilt – an act of intellectual integrity among a supposedly inferior people that so astonished Cortés he reported it to his king.

Further south, the entire Mayan city of Chichén Itzá in south-eastern Mexico was devoted to astronomical observation, the windows of its Spiral Tower being set towards solstices and equinoxes. While travelling in Peru, I saw temples of the Sun in Machu Picchu, Sacsahuaman and Cuzco, all similarly disposed. Cuzco and Chichén Itzá both boast rows of pillars that divide the Sun's movements into periods roughly corresponding to months, helping the farmers to determine the time for sowing.[13] I have also seen solar markers on a hillside overlooking the Island of the Sun in

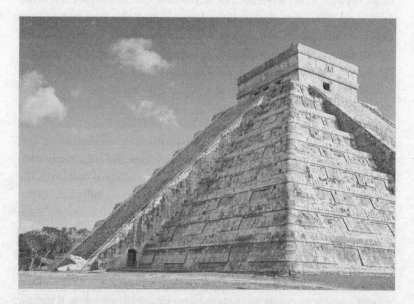

The pyramid at Chichén Itzá ('City of the Water Magician'): on the spring and autumn equinox, 21 March and 22 September, at the rising and setting of the Sun (in Mayan, 'Kim'), corners of the pyramid cast a shadow along the west side of the north staircase in the shape of a plumed serpent, and as the shadow moves the snake seems to slither down the stairs

Lake Titicaca. No wonder that in the last thirty years there has arisen the discipline of astroarchaeology, which combines archaeology with astronomy; most of the research so far published tells the same story, of Sun worship united with practical scientific purpose and cultural celebration.

'To you, to me,' wrote W. H. Auden, 'Stonehenge and Chartres Cathedral are the work of the same Ancient Man under different names: we know what He did and even what He thought he believed in, but we don't know why.' The 'why' is becoming clearer: the desire that places of worship should also serve a practical function. What is noteworthy is how societies living on the margin pour so much of their resources into such constructions, exacting the colossal common effort normally associated with major wars. (Around 1967 a NASA scientist declared that getting man to the Moon was the national integrative effort of modern times, parallel to Egypt's turning itself into one nation by building the pyramids.)

Worldwide, many ancient stone constructions have been so carefully crafted that they have lost nothing of their accuracy. But, because the Earth shifts its position over time (about one degree every seventy-two years), a temple or instrument originally aligned with the Sun will no longer face in precisely the right direction. Travelling in India in the autumn of 2006, I visited the famous Jantar Mantar ('Instrumental Calculation') at Jaipur, one of five observatories erected by the Maharaja Sawai Jai Singh II (1688–?) and the grandest of them all, consisting of sixteen huge instruments, in pink and yellow limestone and marble, set down in a vast park like a giant's playground. Throughout the observatory, the instruments still function effectively, although the one that measured the equinoxes was slightly off, because of the Earth's precession:* the Sun's shadow should have crossed from one

---

*The Earth's axis is engaged in a dance, and as it makes its longer loop, taking about 25,800 years to complete each circuit against the sky, this wobble created by the poles shifts the Earth's incline – only slightly over any human lifetime, but enough over the course of the millennia to change the apparent positions of the stars at any given season of the year. There are also seasonal shifts within the year. If a particular star or group of stars is directly overhead at, say, a given midnight, one finds that they gradually change position as the days, weeks and months go by, until after a year has passed they are directly overhead once again: a process known as 'precession'. In the same way, the Sun's position changes, relative to the stars, until after a year it too returns to the same point. Its annual movement makes it appear to be slipping back from west to east compared with the stars, which move in the same direction – which is why the sidereal day, i.e. the time required for a complete rotation of the Earth in reference to any star, is about four minutes shorter than the solar.

side of its wall to the other on 21 September, but while I was there it did not do so for two more days.

☼

Samuel Johnson's hero Rasselas describes the pyramids of Egypt as 'the greatest work of man, except the wall of China'.[14] To another visitor they appeared 'the Sun's rays made stone'; but however described they are the supreme manifestation of a hierarchy whose kings were inextricably linked with the Sun, and who commissioned these astonishing structures as tombs for themselves and their families. Each tomb symbolized the primeval mound that grew up from the waters during the world's creation: it was believed to form a gigantic stairway to the heavens, where a pharaoh's soul would become one of the 'immortal' stars. The two great moments of solstice, towards which the earliest pyramids were aligned, were seen as gateways for the royal souls on their journey into and out of life.

Just the empiric facts take one's breath away. The Great Pyramid of the Pharaoh Khufu at Giza, on the outskirts of where the ancient metropolis of Memphis once stood (Cairo now spreads over the site), is the oldest and sole survivor of the seven wonders of the ancient world, and ranked for millennia as the tallest structure on Earth. Napoleon, fascinated by all things Egyptian, calculated that there was enough stone in the three Great Pyramids of Giza to build a wall three metres high (he neglected to say how thick) around the whole of France. Although there were imitations at Thebes and elsewhere, the first ones were built in Middle Egypt. The Egyptians knew next to nothing of engineering – they simply multiplied wedge, lever and inclined plane by enormous manpower – yet the maximum inconsistency between side lengths of the Great Pyramid is astonishingly less than 0.1 per cent. Herodotus asserts that the Great Pyramid was built by 100,000 slaves (a more credible figure is 25,000) fed on onions and garlic, who laid an average of 340 blocks a day over two cruel decades.

The most intense period of pyramid-building stretched from 2686 BC to 2345 BC. Beyond the Great Pyramids, there were about eighteen primary sites and some twenty-nine others, each part of what is a single architectural complex. About a hundred of these monuments were built,

The Great Pyramid (the tallest) at Giza, built c. 2800 BC
out of 2.3 million stone blocks

put together from polished, pearly-white limestone, red granite, quartzite, alabaster, mud brick and clay-lime mortar. The largest rests on a base 756 feet square, and the tallest reaches 481 feet, at an inclination of just over 50 degrees.*

The very first pyramid – which is also the world's oldest existing construction in stone – was built at Saqqara, sixteen miles south of Cairo, at the beginning of Dynasty 3 (2690–2610 BC). The cult of Ra reached its zenith under Dynasty 5 (2494–2345 BC), by which time the pharaoh (which means 'great house', a reference to his final residence) had become identified as son of the Sun god. Neuserre (c. 2465–c. 2325 BC),

*The true derivation of the word 'pyramid' has various contenders. The ancient Egyptian term 'pir e mit' has been variously translated as 'division of ten', 'division of number' and 'division of perfection'; also as 'fire in the middle'. The Greeks, possibly jealous, are said to have taken up a derisive term for the structures, *pyramides*, meaning either the little wheaten cakes sold on street stalls or 'storehouses for grain'. There would have been some justification to the latter barb. Christian Europe during the Middle Ages based its idea of Egypt mainly on the Bible, assuming that the pyramids were Joseph's grain storehouses as described in Genesis 41–42, and calling them 'the Barns of Abraham'.

sixth king of Dynasty 5, even had his burial pyramid at Abu Jirab in Lower Egypt made smaller than his Sun temple, as a mark of respect – and possibly fear.

During the reign of Akhenaten and his wife Nefertiti, between 1379 and 1362 BC, the Sun was the all-embracing deity of the Three Kingdoms. After Akhenaten's time, it was demoted from supreme godhood but was still an object of worship: Ramses II (1279–1213 BC) had two temples cut out of the living rock on the west bank of the Nile, south of Aswan, in what today is north Sudan. The taller edifice is 119 feet from base to summit, with four colossal stone gods 67 feet high, atop which stands a row of carved sacred baboons, the Watchers of the Dawn, who as allies of Ra in casting back the forces of night are depicted with arms raised in adoration of the rising Sun. The temple is oriented so that, on each 22 February and 22 October (interestingly, eight months apart, not six), the morning Sun fills the entire length of the elaborate passage chamber – just as at Newgrange – and illuminates four of the five gods against the back wall of the innermost shrine, but not that of Ptah, the primal creator, who dwells in the dark.*

The pyramids have been subject to almost as many theses as Stonehenge. Like that great monument, they clearly had a powerful connection to the solstices. Their basic orientation, established by wooden stakes and ropes, was at first north to south (the solstice axis), then, with those built from the Fourth Dynasty on, east to west (the axis of the equinoxes), even though the entrances remained on the north face. The Egyptologist Martin Isler explains what had to be accomplished:

> The linkage between the rising Sun and a pyramid oriented by its sides to the cardinal points is straightforward. As the Sun travels from winter to summer solstice, it moves along the horizon in an arc of about 50 degrees from north to south at the latitude of Giza. To equate the king's pyramid with the Sun . . . required that the sides of

*Because they lay so far upriver, the temples were unknown until their rediscovery in 1813. With the enlargement of the Aswan Dam in the 1960s, they seemed doomed to being flooded under Lake Nasser. Between 1964 and 1966 UNESCO and the Egyptian government took both temples apart block by block and reassembled them on top of a cliff 200 feet above their original site.

the structure face due east and west. Due east is the midway point for the Sun as it moves between the summer and winter solstices. It is also the only position in which a structure square in plan can have one side face the birth of the Sun in the east, another face its death in the west, and still have its north side directed at the circumpolar stars.[15]

Most pyramids oriented to the equinox were aligned so that, on that date, they seemed to swallow the setting Sun, which must only have increased the witnesses' awe. Their orientation also allowed them to be used to mark the Sun's course (an ancient text refers to 'the shadow of Ra' and 'the stride of Ra') – so that they served as both solar temples and seasonal clocks. Along with their adjoining temples, the pyramids are the surest and grandest evidence of how worship and astronomy have gone hand in hand.

Whole civilizations have married their concepts of time and space with their relationships to their gods, creating a bridge between heaven and Earth. However, the religiously oriented commemorations, solstice points and other solar phenomena are not the work only of ancient or pre-civilized societies. The great cathedrals of Europe face east to declare their sacredness. Although Pope Leo the Great (440–61) forbade Christians to worship the rising Sun, its veneration was deemed acceptable so long as the Sun was seen only as a symbol of Christ entering his church – a doctrine more easily decreed than enforced. To prevent sunlight (which was both distraction and rival) from streaming through open church doorways, Pope Vigilius (537–55) commanded that all places of worship face their apses, rather than their main entrances, east. Only occasionally was dispensation granted for a cathedral to have its doorway aligned with the sunrise of the winter solstice and the setting Sun of the summer solstice, or for it to lie with its main axis facing the rising Sun on a day of special significance, such as its patron saint's festival.*

---

*Medieval man would talk of 'walking widdershins', meaning that it would bring harm to walk around a church in opposite movement to the Sun; walking sunwise around free-standing megalithic monuments survived from the remotest past, and even in the seventeenth century, in parts of Scotland, local people walked sunwise around a stone called 'Martin Dessil' – the Gallic word for clockwise being 'deasil'.[16]

No matter that the Church goes through such periodic renunciations of the Sun. The fact remains that its holy days are rooted in pagan solar festivities, and the astronomers who study the Sun have for centuries been in the employ of the Church. I have walked across a 1798 gnomon – a marked surface that shows the time by registering the shadow cast by the Sun – outside the main church of Bergamo, in northern Italy. Another gnomon, one that establishes the spring equinox, stands at the end of the brass strip, or 'rose line', that runs across the nave of St-Sulpice, in Paris, made newly famous in *The Da Vinci Code*. There is a long history of telescopes being placed in Italian cathedrals, and during the seventeenth and eighteenth centuries the cathedrals of Bologna, Rome, Florence and Paris all served as solar observatories.[17]

Across the Atlantic, St Mark's Church-in-the-Bowery was built and consecrated in 1799 at what is now Stuyvesant Street and lower Second Avenue in Manhattan. After a fire in 1978, restorers uncovered a rose

The bridge at Turtle Bay in Redding, California, by Santiago Calatrava.
It opened on 4 July 2004, and is the world's tallest sundial

window in its south wall aligned with the altar. In 1983 a stained-glass rose window succeeded it, so that at noon on the spring equinox the Sun strews the nave with streaks of coloured light.

As with the sacred, so with the secular. Some two miles north, in the plaza of the McGraw-Hill building on Sixth Avenue at 48th Street, stands the fifty-foot steel 'Sun Triangle', installed in 1973. The steepest side points to the Sun's zenith at highest summer, the lowest arm to its place at noon on the first day of winter, about 21 December. Another recent example is the huge sundial bridge built by the Spanish architect, sculptor and structural engineer Santiago Calatrava between 1996 and 2004: 'tilted like a catapult ready to be sprung',[18] it spans the Sacramento River in Redding, California, its north–south alignment making the pylon's shadow indicate the time of day.

Twice a year, in the borough of Manhattan, the Sun performs a neat trick of its own. Because most of the island is laid out on a grid, and the grid is canted 28.9° east of true north, the rays of the rising and setting Sun are essentially at right angles to its great avenues. This meant that, on 30 May 2009, the setting Sun could be seen shining straight down all the borough's east–west streets at once, from 14th Street (where the city's grid proper begins) on up. In the days immediately after, the point of sunset continued to shift north, right up until the summer solstice three and a half weeks later. Thereafter, that point migrated southwards again, until sunset was once more lined up with Manhattan's cross-streets on 11 July at 8.27 p.m. 'Perhaps,' Neil deGrasse Tyson, director of the Hayden Planetarium, suggests, 'anthropologists in the distant future will presume the Manhattan grid to have some astronomical significance, just as we have found for Stonehenge.'[19]

4

TERRORS OF THE SKY

On the twentieth day, an eclipse befalls. The king is slain, and a nobody seizes the throne. On the twenty-first day, another eclipse. Devastation. Corpses are found throughout the country.

Babylonian Table of Portents, 1600 BC

Nothing can be sworn impossible
. . . Since Zeus father of Olympians
Made night from midday
Hiding the light of the shining Sun
And sore fear upon man.

Archilochus of Paros (714–676 BC),
describing a solar eclipse

Auroras and eclipses, those great celestial dramas, are two of the most awe-inspiring occurrences in nature, and for thousands of years our fear of them has triumphed over whatever the scholars of the ancient world and onwards have told us about their causes.

Auroras are the flares that appear far over our heads, the most spectacular and best known being the Northern Lights (aurora borealis) in the northern hemisphere and Southern Lights (aurora australis) in the southern. They often start off as a mild glow, hugging the horizon; bright patches (called 'surfaces' by scientists) may follow. Next an arc appears, stretching 'like a fluorescent basket handle across the sky'.[1] Finally, the entire heavens will be 'wreathed in celestial fire'. Auroras typically swirl

50 to 200 miles above the Earth's surface, although some have come as close as 35 miles. The flood of particles shot out by the Sun reacts endlessly with atoms and ions in the atmosphere to generate photons – the basic particles of light – and when enough of these collisions occur with sufficient intensity, they set off blazing displays. Given the Earth's rotation (and because the arriving particles are constantly changing direction as they strike the thin upper air), the auroras seem to move across the sky. They are visible mainly at latitudes within the Arctic Circle (most frequently during the equinoxes), and on the shores of the Great Lakes will appear a few dozen times each year; but on at least two occasions, auroras have manifested simultaneously in the northern and southern hemispheres – first on 16 September 1770, when one was observed to the north, while Captain Cook recorded another over the South Pacific; then again in October 2001.

Mercury and Jupiter, like Earth, are magnetized, and any planet with a magnetic field and an atmosphere will probably display auroras. But with the sole exception of Jupiter, which displays even more intense and powerful auroras than ours, none can compare with those found on Earth in amplitude, duration, luminosity or brilliance. One observer has described them as 'Gargantuan, ghostlike arms, chasing and darting, appearing and disappearing spontaneously',[2] while another recorded 'different forms, wavering, many colours diffusing and changing, sometimes far away, sometimes filling the heavens around and above, plunging great dropping spears and sheets of colour earthward towards your very head as though a great hand were dropping colour like burning oil'.[3] Scott, who lost to Amundsen in the race for the South Pole, wrote how overwhelmed he felt before an aurora: 'by its delicacy in light and colour, its transparency, and above all by its tremulous evanescence of form . . . the appeal is to the imagination by the suggestion of something wholly spiritual.'[4] Auroras even make themselves heard, giving off a 'whistling, crackling noise',[5] as in 'the waving of a large flag in a fresh gale of wind'.[6] The late-eighteenth-century American historian Jeremy Belknap compared their sound with a muted rustling or swishing, 'like running one's thumb and forefinger down a silk scarf'.

The giant plumes of gas that burst on the edge of our atmosphere were first given poetic life in the Apocryphal Books of the Maccabees ('horsemen charging in mid-air, clad in garments interwoven with gold').[7] In Norway, they were sometimes seen as Valkyries ('choosers of the slain'). The Inuit believe they are the dead reaching towards those left behind, or spirits playing football with a walrus skull. In the Middle Ages the lights were taken as the cause of plague or disaster, while during the English civil wars of the seventeenth century people saw 'a fiery shaking sword with a blue handle', and as recently as 1802 Walter Scott wrote: 'He knew, by streamers that shot so bright, / That spirits were riding the northern light.'[8]

After the aurora borealis of 14 November 1837, one scientist in New Haven reported that at about six in the evening the sky was thick with falling snow when everything 'suddenly appeared as if dyed in blood'.[9] Trees, housetops, the entire atmosphere, were tinged with scarlet, so much so that people thought the town on fire.* The aurora lasted for half an hour before diffusing 'like the light of an astral lamp'. Elsewhere in the US, great streamers appeared – bright red vertical columns alternating with white clouds – before fusing at their apex. Then, at about 9 p.m., innumerable arches of brilliant scarlet shot up from the northern arc of the horizon, like the striped inside of a gorgeous circus tent. Similar phenomena were reported that night in Pennsylvania, Maryland, Georgia and Ohio – sometimes accompanied by what observers called 'merry dancers', great plumes of colour liberated into movement. Simultaneously, auroras appeared over many parts of Europe, 'a patch of the most intense blood-red colours ever seen . . .' Such vivid colouring has a simple explanation. Atmospheric nitrogen will make auroras purple and violet, while oxygen makes them green and red (from interactions that take place at higher altitudes). But the lights can also appear in reddish-pink and greenish-yellow, or with a violet sheen intermingled with pastel pinks and greens.

*In AD 37 the Emperor Tiberius saw a vivid reddish light in the sky and, as Seneca reports, 'battalions were hurriedly dispatched to Ostia [the vital seaport at the Tiber mouth], in the belief that it was burning'. Then, in London in 1839, fire brigades rushed north to put out a conflagration that turned out to be an aurora; in January 1938 they were called out again, this time in the belief that fires were sweeping Windsor Castle. The Associated Press reported 'The ruddy glow led many to think half the city was ablaze'. Contrariwise, in Daphne du Maurier's *Rebecca*, Maxim de Winter, driving frantically back to his Cornish mansion, sees the evening sky emblazoned in garish orange. 'That's not the Northern Lights,' he cries, 'that's Manderley!'

An aurora borealis, 1 March 1872 at about half-past nine in the evening

Although auroral glows have been known to reach as far equatorward as Italy and France, they are seen that close only about once every ten years. We still do not know exactly what sets off these spectral light displays, although in 2007 NASA launched five spacecraft, each about the size of a washing machine, to try to solve the mystery. Initial findings show that they follow the snapping of magnetic fields – similar to the blowing of an electric fuse. The scientists involved hope to have more information before the next peak of solar storms arrives, between 2010 and 2012.[10] Meanwhile auroras* continue to fill us with wonder – which may be why so much superstition still surrounds them.

☼

*Very different from any aurora is the famous but extremely rare green flash, usually described as 'emerald green' – 'like copper coins igniting on our retinas,' as one witness remembers it – which has a literature all its own, Jules Verne claiming 'a green that no artist could ever obtain on his palette, a green of which neither the varied tints of vegetation nor the shades of the most limpid sea could ever produce the like. If there is a green in Paradise, it cannot but be of this shade.'[11] These flashes appear at sunrise and sunset, on the horizon, usually over the ocean, when some part of the Sun suddenly seems to change colour, and there may be a few seconds when the sky – the Sun's red light blocked by the Earth's rim – turns green (or even violet). The phenomenon has been seen in the Caribbean, the Mediterranean, in Japan, over the waters of the Antarctic, even in the equatorial Peruvian jungle at sunrise.

The most economical definition of an eclipse comes from an Italian book, *Eclipses: An Astronomical Introduction for Humanists*, and runs: 'An eclipse occurs when three celestial bodies are in the same direction, so that one of them obstructs the reciprocal visibility of the other two.'[13] What we now know to be the plane of the Earth's orbit around the Sun is called the 'ecliptic', because eclipses can take place only when the Sun and Moon are both situated on a plane with the Earth's path. In short, a lunar eclipse is caused by the Earth coming between the Sun and the Moon, blocking the sunlight that usually reflects off the Moon and makes it visible. A solar eclipse occurs when the Moon passes between the Earth and Sun, obscuring the Sun. Our Moon is unusually large – both actually to other moons, and relatively to the planet it orbits – and for a total solar eclipse to take place it has to be in exactly the right place, and the Earth has to be exactly the right distance from the Sun. Sometimes, when the Earth and Sun are at their closest, making the Sun seem slightly larger, the Moon will mask only about ninety-eight per cent of it, leaving a circle of light visible and causing what is known as an annular eclipse (from the Latin *annulus*, a ring) – possibly an even more dramatic sight than a total darkening. Usually the Moon cuts across only a section of the Sun, causing a partial eclipse; on such occasions only about one sixth of the Earth's surface encounters some part of the Moon's shadow.

One day long in the future there will be no total eclipses. Observed from the Earth, the Sun and the Moon (in this epoch, at least) have almost the same apparent diameters, as their sizes are in the same ratio to their distances: the Sun is just over 400 times broader than the Moon

---

In March 2006 a photographer captured the even more rare blue flash: an intense beam shooting from the crest of the solar disc at sunset over East Lomond Hill in Fife, Scotland. For this to occur, the atmosphere had to be extremely clear, with a layer of increasingly warm over extremely cold air on the ground – a double layering that intensifies the atmosphere's prismatic qualities, so that the bent sunlight lost the other colours of the spectrum to produce a flash of blue.[12] The science fiction writer Arthur C. Clarke was once sent a photo showing two suns, side by side. The sender was on a cruise ship in the Red Sea and was unable to explain what had occurred; nor could Clarke. The double image was actually the real Sun plus a 'sun dog' or 'mock sun' – a bright patch to one side of the solar disc and usually manifesting slightly below it, caused by the refraction of light from slow-falling ice crystals, hexagonal shapes, like thin slices of a pencil, that bend the light at 22°. But these were all cases of refraction, not of the interaction of particles.

(866,000 miles in diameter as against 2,160), but just under 400 times further away – an extraordinary coincidence, which over hundreds of millions of years will not endure, because the distances between the Sun and the Earth, and the Earth and the Moon, are very slowly increasing. The Moon initially orbited close to the Earth, perhaps just 16,000 miles distant; today it averages 238,000 miles. The slowing effects of tidal friction on our planet will continue to press the Moon into an inexorable compensatory outward spiral, at the rate of 1.5 inches a year. Eventually, some hundreds of millions of years hence, it will be too far off to cover the Sun from our view.

For as long as we have records, eclipses have awakened fear and end-of-the-world imaginings, independent of mythology or culture. Eclipses as well as other heavenly occurrences that were believed to be almost impossible to predict, such as auroras, comets and meteor showers, are described in a Greek word meaning that the stars are aligned against you: *dis-astra* – 'the stars are against'. 'Eclipse' itself comes from the Greek and means abandonment or forsaking. The Chinese word translates as 'being eaten away' (*rìshí*, literally 'Sun-eat'); in Spanish the word means 'to make dim or indistinct', and in Russian 'to black out'. For the Cakchiquel tribe of North America it glosses as 'the Sun carries sickness' – a belief shared by many cultures. In Asian folklore, children born at the moment of deepest darkness will emerge mute or deaf. Even today, many Thais hoard black joss sticks and jelly and sacrifice black chickens to Rahu, the Sun-devouring demon, much as their ancestors did, and give eclipse-tainted goods as alms to beggars in the belief that their sins will be redeemed by their donations.

It is in India that the most enduring superstitions are found. Only a little over a decade ago, in Jajai, a village in the Punjab, pregnant women were warned not to venture outside during an eclipse lest their babies be born blind or with a cleft palate. Food cooked before an eclipse must be thrown out as impure; anyone wielding a knife or axe is believed to be risking an accident. Cows are kept indoors – this being Hindu India – while children must either go to temple or remain at home.[14]

On 22 July 2009 the longest solar eclipse of the century – six minutes and thirty-nine seconds – was best seen from Varanasi, the holy city in north-central India. Three years before, I had travelled there and met a

Sanskrit scholar from one of the city's five universities. We were sitting in his courtyard with cows and rabbits wandering about, and his children, all nine of them, peeping around corners to take in the strange visitor. 'I was silly, and was away from home during the last eclipse,' reflected Dr Sudharka Mishra, pulling at his worn white dhoti. 'Nobody took the cows inside, with the result that that one' – here he pointed to a light-brown three-year-old – 'was born blind.' Sure enough, great mounds of excess skin rose up from the cow's lower eyelids, making it almost impossible for the poor animal to see.

Dr Mishra had gone on to explain that for Hindus solar eclipses were *kalki*, 'polluting'. The obliteration of the Sun, even for a moment, was a terrible omen. When such an event was due, a special ritual known as a homa was held, which involved an offering to a consecrated fire, followed by the Suryagrahana (solar eclipse) puja rite and the chanting of hymns; pregnant women had their bellies smeared with a mixture of ghee (clarified butter) and cow dung, as stipulated in the Vedic scriptures.* In Varanasi, after an eclipse, Hindus make their way to one particular pond in the centre of the city and immerse themselves. I had seen the pond, about fifty feet square: it must have been packed. Ritual cleansings like this are still carried out all over India.

Because early societies experienced eclipses as a crucial disorder, they tried to incorporate them into their world view as something that could be readily understood. In the Qur'an, they are punishments from God, and one must abase oneself to avoid further divine displeasure. North American Indians believed eclipses were the fault of dogs and coyotes; South American tribesmen blamed the jaguar; the Bakairi of Brazil believed an enormous blackbird spread its wings over the Sun; the list goes on.[16]

---

*Ghee is known to have detoxifying properties, while cow dung contains a substance similar to penicillin. In ancient cultures, dung (the fresher the better) was used to combat a range of illnesses, and its particular power to absorb radiation is described in the Vedas. These beliefs persist. In 2005, after Pakistan tested a nuclear weapon, one Indian company advertised its brand of house paint by announcing that it had been mixed with dung: 'Coat your walls with our paint and be free from atomic fallout!' Following the Chernobyl catastrophe of 1986, the Ukrainian peasantry sealed their huts with dung; and even NASA's manned space capsules are said to be coated with a treated layer of dung as a form of protection against the Sun.[15]

Following an eclipse, any setback – the death of a monarch, defeat in battle – would be carefully noted, in the belief that any future darkening at the same point in the solar year would inflict a similar misfortune. Thus it became important to be able to predict when eclipses might occur, and the early glimmerings of science were put to work to shed light on what the future might bring.

It was the Babylonians who discovered that solar eclipses form a pattern. In the course of a year, the Earth's passage makes it appear as if the Sun is moving through a circle. When the Earth has gone 19 times around the Sun and the Moon 223 times around the Earth, a particular eclipse of exactly the same type will repeat itself; so had there been a lunar eclipse at dawn on (our) 18 May 603 BC, say, one could have forecast that an almost identical eclipse would occur at sunset on 28 May 585 BC . To recognize such a pattern over so extended a period of time was an extraordinary accomplishment for so early a civilization.[17]

By the end of the third millennium BC the Babylonians were maintaining a chart of Sun and weather omens (such as lightning, thunder and earthquakes). With this in hand, they tried to outwit the heavens: when an eclipse was forecast, some lowly individual would be selected to stand in for the king until the event had passed. To make it clear that it was he who was to suffer the impending evils, this 'substitute king' would be forced to listen to a reading of the portents.[18] The true king remained out of sight, undergoing intensive rites of purification. Letters written to the real ruler would address him as 'Farmer', but, once the eclipse had passed, the ministers would enquire of the 'Farmer' on which day the hapless decoy should meet his fate, to be mourned as if he were indeed a fallen sovereign.

The Babylonians would measure the eclipse's degree of totality by stretching out the hand at arm's length in front of the eye; this method was later adopted by the Greeks, whose word for 'finger' or 'finger's breadth' also meant a unit of measurement, twelve fingers (oddly) defining a total eclipse.[19] The first total eclipse for which we have records is set down in the Chinese book *Ch'un-ch'iu* as having occurred on 17 July 709 BC, a date agreeing exactly with modern computations.[20] Ancient China left behind a huge number of records; up until the Qing Dynasty took over in AD 1644, there are about a thousand references to these heavenly darkenings. Exact forecasting was desperately important – hence those who managed it won considerable prestige, and people were not slow to hazard guesses, however wild.*

But forecasting was no easy matter. The Babylonians may have achieved it to a certain extent, but few other cultures did. To predict a solar eclipse in Herodotus's day, for example, one would have had to know certain astronomical facts that came to light only in the third century BC. In the second century BC Hipparchus – according to Pliny – in what one might call a solar legend, forecast the motions of the Sun and Moon for six hundred years; three hundred years later, in about AD 150, Claudius Ptolemy's thirteen-book description of the universe, the *Almagest*, laid out a purportedly scientific system for judging eclipses. Such was his authority that his views, for all their inaccuracy, enjoyed wide and prolonged currency, and some 1,500 years later, in 1652, the volumes dealing with such matters were republished as 'an easie and familiar method whereby to judge the effects depending on eclipses'.

For hundreds of years these dread occasions were predicted and recorded as warning, criticism or confirmation, as appropriate. During times of civil war in medieval Japan, the court astrologers tended to over-predict – if nothing happened, they were rewarded for having prevented any occlusion. Chinese astronomers, on the other hand, would use pre-dictions in highly political ways. If they disapproved of their ruler, they would meticulously record an eclipse that had occurred, which could be taken as a divine rebuke to the government, because eclipses were con-sidered warnings from heaven. They might even announce a totally fictitious one, as happened in 186 BC during the reign of the deeply unpopular Empress Kao Tsu. However, if they approved of their rulers,

---

*There is the story, almost certainly legend, of two court astronomers, the brothers Hsi and Ho, who in 2136 BC got so drunk that they failed to predict an eclipse, and thus were unpre-pared, when one took place, to shoot the customary arrows into the sky and beat drums and gongs. The Emperor Chung-K'ang believed that the land had been exposed to the anger of the gods, and had the brothers summarily executed – a harsh punishment indeed, given that the eclipse was only partial. Joseph Needham, in his history of Chinese science, affirms that these names do not refer to actual individuals but to the mythological being who was sometimes the mother, sometimes the chariot driver of the Sun. The name, he explains, 'in some way became split up and applied to four magicians or cult-masters who were charged by the mythological emperor to proceed to the four "ends" of the world in order to stop the Sun and turn it back to its course at each solstice'. All Chinese documents up to this date were burned shortly before the onset of the Christian epoch.[21]

astronomers might just as readily omit all record of an eclipse that had actually taken place. The belief that eclipses presaged misfortune kept Chinese astrologers close to the seats of power. One might imagine that a shortage of ensuing disasters would have devalued their status, but the fates threw up just enough occasions to support their authority. For instance, almost immediately after the eclipses of both 4 March 181 BC and 18 January AD 121, a dowager empress of China met her death; and less than a year after the eclipse of 28 August AD 360 the Emperor Mu died, aged only nineteen.

Humankind seems to have an unquenchable thirst for superstition, and most societies have assumed that great events in the universe at large are conjoined with those in the human world; in the case of eclipses, they were often believed to be markers of the death of a famous person or of some other momentous occasion. 'Every time after such an accident,' Machiavelli says scornfully, 'you get some such extraordinary event turning up.'[22] Thus, random acts of history seemed to confirm such superstitions. Pragmatism could sometimes triumph over folklore, however. In his *Life of Pericles*, Plutarch recounts that an annular eclipse (now identified as that of 3 August 431 BC) so terrified the Athenian sailors that they threatened to desert an expedition against the Lacedaemonians. Seeing that the pilot of his galley was filled with alarm, Pericles doffed his cloak and cast it over the man's head. 'Is there anything so terrible in that?' When the man gave a muffled no, Pericles asked what the difference was between cloak and eclipse, except size. The fleet set sail.

The Emperor Claudius also deployed common sense before the eclipse of 1 August AD 45, which astrologers had told him would fall on his birthday. When they went on to say that their omens predicted rioting against him as a consequence, Claudius put out a public announcement of the imminent event, describing what kind the eclipse would be (annular) and explaining to the citizenry how such celestial happenings come about.[23] His birthday passed by uneventfully.

Foreknowledge of eclipses could be used to exploit superstition, as well as to defuse it. More than once, a ruler whose priests had predicted a solar eclipse would rely on the panic he knew would ensue to send troops to storm an enemy city. In July 1917, after a long desert trek, T. E. Lawrence and his Arab irregulars closed in on Aqaba, now Jordan's only port, which they planned to take from the Turks. Lawrence knew that a lunar eclipse was due on the 6th and held back his troops until the garrison, alarmed by the loss of the Moon, set about frightening off the devouring monster with the usual assortment of drum-hammerings, klaxons and shouting. That gave Lawrence his moment. Aqaba fell.[24]

Sometimes, though, there really was a consequence on Earth for what happened in the heavens. In 1183 a brutal three-year civil war between the Minamoto and Taira clans was ravaging Japan, but just as the Minamoto were preparing for battle the Sun was put out. Having no

The armies of the Medes and Lydians were locked in battle on 28 May 585 BC, part of a gruelling five-year war, when 'the day was turned into night' and the fighting stopped at once

rational explanation for such a calamity, they took to their heels.[25] At other times the connection between eclipse and event appears to have been fabricated. There is a famous case, datable to 28 May 585 BC, when the Medes and Lydians, more than five years at war, made peace, supposedly because an eclipse put a sudden end to a major battle. Herodotus and Pliny both recorded such an event as taking place, the former writing that the great astronomer Thales had predicted a solar eclipse for that exact year (though he was no more precise than that). Even so, the conjunction is most likely mythical, with both historians guilty of political spin, combining an eclipse and a battle from two different dates. Herodotus also seems to have described an eclipse that never took place at all, describing how one shadowed Xerxes of Persia (519–465 BC) as he set out to conquer Greece; modern computations can find no eclipse at the time.*

Eclipses seem to have unhinged even normally authoritative sources. The *Anglo-Saxon Chronicle*, that register from the reign of Alfred the Great to the mid-twelfth century, records:

> In this year [AD 1135] King Henry went across the seas at Lammas [1 August] and the next day, when he was lying asleep on board ship, it grew dark over all lands, and the Sun became as if it were a three-nights-old Moon . . . People were much astonished and terrified and said that something important would be bound to come after this – and so it did, for during that same year the king died.

Henry I, whose old age the *Chronicle* paints as racked with misery, did indeed die exactly two months later, but the eclipse described actually took place two years before. Another legend has the heavens similarly marking the early death of Anne Neville (1456–85), wife of Richard III: 'Rumour ran that the Sun had hid its face because King Richard had poisoned his Queen.'[27]

---

*An annular eclipse is said to have taken place on 7 March 51 BC, the day Caesar crossed the Rubicon. Homer makes two possible allusions, once each in the *Iliad* and the *Odyssey*, but neither reference can be made historical;[26] he creates a third that is surely fictional when he has Odysseus kill Penelope's suitors during an eclipse. (Similarly, Aristophanes, in *The Clouds*, and Borodin, in *Prince Igor*, both use eclipses for dramatic effect.)

Eclipses have been used to date historic events – thus one is recorded eleven days before Alexander overthrew King Darius at Arbela, and by working out events retrospectively one can mark the battle as fought on 21 October 331 BC. But such claims are made with varying degrees of accuracy. The great Johannes Kepler precisely calculated the world's creation day as 23 July 3999 BC, and coupled it with an eclipse – indeed, these feature in similar ways in the creation stories of several cultures, and in attempts to date important events in religious history. 'The birth of our Lord and Saviour is itself quite positively fixed,' contended one nineteenth-century divine. 'It is said in the Bible that Herod . . . died when Jesus was still a child, and probably a few months after his birth. Now, the Jewish historian Josephus tells us . . . there was an eclipse during Herod's last illness; and as this eclipse was known to have occurred in April, in a certain year of what is called the Julian period, it is quite clearly shown that our Saviour must have been born just about the time the Bible states he was.'[28]

An even more ominous darkening, the only one to shadow Jerusalem during Christ's public ministry, is said by Church historians to have occurred on the day of his crucifixion. They describe it as being total, beginning at noon, and passing across the Black Sea from near Odessa on to the Persian Gulf. Matthew, Mark and Luke (probably using a common source) all invoke it in their narratives. According to all three, 'From the sixth hour there was darkness all over the land until the ninth hour', with Luke adding, '. . . while the Sun's light failed'.[29] But Jesus was crucified at Passover, a spring full Moon festival, when the Moon had to be on the opposite side of the Earth. And no eclipse lasts anywhere as long as the three hours that the Gospel narratives proclaim.

Everywhere in the world people have reacted to eclipses with profound fear, yet a simple explanation of what they were was put forward as far back as the sixth century BC, when Thales (c. 624–c. 546 BC) proposed an answer. A century later, his fellow countryman and fellow astronomer Anaxagoras (500–428 BC) was to write that 'The Moon has not any light of its own but derives it from the Sun . . . Eclipses of the Moon are due to its being screened by the Earth, or sometimes by the bodies beneath the Moon; those of the Sun to screening by the Moon when it is new.'[30] For this, however, Anaxagoras was indicted for atheism and condemned to death (he chose exile instead), and the link between eclipses and misfortune continued unbroken for another two thousand years.

Certainly it was still going strong in the early ninth century, when the Archbishop of Lyons appointed by Charlemagne witnessed three lunar eclipses in a single year and recorded that each time his terrified flock took to blowing horns, hammering on metal instruments and screaming, '*Vince, luna,*' urging the Moon to fight back.[31] The Church officially forbade such behaviour (as it would repeatedly do in the centuries to come) but was no match for the strength of superstition – after all, the monster *did* always retreat. Even the great emperor himself was not immune: in 810 Charlemagne was just over seventy when solar and lunar eclipses fell within two weeks of each other (30 November and 14 December) and alarmed him into immediately preparing his will; by the end of January he was dead. Almost thirty years later, on 5 May 840, a solar eclipse lasting nearly six minutes so frightened Charlemagne's son Louis the Pious that he too fell ill and died before June was out. (The bitter struggle for his throne led to the historic Treaty of Verdun, dividing Western Europe into the three major areas we now know as France, Germany and Italy.)

Popular attitudes were little changed even into the seventeenth century, and the ominous conjunction of full solar and lunar eclipses within weeks of one another, albeit rare, was particularly alarming. When this happened in October 1605, Shakespeare used it as a powerful metaphor for a sense of impending doom, incorporating it into the first act of *King Lear*, in which Gloucester invokes the recent horrors:

These late eclipses in the Sun and Moon portend no good to us: though the wisdom of nature can reason it thus and thus, yet nature finds itself scourged by the sequent effects: love cools, friendship falls off, brothers divide: in cities, mutinies; in countries, discord; in palaces, treason; and the bond crack'd 'twixt son and father. This villain of mine comes under the prediction; there's son against father: the king falls from bias of nature; there's father against child . . .

He wanders off, leaving his bastard son Edmund alone on stage. Gloucester is a good man yet a gullible one (as we are soon to see again with his other son, Edgar), and a creature of his time in his acceptance of superstition. Edmund, on the other hand, may be the play's villain but he is also the supreme rationalist, and here has Shakespeare's tongue*:

> This is the excellent foppery of the world, that when we are sick in fortune – often the surfeit of our own behaviour – we make guilty of our disasters the Sun, the Moon and the stars: as if we were villains by necessity, fools by heavenly compulsion; knaves, thieves and treachers, by spherical predominance; drunkards, liars and adulterers, by an enforced obedience of planetary influence; and all that we are evil in, by a divine thrusting on: an admirable evasion of whoremaster man, to lay his goatish disposition to the charge of a star!

Shakespeare's scepticism was shared by his king. That same year, after a ninety-per cent eclipse, James I wrote a satirical letter to Lord Salisbury enumerating all the evils that his subjects might ascribe to the Sun's temporary occlusion. For James, the superstitions attached to forecasting were the devil's works, and he wanted them outlawed. Monarch and playwright alike proved singularly unsuccessful in changing opinion.

John Milton was born three years after the first production of *King Lear*. In 1638 he published *Lycidas*, which laments the death of his Cambridge contemporary Edward King, cast away when his ship foundered off the English coast:

> *It was that fatal and perfidious Bark*
> *Built in th' eclipse, and rigg'd with curses dark,*
> *That sunk so low that sacred head of thine.*

The same year he published *Lycidas*, Milton took a Continental tour, during which he visited Galileo, under house arrest in Florence. The

---

*Shakespeare writes about eclipses in *Julius Caesar, Antony and Cleopatra* (III, ii), *Hamlet* (I, i), *Macbeth* (the witches' cauldron in IV, i , ll. 27–8 contains 'slips of yew /Sliver'd in the Moon's eclipse'), *Othello* ('O, insupportable! O heavy hour! /Methinks it should be now a huge eclipse /Of Sun and Moon, and that the affrighted globe /Should yawn at alteration'; V, ii) and *Henry VI Pt. I* (in the play's opening line, 'Hung be the heavens with black, yield day to night . . .'). They also appear in Sonnets 21, 35, 60, 107 and especially 33, in which he extends the metaphor of a solar eclipse through several lines.

great astronomer was by then frail and blind, yet the two men spoke for some time, and one would like to think that the causes of eclipses would have figured in their conversation; but superstition held sway even over a mind as sophisticated as Milton's. In *Paradise Lost*, his description of the faded splendour of the fallen archangel captures the state of unease that most people still felt in the face of an eclipse:

> *As when the Sun, new ris'n,*
> *Looks through the horizontal misty air,*
> *Shorn of his beams, or from behind the Moon,*
> *In dim eclipse, disastrous twilight sheds*
> *On half the nations, and with fear of change*
> *Perplexes monarchs.**

By the time Milton finished his great work, mentioning eclipses in Book II, lines 663–6 and Book X, lines 410–14, he too was completely blind. He had struggled with poor sight since childhood, and as early as 1647 had already lost most of the vision in his left eye. For the next five years he tried various remedies to stave off his fate, but by March 1652, at the age of forty-three, he had to accept that both eyes had failed him, 'without any hope of day'.[32] In *Samson Agonistes* (1671), he portrays Samson railing in words that might have been his own:

> *O dark dark dark, amid the blaze of noon,*
> *Irrevocably dark, total eclipse*
> *Without all hope of day!*
> *O first created Beam, and thou great Word,*
> *Let there be light, and light was over all;*
> *Why am I thus bereav'd thy prime decree?*
> *The Sun to me is dark*
> *And silent as the Moon.*

Paradise Lost was not completed until 1665, when censorship was severe. One of the chaplains through whom the Archbishop of Canterbury gave or refused publishing licences found this passage objectionable; following the death of Cromwell and in the turbulent days after Charles II returned to power, the lines were seen as dangerous, imputing to monarchs 'perplexity' and 'fear of change'. The poem was allowed to see the light of day only through the intercession of an influential friend of the poet's.

Coincidentally, 29 March 1652 (by the old calendar) – the very same month that Milton finally lost all sight – saw 'Black Monday', a solar eclipse that panicked all Britain. The rich loaded up their coaches and fled from London, and there was a thriving trade in a variety of cordials supposed to allay ill effects. At Dalkeith in Scotland, the poor threw away their possessions, 'casting themselves on their backs, and their eyes towards heaven and praying most passionately that Christ would let them see the Sun again, and save them'.[33] It was left to the diarist John Evelyn to maintain the sceptical spirit: 'Was that celebrated Eclipse of the Sun so much threatened by the Astrologers, & had so exceedingly alarm'd the whole Nation, so as hardly any would worke, none stir out of their houses; so ridiculously were they abused by knavish and ignorant star-gazers.'

Not everyone reacted with such alarm. Samuel Pepys mentions that during a later (spring) occlusion his doctor's daughter, having risen at dawn, was concentrating so hard on composing a letter that not until about nine o'clock – although she was in the midst of an almost total eclipse – did she even notice that 'The light of the Sun look[ed] somewhat dim'.[34] But she was the exception. In Paris, people flocked to confession in such numbers over the days leading up to an eclipse forecast for 21 August 1664 that one priest felt obliged to declare that the event had been put off for two weeks, in order to manage his crowds of penitents.

In the cities of Western Europe, however, the days of superstitious ignorance were numbered, for the march of science was accelerating. Thus, the great astronomer Edmond Halley was able to predict the total solar eclipse of 3 May 1715 (which covered almost the entire south of England for more than three minutes), and so carefully observed it that he could estimate the width of its track. He had readings taken from fifteen different points along its path, which enabled him to calculate the Sun's diameter with greater accuracy than ever before. Thomas Crump, in his wide-ranging survey of the history of eclipses, declares that Halley's observations 'can be taken to be the beginning of modern eclipse astronomy'.[35] Fear and awe remained, but they were now augmented by knowledge.

PART TWO
DISCOVERING THE SUN

Bronze-age petroglyphs from all over the world connect the Sun with fecundity,
as in this drawing from the Camonica Valley, in the Central Alps of North Italy,
of a male figure with the solar disc attached to his phallus

THE FIRST ASTRONOMERS

Of all tools, an observatory is the most sublime . . . What is so
good in a college as an observatory? The sublime attaches to
the door and to the first stair you ascend, that this is the road
to the stars.

Ralph Waldo Emerson, 1865[1]

Astronomy: a fine science. Useful only to sailors. While on the
subject, laugh at astrology.

Gustave Flaubert, c. 1870[2]

In 1894 Sir Norman Lockyer (1836–1920), one of Britain's foremost
living scientists – the man who had discovered helium in the Sun before
it was known on Earth – published *The Dawn of Astronomy*, in which he
broke down the work of his ancient predecessors into three distinct
phases.[3] Civilizations first went through a period of worship, where celestial phenomena were construed only as the actions, moods and warnings
of the gods; next, people began putting astronomy to practical uses, such
as agriculture or navigation; and finally the heavens were studied solely
for the pleasure and purpose of pure knowledge. Though Lockyer's divisions are seen now as too schematic – for instance, we may question
whether the reverential stage is ever completely outgrown, and his definition of the third stage ignores the sense of wonder that often drives
astronomers – his categories still offer a useful yardstick.

Long before mankind gathered into cities, there was astronomy, or at least there were people studying the Sun. We have evidence of observatories in several parts of the ancient world; however, one city above all others deserves to be called the birthplace of an organized and recorded science of astronomy: Babylon.[4]

The country was settled by a people of unknown origin between the Tigris and Euphrates rivers, in what is now the heart of modern Iraq. The surrounding land was called 'Sumer', although another name for this fertile flat area was 'Eden'. The Sumerians began farming about 9000 BC, and roughly 2,500 years later they invented the wheel, cart and plough, transforming every aspect of life – from travel and transport to warfare and industry – and reducing the number of workers required to raise a given amount of food. These were crucial steps in freeing some people to become full-time priests, scholars and merchants. Around 3100 BC they invented a system of writing that incorporated astronomical concepts into two of its characters, one representing a star, the other 'the Sun over the horizon'.[5] By about 2350 BC the Sumerians had been subdued by their neighbours to the north, but it was the conquerors who were transformed: Sumerian remained the language of religion and scholarship, and divining the future by observing the stars became a pre-eminent intellectual discipline.

At first, the Sun was cast as subordinate to the Moon, and worshipped in only two minor cities, Larsa and Sippar. By 1800 BC the growing metropolis of Babylon (about seventy miles south of present-day Baghdad, taking its modern name from the Greek rendering of the Akkadian 'Bâb-ilâni', or 'gate of the gods')[6] had become pre-eminent in the region. Ruled by the Amorite king Hammurabi, who encouraged the careful recording of the movements of the stars, the Babylonians believed the Sun to be a baleful divinity that spied for the high gods and was an unforgiving judge of mankind. A precursor of future images of the Jewish and Christian God, the Sun is depicted as an old man with a long beard, beams radiating from between his shoulders. This figure overtook the Moon in importance and metamorphosed into the Sun Royal, exercising the attributes of earthly divinity and kingship.

The next to subjugate the land, around 750 BC, were the Assyrians, whose kings called themselves 'Suns of the World'. During their campaigns of conquest – in which they would line the walls of each fallen city with the skins of its inhabitants – Babylon sank into political and economic decay. Eventually, a series of revolts culminated in the Assyrians' overthrow in 606 BC, and although the new masters, the Kaldi, or Chaldeans (a loose federation of five tribes of West Semitic origin, from among the southern swamps and coastal plains), held sway for one of the briefest ascendancies in Mesopotamian history, their great king, Nebuchadnezzar (who ruled from 605 to 562 BC), oversaw a renaissance that turned Babylon into the cultural centre of a massive empire. The position of astronomy was assured; special schools flourished not only in Babylon but also in other large towns. Exchanges of ideas among Persians, Jews and Greeks living in the same cities were eased by a lingua franca written in simple characters, Aramaic.

Initially, the Chaldeans – their name now all but synonymous with 'Babylonians' – were content simply to watch the skies, but by around 550 BC they felt curious and confident enough to apply mathematical techniques to their records, while dividing the heavens into zones, the most important being what we now know to be the celestial equator (in later years this would be identified as the path of the Earth's ecliptic journey). Along this they constructed the zodiac (from the Greek 'zo-idion', 'living creature'), a diagrammatic representation of the great figures that they imagined to be marking the Sun's movements. The following mnemonic lists them in order, starting with the vernal equinox in March:

> *The Ram, the Bull, the Heavenly Twins*
> *And next the Crab the Lion shines,*
> *The Virgin and the Scales;*
> *The Scorpion, Archer, and He-Goat,*
> *The Man who bears the Watering-Pot*
> *And the Fish with the glittering tails.*

The Chaldeans' advances were made amid great upheavals. Around 689 BC Babylon had been destroyed by the Assyrian king Sennacherib. The city was then rebuilt in even greater splendour (according to Herodotus, its walls were 56 miles in circumference and 335 feet high), but collapsed once more under the onslaught of the Medes and Persians, to be absorbed into the empire of Cyrus the Great. By the time the

Chaldeans were overthrown in 539 BC they had become famous for their expertise in studying the stars. According to the Greek academic Strabo of Amasia (64 BC–AD c. 23), a special district of the kingdom was 'set apart for the local philosophers, the "Chaldeans", as they were called, who were concerned mainly with astronomy; but some of these, who were not approved of by the others, professed to be writers of horoscopes'. While most of what they recorded was purely descriptive, astrology became the business of interpreting the heavens – not for ordinary individuals, but for king and country. Several branches of this pseudoscience grew up, the first and most important being the art of foretelling a subject's life from the position of the stars, as well as the Sun, at their birth or conception. The Chaldeans would draw on astronomical information to predict human destinies, work out where certain planets were in the sky at a child's birth, and from these circumstances draw conclusions.

Some aspects of the stars' influence were obvious: for example, the

A nineteenth-century vision of Babylonian astronomers observing a meteor trail

Sun's position on the ecliptic was tied to the seasons. If the heavens could tell you the most propitious times for planting or harvesting, why shouldn't they also indicate the best time to marry or to undertake an unavoidable but dangerous journey? Understandably astrologers took to drawing such conclusions about the influences of the celestial bodies, even if these were without scientific basis. Yet without the promptings of astrology, naked-eye astronomy would never have reached so deeply. At first there was no division between the disciplines: one practised astronomy for the purpose of astrology. It was a fruitful partnership. By the fifth century BC astronomers had established a year ahead the expected intervals between sunrise and sunset for various days, the monthly apparent movements of the Sun and much else. They developed the arithmetic required to predict the motions of the Moon and planets, and from there to construct a calendar. Even more important was their discovery of a system to predict the passages of the Moon and the Sun, enabling them to forecast eclipses.

The Chaldeans maintained highly accurate and detailed records, but frustratingly we lack any explanation as to how they constructed their tables. They made their observations from step pyramids called ziggurats, which were attached to temples often the size of fortresses and painted the seven colours of the rainbow. The tower of Babel – portrayed in the Bible as a vainglorious attempt to build a stairway to the stars – is the best known, but the most commanding in all Babylon was another ziggurat that rose over four hundred feet in seven gleaming enamelled levels, its zenith crowned by a shrine that held a table of solid gold and an ornate bed on which each night some young woman awaited the pleasure of the god, or his more assiduous earthly representative.

While much has been made of a climate allegedly perfect for observation – Plato lauded Mesopotamia's cloudless atmosphere and Herodotus wrote that 'the air in these regions is constantly clear, and the country warm through the absence of cold winds' – this was likely based on third-hand accounts. Sandstorms and mirages would have been serious obstructions to following phenomena close to the horizon. Even in the best conditions, it was not easy to distinguish between a planet and a fixed star when gazing on the myriad points of heavenly light.

Observation by itself, then, was not enough. The Chaldeans began to analyse and interpret what they had seen, writing their commentaries in journals, almanacs and star catalogues. By a conservative estimate they recorded 373 solar and 832 lunar eclipses during their brief dominance. A wide range of daily happenings – meteorological events, river levels, fires ravaging the city, the deaths of kings, epidemics, thefts, conquests, plagues of locusts, famines and, of particular importance, fluctuations in the prices of barley, dates, pepper, cress, sesame and wool (always in that order) – were scrupulously transcribed on to clay tablets.[7]*. The Chaldeans also inscribed their calculations of planetary positions on tablets that in AD 509 the Greek writer Heliodorus would name 'ephemerides' ('matters specific to the day') – in effect, working journals, showing the consecutive appearances of the same objects in the sky. About seventy tablets from the main series survive; and of seven thousand notations fully a quarter involve the Sun.

Divination, observation, calculation: in Babylon, anyone who could read studied mathematics. The average schoolchild knew far more about the heavens than do most of us today, who rarely examine the skies. In an age far removed from our modern benefits of gas or electricity, the lights of heaven had no competition, so that a horseman did not carry brands or tapers at night, but rode on under the light of the stars. Yet for all its accomplishments, organized astronomy remained an elitist activity. A college might contain just two or three people genuinely taken up by the heavens – and these were hampered, as astronomers are to this day, by the limits of their technology.

One powerful mathematical tool gave Babylon much of its mystique

---

*The Assyrian capital of Nineveh stood only a few hundred kilometres from Babylon. At the great library established by Ashurbanipal (668–627 BC), at least 22,000 astronomical tablets were collected. In 612 BC, when the Assyrian Empire finally came crashing down, Nineveh was sacked and the entire library destroyed, except for the tablets, which were buried – and thus preserved – under the collapsed walls. Without such a disaster these records would not have survived – to resurface during the 1870s, when a French official in Baghdad, his curiosity piqued by some large fragments of sculpture, began to dig there. It seems extraordinary that such vital sources of information were overlooked for so long. Muslims may have kept their distance because the Qur'an states that the tablets were bricks inscribed by demons and baked in hell, while Christians possibly thought it foolhardy to dig for the buried knowledge of a people so excoriated in the Bible. Another 50,000 tablets, written between 3000 and 450 BC, were excavated from the temple library of Nippur. There may be as many as half a million scattered around the museums of the world, only a fraction of which has ever been deciphered, while countless more still lie under the Iraqi sand.

in the sciences: the place/value numerical system. There were two versions: one used a decimal base, and was the method for day-to-day business and for calculating lunar cycles; the other used a base of sixty, and was for measuring solar cycles. The latter, although more cumbersome, would be employed by various societies for almost two millennia – until the time of Copernicus – and from it sprang the division of degrees into sixty minutes and minutes into sixty seconds.[8] Such a method was far superior to any other in the ancient world, and is curiously close to that of the modern computer.* Wonder was acquiring number as its servant, a crucial development for any science.

Babylonian mathematics – sexagesimal arithmetic, algebraic operations, geometric rules – spread over a far greater area than their astronomy ever reached, and indeed preceded systematic astronomy by more than a millennium. But major shortcomings remained. The early mathematical observers treated time not as a measure but almost as a quality; nor were they interested in quantifying space – there was no concept of absolute distance. No trace appears in Babylonian astronomy of any desire to construct a comprehensive scheme to account for all heavenly phenomena, to investigate their nature and causes, or to take a synoptic view of the universe as a whole. The Babylonians had their lunar calendar, and were the first to divide the day into two twelve-part halves, but their divisions of daylight had to vary with the changing seasons, and they were unable to determine the Sun's exact apparent orbit around the Earth. This shortcoming in turn hampered attempts to work out the shape of the Earth; they never learned that it was roughly spherical. The gnomon (at its simplest, a vertical stick whose shadow is used to tell the time) was beyond them, not least because they had no concept of north, south, east and west as cardinal points: anything approaching compass-like clarity had to await the third quarter of the fourth century BC. They had not even established the sequence of the planets outwards from the Sun, but then no one knew any better. That Babylonian astronomy has its current high reputation may be due to the fact that so many Greek records have perished: clay lasts better than

*In the 1790s, during the French Revolution, radical French mathematicians attempted to grade the circle by dividing each quadrant into a hundred parts, each then decimally subdivided; but the system never caught on. Less scientifically, but more enduringly, the French word for eighty – *quatre-vingt*, four-twenty – survives as a reminder that our ancestors counted on their toes as well as their fingers.

papyrus. However, overall I share Noel Swerdlow's view: 'That the complete development of this science took hundreds of years shows how difficult it was, far more difficult than any other science in the ancient world, both because of the magnitude and complexity of its subject . . . and because it was the first empirical science.'[9]

Around 330 BC the region fell to Alexander, on whose early death his marshal Seleucus took over the greater part of the conquered lands. Sixty years later, the mighty dynast Antiochus I forcibly moved most of the city's inhabitants to the newly founded metropolis of Seleucia, sixty miles north. For the astronomer-astrologers of the old city this was an indignity too far. They kept up their practices in the temple of Bel almost to the end of the first Christian century, but it was their dying gasp. 'Thus ended,' writes James McEvoy, 'the two-thousand-year Babylonian cultural development of watching the heavens.'[10]

☼

Even before Babylon was relocated, its learning had spread to Egypt, Greece and Rome to the west, to India and possibly as far as China to the east. The Egyptians adopted the Babylonian zodiac while rejecting their world view, with its emphasis on portents and fateful predictions and the universe as bad news. Herodotus refers to Egypt as 'a gift of the river': possibly the Nile's annual flood in a cycle ideal for growing crops encouraged a sense of well-being. Their priest-astrologers knew that the Nile would overflow its banks around the summer solstice, when bright Sirius rises, and would make their calendrical predictions accordingly.[11]

From the beginning, the Nile's yearly flooding yielded an unexpected benefit: as the government's main source of income was a land tax, there was a constant need to re-establish boundaries and settle disputes caused by the floods, giving rise early on in Egypt's history to a cadre of skilled surveyors.* The Egyptians, an exceptionally practical people (even if

---

*'Geometry', the pure mathematics of points and lines, curves and surfaces, takes its name from the Greek word for surveying, *ge*, earth, and *metrein*, to measure. 'Algebra', on the other hand, derives from an Arab word meaning 'compulsion', as in compelling the unknown *x* to assume a numerical value. A more colourful notion advanced by some historians is that 'algebra' means 'putting back into its proper place', or 'bonesetting'. The bonesetter in *Don Quixote* is known as *algebrista*, but whether because he resets or because he compels Cervantes never makes clear.

they remained ninety-five per cent illiterate, they had invented writing slightly ahead of the Babylonians, around 3200 BC), devised sophisticated ways of measuring that would eventually be joined to their building skills to create the pyramids. But this was only one of many advances. They tried various ways, for instance, to gauge the Sun's diameter, using gnomons, water clocks and even horses. In this last instance, the horse would be galloped across level ground from the moment the Sun's upper edge appeared on the horizon until the entire disc had risen clear above it – a period of approximately two minutes, in which time the animal was reckoned to cover ten stadia (the term is a corruption of the Egyptian word for 'racecourse'), and it was inferred that, as the motion of the Sun must be equal to that of the horse, its diameter must be ten stadia too. This was hardly good geometry.*

The Egyptians had their own mathematical system at least as early as 2800 BC, but it was crude, and, unlike the Babylonians, they never progressed beyond a basic arithmetic – although they did manipulate this to produce both multiplication and division. With the exception of ⅔, the only fractions they used were in the form of ⅟ₙ. Although the mathematical papyri of around 1800–1600 BC bring off some fairly sophisticated calculations, such as the volume of a truncated pyramid, their geometry could not transact complex computations; nor could they manage the type of calculation necessary to progress from observation to prediction.[12] There is no evidence that they understood the concept of latitude or ever recorded eclipses with any regularity. What did emerge, in the intricate development of their religion and the pronouncements of their priest-astrologers, was a canon of times to be especially careful. And while they derived few predictions from what they saw, they did map stars, planets and constellations and record their movements.

Their understanding of what astronomy could accomplish might be summed up by an inscription on the pedestal of a statue of the

*A modern stade is a tenth of a nautical mile, about 600 feet; but lengths in the ancient world – and indeed on into the seventeenth century – were rarely uniform. A stadion could vary considerably, even within a single society: some say it represented the distance a man in ancient times could cover running flat out without taking breath (reckoned at around 174 metres), and that this governed the length (and naming) of sports stadiums. Others say it originated as a standard length of a ploughed furrow, 600 *pedes* (from *pes*, 'foot'); but the length of a *pes* varied, and in consequence so did that of a stade.

astronomer-astrologer Harkhebi, which dates to the early third century BC: while it tells us little about the man, it does describe his duties. He was to determine the risings and settings of the constellations of the zodiac; to establish the high points of each planet; and from this predict the risings of other bodies. He was also expected to foretell the solstices – 'Knowing the northing and southing of the Sun, announcing all its wonders and appointing for them a time, he declares when they have occurred'[13] – the dates for celebrating the gods' feast days and the exact relation of the spirits of the gods to the pharaoh.

Like most pre-industrial peoples, the Egyptians continuously struggled to impose order on their universe – viewing the Sun itself as needing protection against those forces always lurking to disrupt the continuity of things. In a world characterized by disorder and the frequent calamities of nature, religion joined forces with science, its priests toiling to protect

The family of Akhenaten (Amenophis IV) offers sacrifice to Aton the sun god.
A relief from Amarna, Egypt (1350 BC)

the people through ritual and cult and to create a body of anthropo-morphic explanations for the make-up and workings of the cosmos.[14] The sky was not an empty vault, but a goddess who each night reconceived the Sun – an early example of an immaculate conception – and each morning gave him birth. Even the dead space that separates the heavens from the Earth was a god. In this universe, creation and existence were the product not of impersonal forces but of individual wills and actions; and the various divinities the Egyptians created embodied a network of relationships, at the centre of which basked the Sun.

Over time, their philosophy of creation became yet more complex. Atum, the creator, the 'pre-existing being in whom all existence was inherent and through whose self-realization all creation evolved', brought all matter out of himself – creation was the result of nothing other than his masturbation, a notion to us extraordinary, even ridiculous, but for the Egyptians the self-realization of Atum's own substance.[15] Often Atum would change places with the Sun: the star was his ultimate incarnation. At other times he was the conscious void within which the Sun first arose; the moment of orgasm was as useful a way as any to describe the way in which the universe came into being: a cosmology that suggests that the ancient Egyptians were, by several millennia, the first proponents of a big bang theory.

Such beliefs, and the Egyptians' first struggles towards scientific knowledge, developed in parallel. In these early centuries the Egyptians, like the Babylonians, never attained Lockyer's third stage: it would take millennia before the Sun was studied solely to gain knowledge. But close by was a society that valued knowledge perhaps more than any other in history.

6

ENTER THE GREEKS

Astronomy? Impossible to understand and madness to investigate.

Sophocles, 496–406 BC

Don't confuse progress with perfectibility. A great poet is always timely; a great philosopher is an urgent need. There's no rush for Isaac Newton. We were quite happy with Aristotle's cosmos. Personally I preferred it. Fifty-five crystal spheres geared to God's crankshaft is my idea of a satisfying universe.

Bernard Nightingale, the literary Byron-lover
in Tom Stoppard's Arcadia

The earliest references to astronomy among the Greeks appear around 800 BC in the poems of Homer and Hesiod. By that time their countrymen, like the Babylonians and the Egyptians, had put names to various stars and quantified the risings and settings of the Sun.[1] However, they were not content with merely recording the movements of the heavens: once they took to astronomy in earnest, they enquired into what the skies were made of; the shapes and sizes of the Sun, the stars, the planets and the Earth itself; how far these were from one another, and what revolved around what, in orbits of what form; the number of stars and whether they could be located with systematic accuracy. To this they added a raft of further questions about their immediate world. Exactly how long was the year or the month? When were the equinoxes, and could one pinpoint the exact moment of the solstices? The one matter on which most agreed was that the Earth itself was stationary and central – a belief that would bedevil science for millennia.

Despite that crucial misapprehension, a formidable line of Greek students of the stars extends for over a thousand years, from the time of Hesiod until at least the death of Ptolemy. The muster of ancient Chinese philosophers may have stretched twice as far, but the Greek roll-call is the more impressive. The great German scholar Otto Neugebauer lists 121 astronomers of note – and even that tally excludes figures such as Pherekydes of Syros (the teacher of Pythagoras), Zeno of Elea, Plato and Epicurus, who, while not primarily astronomers, made significant contributions. Many familiar names appear – Aristotle, Euclid, Archimedes, Ptolemy – and others who were giants in their time but less well known now, such as Parmenides, Anaxagoras, Eudoxus, Heraclitus and Aristarchus.

Of course, they built in part on the foundations laid by the Babylonians. Herodotus (c. 484–c. 425 BC) in his only book, *The Enquiries* (more often transliterated as *The Histories*), describes how his countrymen not only copied the Babylonian records and calculations but also borrowed several of their measuring devices, such as that for reckoning the motion of the Sun along the ecliptic. And the first glimpses of an astronomy that went beyond pure observation came from the Egyptians. In the early days of Greek astronomy, however, its practitioners did not even seem to consider the prediction of heavenly phenomena particularly valuable. In the *Dialogue of Phaedrus*, Plato blames his countrymen for not being interested enough in the planets; indeed, astronomy did take some time to put down roots into Greek culture. This mindset changed after the Greeks came into contact with the observation tablets of the Babylonians during the creation of the Macedonian/Greek Empire under Alexander, who dispatched literally camel-loads of Babylonian astronomical tablets to the Greek cities along the Adriatic coast.[2]

By around 650 BC the old aristocratic order had been overthrown, supplanted by a succession of tyrants (then meaning rulers who had seized absolute authority, not necessarily cruel despots) who encouraged trade and economic reform, so that for the first time there was an opportunity throughout the Greek states for a different style of life – enjoying the means to reflect, rather than having to be solely concerned with survival. It was from this point on that, to quote the great scholar of the period D. R. Dicks, 'the Greek passion for rational investigation along mathematical lines . . . converted a mass of crude, observational material into an exact science'.[3]

The first great practitioner was Thales (c. 625–c. 547 BC) of Miletus, a prosperous port serving the whole Aegean. A nobleman by birth, he was also a statesman, engineer and merchant, canny enough, the story goes, quietly to corner the olive oil presses of Miletus and neighbouring Chios just ahead of what he knew would be a bumper harvest, thereby amassing a considerable fortune: proof, he said, that philosophers could make money if they chose.

Thales had travelled to Egypt to learn the elementary principles of practical geometry, a lesson that would inspire him to make astronomy a deductive science. He was the first to posit that the Moon eclipses the Sun when it moves into a direct line between it and the Earth, and the first Greek to maintain that it shone by reflected sunlight. He also is said to have established the times and sequence of the equinoxes. He is amply quoted in Plutarch, Pliny, Cicero and Diogenes Laertius – which is fortunate, as none of his writings survive, and indeed both ancient and modern scholars have built him up into a scientific hero on very thin evidence.

According to Diogenes (b. c. 412 BC), Thales was the first to gauge the Sun's diameter in relation to its apparent orbit around the Earth with any accuracy, estimating it as 1/720 thereof, a ratio that is an exceptionally close estimate of Earth's actual orbit around the Sun. Less impressively, he argued that water was the prime element of the universe, that the Earth floated like a cork on a great sheet of this water and that the Sun was made up of a burning earthly substance. He was on firmer ground when postulating that the duration of the Sun's circuit between solstices was never even, and that the solar year lasted 365 days; but again we know too little of his writings to be certain what he actually said. Ironically, he is best remembered for something he never did: predict the solar eclipse of 28 May 585 BC.*

*Plato adds a less dignified anecdote of the astronomer in his youth being so caught up in his stargazing that on one occasion he tripped and fell down a well, to be promptly teased by 'a clever and pretty maidservant from Thrace', who asked: 'How do you expect to understand what is going on up in the sky when you don't even see what is at your feet?'[4] The first absent-minded-professor joke.

☼

One giant of this time was Heraclitus (535–475 BC), a mystic poet who rejoiced in the nickname of 'He who rails at the people' for his arrogant misanthropy. The 130 fragments of his writings that survive emphasize two aspects of the cosmic order: its continuity and its periodicity, especially that of the Sun. Aspects of his thinking can still be seen in the writings of Darwin, Spencer and other masters of the nineteenth century. He could be absurdly wrong: for instance, he suggests that the Sun is just a foot across, about the size of a shield, and that there is a new Sun every day. But his belief that the universe is composed of worlds that are being created and destroyed in a continuous cycle would have modern physicists nodding in agreement. He understood that nothing is constant – hence his dictum 'One cannot step twice into the same river'.

Another in the line of great Greek mathematician-astronomers was Pythagoras of Samos, who flourished during the second half of the sixth century BC. He is said to have spent twenty-two years travelling in Arabia, Syria, Chaldea, Phoenicia, Greek Gaul (better known today as the French Riviera) and possibly India before settling in southern Italy when over fifty years old. (He also spent some time in Egypt studying astronomy, geometry 'and', as the popular historian Will Durant puts it, 'perhaps a little nonsense'.)[5] An astronomer of exceptional originality, once in Italy he invested his energy in founding an influential religious brotherhood and studying mathematics and theoretical geometry, where he discerned mystical properties in specific numbers. He created a new word to cover such enquiries, which translates as 'love of wisdom' – in Greek, 'philosophy'. In the sixth century 'philosopher' and 'Pythagorean' were synonyms.

Pythagoras saw number as the essence of all things, harmony as the most beautiful and the universe as fundamentally well organized. He pioneered the science of acoustics, which he extended to a hypothetical 'harmony of the spheres', believing that, as with the strings on an instrument, each orbiting planet sounded a characteristic note, its pitch determined by its size and speed. The five planets then known, together with the Sun, the Moon and the Earth, formed an octave. (No one seemed worried by the fact that the Earth, believed to be stationary, would therefore not be vibrating.)

He also believed that those bodies in the sky moved in two separate circular motions, so that the Sun circled the Earth every twenty-four hours in one revolution, but also made a second journey around the Earth along a different circle, at an angle to the first, once every year. He was attempting to rationalize the apparent motions of the stars in a universe deemed geocentric, and his theory broke down when applied to the planets; but this was the first serious attempt to explain their motions. All of modern science stems from Pythagoras's core insight – that there are patterns in nature and that those patterns can be described mathematically.

Pythagoras quickly passed into myth, and legends about him were already circulating by Aristotle's time. He had many followers, several of whom advanced our knowledge of the Sun. We know that one of them, Parmenides, worked out that the bright side of the Moon shone because it was turned towards the Sun, and he was also the first to pronounce that the Earth was spherical. Once this idea had established itself, Pythagoreans went on to reason that the heavens too must be spherical, and so spherical coordinates made their appearance. Philolaus (c. 470–c. 385 BC, a

A French artist's recreation of Pythagoras (c. 580–c. 490 BC) discussing the heavens with Egyptian priests during his prolonged studies in Alexandria

contemporary of Socrates) even pre-empted Copernicus by moving the Earth from the centre of the cosmos, making it a planet. However, in his system the Earth still did not orbit the Sun but rather circled an unseen 'central fire'. All else was spread out across space – including the Sun, which was not a body with its own light but a reflector of this central fire. The universe centred on fire because this was the noblest of the elements, and the centre the most honourable place – a theory adopted in most of the leading schools in Athens.

Some of the finest solar astronomers, however, came from outside Athens. Among them were the three 'Anaxes' – Anaximander, Anaximenes and Anaxagoras – who, like Thales, all hailed from Miletus. As indicated by the prefix 'Anax', which means 'lord', each was well born. Anaximander's pupil was Anaximenes, who taught some thirty years before Anaxagoras, the greatest of the three.

Anaximander was already sixty-four when, c. 560 BC, he wrote one of the first works of natural philosophy. He may also, following the Egyptians, have introduced the gnomon, and appears to have marked down in some form the motions of the planets, the angle of the ecliptic and the dates of solstices, equinoxes and seasons. He engraved the first 'world' map into a tablet of brass, and with his geometric model of Earth, Sun, Moon and stars sought to depict the heavens in non-mythological terms.

Anaximenes (d. 528 BC) was less original, and is best known for his doctrine that air is the source of all things, in contrast to Thales's view that everything comes from water. Although eminent in his time, he had little to say about the Sun – only that it didn't travel under the Earth but circled around it, hidden by mountains at night – in sharp contrast to the third member of the group, Anaxagoras (494–428 BC). Although his name translates as 'lord of the marketplace' – his parents evidently hoped he would turn out a merchant prince – when Anaxagoras moved to Athens from Miletus, at the age of twenty, he gave away his inheritance in order to devote himself to studying the skies.

One of his early pupils – and later close friend – was Pericles (c. 495–429 BC), who grew up to become the most influential person in Athens, at the height of Greece's golden age. This period began after the Greeks had repulsed the Persians in 479 BC and endured until c. 399 BC. During this time Athens became famous, evolving under the guidance of

Pericles into the great intellectual centre of the Mediterranean world, a place where the work of Anaxagoras and his fellow astronomers could flourish. Skywatchers of that era still had only the most primitive technology to help them, and even their observatories were probably no more than flat-roofed buildings. What they did have was mathematical geometry, even then closely allied with astronomy, and the same men were usually active in both fields. Yet solar astronomy continued to be a mix of the utterly wrong (as would come to be seen) and the formidably acute. For an instance of the latter, Oenopides of Chios (active c. 460 BC) not only came up with the idea of the ecliptic, but went on to measure the angle of the Earth's axis against the ecliptic plane and figured it to be 23° 45' – less than three tenths of a degree off from the presently accepted value.

Among his many accomplishments, Anaxagoras was even able to explain lunar eclipses, but this insight was not taken up; he was too far ahead of his time. Around 440 BC a meteorite over a foot in diameter fell on Thrace in broad daylight. Anaxagoras, visiting and seeing it on display, concluded that it had fallen from the Sun, which must, he reckoned, be a mass of red-hot iron; thus heavenly bodies were not divine beings but material objects. The authorities charged him with atheism, and he retired into exile at Lampsacus, where he died, much honoured by his new neighbours, in 428 BC.[6]

About 432 BC Meton of Athens, a leading geometer and engineer, made a detailed study of the summer solstice, probably employing a gnomon-like device known as a heliotropion ('sun-turner'), a pillar fixed on a level platform, which enabled him to measure the ratio of the noon shadow. He also constructed a parapegma, or simple engraved list of astronomical events, and merged the traditional agricultural calendar into one for city use. He became famous – indeed, a celebrity of sorts – his enthusiasm for geometric astronomy such that he found himself publicly held up to ridicule in Aristophanes's comedy, *The Birds*:

PITHETAERUS: In the name of the gods, who are you?
METON: Who am I? Meton, known throughout Greece and at
    Colonus.
PITHETAERUS: What are these things?

METON: Tools for measuring the air. In truth, the spaces in the air have precisely the form of a furnace. With this bent ruler I draw a line from top to bottom; from one of its points I describe a circle with the compass. Do you understand?

PITHETAERUS: Not in the least.

METON: With the straight ruler I set to work to inscribe a square within this circle; in its centre will be the marketplace, into which all the straight streets will lead, converging to this centre like a star, which, although only orbicular, sends forth its rays in a straight line from all sides.*

The same summer that Meton was making his solstice calculations, Athens became embroiled in a struggle against a coalition led by its arch-rival Sparta, a conflict that endured, with one brief interlude, for twenty-seven years. When it finally wore itself out, the political structure of the 'tyrant city' was in ruins. A plague that lasted for three years had killed over a third of the population. Into this turmoil, around 428 BC, a great figure was born. He was named 'Aristocles' – 'best and renowned' – but his prowess as a wrestler at the Isthmian Games led to the nickname 'Platon', or 'broad'. Unlike the fathers of Pythagoras and Socrates, the one a merchant seaman and the other a stonemason, both Plato's parents were wealthy, each coming from one of the most ancient and aristocratic families in Greece.

Athens was a state of only about 250,000 people, half citizens and their families, the rest slaves and resident foreigners, its capital numbering about 75,000. The city was never more than a narrow margin away from serious hunger and was devoid of almost any medical services – a place where even the people who owned other human beings had a standard of living greatly below the working class of modern industrial democracies. (The term for 'burglar' was 'wall-piercer', because houses were so fragile.)

---

*Meton got off lightly. Socrates, whose teaching contained a strong holier-than-thou element, so angered Aristophanes that he ridiculed the former's methods in *The Clouds*; in one scene an old gentleman finds his way to the 'School of Very Hard Thinkers', where he finds Socrates suspended from the ceiling in a basket, engrossed in thought, while his students are bent down with their noses to the ground. He asks one: 'Excuse me, but – their hindquarters – why are they stuck up so strangely in the air?' The student replies: 'Their other ends are studying astronomy.'

Having survived his home city's crushing defeat and a brutal oligarchic revolution in 404 BC, a confused restoration of democracy, then the trial and
death of Socrates in 399 BC, Plato fled to Megara, a trading port some
twenty-six miles to the north-west, thence to Egypt. After various adventures – at one point he was sold into slavery, although quickly ransomed –
by 395 BC he had returned to his home city. Nine years later, following further travels in Italy and Sicily, he borrowed money from friends to
purchase a patch of land on the outskirts of Athens named after the local
hero, Academos. This he would make his Academy, where he was to teach
for forty years: little could he have imagined that it would remain the intellectual centre of Greece for the next five hundred.

Plato made astronomy a branch of applied mathematics at his
Academy, the inscription over its doorway, according to legend, reading,
'Let no one enter who is ignorant of geometry'. No record exists of the
Greeks using algebra before the time of Christ.[7] Astronomy, which Plato
sometimes called 'Spheric', occupies a comparatively small place in his
writings, except in the dialogue *Epinomis*, which describes the work of an
astronomer. Having embraced the concept of a spherical Earth, stationary at the centre of a spherical universe, he encouraged his students
to investigate this system more carefully, but this soon raised some potentially disturbing questions. Plato believed the stars and planets to be
visible images of immortal deities whose movements were part of a transcendent order. However, some celestial bodies did not move with the
same regularity as the rest, but 'wandered' (the Greek word for planet,
planetos, originally meant 'wanderer'). How to explain these motions
within the context of his belief in that transcendent order?

For Plato, the task of philosophers was to comprehend the reality
underlying creation. If direct observation conflicted with understanding
of that reality, he cautioned that things were not necessarily what they
seemed – a valuable enough axiom in scientific study generally, but a
douche of cold water for would-be astronomers in the Academy. But
Plato's pupils were not put off, because even though he never understood the discipline, his Academy made their deliberations possible.
When he died in 347 BC, one of his students erected an altar to him and
gave him a funeral almost befitting a god. The student's name was
Aristotle. He didn't even like Plato very much.

☼

Aristotle (whose name translates as 'best objective') was born in Stagira, to the north of Greece, in 384 BC. By the time he was ten both his parents were dead, and at seventeen he was sent to the Academy, staying for almost twenty years and becoming so much its star that Plato called him 'the mind of the school'. Possibly inspired by his father, who had been court physician to the king of Macedon, he first concentrated on medicine, biology and zoology, soon advancing from pupil to teacher and lecturing on almost every subject. When Plato died, he was succeeded by his nephew Speusippus, whose intellect Aristotle disdained. Possibly piqued, maybe keen to widen his horizons, and certainly aware of the mounting animosity towards the Macedonians, who had so recently conquered Athens, he chose exile at a school on the Aegean island of Assos. When his patron, the ruler of Assos, was betrayed and crucified by the Persians, he fled to nearby Lesbos, until he was invited to tutor the unruly son of Philip II of Macedon, the future Alexander the Great. About 335 BC he returned to Athens and, possibly with money from Alexander, founded his own school, the Lyceum, near the temple of Apollo Lyceius, god of shepherds.

His followers were known as the 'Peripatetics', taken from their master's habit of strolling up and down the covered walkway, the *peripatos*, while he lectured. The students came from prosperous merchant and landowning families and keen rivalries developed between the Lyceum, the Academy (with its largely aristocratic student body) and the school of the political theorist Isocrates, which catered to Greeks new to the city. Isocrates's strength was rhetoric, the Academy's mathematics, politics and metaphysics, the Lyceum's natural science.

By this time the Pythagorean notion of two sources of light, a central fire with the Sun its mere reflector, had been discredited, but its sister idea of several interconnecting crystalline spheres revolving around the Earth had been seized on and developed. Eudoxus of Cnidus (in what is now south-west Turkey), who was teaching in Athens when Aristotle arrived, argued that the universe comprised not two spheres but twenty-seven. This elaboration enabled him to explain why four of the planets sometimes came to a stop, then reversed course and moved westwards (the phenomenon of 'retrograde motion'). Each planet's path, he argued, was that of a hippopede – a name taken from the figure-eight-shaped rope for hobbling horses' hooves. This theory also made sense of the Moon and the other planets' travelling along approximately the same path as the Sun but then deviating north and south.

Invited to lecture at the Lyceum, Eudoxus postulated that the stars and the planets were fixed to the shells of these twenty-seven spheres, which, transparent and thus invisible, encircled the Earth. One carried the Sun, which completed its transit every twenty-four hours, to bring about the day. Another slowly rotated the Earth about its axis, by which it was attached to a larger sphere, this rotation establishing the year. The axis of the newly posited sphere, tilted relative to its outer neighbour, was what moved the Sun down the sky in winter and up it in summer.

Aristotle went even further, proposing that the heavens were made up of *fifty-five* such bodies, all rotating around the Earth at different but constant speeds – which makes one think of some circus high-wire acrobat spinning plates. His idea was that, since there could never be an empty space in the heavens ('Nature abhors a vacuum'), and the spaces between the spheres in this universe had to be filled, the greater number, each sphere nestling with those immediately above and below it, would do the job nicely. The Earth was completely surrounded by this rotating nest of spheres, like the core of an onion. Beyond all these spheres Aristotle envisioned yet one sphere more, the Prime Mover – not to be confused with the creator of the universe, but still the dynamic force that drove all things. This tortured web of elaboration satisfied the world's astronomers for centuries, until the invention of the telescope.

The whole structure was based more on a priori philosophic speculation than empirical observation, so Aristotle underpinned it with logic as best he could. The universe is spherical, he reasoned, because, of all shapes, a sphere, which looks the same no matter from where it is viewed, is the most perfect. Earth too has to be a sphere, as proved by the Moon's face being overrun by a curve of darkness during an eclipse. Furthermore, travellers north or south do not see the same stars at night as those who stay at home, nor do the stars appear to occupy the same positions in the sky: so terrestrial journeys must be over a curved surface.

Why the Earth should be at the centre of the universe was explained by its uniqueness and immobility. While an understanding of gravity was millennia away, Aristotle argued that the Earth's heavier elements – rock, soil and water – were of their nature drawn towards the centre of the universe, where they came together in the shape of a sphere. The lighter elements – air and fire – intrinsically moved up and away. Echoing Plato, he added that heavenly bodies were not made up of these four elements but of ether, which he named (because it was the fifth) 'quintessence . . . more divine, and prior to . . . the four in our uniquely changeable world that lies under the wandering Moon'.[8] This ether was neither heavy nor light but 'ageless, unalterable and impassive [i.e. free from suffering]', and – key to the construct of Greek astronomy – passed in an unceasing circular motion, so bringing each heavenly body back to its starting point, enforcing an eternal recurrence.

There were obvious flaws in this explanation. For example, each planet – and the Sun – attached to its sphere must always be the same distance from the Earth; yet the Moon appeared to change in size by up to almost a sixth: did not this indicate that its distance from the Earth was changing? Another indication that the planets' distance from the Earth was not constant was that their brightness – especially that of Mars – was variable. Transient mutable phenomena such as comets and meteors were assumed to be atmospheric occurrences, taking place below the orbit of the Moon. Eudoxus and Aristotle for some reason ignored all this, as well as the fluctuation in the apparent velocity of the Sun. But what Eudoxus did discover, and Aristotle appreciated, was that the motion of the planets could be explained by combining the uniform rotations of concentric spheres on inclined axes. At least in respect of geometrical model-building, this marks the beginning of scientific astronomy.

Following Alexander's death in 323 BC, anti-Macedonian agitation drove Aristotle from the city a second time, and he died the following year in Chalcis, on Euboea, leaving behind the prototype of a modern university library, a zoological garden, a museum of natural history and an unparalleled body of work (he is said to have spent his honeymoon collecting specimens of marine life). His strengths were not in mathematics or physics and he undertook no astronomic observations, yet his influence on this as on the other sciences was profound. He systematized the whole approach to learning and settled the principles governing research, collecting a mass of data from which deductions could be drawn.

Even as he advanced his beliefs, however, his almost exact contemporary, Heracleides of Pontus (390–322 BC – they died within days of each other), who had studied under Plato but lived on the southern coast of the Black Sea, is said to have suggested that Mercury and Venus might revolve around the Sun, not the Earth, and also that the Earth might rotate daily on its axis, despite appearances to the contrary. He went on to assert that the Sun caused winds, which in turn brought about high and low tides. A younger astronomer, Theophrastus ('speaking like a god' – a nickname given him by Aristotle) of Eressos in Lesbos, discovered sunspots, though we have no idea how. Aristotle's will designated him his successor.

Early in the third century one of the most remarkable of all Greece's astronomers, Aristarchus of Samos (310–230 BC), was the first to discover precession. He also calculated that the ratio of the Sun's diameter to the Earth's was between 19:3 and 43:6. How could so large a body revolve around one so much smaller? A few years later he made the world-shattering – or at least Earth-demoting – assertion that our globe must go around the Sun, not the other way around, and that, apart from the Moon, the other planets did too. The Earth, not the heavens, turned daily, and made a complete circle of the Sun in the course of a year. The Sun, along with the fixed stars, did not move.

A Stoic astronomer in Athens promptly branded him an atheist and circulated a tract charging him with having set 'the hearth [sic] of the cosmos in motion'. Otherwise, this new theory, so momentous in its implications, was buried. This may seem odd, given that Aristarchus's description not only built on discoveries and investigations that had been in train since the sixth century BC, but also resolved a whole array of problems, such as retrograde movement. However, if the Earth truly moved, then Aristotle's theory of falling bodies would be undermined, without any compensating theory to take its place; and, if we were living on just another planet, what would become of the terrestrial–celestial dichotomy implicit in the majesty of the heavens?

As Aristarchus himself had to admit, there was nothing to be seen in the world around us to suggest that the Earth moved: surely if it did it would hurl people against one other, clouds and birds would be left behind, things would just be pitched off. Common sense dictated that he was wrong, reinforced by the prejudice that man must surely live at the centre of creation. Then there were the doctrines of astrology – at that time still respected as a science – which also required a fixed, central Earth. All in all, there seemed many good reasons for keeping the Earth at the centre.

Whether or not the hypotheses of this solitary astronomer, working on his own, ever reached them, the so-called Hellenistic philosophers (i.e. those who practised in the centuries after Aristotle's and Alexander's deaths) decided to leave matters (and the Earth) where they were. The Sun was important, but the Earth much more so. If difficulties emerged to which science had no answer, they should not threaten the status quo. One wonders whether some harboured doubts. For most astronomers of that time, science existed to support the conviction that humanity inhabited a universe of manifest order and design. But, as Durant writes,

> Since the organization of a religious group presumes a common and stable creed, every religion sooner or later comes into opposition with that fluent and changeful current of secular thought that we confidently call the progress of knowledge. In Athens the conflict was not always visible on the surface, and did not directly affect the masses of the people; the scientists and the philosophers carried on their work without explicitly attacking the popular faith, and often mitigated the strife by using the old religious terms as symbols or allegories for their new beliefs; only now and then, as in the indictment of Anaxagoras . . . did the struggle come out into the open, and become a matter of life and death.[9]

The next figure of substance was Hipparchus of Nicaea, who has been called the greatest astronomer of antiquity (even though he stayed firmly within geocentric orthodoxy). Operating c. 130 BC from his observatory on the island of Rhodes,* he determined the length of the solar year to

---

*The centre of a flourishing Sun cult, Rhodes devoted to it a quadrennial festival with athletic games, the most important being a chariot race, and erected the famous Colossus, one of the seven wonders of the ancient world, in 284 BC. This figure of the Sun god towered over 105 feet only to collapse in an earthquake in 218 BC. It took 900 camels to haul away the wreckage.

Hipparchus (c.190–c.120 BC) at the Alexandria Observatory.
At his left is the armillary sphere he invented, while he is looking through,
not a telescope, but a sector-defining tube

within six minutes of accuracy and made numerous calculations of the
Sun's diameter. He mapped some 850 stars (crucial, as without defining
positions in the sky, mathematical astronomy is impossible), and devised
a scale, divided into six ranges of intensity, that measured their brightness: even today, no one has been able to improve on it.

So accurate was his mapping that he had to accept Aristarchus's discovery of precession when his calculations indicated that the stars were
not quite fixed in relation to the Sun. Ever the committed Aristotelian,
he set out to explain the apparent circular motions of the main celestial
bodies in a way that would make sense without upsetting the geocentric
theory. Aware that the seasons were of unequal length, he had the Sun
revolve at a uniform speed, but moved the Earth from the centre of its
path, then suggested that it was how the plane of the Sun's motion lined
up with the Earth's axis that determined the timing of both the solstices
and equinoxes. He remains the best early example of how astronomers
tortured their reasoning to keep the Earth where the ancient world
wanted it.

Hipparchus's long and productive life ended around 120 BC, and with him the long Greek tradition of astronomical observation and speculation. Rome was in the ascendant, and uninterested in the heavens. It would take over 200 years for another outstanding astronomer to appear.

☼

Long before Hipparchus, the great successor states of the Macedonian Empire, from Greece to Iraq, had been conquered by the Roman Empire. Most members of the Roman elite were suspicious of Greek scientific learning (its medicine aside). By the last years of the Roman Empire, a measure of astronomical knowledge was at least considered a basic aristocratic educational requirement, but its study was accepted only in so far as it could be applied to literature. Did it help towards a better understanding of a poem? In an empire of some 50 million people, the number of natural philosophers in Rome itself became too small for any fruitful collaboration,[10] leading to the decline in scientific knowledge generally.[11] There was virtually no solar astronomy of great value by an Italian-born scientist for over 900 years. The historian of science Timothy Ferris says of the Romans:

> Theirs was a nonscientific culture. Rome revered authority; science heeds no authority but that of nature. Rome excelled in the practice of law; science values novelty over precedent. Rome was practical, and respected technology, but science at the cutting edge is as impractical as painting and poetry . . . Roman surveyors did not need to know the size of the Sun in order to tell time by consulting a sundial; nor did the pilots of Roman galleys concern themselves overmuch with the distance of the Moon, so long as it lit their way.[12]

It is an irony of history that while Rome was turning its back on the heavens, one part of its empire was celebrating the last major figure of this period. Ptolemy – Claudius Ptolemaeus (c. AD 90–168) – was an Egyptian geographer and astronomer who flourished for some forty years at Canopus, some fifteen miles east of Alexandria, nicely rendered by that iconoclastic historian Dennis Rawlins as 'an infamously licentious town which was an ancient combination in one of Hollywood, Lourdes and Las Vegas'.[13] He left four works, any one of which would place him among the most important authors of the ancient world: *Syntaxis mathematica*, better known by its Arabic title, the *Almagest*, a thirteen-volume 'book' of data about the stars; *Tetrabiblos* (the 'Bible of the astrologer', which he begins by distinguishing between the two modes of studying the heavens, mathematical astronomy and horoscopic astrology); *Harmonics*, which relates musical harmonies to the properties of mathematical proportions derived from the harmonies he considered to exist throughout the universe; and *Geographia*, a compilation of what was known about the world at that time.*

In the *Almagest* Ptolemy suggested that the planets moved in circles within other circles, with the Earth at their very centre, all the while assuming that their actual movements were unknowable. He concluded that the Sun was about 1,200 Earth radii from us (one-nineteenth of the actual figure), essentially the same estimate accepted during the whole of the Middle Ages. At one point he does consider a Sun-centred universe, but discards it: after all, where was the evidence? But he does hold that one of the two main composite motions of each planet is governed by its relation to the Sun, which made the conversion to heliocentricity much easier when it came.

When the *Almagest* appeared, critics charged that he had lifted large sections from Hipparchus and that several observations were fabricated – at one point he does unwittingly assign two quite different dates thirty-seven days apart to the same celestial event, and there are many other instances of sloppy research and borrowed ideas.[14] But if he was a fraud he did a hopeless job of covering his tracks, and his achievement is better seen from a different perspective.

In his history of astronomy, Colin Ronan points out that during these years scholarship was 'primarily in a reminiscent mood, collating and assessing the achievements of previous generations'.[15] Ptolemy was the

---

*'Books' then were more like scrolls, of necessity brief affairs, corresponding to a modern-day substantial chapter. 'Liber' meant a unit of writing, not a unit of creation (a book in our present-day sense), so it is not impossible for someone to have written several dozen books in one lifetime. Livy authored one book of 127 chapters, Caesar's opponent Varro many books, totalling 490; yet in modern terms just 200 to 250 pages. By contrast, Henry James authored some forty books, few of them slim. But then every generation throws up at least one person who cannot stop writing.

supreme collator; although the *Almagest* contains errors not corrected until the seventeenth century, the huge number of tables he created was sufficiently accurate to be taken up by Copernicus (never a sophisticated observer of the skies). The *Almagest* shares with Euclid's *Elements* – which laid down the basis for geometry – the distinction of being one of the two mathematical texts longest in use. It was Ptolemy's thinking that set the course of astronomy for the next 1,500 years.

Ptolemy regarded his work as part of an ongoing enquiry, but it was treated by his successors as definitive. If, as seems most likely, he came up with his theory of orbits within orbits in order to maintain the Earth at the centre of the universe, then in turn had to shape his 'observations' to fit his theory, he might almost not have bothered. Rulers and ever more powerful Church dignitaries – their views reinforced by the rising importance of the Christian idea that doctrine counted for more than knowledge – were becoming sceptical of the value of looking into the heavens. With Ptolemy's death, as Ronan has it, 'the light of astronomical research went out [throughout Western Europe] for a thousand years';[16] no astronomers from that civilization made any significant advance on what he had done. Conditions for scientists began to worsen, and not until far on into the Middle Ages were Ptolemy's ideas revived, as the philosophy of Aristotle was wedded to medieval theology in the great synthesis of Christian faith and ancient reason undertaken by such philosopher-theologians as Aquinas. The 'Prime Mover' of Aristotle's universe (while never conceived by him as the cause of all creation) fed into the God of Christian theology; the outermost sphere of the Prime Mover became the cosmological embodiment of the Christian vision of heaven; and the central position of the Earth was interpreted as a sign of the Christian God's concern for mankind. Thereafter it served the Church not to study the heavens too carefully.[17]

Astrology was the leading discipline of the time; fortune-telling became an obsession that lasted through the centuries, and its connections to alchemy and number symbolism were to become important elements in both Christian and Muslim Arab thinking. As Nietzsche was to ask: 'Do you believe that the sciences would ever have arisen and become great had there not been beforehand magicians, alchemists, astrologers and wizards, who thirsted and hungered after obscure and forbidden powers?'

Rome's leaders were no exception. In the days before he accepted the
sobriquets of 'Augustus' ('the Increaser') and 'emperor' ('victorious general'), Caesar's great-nephew and successor, Octavian, was converted to
astrology when a leading practitioner examined his horoscope and
immediately knelt in worship before the man he saw as his future sovereign. Although the adopted son of Augustus, Tiberius (42 BC–AD 37),
banished all astrologers from the capital, he privately continued to
depend on them. Nero (AD 37–68) too was officially sceptical, but had
one Barbillus read the heavens to discover his enemies, whom he
promptly executed. These years and the centuries that followed were a
period of regression, as astronomical observation and investigation
became subservient to sun worship and false prophets abounded.
Tertullian the Convert (c. AD 155–245), a lawyer from Carthage, wrote:
'As for us, curiosity is no longer necessary.'

Here astronomy came to theology's aid. Ptolemy's theories survived
the collapse of the Roman West for their technical quality, their coherent picture of the cosmos, their usefulness to astrology and their
harmony with Christian teaching. The system outlined by Pythagoras,
refined by Plato and Aristotle, given final form by Hipparchus and
recorded by Ptolemy promoted, as Franz Cumont says, 'a divinity unique,
almighty, eternal, universal and ineffable, that revealed itself throughout
nature, but whose most splendid and energetic manifestation was the
Sun'.[18] The way these astronomers described the Sun chimed in almost
perfectly with how the Church, appropriating solar imagery, wanted its
believers to see the Christian God.

This cumulative reading of the universe afforded far more to
Christianity than merely a legacy of pagan festivals and symbols. 'To
arrive at the Christian monotheism,' continues Cumont, 'only one final
tie had to be broken, that is to say, this supreme being residing in a distant heaven had to be removed beyond the world.' The various cults
devoted to the Sun not only made straight the roads for Christianity; they
heralded its triumph. Little wonder that for the next 1,400 years Church
and state should combine so effectively to keep man's understanding of
the universe exactly as the Greeks had left it.

GIFTS OF THE YELLOW EMPEROR

Among the Greeks, the astronomer was a private person, a philosopher, a lover of truth ... in China, on the contrary, he was intimately connected with the sovereign pontificate of the Son of Heaven, part of the official government service, and ritually accommodated within the very walls of the imperial palace.

André de Saussure, c. 1910[1]

The people of our humble land have always understood heaven.

Zou Yuanbiao, writing to Matteo Ricci, c. 1597[2]

So much of our understanding of China in general and its solar science in particular depends on the work of one Western scholar – the remarkable British polymath Joseph Needham, whose life spanned the twentieth century (1900–95). He has been compared with both Darwin and Gibbon in bestriding his area of study. A contemporary went so far as to claim that one would have to reach back to Leonardo to find so much learning in a single individual – Needham's knowledge embraced mathematics, physics, history, philosophy, religion, astronomy, geography, geology, seismology, mechanical and civil engineering, chemistry and chemical technology, biology, medicine, sociology and economics.

He spoke eight languages with comfort, wrote poetry throughout his life and was a profoundly committed Christian and Communist, while other interests ranged from morris dancing to steam trains, accordion-playing and nudism.[3] He spent most of his life in Cambridge, initially specializing in embryology and morphogenesis, and in 1924 marrying a fellow don, the biochemist Dorothy Mary Moyle. Then in 1937 a Chinese postdoctoral student, Lu Gwei-djen, arrived at the department of biochemistry: Needham fell in love, and totally redirected his intellectual pursuits. He taught himself enough Mandarin to get by, and became obsessed with two questions: why were the Chinese so unaware of their past scientific achievements? And why had not the scientific revolution that swept through seventeenth-century Europe taken place in China first?

Just before the outbreak of the Second World War, Needham signed a contract with Cambridge University Press to write a one-volume history, 'taking up the fundamental question of why modern science originated in Europe and not in China'.[4] Going out in 1943, he spent two years with the British Scientific Mission in Chungking (Chongqing), the Nationalist Party capital of China, and for the next six years travelled the country by whatever conveyance was available – jeeps and junks, camels, wheelbarrows, litters and goatskin rafts – collecting rare works on traditional learning as he went. He returned home, his 'short study' having grown into three volumes. The first appeared in 1954 and included a table of contents for the entire enterprise, which at that point he was projecting to seven volumes. Eventually, those seven would become twenty-seven, mostly written by Needham, who worked at them almost until his death at ninety-five. The chapters on astronomy alone fill 300 pages.[5] While his remarkable work has been augmented over the years, and sometimes corrected in its detail, it has never been replaced.

☼

In one central myth of China's beginnings, the early king Fuh-hi and his four successors, together known as the 'Five Sovereigns', are said to have founded the Chinese Empire, ruling from c. 3000 to 1600 BC and expanding the kingdom from what is now north-central China to the eastern sea – a huge territory, with an 8,700-mile coastline. A competing myth presents a certain Xuan Yuan, or Huangdi, the Yellow Emperor, who became the leader of all the northern tribes after decisively defeating his enemies in 2698 BC. He could move through the air at enormous speed, riding a dragon as high as the Sun – this beast having come from 'the land where Suns are born'. Legend holds that he believed that numbers had philosophical and metaphysical properties and helped to 'achieve spiritual harmony with the cosmos'. Both Xuan Yuan and Fuh-hi, like the Babylonians, are said to have devised a numerical system based on sixty as part of a method of timekeeping known as the 'Heavenly Stems' and the 'Earthly Branches'.

Mythical though these figures were, there is some truth in what the myths describe. The first dynasty to appear in Chinese records is the Xia, running from 2033 to 1562 BC. During the subsequent Shang or Yin dynasty (c. 1556–1045 BC) the Chinese developed a calendar, learned how to write and became highly proficient in bronze metallurgy. This era was a critical time in Chinese intellectual history. We know that they were investigating the heavens by the fifteenth century BC, and inscriptions suggest that by the fourteenth – well before the earliest known Sumerian calendars – they had established a solar year of 365¼ days, formulated systems for measuring daily time, incorporated the cardinal points of direction into their cartography and mapped the paths of the Sun and the Moon accurately enough to predict eclipses.

Astronomy was of vital interest to the Chinese, and considerable resources were devoted to it. Early on in his account, Needham makes it clear that until at least the sixteenth century AD they connected celestial events with the performance of their rulers, whom they saw as divinely appointed. The Earth, its emperors and the entire cosmos were one, and so long as a sovereign governed well, the celestial bodies would follow their appointed courses without deviation or surprise. Were his reign unjust or otherwise defective, comets or new stars would blaze across the sky.[6] Since unusual heavenly events either presaged disaster or pointed to inadequacies of administration, a ruler had good reason to keep such portents secret, and records were closely guarded. While major happenings were visible to all, many lesser ones were not noticed and could be passed over or publicized at will; thus many astronomers gained considerable power, even influencing policy, and were central figures of court life. They developed so-called 'techniques of destiny', which included inscribing predictions on the shoulderblades of oxen or deer or on tortoise shells. These 'oracle bones' are our main source of information about the period from the fourteenth to the eleventh century BC.

Given the discipline's potential as an instrument of intrigue (independent astronomers might be tempted to calculate horoscopes to the advantage of a rival would-be dynast), the emperors made it a crime to leak astronomical information – just as Tiberius and Nero would later do in Rome. A typical edict declares:

> If we hear of any intercourse between the astronomical officials or their subordinates and officials of other government departments or miscellaneous common people, it will be regarded as a violation of security regulations . . . astronomical officials are on no account to mix with civil servants and common people in general. Let the Censorate look to it.[7]

Even to study astronomy in private was punishable by a two-year prison sentence.[8]

The astronomer Xi Chu (AD 1130–1200) using a gnomon to determine the summer solstice. The animals may represent the Chinese zodiac

The most startling difference between Chinese and Western astronomy was the method the Chinese used to track the planets through the night sky. They named the ecliptic the 'Yellow Path' (*huang-tao*), with a set number of 'houses' – the Ten Heavenly Stems – positioned along it. These combined with both the Twelve Earthly Branches and another sequence, the Five Elements, to create cycles of sixty hours, days and years. To make the construct easy to remember, each Branch was assigned an animal symbol.* 'So far as the historian of science is concerned,' Needham comments tersely, 'whoever invented the animal cycle is welcome to it.'[9] As he notes, a Western astronomer would hardly recognize a map of the constellations produced by a Chinese astrologer – a fair point, as this famous menagerie does tend to obscure the real advances that the Chinese made.

The recording of astronomical observations and cultural practices tended to go together. The *Li Ki*, or Book of Rites, for instance, a compilation of religious practices from the eighth to the fifth century BC, describes the Son of Heaven (the king/emperor) as a stargazer whose function was to predict the moment when his people were to participate in the rites of heaven and Earth by sowing the new season's grain. Stargazing (as with the ancient Greeks) was seen as part of everyday life – as the late-Ming scholar Gu Yanwu wrote: 'In the Xia, Shang and Zhoui dynasties, everyone was an astronomer.' 'Astronomy' was *tian wen*, literally 'the pattern of the skies', and encompassed the systematic study of all heavenly events – the waxing and waning of the Moon, the movements of the planets and the type and colour of comets. The project of mapping the heavens was not unique to the Chinese, but, along with the Babylonians, they were the most persevering and accurate observers until the Arabs, and, from the middle of the fourteenth century BC on, they recorded 900 solar eclipses over the next 2,600 years. The Chinese of this era also noted the appearance of sunspots – they are mentioned at least 120 times in the official histories. Although

---

*The myth developed that the Buddha invited all the animals to celebrate the New Year, but only twelve turned up: mouse, ox, tiger, rabbit, dragon, snake, horse, sheep, monkey, rooster, dog and pig, in the order they now take in the zodiac. Out of gratitude, he decided to name the years after them in the order of their arrival, people born in a given year inheriting the personality traits of that particular animal.

observations of the shape and splitting of such spots did not come until much later – these were first noted in 28 BC – this was long before they were written up in the West. One wonders how the Chinese saw spots at all, when their skies were not especially clear, and even given good conditions a spot has to be more than 50,000 miles in diameter to be visible to the naked eye.

The sixth century BC saw the standardization of weights, measures and other practicalities such as the width of roads. Even gnomons were officially set at eight *chi* (2.4 metres). By the fifth century the Chinese understood what caused eclipses and were the first to record the appearance of what we call Halley's Comet, in 467 BC. Between 370 and 270 BC, two of the greatest Chinese astronomers, Shih Shen and Kan Tê, along with a colleague, Wu Hsien, drew up the first star catalogues – to give some idea of how advanced this was, Hipparchus would not produce comparable work for another two centuries.

In all this the Sun operated as one force among many (Chinese painted screens depicting the universe often omit the Sun altogether): to give it primacy would have been to disturb the balance of nature. The Chinese saw the heavens as a bowl turned upside down, resting on a square Earth, with four cardinal points of terrestrial space, north, south, east and west, and a fifth, the centre, heart-point of the Celestial Realm. The Sun was essentially a circumpolar star, illuminating first one zone of the Earth, then another. The Sun and Moon were attached to heaven, which moved at great speed, carrying them along with it.

From c. 200 BC on, the main observatory under the Ch'in dynasty employed a corps of over 300 astronomical experts. They divided the sky into twenty-eight unequal sectors, radiating from the north celestial pole in the same way that lines of longitude radiate from the poles of a terrestrial globe, derived from the Moon's apparent motion set against the stars – which allowed them to calculate a 'sidereal month' of 27.32 days. These astronomers employed a cosmograph, or *shipan*, with a rotating disc (heaven) above a square, fixed plate (Earth). Around the disc's edges were marked the names of the twenty-eight constellations, while in the disc's centre was a representation of the Plough. The instrument enabled the user to locate a star at any time of the year, and also functioned as a clock. Once the astronomer had calculated what star was passing the meridian at sunset, he could tell which constellation had reached its zenith at noon or midnight – and, knowing that a particular constellation had risen at the vernal equinox, which stars were reaching their high point, or alternatively were on their way down. Some time before the second century BC, this cosmograph would evolve into the compass, its function shifting from the celestial to the terrestrial.

Chinese astronomy was thus predicated on a very different system from that of the Greeks or later Europeans. The Egyptians and Greeks charted the Sun's risings and settings and those of the other stars near the ecliptic, so for instance the risings of Sirius, measured against successive sunrises, would announce when the Nile was about to flood. 'Such observations,' wrote Needham, 'required no knowledge of pole, meridian or equator, nor any system of horary measurement ... Attention was concentrated on the horizon and the ecliptic.'[10] The Chinese focused instead on the Pole Star and its circumpolar brethren, basing their system on the meridian – the great circle of the celestial sphere passing through the Pole Star at the observer's zenith – and calculating the culminations and lower transits of the Pole Star's immediate neighbours.* (Not that the two systems were entirely separate. Homer, for one, was aware of the significance of the Pole Star as a mark of the north; and the sentinels at the siege of Troy changed their guard according to the vertical or horizontal positions of the tail of the Great Bear.)

Such an approach had a clear scientific basis, but for most Chinese their understanding of the heavens – and thus of the Sun – was bound up in their overall way of looking at the world. Chinese cosmology affirms a life of balance, of virtue, of performing one's duty: and their philosophy of what occurs in the heavens reflects whether or not that balance is achieved – particularly as understood through their three key terms, ch'i, yin and yang. Before about 300 BC, ch'i in its narrowest sense

---

*A circumpolar star is one that, as viewed from any given latitude on Earth, never sets (i.e. never disappears below the horizon), due to its proximity to either of the celestial poles. Such stars are therefore visible from that location for the entire night on every night of the year – and would be continuously visible throughout the day too, were they not overwhelmed by the Sun's glare.

means 'life energy', but thereafter it was used to cover a multitude of phenomena: air, breath, smoke, mist, fog, the shades of the dead; also cloud forms, more or less anything perceptible but intangible; the physical vitalities; cosmic forces and climatic influences that affect health and the seasons (many of the Sun's effects were manifest in *ch'i*); flavours, colours, musical modes. *Ch'i* could be benign and protective or pathological, an agent of disease, and destructive.

One of the primary schools of Chinese thought, Taoism, is a philosophy of duality and balance – thus the doctrine of *yin* and *yang*, which embrace almost all opposable forces, *yin* being cool, wet and female, *yang* hot, dry and male. Other antitheses include day and night, light and dark, honest and deceitful.[11] Taoists propound the idea that, as *yin* and *yang* surge through the Earth, they cause the periodic changes that bring about tides and seasonal heats and colds. They believed that numbers were key to maintaining balance, five being particularly auspicious – each of the five prime locations of north, south, east, west and centre having a corresponding colour, animal, element and flavour. Nine was also important, not so much auspicious as a number to be heeded, so that on the ninth day of the ninth month – the double *yang* – the people of Beijing would assiduously set out to climb two of the city's main hills (the Wall of Pure Metamorphosis and the Destiny of Trees Surrounding the Gate of Ch 'i) in order to avoid disaster.[12]

Yin and *yang* modified *ch'i*: they were the complementary aspects of any configuration of space or time, what Needham calls the 'two fundamental forces or operations in the universe'. The Sun was of a fiery *yang* nature, the Moon and the Earth were *yin*, air simply a vacuum.* The distinction was between bodies that shone by their own light and those that shone by reflected light. Furthermore, there were five phases (*wu hsing*, the five powers, or the five travellers, because in continual movement) of five life processes: wood, fire, earth, metal and water. By the second century AD no one was sure what the unifying

---

*As in most cultures, the Sun had several synonyms, including Radiant Numbers (*yao ling*), Vermilion Luminosity (*chu ming*), Lord of the East (*tung chun*), Great Luminosity (*ta ming*), Crow of Yang and Fluid Pearl. Needham comments that the Chinese elite used many synonyms 'precisely in order to bewilder the uninitiated'.[13]

principles of these materials were, but despite such uncertainty a common language emerged to describe the cosmos and its analogues on Earth for both state and individual: a dynamic harmony existed, leading to one true way. This classifying system originated in humanistic thought, then passed into state cosmology, and from there was taken up by scientists.

✧

In 221 BC the eastern kingdoms were forged together to become China, the great central state of Ch'in. 'China' means 'land of the house of Ch'in' but also translates as 'middle': Ch'in was the Middle Kingdom. The long distances and physical barriers that had isolated it from other civilizations gave its people the impression that their country was the centre of the Earth, the sole source of civilization – an idea that would persist for millennia.

However, China's expansion into central Asia in the first century AD would expose its astronomers to Hindu and Persian ideas, provoking intense speculation in cosmology and astronomy. Some of these doctrines were almost too far ahead of their time: in AD 5 Wang Mang (45 BC–AD 23), regent and later emperor, argued from what he had learned from these foreign influences that the workings of the heavens happened 'of themselves', and natural disasters were not sent to warn or punish, a thesis that went largely unheard. Another theory that won wide acceptance, associated with the Later Han Dynasty (AD 25–220), pictured the universe as unending empty space in which floated the Sun, Moon, planets and stars – notable as the first time anyone from any country had suggested an infinite uncentred universe.

Much of Chinese mathematics (from, say, c. 100 BC on) was devoted to calculating the calendar and predicting the positions of heavenly bodies. (As it was, Chinese calendric science produced a strong tradition of algebraic-arithmetical astronomy, in contrast to the Western emphasis on geometry.) In 132 China's greatest 'Grand Astrologer' Zhāng Hêng (AD 78–139), basing his insights on such mathematics, invented the world's first seismoscope for registering earthquakes: a cylindrical device with eight dragon heads around the top, each with a ball in its mouth, and eight frogs, one directly beneath each dragon, on

A reconstruction of the water-driven astronomical clock tower of *Sū Sòng*
(AD 1020–1101) at Kaifeng, in north-east China. The clock tower
employed 133 different clock jacks to indicate and sound the hours

the bottom. When a shock wave came, a ball would clatter out of a dragon's mouth into a frog's. Zhāng also wrote that moonlight is only reflected sunlight, and that lunar phases are the result of the Moon's changing reflections of sunlight from different points on its course. He invented the world's first water-powered armillary sphere; documented about 2,500 stars in his extensive star catalogue; and invented an astrolabe with graduated rings and a sighting tube.* While his fellow Chinese were reacting to lunar eclipses by pounding gongs, Zhāng was explaining their real cause. Needham believes he set 'the standard for

*Astrolabes show how the sky ideally looks at a specific place and time. This is done by draw-
ing the constellations on their faces and marking them so positions in the sky are easy to
find. Once set, the instrument's face has a complete representation of the sky, allowing
many astronomical problems to be solved visually.

excellence in astronomy', although even Zhāng was partly conditioned by the thinking of his time: while he was the first in China to construct a celestial globe that actually turned, it bore little or no relation to reality, but rather reflected his vision of the universe as expressed in *Hun-i Chu*: 'The heavens are like a hen's egg and as round as a crossbow bullet; the Earth is like the yolk of an egg, and lies along the centre. Heaven is large and Earth is small.' And, in a passage not translated by Needham, Zhāng adds: 'Heaven takes its body from the *yang*, so it is round and in motion. Earth takes its body from the *yin*, so it is flat and quiescent.'[14]

Between the second and eleventh centuries AD little changed in Chinese understanding of the Sun – although their astronomical literature was equal in extent to those of botany, zoology, pharmacy and medicine taken together. By the twelfth century the imperial library contained 369 books on astronomy and allied subjects, and by the thirteenth the empire's astronomical instruments were far superior to anything in Europe; however, the non-geometrical nature of its mathematics left it still unable to chart the skies accurately and hampered further progress. This was one consequence of its isolation. Until about AD 400 the country was almost completely shut off from all but its immediate neighbours. Christian missionaries made the arduous journey overland as early as the following century, but there was little if any cross-fertilization of ideas, and the country did not become an object of general curiosity in the West until 1250, when Pope Innocent IV sent Franciscan monks into Asia – to discover a civilization that was in many ways remarkably advanced.

The Franciscan reports were augmented several decades later by Marco Polo (1254–1324), who, writing about a period towards the end of the thirteenth century, described a land 'of immense extent, with the largest cities, widest rivers and greatest plains ever seen, where gunpowder, coal and printing were in general use'.[15] One might add to that an economy based on paper money, cities with over a million people, a genuine bureaucracy and such items as silk, tea, porcelain, herbal medicines, lacquer, playing cards, rockets, astronomical clocks, dominoes, wallpaper, the kite and even the folding umbrella. Marco Polo was describing China as it was under Kublai Khan, during whose reign the Chinese Empire stretched from the Pacific to much of Eastern Europe, and China was open to foreign influence as never before.*

Under the Ming Dynasty (1368–1644), as the country again sealed itself off for more than a hundred years, its astronomical undertakings sank into the general deep sleep of science – a decline so precipitous that it is hard to believe it happened; but it may have been that the Chinese 'believed that because the good life had been discovered their education need never change'.[19] The situation would alter only with the coming of the Jesuits. They found a culture profoundly cut

---

*How far, if at all, was ancient Chinese astronomy influenced by the West? In 1986 the Palace Museum in Beijing sent two horse bones to the British Museum asking for help. Both bore fragments of cuneiform script, and one was identified as a poor copy of a cylinder (of which the BM has an example) that describes the conquest of Babylon by Cyrus II of Persia in 539 BC. At first it was presumed that the horse-bone script, with its square structure and pictographic aspects, was a forgery, but after prolonged study experts decided it was genuine. How had it suddenly turned up in Beijing? In 1987 an article appeared in a Chinese historical journal by one Xue Shenwei, a doctor of traditional medicine who had died well into his eighties the year before. Travelling in eastern China in 1928, he had learned of the bones from a local scholar. In the years that followed he monitored their progress, and over a decade later bought them from an antique dealer, who told him that his purchases came from the north-west of Xinjiang, in the desert around Gu'erban: the villagers there called such pieces dragon bones and used them as spindles. During the destructive excesses of the Cultural Revolution, Xue Shenwei buried the two bones; when he judged it safe he dug them up, and just before his death gave them to the Palace Museum.[16] Xinjiang probably lay on the line of an old trade route, so fourth-century-BC visitors from the West could indeed have carried the bones with them. I asked a senior member of the BM whether this discovery was not rather important. 'I suppose,' he mused, 'we do tend to keep some things to ourselves.'[17]

Six years after Beijing received its bones, about 113 mummified corpses, dating back at least 3,800 years, were excavated at Qizilchqa, known as 'the Land of Irrevocable Death' – through which runs a key stretch of what would become the Silk Road, the 4,000-mile network of interconnecting beaten paths and caravan tracks that served from around 500 BC to AD 1500 as the main highway between Europe, the Middle East, India and China. The bodies, although their skin is parched and blackened, have long noses, long skulls and deep-set eyes, typical of Caucasians – even blond hair. Qizilchqa, in the Taklamakan desert of the Tarim Basin (modern-day Xinjiang), lying between the Gobi Desert to the north-east and the Himalayas to the south, is one of four sites where such corpses have been found, not in ones and twos but groups of thirty or more, suggesting permanent settlements and not just a few stray travellers. Victor H. Mair, Professor of Chinese Studies at the University of Pennsylvania, believes that there was a substantial two-way traffic between East and West beginning at least 3,980 years ago. Would such caravans have brought scientific knowledge with them? We do not know, and the present Chinese government has impeded further research, but at least two of the 'Tarim mummies' were still wearing their original silk clothes – decorated with a star map and an ancient text, possibly in Phoenician, on astronomy.[18]

off from much of the knowledge of the West, including its geography as much as its astronomy. When one of the missionaries hung up a map of the world in his hut and invited his Chinese guests to examine it, he reported:

> To them the heavens are round but the Earth is flat and square, and they firmly believe that their empire is right in the middle of it. They do not like the idea of our geographies pushing their China into one corner of the Orient. They could not comprehend the demonstrations proving that the Earth is a globe, made up of land and water, and that a globe of its nature has neither beginning nor end. The geographer was therefore obliged to change his design and . . . left a margin on either side of the map, making the Kingdom of China appear right in the centre. This was more in keeping with their ideas and it gave them a great deal of pleasure and satisfaction.[20]

Chinese geography may have been backward, yet it was three Chinese inventions – the magnetic, south-pointing compass, the true rudder, and ships that could sail to windward, with a compartmentalized hull to make them less liable to sink when broached – that made it more feasible for Europeans to sail eastwards by sea. Even then, when in 1551 the first of the great missionaries, Francis Xavier, set out from neighbouring Japan, it took him nine months of negotiations to establish himself on an island seven miles offshore, and a further nine weeks to set foot on the mainland. Some thirty years later, an Italian member of the Society of Jesus, the mathematician and astronomer Matteo Ricci, took four months to reach the Portuguese outpost of Macao, enduring illness, shipwreck and imprisonment before finally entering Beijing.

Ricci was the most notable of all the Jesuit missionaries to China, the first Westerner to be admitted to Beijing and then the Forbidden City. While baptizing many high officials, he acquainted his hosts with the advances of the Renaissance – although he and his fellow Jesuits also imported many of the current scientific errors of Europe, and indeed would ridicule the notion of 'the emptiness of the heavens'. Yet, as Needham observes, 'The vision of infinite space, with celestial bodies at

An eclipse of the Sun in 1688, viewed by Jesuit missionaries in the
presence of the King of Siam

rare intervals floating in it, is far more advanced ... than the rigid
Aristotelian-Ptolemaic conception of concentric crystalline spheres.'[21]

Soon after his arrival in 1584, Ricci set about expounding to his hosts
what he had learned in his European studies – even providing them with
a world map that, among other things, proves that the Sun is larger than
the Moon.[22] The Chinese resisted, as to their minds they were the intellectual masters of the world, certainly in the sciences, until Ricci
produced a chiming clock that he said replicated the movements of the
stars. To test it, his hosts set about predicting a solar eclipse by their
methods, while Ricci and his European assistants used their device. The
Chinese-predicted hour came and went in the full light of the Sun; the
Western-predicted eclipse fell precisely on time.

As Ricci's renown grew, he was invited to the country's most important
observatory, in Nanking. Instead of the elaborate semi-magical charts he
had expected, however, he found magnificent instruments of cast
bronze, embellished with imperial dragons, more sophisticated than any
he had seen in Europe: a sphere for determining eclipses marked with
degrees according to the European system, but in Chinese characters; a

The Imperial Observatory in Peking, as it was in the late seventeenth century.
From here sunspots were discovered and records of comets kept. Large bronze
spheres (lower frame), a quadrant (upper left) and a sextant (upper centre)
are all identifiable

gnomon; and, largest of all, a nest of four astrolabes. The eunuch in charge* explained that the instruments had been cast by a Muslim astronomer some 250 years before.

Ricci examined one of the spheres and pointed to the graduated arms. 'Is that where you read the shadow?' The eunuch looked puzzled, then nodded, before trying to pull the visitor away, but Ricci continued his inspection, noticing that degrees were indicated by a system of knobs that could be read at night by touch. Then he made a startling discovery.

'Surely Nanking stands at thirty-two degrees?'

---

*'In the later periods of the Byzantine Empire' writes the historian Beverly Graves Myers, 'the heavenly court found an earthbound reflection in the imperial court. The eunuchs who surrounded the emperor – who was considered God's representative – with their brilliant white clothing and glowing beardless faces served as the equivalent of angels.'[23] Sun Yaoting, the last eunuch of the imperial court, died as recently as 1996, aged ninety-three.

The eunuch agreed.

'But these instruments are set for thirty-six degrees.' His guide was silent and tried again to divert Ricci's attention. Intrigued, his guest posed two or three simple questions about the gnomon – all of which received evasive answers until at last the chief eunuch confessed.

'They're beautiful instruments, but we don't know how to use them. Two sets were made, one for Beijing, the other for Pingyang. These are from Pingyang.'

Now Ricci understood. The town of Pingyang stood at thirty-six degrees, and the instruments had never been adjusted. They were no more than objects of magic. He gradually learned that some astronomical knowledge had come from the Arabs – so China had acquired its preliminary understanding of the Ptolemaic system from the same source and at about the same date as Europe. In the reviving West, such information had been enthusiastically received and elaborated; but in China, where science was not judged conducive to the good life or to good government (as opposed to astrology's value as the source of predictions), it was not included in the approved list of studies.[24]

Ricci did his best to dispel such ignorance, but considerable opposition to his teaching remained. Not until three years after his death, when the Chinese Board of Astronomers seriously mistimed an eclipse, did the throne finally decree that the calendar be brought up to date and European works on astronomy translated. However, China was almost immediately invaded by the Manchu, order broke down nationwide and the country was once more closed off. Isolated it may have been, yet well into the eighteenth century it was held in awe by many Europeans as the most prosperous and sophisticated of any civilization on Earth. Samuel Johnson's first prose work (1733) refers to the Chinese as 'perfectly, completely skilled in all sciences'.[25] Not until 1850 was their astronomy merged with that of the rest of the world.

Today the Imperial Observatory has its own subway station.

8

THE SULTAN'S TURRET

Awake! For Morning in the Bowl of Night
Has flung the Stone that puts the Stars to Flight;
And Lo! The Hunter of the East has caught
The Sultan's Turret in a Noose of Light.

Opening verse of the *Rubáiyát* of Omar Khayyám

And when the night overshadowed him he saw a star and said,
'This is my Lord!' And when he saw the Moon beginning to
rise he said, 'This is my Lord!' And when he saw the Sun
beginning to rise he said, 'This is my Lord, greatest of all.'

Qur'an[1]

Back in 1959 the novelist and cultural historian Arthur Koestler published *The Sleepwalkers*, a remarkable account of scientific discovery through history. His summary of what Islamic civilization added to astronomy is harsh. The Arabs, he concludes, were

the go-betweens, preservers and transmitters of the [Graeco-Indian] heritage. They had little scientific originality and creativeness of their own. During the centuries when they were the sole keepers of the treasure, they did little to put it to use. They improved on calendrical astronomy and made excellent planetary tables; they elaborated both the Aristotelian and the Ptolemaic models of the universe; they imported into Europe the Indian system of numerals based on the

symbol zero, the sine function and the use of algebraic methods; but
they did not advance theoretical science. The majority of the scholars
who wrote in Arabic were not Arabs but Persians, Jews and Nestorians;
and by the fifteenth century the scientific heritage of Islam had largely
been taken over by the Portuguese Jews.[2]

While his comments are correct in one sense, they do not take into
account that during Europe's Dark Ages any holding operation – if that
adequately describes the Islamic contribution – had great value in itself.
Just to retain knowledge and incrementally advance it was enormously
important, not least in solar astronomy. Tellingly, the great Belgian-
American scholar George Sarton (1884–1956) divides his magisterial
five-volume *History of Science* into half-centuries, each of which he asso-
ciates with one central figure. He begins in 450 BC with Plato, followed
by Aristotle, Euclid and Archimedes; but from AD 750 there traces an
unbroken line of Arab and Persian scientists – Jabir, al-Khwarizmi, who
worked on calendars, the true positions of the Sun, spherical astronomy
and eclipse calculations, al-Razi, al-Masudi, Abul-Wafa, al-Biruni and
Avicenna – ending with Omar Khayyám (AD 1050–1100).

Ghiyath al-Din Abu'l-Fath Umar ibn Ibrahim Al-Nisaburi al-Khayyámi –
the man the West calls Omar Khayyám – was largely unknown to it until
1859, when an effete Suffolk gentleman scholar, Edward Fitzgerald, published an anonymous translation of the 4,000-line *Rubáiyát* (literally
'Quatrains' – stanzas of four lines). By that time Fitzgerald was well into
middle age, and his previous work had aroused little interest. He was
probably unsurprised when this too seemed to meet the same fate. Two
years after publication, however, a London bookseller sold off his
remaining stock from a penny bargain box outside his shop near
Leicester Square, and another poet, Dante Gabriel Rossetti, picked up
a copy and was soon declaring the work a masterpiece: by the end of the
century, aided by the long-running enthusiasm for Persian culture in
nineteenth-century Britain, it would become the most popular poem of
its time. Yet the *Rubáiyát* turns out to be only one of Khayyám's many
accomplishments.

Khayyám lived in Naishápur (now in north-eastern Iran), close to the
Afghan and Turkestan borders. His last name means 'tent-maker', and

in one of his verses he writes punningly of 'Khayyám, who stitched the tents of knowledge' – which is precisely what he did. He was the author of the well-received *Treatise on the Demonstration of Problems of Algebra* and was the first to come up with a general theory of cubic equations; he also helped reform the Persian calendar, basing it on the solar year, not the lunar – the result, according to Gibbon, being 'a computation of time which surpasses the Julian, and approaches the accuracy of the Gregorian style'.[3] He measured the solar year at 365.24219858156 days, an extraordinarily accurate computation, given that at the end of the nineteenth century it was measured at 365.242196, now refined to 365.242190. He constructed a star map and helped build a major observatory.

Khayyám and his seven fellow scientists are Sarton's group of choice, but they were abetted by many others. Faced with Koestler's dismissal, one might ask what made Arab science, in particular its mathematicians and astronomers, so outstanding for more than 350 years, and what is a fair estimate of Islam's contribution to solar astronomy? These two questions take up the first half of this chapter, and it turns out that the answer to one is tightly bound up with that to the other, because Islam valued knowledge not only for its own sake but also, in the case of astronomy – and specifically the study of the Sun – for its contributions to religious observance. Thereafter we will look at how, from around AD 750 through to the coming of Copernicus, studying the Sun was evolving in other cultures – most notably in India – and back in Europe.

It all started in AD 630, when the Prophet Muhammad (Muhammad ibn 'Abdu'llah ibn 'Abdu'l-Muttalib) marched on his birthplace, Mecca. The city surrendered without a fight and became the centre of pilgrimage for Muhammad's new religion of Islam. In the moment of his next victory, over the Jewish settlement of Khaibar, he and some of his men were given dinner by a local woman who spiked their meat with poison. He took over two years to die, finally succumbing eight days short of his sixty-third birthday.

By then he had already established Islam as a social and political force on a par with Judaism and Christianity. Within a few decades his successors would unite Iran, Iraq, Egypt (taking Alexandria and burning what remained of its great library), Syria, Palestine, Armenia and much of

North Africa. By AD 750 the rest of North Africa had followed, along with the entire Iberian Peninsula and much of central and large stretches of south Asia. In the tenth century Islam moved even deeper into Africa, as well as spreading across what had been the mainly Hindu principalities east of the Indus in what are now Pakistan, India and Bangladesh, then onwards as far as the frontier provinces of China.

The Muslim theocrats could not have been more unlike their Christian or Chinese counterparts in that they tended towards tolerance – and so, during its golden age, Islam was eager to embrace learning of all kinds, its scholars regularly exchanging ideas with astronomers and natural historians from other countries. During the years AD 661 to 751 the caliphs,* ruling from Damascus, strove systematically[4] to increase the store of knowledge.

The reason for such passion for learning lay in the Muslim holy book. More than a fifth of the Qur'an – over 1,100 verses – is concerned with natural phenomena.† Along with the teachings of Muhammad, the Qur'an forms the great motivating force both for individuals and the state to concentrate their energies on such subjects as mathematics and astronomy. As the Islamic scholar M. A. R. Khan has noted:

> Some of the most eloquent passages in the Qur'an refer to the grandeur of the stellar world, the regularity of solar and lunar movements among the constellations, the phases of the Moon and the dazzling brilliance of the restless planets. No wonder that the Arabs and the later converts to Islam from other nationalities and civilizations took to astronomy so enthusiastically and left their permanent mark on it.[6]

The successors to the caliphs of Damascus, known as the Abbasid Dynasty, came to power in AD 751 and ruled for more than five centuries, choosing Baghdad over Damascus for their capital and dedicating

---

*'Caliph' comes from the Arabic *khalīfa*, meaning 'vice-regent' or 'successor', specifically the successor to Muhammad. It also defines what such a person's role should be: to govern according to the Qur'an and the practice of the Prophet. It is a word seldom used for anything other than the leader of the entire Sunni community.

†A nice echo of the Qur'an's place in things appears in Zadie Smith's first novel, *White Teeth*, when the twins Millat and Magid Jones meet in a university lecture room: 'Millat arranges the chairs to demonstrate the vision of the solar system which is so clearly and remarkably described in the Qur'an, centuries before Western science.'[5]

themselves to building on the scientific achievements of the first caliphs. Some twelve years after their move to Baghdad, they began to reconstruct that city, working from a design created by a philosopher and an astronomer. Within fifty years of its planning, this new metropolis became the pre-eminent cultural and scientific city of the world. At a time when the Church forbade the practice of intrusive surgery, Arabs were giving anaesthetics and performing complex operations. Mathematics exploded: they introduced the tangent in trigonometry, posed and solved cubic equations, conducted extensive studies of cones, replaced chords with sines and laid down the basic theories for the solution of trilineal figures (a rarefied area of higher mathematics). They brought the world its first international banking system, and also the pendulum and the windmill. Astronomy flourished, along with the rest of the sciences.

While the centre of observational astronomy was Baghdad, the renaissance went far beyond one city, thriving in Muslim capitals from Spain to central Asia. The caliphs, particularly al-Mansur (who ruled from AD 754 to 775), were great patrons. In a typical example, a scientist from India arrived at court bearing a copy of the *Sūrya Siddhānta* ('Solar Principles'), written by the Hindu astronomer Aryabhatta 400 years before. The caliph had it rendered into Arabic – Muslims were familiar with Indian and Persian writings some time before they read the great Greek texts. The caliphate of al-Ma'mun (AD 813–833) was even more progressive. Himself an astronomer, he founded a celebrated academy, the House of Wisdom, and paid his savants famously well: one senior librarian, on his retirement, was given gold equal to the weight of all the books he had translated. One of Sarton's eight, al-Khwarizmi, belongs to this period, as does al-Kindi, whose massive work *Optics* would become the basis for many of Newton's theories.

It was al-Ma'mun who introduced the best Greek works on astronomy to his people – Ptolemy's *Almagest* was translated at least five times in the late eighth and early ninth centuries, once into Syriac and four times into Arabic. He built two observatories, one in the desert outside Riyadh and the other at Qasiyaun, west of Damascus: over the next seven centuries a dozen observatories would be set up at royal expense, and many more were financed by the astronomers themselves. Much of this research was Sun-related: calculating the rates of precession and the length of the tropical year, and advancing reasons for why the Sun appears larger near the horizon than at the zenith.

Their work was also specifically bound up with Islam. Muslims seek to pray facing Mecca (a mountain-set city about fifty miles inland from the Red Sea). It is not only Muhammad's birthplace: it is also revered as the first part of Earth to be created, where Ibrahim (Abraham) and his son Isma'il built the Kaaba ('cube') that Muhammad was to adopt as the focus of all Islamic worship, a physical revelation of the presence of God, and which is tended by Muhammad's tribe, the Quraysh. Such ritual acts as the call to prayer or the halal method of slaughtering animals for food must also be performed facing the Kaaba; and Muslim graves and tombs were laid out so that the body lies on its side facing the holy wall. (Modern burial practice differs slightly, but the body remains oriented towards the Kaaba.)[7] Since Muslims direct their prayers and other acts of devotion towards Mecca, they need to know *exactly* where the Kaaba lies.

How do worshippers spread across such a vast part of the globe determine the *qibla*, the direction that they must adopt to be facing towards the holy city?* In the beginning the faithful used the risings and settings of the Sun and the positions of the fixed stars – the earliest mosques faced Mecca only roughly. Out of a desire for a more faithful precision, the caliph al-Ma'mun commissioned a team of scholars to establish the coordinates of Mecca and Baghdad. Similarly inspired, the mathematician al-Biruni wrote a treatise on mathematical geography, *Tahdid*, expressly for the purpose of determining the *qibla* from Ghazna, the walled city along the route between Kabul and Kandahar. During the ninth century, according to the distinguished Arabist David King, 'Different schemes of sacred geography were developed . . . in which the world was divided into sectors about the Ka'ba[,] and the *qibla* in each sector was defined in terms of the rising or setting of the Sun or a certain fixed star.'[8] Muslim mathematicians compiled tables to express the *qibla* as an angle to the meridian for each degree of difference in terrestrial latitude and

---

*The Kaaba is actually a pre-Islamic pagan shrine of uncertain origin and date, standing about fifty feet high by thirty-five feet wide in the centre court of the great mosque of al-Haram. Embedded in its walls is a large black stone, held by scientists to be a meteorite, that all pilgrims are expected to kiss.

longitude, using geometric and trigonometric solutions. The extremely complex calculations involved led to the production of astronomical tables that made it possible for believers, wherever they might be, to know precisely which way to face. By the beginning of the tenth Christian century, mosques were being aligned towards Mecca with much greater accuracy: those in Egypt and Andalusia, west of the City of God, face the Sun as it sets at the midwinter equinox, while mosques to the east, in Iran, Iraq and central Asia, face the rising Sun.

It was not just a question of the direction to pray, but when. Muslims required accurate knowledge of the Sun and Moon so that they could determine the exact hours for prayer. According to the twentieth-century historian of science Mohammad Ilyas, 'so important was this role that only the major astronomers were assigned to the principal mosques in the different parts of the Muslim world during the medieval period'.[9] The Qur'an does not specify particular times for prayer; rather, the faithful are encouraged to use the passage of the Sun for guidance. In the early years of Islam, the times were set by simply observing shadow lengths by day and twilight phenomena in the evening and early morning. Muslims begin their day at sunset with *maghrib*, moving on to their second prayer, the *isha*, as night falls; then come the *fajr*, or dawn prayer, and the *zuhr* at noon, to commence shortly after the Sun actually crosses the meridian. Last comes the *asr*, or afternoon prayer, which begins when the shadow of any object has gone beyond its midday minimum by an amount equal to its own length. In some Muslim communities a sixth prayer is added, the *duha*, begun at the same time before midday as the *asr* begins after it.

While some early mosques were built without consulting an astronomer, young scholars were encouraged to study astronomy and mathematics, not just to prepare a calendar (the Islamic version is strictly lunar) but also as a crucial aid to religious practice. That the long line of great medieval scientists from the Muslim world coincided with the development of a dynamic and coherent Islamic liturgy is no accident.

☼

The introduction of the plane astrolabe – Greek for 'star-taker' – possibly as early as the eighth century by the Islamic mathematician and

Two astronomers stargaze through a seventeenth-century telescope

astronomer Muhammad al-Fazari, who died c. 796–806, only increased
Arab interest in studying the heavens, particularly such spectacular
manifestations as comets and eclipses. A highly versatile instrument, the
astrolabe could measure altitude and answer so many of the key questions about spherical astronomy that European scholars dubbed it 'the
king of instruments', writing dialogues in its honour. The best craftsmen
were specifically named *Alasterlabi* – makers of astrolabes – such was the
demand for the product. Most such devices were fashioned in the royal
workshops (*buyūtāt*); to make and decorate one's own instrument was a
significant achievement.[10]

When compared with the Greeks' speculative approach, the use of
these instruments – astrolabe, meridian quadrant, armillary sphere and
celestial globe – transformed the practice of astronomy, providing the
foundation on which scientists build the equipment and laboratories of
today. In 992 the obliquity of the ecliptic was measured with a quadrant
whose radius was forty cubits – about fifty-eight feet; at the great observatory at Samarkand the gigantic meridian arc had a radius of over a

hundred and twenty. These magnitudes alone gave Arabian astronomers considerable advantages in accuracy and would be their greatest contribution to our understanding of the Sun.

By the thirteenth century the Abbasid caliphate, harried by Mongol invaders, was in disarray, and east and west broke into semi-independent states. In 1220 Genghis Khan, at the head of some 200,000 horsemen, swept across Central Asia and Afghanistan, overwhelming Samarkand, Bokhara and Balkh, and pressing on into eastern Turkey. Genghis died at the age of seventy-two, by then having converted to Islam and changed his passion from destruction to building, but not before he had managed to erase, in the space of twenty years, one of the world's most advanced civilizations. Meanwhile, the Crusaders were again wreaking havoc on the western frontiers of the caliphate. The death blow came in 1258 when the Mongol prince Hulāgū Khan, Genghis's grandson, sacked Baghdad, destroying its palaces and public buildings. The libraries and collections of fine art that generations of caliphs had so assiduously assembled were pillaged and burned. Some believe that only one book in every thousand survived.

Arab astronomy, however, would live on. Three years before he sacked Baghdad, Hulāgū had succeeded in taking the vital mountain stronghold of the Assassins.* Among the prisoners was the astronomer Naṣīr al-Dīn al-Ṭūsī, who had been held as an honoured astrological adviser by the Assassins' grand master. Al-Ṭūsī now transferred his allegiance to Hulāgū, and in 1256 a huge observatory was begun under his direction at Maragha, south of Tabriz, and continued to function for over eighty years, becoming one of the most important of its time. Al-Ṭūsī's criticisms of Ptolemy's theory of planetary motion would anticipate the Copernican revolution.

The Mamluk sultans ruled Egypt and Syria from the mid-1200s to the early 1500s. Moving northwards to Turkey, westwards to north-west Africa and southern Spain and eastwards to India, they effected a renaissance

---

*These were a particularly terrifying and bloodthirsty religious sect founded in 1090 violently opposed to the Abbasid caliphate. Amin Maalouf, in his history of the Assassins, writes: 'The serenity with which the members of the sect accepted their own death led their contemporaries to believe that they were drugged with hashish, which is why they were called hashashûn, or hashîshîn, a word that was distorted into "Assassin" and soon incorporated into many languages as a common noun.'[11]

of sorts – new centres of learning grew up around the great observatories at Maragha and Samarkand, the latter built under the patronage of Ulugh Beg (1393–1440), a grandson of King Tamur the Lame (Marlowe's Tamburlaine). Ulugh Beg ('great prince') devoted himself to astronomy and compiled star tables so precise that for generations of Arabic astronomers they superseded even Ptolemy's. But the resurgence was transient – the only part of the Samarkand observatory that still survives is a giant underground marble sextant.

Various revivals were attempted in the years that followed, but none lasted long; after Ulugh Beg was assassinated in 1440, science in the then Muslim world withered away. But for 350 years Muslims had put astronomy to work in the service of Islam, and by so doing kept it alive.

☼

In 1068 a scholar of the time picked out a group of eight nations according to their perceived contribution to science: 'the Hindus [Indians], the Persians, the Chaldeans, the Hebrews, the Greeks, the Romans, the Egyptians [the scientists at Alexandria] and the Arabs.'[12] The list is a simplification, ignoring the many interactions among them; it gives no credit to China, which, although it may have been a world unto itself, should not have been discounted; and imperial Rome is an unexpected and perhaps unwarranted inclusion. But that list will shape the rest of this chapter, which looks at what Hindu scientists gave to solar astronomy and assesses how Western Europe fared in the years before Copernicus.

The Vedic religion, the precursor to modern Hinduism, was one of the earliest to have its beliefs and rituals set down in written form (Sanskrit); the Vedas as a whole (the earliest of which date back to 1500 BC) are filled with hymns praising the Sun as the source and sustainer of all earthly life, and one of its associated texts, the *Jyotisa Vedānga*, refers to the Sun ('Surya' or 'Aditya') as 'a storehouse of inexhaustible power and radiance'.* The first Indian study devoted to astronomy, the Vedas were already an influential work by the fourth century BC.[13]

---

*One word for 'bright' in the *Jyotisa Vedānga* is 'citrus'. It seems reasonable to suppose that the fruits we know as 'citrus', such as oranges and lemons, gained a generic name from their Sun-like colour. When the strife goddess Eris set out to precipitate the Trojan War, she cast down not an apple but an orange, for which the Greek is *chrisomilia* – 'golden apple'.

Perhaps as a result of that early interest, the influence of Indian science was pervasive – and not only in China, where Indian astronomy made inroads within a century or so of the Vedas being written. That great authority on early science Otto Neugebauer writes that Indian astronomy constitutes 'one of the most important missing links between late-Babylonian astronomy and the fully developed stage of Greek astronomy represented by the *Almagest*'.[14] It is possible that Indian astronomers exchanged ideas with their Greek and Babylonian counterparts at least from the time of Ptolemy, if not well before: key concepts already present in the Rig-Vedic texts turn up in Babylonian astronomy around 700 BC. We do know that Indian astronomy was a felt presence in China, along with Buddhism, during the Tang Dynasty (AD 618–907), when many of their skywatchers took up residence in the Chinese capital. By AD 718 Indian astrology had also taken root, while the Chinese adopted the Indian calendar, organized around the twenty-eight lunar mansions. Next to be influenced was Persia, whose Abbasid caliphs were so enrapt by Indian science that they worked to spread its methods into central Asia.

From the fifth century AD Jainist doctrine (with its emphasis on rebirth and reincarnation but also on supernatural forces issuing as karmas)* began to assert itself through most areas of India. It decreed Mount Sumeru – supposedly situated in the Himalayas – to be the Earth's centre, around which revolved the Sun, Moon and stars. By day, the Sun was thought to travel south from the mountain, while at night the Himalayas hid it from view. Jains had no notion of daylight in one hemisphere and night in the other, but imagined each to have its own Sun.

The Indian picture of the cosmos derived primarily from religious doctrines, and none of their ancient records of the Sun or of sunspots or eclipses has been found. Instead, Indian astronomers took the statistics compiled in Babylon and Egypt and adapted them to their own needs – calendar computation, timekeeping, the casting of horoscopes

---

*Jainism, or Jain Dharma, began in the sixth century BC as a revolt against Hinduism. It regards every living soul – even those of animals and insects – as potentially divine and advocates non-violence. Unusual in being an atheistic religion, it teaches among other things that one should not eat, drink or travel after sunset. Currently there are 5.2 million Jains in India, found in all but one of the country's thirty-five states.

and the prediction of eclipses. Yet it was the Indians who reckoned the age of the Earth as 4.3 billion years, when even in the nineteenth century many scientists were convinced it was at most 100,000 years old (the current estimate being 4.6 billion). Indian civilization most influenced Western astronomy not through its theories but through its development of mathematical tools: brilliantly innovative spherical trigonometry, for instance, and their invention of the decimal base, with the concept of zero functioning as a number.[15]

☼

The fall of Rome extinguished the essentially bilingual civilization of the West, resulting in the almost total disappearance of the Greek language outside the eastern empire, with most major works of Greek science meeting the same fate. Astronomy moved philosophically from a pagan to a Christian milieu, and geographically from one intellectual centre in Alexandria to another in Constantinople, but for between 300 and 700 years after Ptolemy the discipline, both observational and theoretical, was almost entirely neglected. The vicissitudes of wars and faiths subjected learning to long hiatuses – disastrous for a highly technical science dependent on competent and consistent instruction, handed down from one generation to the next. In 398 a Christian mob wrecked the library of the Alexandrian Museum.* In 415 another mob lynched the leading astronomer, the Lady Hypatia. Any writings from 650 to 800 were negligible, partly the effect of Iconoclastic persecutions in the eighth century.

Throughout much of the Old World, most people were living on the absolute margin, their lives threatened by barbarian invasions, calls to war and ever more frequent migrations, famine, plagues, floods and other disasters. In 526 Antioch was wrecked by an earthquake; in 542

---

*Of course, solar astronomy was also practised outside the main cultures we have examined, but either of a basic kind or not essentially differing from that of the countries we have looked at. The main contributions of South American and Mexican astronomy are discussed in Chapter 20. Otherwise one may take as typical the customs (at least up to the early years of the twentieth century) of New Zealand's Maori, where an elite drawn from the *tohunga* (or 'adept') class would spend hours each night contemplating the *ra ririki*, or little Suns (the stars), and were relied on as weather forecasters. Travellers and fishermen consulted them before any journey, and the power of the *tohunga kokorāngi* stretched to foretelling the general aspect of future seasons, their fruitfulness or otherwise. Such men were valuable observers; but their contribution to solar astronomy itself was minimal.

Indian folk drawing depicting Surya Mandala with a cobra,
symbolizing the Sun, water and vegetation

Constantinople was scourged by bubonic plague; nine years later an earthquake and tsunami killed tens of thousands in Lebanon. The year 856 saw a major earthquake in Greece, 1069 a famine in Egypt. The tenth century alone brought twenty severe famines, some lasting two to three years. The eleventh century was no better: France endured at least twenty-six famines, England a famine every fourteen years or so. Earthquakes in 1138, 1268 and 1290 killed upwards of half a million. In 1228 a great flood swept away 100,000 people and a substantial portion of the land in the Low Countries. Then, in the middle of the fourteenth century, the bubonic plague returned in the form of the Black Death, raging for six years and carrying off well over a third of the people of Europe as well as multitudes in Asia and North Africa, probably 25 million in all. It was achievement enough to survive, let alone map the heavens.

At the end of the ninth century the West was just beginning to emerge from its Dark Ages. Arab astronomers had streamed into Spain, and by the tenth century Cordoba, which boasted half a million inhabitants, several libraries, a medical school and a large paper trade, was a substantial centre of learning. While trade routes between the West and the Arab world were by this time long established, the exchange of scientific knowledge began only when Christian scholars visited Spain in the eleventh century, not only imbibing Arab ideas but also reclaiming their own heritage as a result. By the twelfth century Toledo was the hub of translating from Arabic into Latin, Hebrew and Castilian, as well as many other languages. The *Almagest* was rendered by an anonymous Sicilian into Latin in 1160, although it remained a difficult work: some words were not understood but simply transliterated – such as 'nadir' and 'zenith' – with the result that a number of simplified and often inaccurate versions appeared. Like a game of Chinese whispers, each translation became progressively more removed from the original.*

Even so, by 1200, thanks to the Arab translations in Cordoba and Toledo, reasonably accurate Latin versions of Ptolemy and the main works of Aristotle, Plato, Euclid, Archimedes and Galen were circulating. Ptolemy's writings became almost synonymous with astronomy and were rarely questioned – although the leading Arab astronomers certainly had their doubts, discerning yet more errors in his system, particularly in his theories of planetary motion. How to explain, for example, the great variation in the brightness of Mercury and Venus by a Ptolemaic reading? The greatest of a new generation of Arab scientists, Ibn Rushd (1126–98), known as Averroës, whose massive commentaries on Aristotle earned him the sobriquet '*the* commentator', wrote: 'The astronomical sciences of our days offer nothing from which one can derive an existing reality. The model that has been developed in our times accords with the computations – not with the way things actually are.'[16] King Alfonso X of Castile (1221–84) is said to have remarked, on being introduced to the Ptolemaic model, that, had he, Alfonso, been present at the Creation, he might have given the Almighty some better advice.

From the dissolution of the last vestiges of imperial authority at Rome in the late fifth century until the final years of the Renaissance, the Roman Catholic Church dominated the West. During the first half of this

---

*The great scientist's influence endures: in the early months of 2008 a New York couple sent their young son to a small nursery school in Brooklyn. *Seven* of his fellow pupils were named Ptolemy.

period, virtually the only schools were run by monks: pupils had to make do with limited and second-hand sources, their teachers offering little encouragement to original enquiry or even observations. Scientific knowledge was for the most part regarded as little more than an aid to understanding the Bible. St Augustine of Hippo, one of the largest-minded of Christian thinkers, observed:

> It is not necessary to probe into the nature of things, as was done by those whom the Greeks called *physici*; nor need we be in alarm lest the Christian should be ignorant of the force and number of the elements – the motion, and order and eclipses of the heavenly bodies; the form of the heavens . . . It is enough for the Christian to believe that the only cause of all created things, whether heavenly or earthly, whether visible or invisible, is the goodness of the Creator, the one true God.[17]

A Dutch drawing of 'the astrolabe of Lansberg', based on Ptolemaic principles according to the original caption, though it looks more like a sundial that the man is consulting

Professor John North of Trinity College, Cambridge, a punctilious cus-
todian of the period, adds: 'Generally speaking, Christian man saw
himself as a degraded and miserable creature whose only hope was in
prayer and penitence, and as one to whom a rational understanding of
the motions of the planets was supremely irrelevant.'[18] It is hardly surprising, then, that astrology continued to be an integral part of everyday
life, those few astronomers who remained being relegated to accumulating data and improving the precision of formulas and parameters.
Astrology became the patron of astronomy, providing a market for treatises and tables and so contributing to the survival of works that would
otherwise have perished.[19]*

Gradually, however, the propagation of Christianity, centred as it was on
an all-creating single deity, led the way to a scientific approach to the workings of the natural world through its rejection of the plethora of charms,
omens, dreams and suchlike foolery that went to make up astrology. This
at first hardly helped astronomy. It is true that no medieval student could
get his master's degree without understanding the rudiments of how to
study the heavens; on the other hand, medieval Christian universities,
taking their lead from the Church, never gave astronomy a prominent
position: it may have been part of every introductory course, but was
taught at an elementary level, with the main deficiencies of the Greek
world view persisting.

As the cultural climate warmed, and the first Christian universities in
Europe were established in cities such as Oxford (1096), Paris (between
1150 and 1170) and Bologna (1158), the *Almagest* found its way into the
curriculum of mathematical education both in Europe and the Near and
Middle East, being taken as a natural sequel to Euclid's *Elements* and the
treatises on spherical astronomy by Autolycus and Theodosius; even
then, most students found it too difficult and resorted to interpretive
texts that dealt with basic solar questions.

With the formation of universities there arose a figure who would
come to dominate the intellectual landscape of Western Europe. The
Franciscan friar Roger Bacon (c. 1214–1294) was within his lifetime

*The famous 'All the world's a stage' speech in *As You Like It* (1599) is in part based on an
astrological reading of man's seven ages, each of the so-called 'planets' guiding a particular
period of time. The Sun oversees 'young manhood' – the years twenty-three to forty-one.[20]

hailed as 'Doctor Mirabilis' (the Wonderful Teacher) for his breadth and depth of understanding, and held by some to be the first true scientist.[21] When Oxford was given its charter, he fought to have science made part of the curriculum, insisting that it was complementary, not antagonistic, to the faith. 'Everywhere,' he declared, there was 'a show of knowledge concealing fundamental ignorance.' Pope Clement IV, who as Guy de Foulques had come to know Bacon while serving as papal legate to England, now wrote to his old friend asking him to undertake a book on how the sciences should be taught.

Within eighteen months Bacon had completed three large volumes, in which he offered not only his ideas for better scientific methods (for instance, that laboratory experiments be part of one's education) but also a vehement attack on the vices of clergy and monks. Over the next ten years, going far beyond his commission, he would anticipate an extra-ordinary range of discoveries and inventions: the magnetic compass needle, a method for constructing a telescope, reading glasses, the steamship, the aeroplane, the workings of radiology, the cause of rainbows, the principles of a camera obscura to project pictures – even television. He also wrote several essays on astronomy. But Pope Clement died before he could read any of this, and his successor, Nicholas IV, condemned Bacon's writings as heretical. By 1278 the Wonderful Teacher had been thrown into prison (most likely, an unpleasant form of house arrest), which he was made to endure for the next fourteen years, dying at age eighty or thereabouts in 1294.

Over the next 200 years astronomers remained preoccupied with the adaptation and manipulation of a bewildering profusion of tables, and for all the power of Bacon's reasoning the universities continued to view science as a way to establish the mathematical relationships between the phenomena of nature rather than conducting an open-minded enquiry based on empirical research and experiment. As a result, a scholar proved his prowess through the mastery of the most difficult predictions (eclipses) and of the geometrical principles underlying the interminable tables.

The one exception to all this was a revival of fascination with Aristotle. Before the twelfth century his works were largely unknown in Christian Europe, but now they came back into fashion with a vengeance. In particular, people quoted *On the Heavens*, which proposed that science should seek out the *causes* of phenomena. According to Aristotle, a stone, for instance, falling towards the centre of the Earth increased its speed because, being of earthly nature, it was impatient to get home: all motion was a transition from 'potency' into 'act', the realization of what may exist in any object's basic nature. This might be physics back to front, but it won a wide following. (Arthur Koestler would remark that, in moments when we curse an obstinate gadget or a temperamental machine, we all revert to Aristotle.)[22]

The end of the twelfth century saw two main schools of thought: the strict Aristotelians and the mathematical astronomers, who distrusted any theory that could not be supported by elaborate tables and calculations. Bacon had helped resolve the conflict between Aristotle's and Ptolemy's theories by introducing a new explanation for the ways in which the various celestial spheres might interact. Full of notions of outer convex and inner concave surfaces, his idea was no better than its predecessors, but it kept the peace – and kept both Greek thinkers in the classical canon. Aristotle and Ptolemy continued to be taught, and their incompatibilities, with each other and with observable phenomena, were tacitly accepted. Then in 1277 the Bishop of Paris condemned 219 propositions of Aristotle's, declaring several of his doctrines (such as the eternity of the world) to be heresies. The Aristotelians retreated.

A century later, in 1377, another Parisian cleric, Nicole Oresme, wrote a commentary on Aristotle dismissing his arguments for the immobility of the Earth as purely relativistic. Although Oresme (later Bishop of Lisieux) could not quite persuade himself that the Earth actually did perform a daily turn about its axis, he at least raised the possibility that it did so. This tentative probing of the ancient assumptions about an unmoving globe was, like the Arab criticisms, but one more dent in the bodywork of a trusted vehicle, and the Greek view of the heavens continued to trundle happily along, its engine carefully lubricated by both Church and science. Heavenly bodies, the Sun included, still moved in uniform circles and the Earth lay immobile at the centre of a rotating universe.

But that was about to change.

9

THE EARTH MOVES

In the centre of everything is the Sun. Nor could anyone have
placed this luminary at any other, better point . . . Not without
grounds do some call it the World's Lamp.

Nicolaus Copernicus

People give ear to an upstart astrologer who strives to show
that the Earth revolves, not the heavens or the firmament, the
Sun and the Moon . . . He's a fool who wants to turn the art
of astronomy on its head.

Martin Luther[1]

While still a teenager, Groucho Marx took it upon himself to be his
brothers' teacher, and one day asked Harpo, two years his senior, how
the world was shaped. Harpo confessed that he wasn't sure.

Groucho gave him a hint. 'What is the shape of my cuff buttons?'

'Square.'

'I mean the cuff buttons I wear on Sunday – not every day. Now:
what's the shape of the world?'

'Round on Sunday, square on weekdays,' replied Harpo, who shortly
after took his vow of professional silence.[2]

The tale has been around for a long time, and the question for even
longer. What shape is the world? In the first century AD Pliny the Elder
concluded it was a globe, but he also had a hedging theory that it was
'shaped like a pine cone'. The encyclopedist Isadore of Seville would also

declare the world circular, but flat, like a wheel. The Venerable Bede thought it a sphere, but could not imagine that people could live in the inverted antipodal ('feet-opposite') regions. These sages were mostly on their own, as from the fifth through the end of the tenth century Europe returned to a belief in a rectangular world: mapping orthodoxy put Jerusalem at its centre; but when in 999 the astronomer Gerbert became Pope Sylvester II, he adopted Pliny's globe as Church doctrine. In 1410 the appearance of Ptolemy's *Geography* in Latin confirmed the by then generally accepted notion that the world was round, and by the century's end, as Columbus set out for the Indies by sailing west from Spain, sailors and scientists alike embraced the idea of a spherical Earth, even if they were unsure of its size. It took a far later generation to show that its rotation causes bulges at the equator, so it is not a perfect sphere after all.*

But where was our planet in relation to the rest of the universe? 'Among the masses,' writes William Manchester, 'it continued to be an article of faith that the world was an immovable disk around which the Sun revolved, and that the rest of the cosmos comprised heaven, which lay dreamily above the skies, inhabited by cherubs, and hell, flaming deep beneath the European soil. Everyone believed, indeed *knew*, that.'4 Well, not quite everyone: an increasing number of astronomers were finding that they could not make a geocentric universe fit all the facts. Muslims and Christians might endlessly revise the *Almagest*, but none of the variants they came up with could compute the planetary positions in a way that would square with the increasingly accurate observations of the sixteenth century. It was not that no one had figured it out: 400 years before Aristarchus of Samos advanced his notion of a heliocentric cosmos, philosophers in northern India argued that if the Sun were the largest object in their universe, and if gravity were holding it together – they believed both to be true – then the Sun had to be at its centre. There were others: early in the eleventh century al-Biruni had come to

---

*In 1956 Thomas Bailey's influential *The American Pageant* stated without a line of evidence that Columbus's 'superstitious sailors . . . grew increasingly mutinous . . . because they were fearful of sailing over the edge of the world'.3 Both Tolkien (with Arda) and C. S. Lewis (with Narnia) envisaged a flat world at least once in their fiction, while about 1910 a schoolmaster in Mississippi was dismissed for teaching that the Earth was round. Otherwise one is left with a Monty Python sketch, 'The Crimson Permanent Assurance', in which a pirated office building plunges off the edge of the world.

roughly the same conclusions, and the German theologian (later cardinal) Nicholas of Cusa (1401–64) had speculated on the possibility. But before the advent of the printing press and with no one to record their thoughts, such musings mostly died in the dark.

When in 1450 Johannes Gutenberg finally invented movable type, not only could these theories be preserved in print but they could be widely disseminated. Readers could enjoy more information in less space, along with pagination, indexes, standardized punctuation, tabulated data and accurately reproduced illustrations; while scholars could now put together personal libraries and have access to the same printed text that they could discuss by letter with colleagues, no matter how far distant. By 1500 printers' workshops could be found in every important Western European city, and there were some 6 to 9 million copies in print of more than 35,000 published titles.

In 1465 a thirty-year-old astronomer from Königsberg, in Bavaria, Johann Müller (1436–76), began to write and commission works on astronomy, along with almanacs and tables. These books and papers, given the need for tools to aid navigation and exploration and the ever-present enthusiasm for astrology, proved highly popular. Within five years he had established himself in Nuremberg with his own observatory and printing shop, and by the century's end nearly every astronomical manuscript of value had been disseminated throughout the West.

Enter Nicolaus Copernicus (1473–1543), a shy and retiring cleric in Polish-ruled East Prussia, who had become curious about the doctrines of heavenly spheres during ten years' study in Bologna and Padua, only to find himself, like so many before him, having to make adjustments to accommodate the contortions of Ptolemy.[5] How could Mercury and Venus never seem to stray far from the Sun, while Mars, Jupiter and Saturn sometimes moved backwards? He was compelled to the idea of a heliocentric system, which fitted the evidence far better than Ptolemy's solution – though still imperfectly, since like everyone else he continued to assume that the planets travelled in circular, not elliptical orbits. He included Earth as another orbiting planet – a notion that he tried to dismiss as absurd, but that wouldn't go away.

Now thirty, he returned to the home of his bishop uncle (that 'far corner of the Earth', as he sourly dubbed it) where he built a primitive observatory in one of the turrets in the wall surrounding the cathedral precincts. Following directions in the *Almagest*, he set about constructing the same crude wooden instruments that had been used for centuries, and spent his nights gazing at the stars. Around 1514 he began circulating copies of what later scholars would title *De hypothesibus motuum coelestium a se constitutis commentariolus*, or 'Little Commentary', which laid out his vision of a Sun-centred system, and asked his friends and fellow scholars for their opinions. At this particular moment the Catholic Church was imbued with a growing enthusiasm for learning, and was happy to encourage original scientific research, as long as it did not openly challenge doctrine. When Pope Leo X sent an encouraging note, and liberal members of the Curia let it be known that they too approved, plans were made to print a more substantial version. One would have thought that the basic thesis was self-evidently controversial, yet at this stage there was no hint from official channels of the storm to come.

In 1532 Copernicus's new system was given its first presentation, by no less than the Pope's private secretary, to an invited audience in the Vatican gardens. The guests were favourably impressed. Copernicus began to win a degree of fame without actually having committed anything to public print. Yet still he hesitated – partly, no doubt, because he was unable to find any direct evidence that the Earth *did* move, let alone round the Sun, partly because he must have suspected that Protestants, with their literal interpretation of the Bible, would be far less sympathetic than his friends in the Vatican. Almost thirty years would elapse before he allowed an extract from the work in progress to be included in a book on trigonometry.* The whole came to 212 folio pages and appeared in 1543 in an edition of a few hundred copies under the title *De revolutionibus orbium coelestium*, 'On the Revolutions of the Heavenly Spheres'. By this time the ageing canon was seventy, afflicted with apoplexy and paralysed 'from ear to heel': he saw the first copy to come off the press only hours before he died.

'The scorn which I had to fear,' Copernicus writes in *De revolutionibus*, 'on account of the novelty and absurdity of my opinion almost drove me

---

*The book was mainly the work of Copernicus's friend and sole pupil Georg Rheticus (1514–74). It was printed at Wittenberg, the same university to which Shakespeare sent that supreme procrastinator Hamlet; but Copernicus was hardly unique in his reluctance to publish. Darwin let *The Origin of Species* languish in a desk drawer for fifteen years.

to abandon a work already undertaken.' That said, he was not bashful in claiming credit. It is true that in a dedicatory letter to Pope Paul III (the pontiff who, Dava Sobel reminds us, excommunicated Henry VIII and extended the remit of the Inquisition)[6] he acknowledges that he first learned about the possibility of a moving Earth from reading Cicero, and had discovered from Plutarch that others had shared such a view. But while he initially acknowledged the contribution of Aristarchus of Samos, in his last editing he struck all mention of him.[7] He outdid Aristarchus, however, in offering detailed models of the planets' passage around the Sun – Mercury's in approximately eighty days (our current figure is 87.97), Venus's in nine months (our figure is 224.7 days), the Earth–Moon system in a year, Mars in not quite two Earth years (1.88), Jupiter in twelve (11.86) and Saturn in thirty (29.4): all reasonable estimates. He also managed to determine within five per cent of the correct values the planets' least and greatest distances from the Sun.

He proposed that the Earth orbited the Sun, rotating once every twenty-four hours (thus making the stars appear to revolve in the opposite direction). He placed the six known planets in their correct sequence from the Sun – Mercury, Venus, Earth, Mars, Jupiter, Saturn – but that is as far as he got; the spheres of the stars remained fixed, and the trajectories of the planets circular, not elliptical. Copernicus did not get everything right but he delivered more than enough to stir up bitter argument. Martin Luther (1483–1546), one of several leading churchmen who viewed the new theory as not only shocking but gravely in error, asked plainly: 'Who will venture to place the authority of Copernicus above that of the Holy Spirit?' After all, he continued, Joshua commands the *Sun* to stand still, not the Earth.[8]

As long as Copernicus's ideas were represented as hypothetical, Rome herself remained silent, and the intellectuals in her flock followed suit. For fifty years after his death, most astronomers privately acknowledged the importance of what he had revealed, while publicly maintaining their belief in a stationary Earth. 'The acceptance of the heliocentric theory was far from general right up to the time of Newton,' the astronomer Patrick Moore has written, 'and in countries far away from the sphere of Mediterranean culture the old ideas persisted for even longer.'[9] In China and Japan, the Jesuits continued to teach the received cosmology. The

notion of a huge and unknown universe of which the Earth was but one
minute part would put much of the Christian apparatus of belief at risk,
and that was not a thing to be undertaken lightly. The reaction of the
baptized world recalls the wife of a Victorian canon of Worcester
Cathedral who, informed that Darwin's theory implied the descent of
man from ape-like creatures, cried out, 'Descended from the apes! My
dear, we hope it is not true. But if it is, let us pray that it may not become
generally known.'[10]

Copernicus, ever the reluctant revolutionary, had aimed to revitalize
the Ptolemaic tradition of mathematical astronomy, not to supplant it.
Again and again his calculations fall into place alongside Ptolemy's – the
very structure of *De revolutionibus* mirrors that of the *Almagest*, and its
tables were no more accurate than those of the old orthodoxy.* 'Had it
not been for Tycho Brahe† and Kepler,' concludes Otto Neugebauer
acidly, 'the Copernican system would have contributed to the perpetua-
tion of the Ptolemaic system in a slightly more complicated form but
more pleasing to philosophical minds.'[11] Arthur Koestler is even more
damning: in almost 400 pages devoted to the period from Copernicus to
Galileo, he buries the Polish canon's reputation – in character 'a stuffy
pedant, without the flair, the sleepwalking intuition of the original genius;
who, having got hold of a good idea, expanded it into a bad system,
patiently plodding on, piling more epicycles and deferents into the drea-
riest and most unreadable among the books that made history'.[12] Yet
Copernicus did strike the original blow in a genuine revolution. He was
the first to take the idea of an Earth that rotates on its axis every twenty-
four hours around a central Sun and to work it into a comprehensive
theory.

---

*Despite this judgement, Copernicus continues to exercise a fascination on writers.
Currently Dava Sobel is writing a play about him, and Dennis Danielson, a Vancouver professor, has recently published a biography of Georg Rheticus, the conflicted German
professor of mathematics who did so much to help prepare *De revolutionibus* for publication but received no acknowledgement. Rheticus is the central narrator of John Banville's
novel *Doctor Copernicus*. Banville, winner of the 2005 Booker Prize for Fiction, has also written novels about Kepler and Newton.

†He is generally referred to just as 'Tycho'. Of other major scientists, only Leonardo da
Vinci is similarly foreshortened – and he had other gifts. Artists are often known by their
first names: Michelangelo, Raphael, Claude, even Inigo (Jones); in Tycho's case his easy
consorting with royalty may have led to the fashion.

☼

In 1546, three years after Copernicus died, there was born an astronomer who would model a cosmos that retained much of Copernicus's heliocentric theory but did not require the Earth to move. Tycho Brahe (his first name is Greek for 'hitting the mark') came from a Danish noble family, and, growing up a Protestant, he inherited a world view framed by a strict interpretation of the scriptures. He saw his first eclipse as a freckle-faced boy of fourteen and was struck that it had been predicted to the day. How could imperfect man anticipate the majestic acts of the heavens? He took up astronomy to find out. At sixteen he observed how the conjunction of Jupiter and Saturn, predicted by the still operative Julian calendar for 25 August 1563, was two days off if one went by Copernicus's tables, over a month if by Ptolemy's. (A 'conjunction' occurs when two objects have the same celestial longitude and so seem to occupy the same space.)* The astronomical system, he decided, was in dire need of overhaul. That would be his mission.

The death of Tycho's foster-father (his uncle Jørgen Brahe) in 1565 brought him into his inheritance, and he soon built an observatory at Herrevad Abbey, near Copenhagen, ransacking Europe for the finest instruments, among them a celestial globe five feet in diameter which cost five thousand *dalers*, the equivalent of eighty years of a schoolmaster's salary.† A massive mural quadrant, a great quarter-circle made of brass, it had a radius of more than six feet and enabled Tycho to see exactly when stars crossed the meridian. He added a huge sextant, which accurately measured the latitude and longitude of celestial objects. The

---

*Jupiter and Saturn are not only the largest of the Sun's planets but also the furthest consistently visible to the naked eye. They align only every twenty years. One theory holds that their conjunction in 7 BC formed the 'star of Bethlehem' from which the Wise Men predicted the coming of the Messiah. They next meet in our skies on 21 December 2020.

†In other respects Tycho followed the pursuits more typical of a young man of his time. During Christmas 1566, while at the German university of Rostock, he attended a dance and, the worse for drink, got into an argument that led to a duel with broadswords in the dark. His opponent sliced off a sizeable chunk of Tycho's nose, and thereafter the astronomer wore a prosthesis made of silver and gold blended into a flesh tone, with an adhesive balm to keep it from falling off. In 1901 his tomb was opened and his remains examined by medical experts who claimed his nasal opening was rimmed with green, a sign that the false nose had been made of copper. Historians have speculated that he wore the lighter copper nose for everyday use and his more precious one on special occasions.

complex also included a paper mill, a glass works, a tubal intercom system, flush toilets, a private jail, an instrument factory and a chemical laboratory – in effect, facilities sufficient to create an entire self-contained universe.

On the evening of 11 November 1572 Tycho was walking back from his observatory for supper when he was brought up short by a brilliant white light outshining Venus. 'Amazed, and as if astonished and stupefied, I stood still, gazing with my eyes fixed intently,' he wrote. '. . . I was led into such perplexity by the unbelievability of the thing that I began to doubt the evidence of my own eyes.' He called several servants, who confirmed what he saw, but fearing this might be a group delusion he went on to stop peasants in the street to ask if they could see what he was seeing. Over the next eighteen months the light, slightly north-west of the three stars that make up the right-hand stroke of the familiar 'W' of Cassiopeia, remained at a point in the sky where no star had been observed before. Lying outside the zodiac, it could not be a planet, nor did it fit the descriptions of comets. Sometimes it blazed so intensely that it could be seen in daylight; but by December it had dimmed to the brightness of Jupiter, and by the following March was but a vivid point, modulating from white to red and finally to a leaden grey. Tycho monitored it until it disappeared around March 1574.

Were this a new object in the sky, that would require a dramatic revision to Aristotelian cosmology, for according to Aristotle change and decay were confined to the world below the Moon: everything above was immutable. But something in the heavens was all too evidently changing, for what Tycho had witnessed was a huge stellar explosion, a star in its death throes – a phenomenon that would one day be known as a supernova.

Tycho was the last astronomer of note to work without the aid of a telescope, but his was an age with a new respect for hard facts, and he became the most gifted observer on record, his naked-eye measurements accurate to the arc-minute – i.e. one-sixtieth of a degree. (The human eye is a superb long-distance instrument, able to discern on a clear night a candle flame seven miles away.) Tycho rigorously catalogued the planets and stars, listing a thousand of the latter, and in 1573 published *On the New Star*, a pamphlet about the supernova that helped establish his reputation. The following year he gave the Latin inaugural oration at

Copenhagen University, with the entire academic community in attendance. 'In our time,' he began,

> Nicolaus Copernicus, who has justly been called a second Ptolemy, from his own observations found out something that was missing in Ptolemy. He judged that the hypotheses established by Ptolemy admitted something unsuitable and offending against mathematical axioms ... He therefore arranged, by the admirable skill of his genius, his own hypotheses in another manner and thus restored the science of the celestial motions in such a way that nobody before him has considered more accurately the course of the heavenly bodies. For although he devises certain features opposed to physical principles, e.g. that the Sun rests at the centre of the universe, that the Earth, the elements associated with it, and the Moon move about the Sun in a triple motion, and that the eighth sphere remains unmoved, he does not for all that admit anything absurd as far as mathematical axioms are concerned.[13]

Tycho knew that the old system failed to make sense, but he was unwilling to let go of the Earth as the centre of everything. He produced what became known as the Tychonic System, which resembled that of Copernicus except that it had two centres, the Earth and the Sun. The planets other than the Earth orbited the Sun while the Sun orbited the Earth – a fine compromise between the Bible and science, if to our eyes untenable. But Tycho was an observer, not a theorist, and pressed on unperturbed.

His account of the supernova both intrigued King Frederick II and led him to worry that his leading astronomer might leave Denmark for more tempting pastures. The king decided to grant Tycho the remote island of Hven, situated midway between Zealand and Sweden, its cliffs a hundred feet high and in clear sight of the great fortress Frederick was raising at Elsinore. With it came a generous endowment consisting of the rents from the island's householders for the rest of his life: at one point in the 1580s Tycho commanded one per cent of Denmark's entire income.

He immediately set to work on two new observatories on Hven; the underground 'Castle of the Stars', unshakeable by any storm, and a larger complex above ground, 'the Castle of Uranus', named after the Greek

The castle and observatory of Uraniborg (in Danish, Uranienborg) on the island of
Hven, in Denmark, constructed around 1584 by Tycho Brahe

god of the skies. The latter, nicknamed his 'Museum' – literally 'Temple
of the Muses' – boasted ramparts, pavilions, botanic gardens, fountains,
an aviary, a print shop, cellars and towers. By the time it was all completed, 'there was no other place quite like it anywhere in the world'.[14]

In November 1577 Tycho was fishing one of the island's many ponds,
his workers hauling in the nets, when he noticed what looked like
another new star. As dusk faded into night, he saw that it had a reddish
tail stretching away from the Sun, the unmistakable signature of a comet
(a word that originally meant 'head of hair' in Greek; Aristotle appropriated it because he thought comets – as we now know, no more than
conglomerates of ice and dirt – looked like hairy stars). Tycho rushed to
one of his observatories, where he determined the newcomer's position
by measuring the angle and distance from its two nearest stellar neighbours. For four months he tracked its fading dash across the skies, and
found that it had no measurable 'parallax' (a phenomenon exemplified
by the seeming movement of objects in the foreground against those in
the background, when one moves one's head).

In a report to the king, he calculated that the object was at a distance
of more than 230 times the radius of the Earth (four times that between

the Earth and the Moon). Even though Tycho's calculation was badly off, this, even more than the earlier supernova, convinced him that Aristotle was wrong: separate spheres could not be spinning above and below the Moon. He postulated instead a common atmosphere pervading the universe, in which novas and comets could burst out anywhere. The old doctrine of multiple spheres, he wrote in 1588, 'a theory that authors have invented to save appearances, exists only in the imagination so that the motions of the planets in their courses may be understood by the mind'.

From his mid-thirties on, Tycho became progressively more eccentric. He kept a dwarf named Jepp, who he believed had powers of clairvoyance, and who sat under his table at dinner (unless required as a jester) feasting on titbits fed him by his master; and also a tame elk, which one night wandered upstairs to an empty room and lapped up so much beer that it crashed back down the stairs, broke its leg and died. His own fortunes would also take a turn for the worse: in 1588 Frederick II died (he too from a surfeit of drink) and, his son being only ten years old, for the next nine years there ruled a series of regents. All were Tycho's friends, so nothing changed; but when the new king, Christian IV, turned nineteen he determined to shut off the favours this rattle-brained stargazer had enjoyed for over a quarter of a century. Tycho quit his island, taking with him his students, servants, instruments and printing press, and settled in Catholic Prague. Here he built his last observatory – at Benátky Castle, about thirty miles outside the city, where, until his death in 1601, he spent his time writing poetry and summarizing his researches. But he published not a word more.*

The year before Tycho left Hven, he was sent the first publication by a twenty-five-year-old German schoolteacher of astronomy and mathematics. Johannes Kepler (1571–1630) was short-sighted, surly, reckless, erratic and from a poor and severely dysfunctional family. (His mother

*It was during this period – actually in 1592 – that two of Tycho's cousins by marriage, Frederik Rosenkrantz and Knud Gyldenstierne (which means 'Golden Star'), travelled to England on a diplomatic mission, in the course of which Shakespeare apparently learned about them and introduced them into *Hamlet* (completed around 1601) as two famously untrustworthy courtiers. It was probably not an unfair portrait, at least in the case of Rosenkrantz, who had years before impregnated a lady-in-waiting, for which outrage he had been sentenced to forfeit his title and lose two fingers, but he escaped that fate by enlisting against the Turks. Both courtiers enjoyed further fame when Tom Stoppard re-immortalized them in *Rosencrantz and Guildenstern Are Dead*.

The frontispiece of Kepler's *Cosmographic Mystery* (1596), showing the planetary spheres to be nested alternately with the five Platonic solids, so determining the sizes of the spheres and limiting their number to six. From this, Kepler estimated the distances of each planet from the Sun

had been accused of witchcraft, and her son fittingly had the devil's own job saving her from the stake.) He was also a genius. His *Mysterium Cosmographicum* (*The Cosmographic Mystery*) did not take one far in the way of explaining the universe, as it held to the notion of concentric spheres, but it did show its author's powers as an original thinker, and Kepler did place the Sun at the centre. Tycho invited him to visit, but Hven proved to be too far from Kepler's home in Graz. Four years later, however, with Tycho now close to Prague, such a journey became practical; and when in 1598 Kepler, a Lutheran, saw his school closed by order of the Jesuit-educated Archduke Ferdinand, the die was cast.

On 4 February 1600 Tycho, fifty-three, and Kepler, twenty-eight, finally met, 'face to face, silver nose to scabby cheek'.[15] They spent three fractious and not particularly satisfying months in each other's company (the Dane suspicious and secretive, the young German obstinate and critical), during which Tycho insisted that, before Kepler returned to Graz to bring order to his family's affairs, he should pledge in writing to reveal nothing about his host's work. But when that same year Tycho's principal

assistant returned to Denmark, Kepler was invited to move in perma-
nently. Although the two men seldom spoke to one another except at
meals, when Tycho died the following October Kepler found himself
appointed to the post of Imperial Mathematician (a position that would
have been Tycho's, had he not been too lofty an aristocrat for such a
job). Kepler immediately set to work drawing on Tycho's vast and precise
body of observations to advance his own theories, several of which (such
as the motions of the planets) were the opposite of his benefactor's.

By 1605 Kepler's main researches were completed – he was to say that
he had worked so hard on his calculations during these years that 'he
could have died ten times'. Interference from Tycho's heirs slowed down
the publication of his findings, but at last, in 1609, Kepler's *The New
Astronomy* appeared. In it he established two cardinal principles: that the
planets moved not in circles but in ellipses, and that their speed varied
according to their distance from the Sun. The Vatican called his views
heretical, as it considered only circular paths perfect; Kepler replied that
celestial imperfection was there to enable God to make better music,
since He had used a musical scheme to establish the positions and
motions of the planets around the Sun.* He was also able to strengthen
the claims of both Copernicus and Tycho that the stars were immensely
further away than anyone had supposed. The universe for Kepler was
vast beyond imagining.

That great philosopher Blaise Pascal (1623–62), who would abandon
science and devote himself to religion, was to respond to these revelations by having the atheistic 'libertin' of his *Pensées* cry out: 'The eternal
silence of these everlasting spaces terrifies me!' It terrified Kepler too,
who was heard to exclaim: 'The infinite is unthinkable.' But his findings
strengthened rather than undermined his religious beliefs. 'Whoever is
too stupid to understand astronomical science,' he declared, 'or too
weak to believe Copernicus without affecting his faith, I will advise him
that, having dismissed astronomical studies and having damned whatever

---

*In this he was returning to the Pythagorean belief in celestial harmony, according to which
the heavenly spheres made music as they spun. Kepler realized that what previously had
seemed the most harmonious scheme for the heavens – uniform circular
motions – could generate only a monotone for each sphere; a heavenly body moving at
variable speeds would generate more notes, which, combined with other bodies, would
create complex polyphonic harmonies – a much richer system.

philosophical opinions he pleases, he mind his own business and betake himself home to scratch his own dirt patch.'[16]

Between them, Copernicus, Tycho and Kepler laid the foundations of the scientific revolution of the seventeenth century. Although virtually no astronomer practising at the time of Tycho's death believed in a heliocentric cosmology, all employed Copernican techniques; and as Owen Gingerich, Harvard's Professor of Astronomy, says, it was Kepler, 'the world's first astrophysicist', who was 'the man who really forged the Copernican system as we know it'.[17] By the time of Kepler's death thirty years later, the entire profession was doctrinally Copernican, and accepted not only Tycho's evidence of how the stars moved but Kepler's theories of how the planets did too. Cosmology had been transformed by these three men's discoveries – and by the appearance of the telescope. But that is a different story, the story of Galileo.

☼

> *In the year sixteen hundred and nine*
> *Science's light began to shine.*
> *At Padua city, in a modest house,*
> *Galileo Galilei set out to prove*
> *The Sun is still, the Earth is on the move.*

So begins Bertholt Brecht's play *Galileo*, originally entitled *The Earth Moves*. It opens with Ludovico, who is courting Galileo's daughter, calling on the scientist at his house in Padua, then part of the Venetian Republic. 'I saw a brand-new instrument in Amsterdam,' he tells his host, who is at this point a professor of mathematics. 'A tube affair . . . It had two lenses, one at each end, one lens bulged and the other was like that.' (He gestures.) 'Any normal person would think that different lenses cancel each other out. They didn't! I just stood and looked a fool.'

Galileo asks if it is a recent invention.

'It must be,' replies Ludovico. 'They only started peddling it on the streets a few days before I left Holland.'[18]

The lines Brecht gives the young suitor reflect what was actually happening in Holland, and Galileo's reaction in the play is also true to history. Ever the entrepreneur, he learned to grind and polish lenses and soon had put together his own, superior instrument. Venice was of necessity an unwalled city, which made early detection of enemy attack vital. On 8 August 1609 he came before the Doge's assembled court, which he had invited to the tower of St Mark's Cathedral to demonstrate the applications of his device. As Brian Clegg writes: 'The elderly senators had to be restrained from fighting over the chance to be the next to climb to the roof and scan the horizon for ships. They were like children with a new toy.'[19] Having shrewdly presented his showpiece, Galileo had his academic stipend nearly doubled, from 520 florins a year to 1,000 (the equivalent of $300,000 now),[20] and his professorial tenure was confirmed for life.

The portraits we have of the early Galileo show him as beefy, ginger-haired, short-necked and coarse-featured: they plainly suggest his stubborn nature and abrasive self-assertiveness. 'A devoted careerist with a genius for public relations',[21] he now claimed to be the original begetter of the instrument (which he preferred to call a '*tubus*'), and such was his skill as a lens-maker that it was not until 1630 that anyone produced a higher magnification; but in fact there were several 'begetters' before him.

Roger Bacon, in 1268, is probably the earliest person to have developed a far-seeing instrument in the West; Chinese astronomers had come up with something similar even before that. Leonardo has left intriguing hints in his diaries and notebooks that he may have discovered the telescope before the fifteenth century was out – 'Make mirrors to make the Moon big' and 'In order to observe the nature of the planets, open the roof and bring the image of a single planet on to the base of a concave mirror. The image of the planet reflected by the base will show the surface of the planet much magnified.'[22] Over the years, however, a Dutch lens-grinder,* Hans Lippershey, has been generally acclaimed as the main inventor. Late in 1608 he claimed to have discovered that the combination of a convex and a concave lens could make distant objects look nearer: in September of that year he presented a pair of biscuit-sized lenses fixed inside a tube to the German Prince Maurice of Nassau.

---

*Magnifying glasses became common in the thirteenth century, but these were cumbersome, especially for writing. Craftsmen in Venice began making small discs of glass, convex on both sides, to be worn in a frame: spectacles. Because these were shaped like lentils, they became known as 'lentils of glass', hence, from the Latin, lenses.

Upper left: Nicolaus Copernicus (1473–1543), upper right: Tycho Brahe (1546–1601),
his false nose clearly visible, lower left: Johannes Kepler (1571–1630),
lower right: Galileo Galilei (1564–1642)

By the following April spyglasses could be bought in spectacle-makers' shops all along the Pont Neuf in Paris. The Dutch prototype magnified by a factor of three; very soon there were instruments capable of magnifying more than twentyfold.

It was Galileo who got the credit for using telescopes to examine the heavens. 'It is hard to think of more surprising discoveries in the entire history of science,' posits Noel Swerdlow. 'In about two months, December and January [1609–1610], he made more discoveries that changed the world than anyone has ever made before or since.'[23] Hyperbole, but understandable. Within seven months of his presentation to the doge, Galileo was describing his findings in terse, declarative style in *Sidereus Nuncius* ('The Starry Messenger', although he himself seems to have intended the title to mean 'A Message from the Stars'). 'I have seen,' he wrote, 'stars in profusion, stars that have never been seen before, and which surpass the old, previously known by more than a factor of ten. But what will excite the greatest astonishment by far . . . is this, namely, that I have discovered four planets . . .' The profusion of stars he refers to was the Milky Way. The telescope suggested, which the unaided eye could not, that the heavens had *depth*. The idea that the Milky Way was not a streak across the sky but a multitude of stars, the crowding together of uncountable brilliant spots of light, was astounding.

The 'planets' that Galileo had seen were Jupiter's four major satellites,* now known as the 'Galileans'. Their discovery proved that planets other than Earth had objects in orbit around them, which strongly supported the Copernican view of the solar system: were Jupiter fixed to a crystal sphere, these moons would have shattered it. The telescope enabled Galileo to produce the first systematic drawings of the Moon, revealed that Venus had phases, and also showed the planets to be much larger than previously thought. There followed a greater discovery yet: the Sun had blemishes. It was 'spotty and impure' – what Galileo had seen were sunspots. Moreover, he had watched them move along with

---

*The word 'satellite' comes from the Latin *satelles* (an attendant at court) and was first used astronomically in a letter Kepler wrote to Galileo in the summer of 1610. Almost immediately afterwards, for no discernible reason, Galileo broke off all contact and refused throughout his life to acknowledge Kepler's law of elliptical orbits. For him, everything in the heavens went around in circles.

Galileo as hero of science: his triumphant visit to the Vatican in 1611

the solar surface, which meant that the Sun itself was revolving. Preoccupied with determining the periods of Jupiter's moons (and protecting his position at court), he did not give as much attention to these markings as to his earlier finds, confessing he 'did not know, and [could] not know, what the material of the solar spots may be', but they appeared to him most like clouds.[24]

Others before him may have seen objects cross the solar disc, but it was Galileo's account that astounded a wide audience and made him famous. In 1611 he made a triumphant visit to the Vatican,* where he was granted a private audience with Pope Paul V. It has ever been a fine art to be able to judge whether Rome would applaud or condemn, and Galileo seems to have reined himself in – perhaps mindful of a quip made by a curial cardinal that the Bible was 'intended to teach us how to go to heaven, not how the heavens go'.

---

*At a banquet there a guest coined the word 'telescope', Greek for 'far-seer'. Tom Stoppard's screenplay *Galileo* has the Jesuit scholar Robert Bellarmine neatly remark to him: 'How cruel to make the Greeks give a name to the instrument you wield against their cosmos.' Galileo's telescope survives in one popular visual aid: opera glasses. They began in their modern form in Vienna in 1823 – two simple Galilean telescopes with a bridge in the centre.

Three other astronomers now came forward to claim that they too had noted such marks. The English scientist Thomas Harriot (1560–1621) and two Germans, Christopher Scheiner (1573–1650) and Johann Goldsmid (his name Latinized to Fabricius, 1587–1616), each issued pamphlets, the first being Fabricius's, an account that he took proudly to the Frankfurt Book Fair in the autumn of 1611. Harriot, a steward in Sir Walter Raleigh's household, had made 199 observations of sunspots between 3 December 1610 and 18 January 1613; he just never published them. Scheiner, a Jesuit priest, had made his observations from 21 October to 14 December 1610, but did not publish his records until 1612. Galileo's first sunspot drawings belong to 3 to 11 May of 1612. It genuinely seems that each of the four men separately detected the phenomenon and was not passing off another's findings as his own.

Whoever deserves the credit, it was Galileo's claim that touched off the ensuing debate. Scheiner's observation that the Sun was blemished was discounted by his fellow Jesuits, but he argued that the spots were not strictly solar, rather the silhouettes of previously undiscovered small planets revolving close to the Sun's surface – after all, he said, reverently echoing Aristotle, the Sun was perfect and could not be blemished. The following year Galileo replied with *Letters on Sunspots,* three public epistles in which he asserted that the marks did indeed lie on the Sun's surface. His evidence was that the spots displayed a peculiar speed-up, slow-down motion as they crossed the Sun, becoming elongated and foreshortened as they neared the solar edges – exactly the behaviour of something fixed to the surface of a spinning globe.*

In the last of his three letters, Galileo (possibly looking for a fight) made his first public endorsement of Copernicus's system. The letter enraged Rome, which had given clear notice of its intentions in the

*Robert Burton (1577–1640) characterizes the confusion wrought by these conflicting views in his classic *Anatomy of Melancholy,* a wide-ranging enquiry into contemporary science and the ills of society first published in 1621. 'And so,' he writes, 'while fools are running away from one vice, they plunge into the opposite ones, as a tinker stops one hole and makes two, he [the latest astronomer to pronounce his views] corrects them and doth worse himself, reforms some and mars all. In the meantime, the world is tossed in a blanket amongst them, they hoist the world up and down like a ball, make it stand and go at their pleasures: one says the Sun stands, another he moves; a third comes in, taking them all at rebound, and, lest there should any paradox be wanting, he finds certain spots and clouds in the Sun, by the help of glasses...and thus they disagree among themselves.'[25]

1590s in the case of the Neapolitan philosopher Giordano Bruno (nicknamed 'the Exasperated'). When Bruno declared that a rotating Earth
in orbit around the Sun was unassailable fact, the Inquisition convicted
him of being the worst kind of heretic, a pantheist who had demoted
God from supreme creatorship, and had him transported on a mule to
Rome's Campo de' Fiori, where he was hanged upside down, stripped
naked, then burned at the stake, an iron spike driven through his tongue
to prevent further blasphemies. For years, Catholics had been forbidden
to read Copernicus unless the crucial nine sentences claiming his ideas
were more than theory had been deleted. Even with these removed, the
Congregation of the Index would finally ban the work – on 15 March
1615 – on the grounds that it defended 'the false Pythagorean doctrine
that the Earth moves and the Sun is motionless'.

Galileo could no longer escape. He was summoned before the Holy
Office, and on 26 February 1616 a report was filed stating that he had
been commanded to give up Copernicanism and 'to abstain altogether
from teaching or defending this opinion, and even from discussing it'.
He still was not condemned, but told to direct himself to other studies.
There matters rested for seven years, until Maffeo Barberini, once an
astronomer himself, was elected Pope as Urban VIII. A long-time friend
of Galileo's, he now invited him to Rome. Over the course of six audiences the two men strolled the Vatican gardens discussing the
heliocentric thesis. Urban is said to have told his old confrere that he
could not revoke the 1616 condemnation, but urged him nonetheless to
develop a formal comparison of the Copernican and Ptolemaic systems –
on one condition: no conclusion was to be drawn as to the truth of either
theory, since God alone understood the workings of the universe.

Galileo set to; it would take him nine years. Finally, with the approval
of the local censor in Florence, he produced his *Dialogue on the Two Chief
World Systems*, writing the first edition in vernacular Tuscan, the second
in scholarly Latin.* 'God pleased to make me the first observer of an

---

*One of Brecht's characters says that Galileo wrote 'in the language of fishwives and mer-
chants'. Maybe: but he knew other tongues and when to employ them. 'The common use
of Latin,' writes H. L. Mencken, 'which continued to be the language of the learned on
to the beginning of the eighteenth century, made for an easy exchange of ideas, and what
was discovered or suggested in one country was quickly taken up in all the others.'[26]

admirable thing kept hidden all these ages.' The weight of his argument was clear: the heliocentric version was scientifically indisputable; the Earth was in motion because mathematics demanded it must be.

Urban, already bruised by the battles of the Counter-Reformation, felt all 'the fury of the betrayed lover'[27] and let loose the Inquisition. In a formal trial the following year, Galileo was condemned 'on vehement suspicion of heresy'. Heliocentrism per se had never been declared heretical either ex cathedra or by an ecumenical council; it was just, as one commentator puts it, that 'Galileo was intent on ramming Copernicus down the throat of Christendom'. Kepler, for one, was furious with his fellow scientist: 'Some, through their imprudent behaviour, have brought matters to such a pass that reading Copernicus's works, which was absolutely free for eighty years, is now prohibited.'[28] In effect, Galileo forced the Church to silence him. It did, banning further sales of the *Dialogo* and ordering all extant copies confiscated. 'It is not given to man to know the truth,' Brecht's examining cardinal instructs Galileo. 'It is granted to him to seek after the truth. Science is the legitimate and beloved daughter of the Church. She must have confidence in the Church.'

Brecht claimed that the scientist was threatened with the rack and other instruments of torture, but of this there is absolutely no evidence (mark it down as an Urban legend). If anything, curial officials took special pains to avoid a collision. During his trial Galileo, by this time seventy, was housed in a five-room apartment overlooking the Vatican gardens, and was even accorded a valet and another servant to look after his food. Happily for the Church, on 22 June 1633 he fell to his knees in the great hall of the Dominican convent of Santa Maria Sopre Minerva and recanted – perhaps because, in the end, he was a devout Catholic and saw the larger point that the Holy See wanted to make: science cannot be seen as the ultimate source of authority* – or perhaps because,

---

*Aristotle had asked: if the Earth did indeed orbit the Sun, then why were not stellar parallaxes observable? That is, there ought to be a shift in the positions of the fixed stars as seen from the Earth on one side of the Sun, then six months later from the other side. Tycho had not resolved the question, and even with his best telescope Galileo would not have been able to detect such changes, as the stars were too far away. This issue was not resolved until 1838, when the German astronomer Friedrich Bessel (1784–1846) published the first reliable measurement of a star's apparent annual shift.

as he was forced to acknowledge as part of his retraction, he did not have incontrovertible evidence that he was right.

When his sentence was imposed, it was that he be confined to his villa in Alcetri, just outside Florence, where he spent the rest of his life at his studies, in particular on the science of dynamics. By 1637 he had lost his sight (though not from observation of the Sun), one eye at a time. The following year he received several visitors, including Hobbes and Milton, the latter of whom, six years on, in *Areopagitica*, recalled discussing cosmology with Galileo, 'grown old, a prisoner to the Inquisition, for thinking in astronomy otherwise than the Franciscan and Dominican licensers thought'. Milton returned to the subject in *Paradise Lost*:

> *Sollicit not thy thoughts with matters hid,*
> *Leave them to God above, him serve and feare;*
> *. . . Heav'n is for thee too high*
> *To know what passes there; be lowlie wise.*[29]

In the same poem, he describes Satan's landing on the Sun as creating a mark like a sunspot seen through a 'Glaz'd Optick Tube' (i.e. a telescope). An endorsement? Anyway, his advice came far too late. At least Galileo, contrary to legend, did not spend a single day in a prison cell. He wrote two comic plays, lectured on Dante and continued his academic studies before dying in 1642 at the age of seventy-eight. 'In a generation which saw the Thirty Years War,' wrote Alfred North Whitehead, 'the worst that happened to men of science was that Galileo suffered an honourable detention and a mild reproof, before dying peacefully in his bed.'[30]

Almost a hundred years later, Benedict XIV would grant an imprimatur (Rome's formal permission to publish) to the first edition of Galileo's *Complete Works* (even though, paradoxically, the ban on Copernicus's work would linger on until 1828). Another 230 years would pass before John Paul II commissioned a study of Galileo's case, in 1979. After more than twelve years' further deliberation, in 1992 the Vatican finally conceded that Galileo had been justified in what he said. Then, in March 2008, Pope Benedict XVI announced that a statue of the great man would be erected within the Vatican walls. The head of the

Pontifical Academy of Sciences (himself a nuclear physicist) declared: 'The Church wants to close the Galileo affair and reach a definite understanding not only of his great legacy but also of the relationship between science and faith.'[31] Even by the standards of this great institution, it was all just a little slow.

☼

There were reasons why these arguments took so long to become accepted. Technologies profoundly affect belief systems, and only with the invention of the printing press could ideas spread widely. Still, where such notions involved a revolutionary new way of thinking, there was bound to be great resistance. Sigmund Freud (1856–1939) popularized the view that Copernicus had caused particular consternation by moving mankind from its pre-eminent position at the centre of the universe. In fact, the distress was even greater in the wake of Galileo's revelations about sunspots. There was something horrible about the Sun's not being perfect – as if the ravages of the human face had been translated to the mighty disc and magnified. Before Galileo, the Sun had been an impeccable sphere. Suddenly it was grubby and mottled.

With the advent of the telescope, seventeenth-century man had to frame things anew, not from imaginative interpretation – Sun gods, chariots in the sky, dragons devouring the solar disc – but by repeatedly qualifying and sieving the latest evidence. It must have been profoundly deracinating to think that this great star was, despite being at the centre of the solar system, only a small part of the Creator's purpose – and a spotted one at that.

10

STRANGE SEAS OF THOUGHT

> I do not know what I may appear to the world, but to myself
> I seem to have been only a boy playing on the seashore, and
> diverting myself in now and then finding a smoother pebble
> or a prettier shell than ordinary, whilst the great ocean of
> truth lay all undiscovered before me.
>
> Isaac Newton[1]

> When an apple ripens and falls – what makes it fall? Is it that
> it is attracted to the ground, is it that the stem withers, is it that
> the Sun has dried it up, that it has grown heavier, that the
> wind shakes it, that the boy underneath wants to eat it?
>
> Leo Tolstoy, *War and Peace*[2]

'I will make an end,' wrote the friendless and isolated schoolboy Isaac Newton (1643–1727) in his Latin exercise book. 'I cannot but weep. I know not what to doe.' His principal biographer describes him as 'a tortured man . . . an extremely neurotic personality who teetered always, at least through middle age, on the verge of breakdown'.[3] He never married – as a character in Tom Stoppard's *Arcadia* jokes, sex was 'the attraction which Newton left out'. One can hardly imagine him enjoying Galileo's description of wine as 'light held together by moisture'. He had no ear for music, dismissed great works of sculpture as 'stone dolls' and viewed poetry as 'ingenious nonsense'.[4] Yet when Sir Isaac Newton died he was acknowledged to be the supreme genius of his age. Famous throughout the Western world, he had served for thirty years as president of the Royal Society (the world's greatest scientific institution, founded under the amateur physicist Charles II); twice as a Member of Parliament; and as a surprisingly energetic Warden of the Mint, in charge of all public coinage. He was laid to rest in Westminster Abbey beneath an ornate marble monument twenty-five feet tall, his effigy recumbent beneath a celestial globe. By his side, cherubs weigh the Sun and planets; the Latin inscription declares: 'Mortals rejoice, that there has existed so great an ornament of the human race.' No one in history had learned 'what to doe' to greater effect.

Before he was twenty-four Newton had started to formulate the principles of gravity – the basis of all modern predictive astronomy – and to prove that objects on Earth and in the sky move according to the same laws. He had already made important discoveries about the nature of optics, particularly the qualities of colour and light, and would go on to develop a law of cooling, articulate the principles of the conservation of momentum, study the speed at which sound travels through the air and theorize about the origins of the Sun. He was the first to explain the nature of tides, wrote innovatively about the construction of telescopes and co-founded calculus – the mathematical discipline necessary for every scientific advance of the twentieth century. 'Let no one suppose,' declared Einstein in 1919, at the time his theories of relativity were bringing him world renown, 'that the mighty work of Newton can really be superseded by this or any other theory. His great and lucid ideas will retain their unique significance for all time.'[5]

☼

Isaac Newton was born on Christmas Day into a marginally genteel farming family in Woolsthorpe-by-Colsterworth, in the eastern coastal county of Lincolnshire. His father's early death and his mother's hasty remarriage left him in the care of his grandmother – not unusual in those days, but probably not emotionally nourishing either. As a child he built clocks and sundials and was known for his skill at telling the time by the Sun; by June 1661 he was studying at the College of the Holy and Undivided Trinity, then as now the largest of the colleges of Cambridge University.

Though Aristotle and the other great Greek philosophers still

Sir Isaac Newton (1643–1727)

undergirded the prevailing orthodoxy, radical innovation had come in the person of René Descartes (1596–1650) who, while keeping much of his research secret for fear of the Church's deadly displeasure, in 1644 did publish *Principia Philosophiae*, in which he asserted that the Sun was just one of many stars, and that each stood at the centre of its own 'vortex'. For Newton, such a universe begged a host of questions. Why did any object tend to descend ever downwards? Or move in a particular direction? Galileo had seen that the Moon had mountains and chasms not unlike Earth's; but if it were made of the same substance as our planet, what held it in the sky? Why did it circle the Earth rather than plummet down, or spiral away? Queen Elizabeth's personal physician William Gilbert may have provided a clue when he asserted in *De Magnete* (1600) that the Earth displayed what we now call a 'dipolar magnetic field' – i.e. its poles have both a positive and negative charge, making it 'a great magnet'. But what *was* a magnet?

All these conundrums Newton pondered in his room between

Trinity's Great Gate and its chapel. But during the severe winter of 1665 plague spread from the Continent, leaping from parish to parish, killing thousands every week. In little more than a year, one out of every six Londoners was dead. Cambridge was not immune; the sixteen colleges shut down (they were still small communities: even by 1800 the university would have only 400 students), and Newton returned to his grandmother's large stone farmhouse. His isolation lasted about nineteen months, during which he effectively created modern mathematics, mechanics and optics. As he himself recalled: 'I was in the prime of my age for invention and minded Mathematics and Philosophy more than at any time since.'[6]

The power of his insight is revealed in the answers he found to two particular solar questions: the Sun's mass relative to that of the Earth, and its relative density: he reckoned it to be 28,700 times more massive (a considerable underestimate, but closer than anyone else had come), and slightly more than a quarter as dense (within two per cent of our current value). The law of gravity, however, proved more elusive: it would be nearly twenty years before he advanced the theory publicly, and by then he was at least in part inspired by others. Around 1639 another Cambridge-educated astronomer, Jeremiah Horrocks (1617–41), had suggested that the Moon's motion, while affected by the Sun, was also influenced by some force emanating from the Earth. Thereafter three members of the Royal Society, Robert Hooke (1635–1703), Edmond Halley (1656–1742) and Sir Christopher Wren (1632–1723), as well as a French priest, Ismaël Bouillaud (1605–94), all posited that this force diminished rapidly the further the object was from the Earth's centre. But it was Newton who both recognized that a general law was in play and first demonstrated how it worked.

He concluded that a body on the Earth's surface is drawn downwards by a force some 350 times stronger than the tendency of the Earth's rotation to fling it outwards – a force he named 'gravity' (*gravitas*, meaning heaviness). The question was not whether such a force existed – Galileo had shown that it did – but whether it extended so far from Earth that it could be the same power that held the Moon in its orbit. Newton calculated that if the reciprocal attraction were proportionate to the size of the attracting masses, and diminished according to the square of the distance between them, then this explained not only how the Moon orbited but how this same force shaped the orbits of all the planets – hence his term 'universal gravitation'. With this formula he had determined how the planets and the stars held their places, what caused the precession of the equinoxes and explained the ebb and flow of the tides. Even now the majestic range of this reasoning ranks among the greatest of all advances in human understanding; but having made the calculation, he set it aside.

Many years later he would recount to at least four people that he had been inspired by an apple in his garden. What if the power that made the fruit fall from the tree was not limited to a certain distance from the Earth but extended much further? He never wrote of any single burst of insight, just that 'I began to think of gravity extending to the orb of the Moon' as a force acting ever outward from its centre.*

☼

In less than two years Cambridge reopened, and Newton, returning, quickly made several important discoveries about light. Plato had thought of vision as the result of particles shot out of the eyes; but was the essence of light outside or inside the observer? Aristotle had understood that light was necessary for colour to exist, and Ptolemy had experimented with angles of refraction – but where did colour come from? Was it a gift from the Sun?

This time inspiration arrived in the form not of an apple but of an ornament. A few miles out of town, on the bank of the river Cam, lay Stourbridge Common, where once a year a large fair was held. Newton bought a prism from a local lens-grinder, a simple triangular block of glass shaped like a Toblerone bar. Back in his rooms, he set down the prism: when the Sun fell on it, different colours emerged. Did the glass change the light, or did the sunlight contain the colours and the prism separate

---

*The story of the apple was created from a series of 'contemporary accounts', the first being Voltaire's, in his *Essay on the Epick Poetry of the European Nations.* '[Newton] was walking in his garden,' he wrote, 'saw some fruit falling from a tree, and let himself drift into a profound meditation.'[7] Similar reports appear in the memoirs of people who knew him, including Catherine Barton, Newton's niece (and Jonathan Swift's romantic fantasy); and in the work of his first biographer, William Stukeley, who unequivocally asserts that 'the notion of gravitation came into his mind . . . occasion'd by the fall of an apple'.[8] Whether or not there is any truth to the story – Newton's bedroom desk did overlook an orchard – one undeniable consequence was the naming of a new variety of apple: Newton's Pippin.

Newton using a prism to break white light into its spectrum, as imagined
by a nineteenth-century French artist

them out? Newton knew that a telescope produced a rainbow effect
around whatever was being viewed, because every lens edge is a prism, but
he was unconvinced by the explanation that white light, as it travelled
through a lens, darkened at the thin areas, becoming red, and darkened
even more at the thick areas, turning blue. He set up an experiment in
which a thin sunray struck the prism, which made the separation of light
even clearer. 'It was at first a very pleasing divertissement,' he wrote,

> to view the vivid and intense colours produced thereby; but after a
> while applying myself to consider them more circumspectly, I became
> surprised to see them in an oblong form; which, according to the
> received laws of Refraction, should have been circular. And I saw . . .
> that the light, tending to one end of the Image, did suffer a Refraction
> considerably greater than the light tending to the other.[9]

The different colours refracted differently, and in rainbow order, the
prism bending them into different directions – which meant that the rays
could be separated into components. Newton identified red, orange,

yellow, green, blue, indigo and violet. Indigo is not really a separate colour and orange is questionable, but Newton, prey to the charms of numerology, organized his observations under the magic number seven.[10]

His next task was to isolate one single colour from white light and pass that through a second prism. He aligned two glass wedges and rotated one to direct first blue light, then red, through the second, to find that the blue rays were refracted by the first prism and slightly more by the next. He also observed that the second never created new colours or altered those received from the first. '*Light*,' he noted, 'consists of *Rays differently refrangible*' – i.e. capable of being turned from one direction to another. So colour was not a modification of light but a basic property; white light, which had been thought to be free of tint, actually contains all the colours the eye can see.

Descartes had thought colour was induced by the rotation of the minute particles that made up light rays. Newton's thesis was that the colours we observe are separated by passing through objects, not by the objects themselves generating new colours. He dismissed Descartes's notion, establishing that – for example – daffodils are not intrinsically yellow, and rainbows are no more than the cumulative consequence of drops of water building up in the atmosphere during a shower, so performing the same function as a prism. The colours of a rainbow – or of anything else – are a function of the way our eyes register individual wavelengths of light.

Culturally at least, Newton was on dangerous ground. Rainbows were icons of nature, closely associated with the Sun.[11] The Greeks had considered them the paths made by messengers between Earth and heaven. Inca armies carried banners of the rainbow, North American Indians believed that the dead dwelt in the 'land of rainbows', the rebellious German peasant hosts of the sixteenth century marched beneath a rainbow ensign, emblem of apocalyptic hope. And now the great arc in the sky was suddenly contracted into no more than a side effect of passing raindrops.* More than a century on, the Romantics

---

*The rainbow flag remains a symbol of progressive causes, among them that of the gay and lesbian rights movement, the different colours symbolizing diversity, while Rainbow Gatherings are conventions of hippies who come together on public lands to promote ideas of peace and community. The only supposedly malign rainbow is caused by the Moon – a 'lunar bow' that appears in different shades of grey on showery nights, always opposite the Moon. Sailors used to believe it presaged a death among the crew.

would be appalled at this demotion, and as late as 1817 Keats would denounce Newton for having virtually destroyed the poetry of the rainbow by reducing it to the spill of a prism. But he drank to the great man's health all the same.

Newton could lose himself in an enquiry that knew few (if any) bounds. In one experiment he stared for as long as he could bear at the Sun's reflection in a looking glass, repeatedly turning to a dark corner of his room to see what spots and colours floated in the darkness as an after-image. He continued with this test over and over again until, fearing permanent damage, he shut himself up in a darkened room to wait for full sight to return. It took three days. In another experiment, intent on proving that colour perception was affected by pressure on the optic nerve, he slid a darning needle around his eye socket until he could poke at its rear wall, noting dispassionately 'white, dark and coloured circles' for as long as he prodded away with 'ye bodkin'. He was never merely curious about a subject – obsession was his second nature.

Since light rays of different colours differ also in their refrangibility, Newton concluded that the indistinctness of an image formed by the main lens of a telescope derived from differently coloured rays coming into focus at different distances. A single lens could not possibly deliver a distinct image, as the refracting telescope – like a prism – would disperse white light into colours, surrounding the images of stars and planets with false tints. So he invented the first practical reflecting telescope (known today as the 'Newtonian' type). Grinding his own mirrors (not a pleasant job: in 1677 the great Dutch philosopher Spinoza would die at forty-four, his lungs rotted from years of inhaling the glass particles produced by his lens-making), he made a superior instrument by widening the mirror – though still only to an inch across. He then cast a two-inch mirror and ground it into a spherical curvature, placing it at the bottom of a tube, which caught the reflected rays in a forty-five-degree secondary mirror, which in turn cast the image on to a convex ocular lens outside the tube – the lens he looked through. In 1671 he sent this little instrument, just over six inches in length, to the Royal Society, where the sensation it caused among the 200 or so members encouraged him to publish *On Colour*, a work that he later expanded into the *Opticks* of 1704. This last laid out his theories in full and ended with what would become a famous set of rhetorical 'queries', expounding his speculations on the nature of the physical world, which he predicted would be answered in generations to come.*

Meanwhile, despite Pope's famous epigram, 'Nature and Nature's laws lay hid in night: /God said, *Let Newton be!* and all was light', some in the academic world were less convinced. Many of Newton's peers were incredulous when they read in *On Colour* that light was made up of tiny particles that create movements in the ether. Deeply offended by this response (as by much else), Newton lapsed into characteristically ven-omous argument, constantly seeking redress for slights real or imagined, responding yet more fiercely to any criticism, often with the object of inflicting highly personal wounds, while refusing to make allowances for those he considered mere 'splutterers in mathematics' (which, next to him, of course, they were).

The most enduring of his enemies was Robert Hooke, surveyor to the City of London, chief assistant to Christopher Wren and Curator of Experiments to the elite cadre who formed the Royal Society. The sources of their rivalry date back to 1672, when Hooke first criticized Newton's theories on light, saying they had not been proved in enough detail. He was powerfully placed, as well as respected for his work on barometers, ultraviolet light, thermometers, wind gauges and the nature of elasticity, and for demonstrating that air was essential to life. He had also written extensively on light in *Micrographia* (1665), an exposition on another of his inventions, the compound microscope.

Newton, furious, declared Hooke incapable of grasping what he had to say. Their reciprocal sniping continued unabated for years, not helped when Newton learned that the great Dutch mathematician Christopher

---

*Newton was not alone in his ponderings. Back in the fourth century AD the Chinese philosopher Ko Hung had written: 'The human eye is colour-blind, and the pupil short-sighted; this is why the heavens appear deeply blue. It is like seeing yellow mountains sideways at a great distance.'[12] It would take until 1870 for the British physicist John Tyndall to discover that blue scatters more easily than red, which accounts for the colour of a sunny sky. In 1810 Goethe, a considerable scientist as well as a literary genius, published a 1,400-page treatise entitled *Zür Farbenlehre* ('On the Theory of Colours'), in which he had reformulated Newton's definition. Newton had viewed colour as a physical problem, which involved light striking objects and entering our eyes. Goethe realized that the sensations of colour reaching our brain were also influenced by the mechanics of vision and by the way our brains process information. He argued that what we see depends on the object in view, the lighting *and* our perception.

Huygens (1629–95) had proposed another challenge to his theory: that light consisted of waves, not particles. In fact, light exhibits the properties of both, but that discovery lay centuries ahead; so the dispute was further exacerbated by each side's being able to show that its opponents were not entirely correct, yet unable to prove its own version conclusively.

Events now move to a coffee shop in London's Strand, where in January 1684 Hooke, Halley and Wren found themselves arguing about the attraction between the Sun and the planets. After much debate, Wren offered a handsome prize – a book of the winner's choice worth up to forty shillings, a good month's income for a labouring family – if any one of them could demonstrate within two months the kind of planetary orbit that would result were the Sun's gravitational attraction to obey an inverse-square law. The time expired, and none of the three had come up with an answer, so Halley was dispatched to Cambridge to put the issue to Newton.* As Halley later told it, the great man immediately replied, 'An ellipse,' adding that he had solved the problem of gravity some time ago but not told anyone, although he would now set to work on putting the solution into publishable form. On hearing this, Hooke proclaimed that he had had the idea at least fifteen years before and had written to Newton back in 1679, arguing for just such a law. He may well have intuited the nature of gravity then, but he lacked the mathematical skills to prove it; and while he probably did feel cheated, history is littered with almost-discoveries.

Newton's mockingly false-modest response, in a letter to Hooke, put it succinctly: 'If I have seen further, it is by standing on the shoulders of

*Newton had few friends, but Halley was one, an outstanding astronomer and physicist in his own right. In 1705, having perceived that the characteristics of the comet of 1682 were nearly the same as those of the comets that had appeared in 1531 and 1607, Halley concluded that all three were the same object, returning not quite every seventy-six years, and forecast it would reappear in 1758, which it did. While neither he nor Newton lived to see it, the fact that the comet maintained a steady orbit dramatically reinforced the case for Newton's formulation of gravity. According to Owen Gingerich, Halley was lucky in that only 'his' comet was bright enough (among the 140 or so periodic ones now known) to be identified without the aid of a telescope. At its closest, it shines with a light equal to a quarter of the Moon's. Halley's calculations enabled the comet's earlier appearances to be tracked back through history. Chinese astronomers observed it in 240 BC and possibly even as early as 2467 BC. A Babylonian tablet records it in 164 BC, and the Bayeux Tapestry in 1066. The most recent appearances were in 1910 and 1986. Halley (as the comet was titled immediately after its return in 1758) will next appear in 2061.

giants.' Their fractious goading would end only with Hooke's death in 1703 – after which, amid much political jockeying, his great rival was elected to succeed him as president of the Royal Society. But success could not alter Newton's nature. Despite the acclaim that met the *Principia* (from which he excised an acknowledgement to Hooke), he withheld the lion's share of his researches,* closeting himself in his room in Trinity for days at a time, careless of meals, toiling by candlelight, retreating almost completely into himself. He was a man ever at odds with his world.

☼

For all the emphasis Newton placed on the notion that the fundamental laws of nature were universally the same, he didn't want a predictable or purely mechanical universe, but rather one that had a place for things of the spirit – thus alchemy, in its early days principally concerned with transmuting 'inferior' bodies, especially metals, into 'superior' ones by means of a 'proper medicine'. By the twelfth century this secretive activity had infiltrated European culture courtesy of the Arabs. At first, little distinction was made between it and chemistry, as both concerned themselves with differences and boundaries in material substances. But alchemy also embraced the exploration of nature in the form of procreation, fermentation, transmutation and transfiguration.

Alchemists recognized seven main metals, each corresponding with a planet, the Sun being identified not only with gold as a substance but with 'philosophical gold', the mysterious power hidden within it. All this was of supreme interest to Newton. 'Noe heat is so pleasant & beamish as the suns,' he wrote. The alchemists' Holy Grail was the so-called philosopher's stone, a perfect balance of elements and forces, which was believed to have the power to transmute any metal into gold and to confer on earthly man an all-seeing omniscience. When Milton's Satan lands on the

*Newton's unpublished writings amount to some 2.7 million words, including hundreds of pages about religion. These he sensibly kept secret: although submissive to the Church of England, he held that to regard Christ as God was a sin and that the Holy Trinity was a blasphemous doctrine – a deeply heretical position that could have imperilled his life. His papers remained in private hands until 1936, when they were sold piecemeal at Sotheby's. One of the buyers was the economist John Maynard Keynes, who acquired most of the writings on alchemy, chiefly to save them, and reportedly read them in taxis going from one Treasury meeting to the next.[13] Keynes told a gathering at the Royal Society that he viewed Newton as 'the last of the magicians, the last of the Babylonians and Sumerians'.[14]

An Alchemist in his Workshop, a nineteenth-century recreation.
Alchemy was full of nonsense, but helped develop chemistry as a science

overpoweringly radiant Sun, he finds it difficult to describe but reports that it might be akin to the philosopher's stone, which is 'Imagined rather oft then elsewhere seen, /That stone, or like to that which here below /Philosophers in vain so long have sought'.[15] This talisman went by many names, including 'Sun', and was deemed to take two main forms: a white stone, which could translate base metals into silver, and the red stone of the solar stage, which could translate them into gold.

In the garden shed outside his room, adjoining the wall of the college chapel and with a special chimney to carry away the fumes, Newton constructed a laboratory where a furnace burned day and night. When he passed a crimson-tinged alloy, cinnabar (red mercuric sulphide, known to painters as vermilion), through the fire, he extracted a liquid metal known as quicksilver, held by alchemists to be 'philosophical mercury', or *Mercurius*, the material from which all things were made. Newton was so enamoured of its power that he filled his room with crimson-upholstered furniture, crimson curtains, crimson cushions, even a crimson mohair bed. In the end, his constant handling of mercury built up the metal inside his body to inflict tremors, sleeplessness and some say paranoid delusions (one recent theory is that he came to suffer from Asperger's syndrome, a form of autism). Yet he was right in thinking alchemy possible; it just couldn't be effected by the chemistry of his time.

☼

The epoch of discovery from Copernicus to Newton is generally known as the 'Scientific Revolution'. 'Science was a minor partner in culture in 1600,' observes James Gleick.[16] 'By 1800 it was *the* major part of culture' – to the extent that fashionable ladies had their portraits painted with sextants and telescopes at their feet. Newton casts such a long shadow that it is easy to underestimate his successors. (Isaac Asimov once remarked that when scientists dispute among themselves who is the greatest scientist in history, what they are actually arguing about is who is the *second* greatest.) While the century after Newton's death saw far fewer discoveries about the Sun, important discoveries there were, including the first quantitative measurements of the speed of light, the notion of planets of other stars, the first reports of dark lines in the solar spectrum and pioneering work on the relation of clouds to the Sun.

One question particularly fascinated scientists and public alike: how far was it from the Earth to the Sun? None of the existing calculations was even remotely convincing. This is where the passing of Venus between the Earth and Sun – the visible rapid transit of a dot in front of the roiling solar surface – became vital.* Kepler's famous third law states that the cube of a planet's distance from the Sun is proportional to the square of the time that planet takes to complete its orbit – thereby giving us each planet's relative distance from the Sun; it does not, however, give its *absolute* distance.

Although Venus has made this passage for millions of years, it was the

---

*Venus crosses the Sun four times within a 243-year cycle. Transits come in pairs eight years apart, followed alternately by gaps of 121.5 years and 105.5 years. Occasionally, a 'double transit' may not occur because Venus does not quite pass between the Sun and Earth – as in the 'missing' transits of 416, 659, 902, 1145 and 1388.

young Lancashire curate Jeremiah Horrocks who on 24 November 1639 (having rushed home from church, where he had been obliged to give two sermons) became the first to observe Venus crossing. He realized that if a transit were observed simultaneously from two locations set far enough apart, the data gained would be enough to work out the distance to Venus, the distance from the Earth to the Sun and finally the size of the whole solar system. Having sent a friend to witness the event from Manchester, Horrocks brought off his own observation and contentedly wrote: 'The object of my most sanguine wishes . . . just wholly entered upon the Sun's disc.'[17] However, his larger ambition was not to be realized, for his friend's point of vantage proved too close to his own position to be useful, and two more transits would have to occur before anyone else effectively ventured such an observation.*

Horrocks's pioneering effort was a heroic beginner's attempt. In 1716 Halley published *A New Method of Determining the Parallax of the Sun, or his Distance from the Earth*, in which pamphlet he advocated a much more sophisticated approach; deploying as many observers as possible worldwide. By 1761 the scientific community was on notice for the next crossing. Joseph-Nicolas Delisle, a Parisian who had built an observatory and school of astronomy in St Petersburg, dispatched astronomers to India, St Helena and other spots likely to provide a clear view of the 6 June transit. At least 120 observers were posted to more than 60 stations. But the event fell in the middle of the Seven Years War, and two of the scientists, the astronomer Charles Mason and the surveyor Jeremiah Dixon (later to demarcate the Mason–Dixon Line), were aboard a ship attacked by a French frigate on their way to Sumatra, eleven of her company being killed. While Mason and Dixon's observation was aborted by war, clouds thwarted others.

Undismayed, the world's astronomers girded their lenses for the next passage, calculated to occur on 3 June 1769. This time seventy-six observational posts were spread around the globe. The Royal Society

---

*Because the Sun's diameter is around 110 times that of Venus, the planet's silhouette in transit, if viewed from different places on Earth, is outlined against different points on the solar surface. If the transit is observed from places a known distance apart, simple geometry yields the distance to Venus: this is 'triangulation', or 'parallax' (from the Greek, for the value of an angle). How do people hundreds of miles apart compare their bearings on the silhouette? They must decide on a particular stage of the transit when everyone will observe; confirm the exact time they do so; and note the exact position of Venus on the Sun.

sent observers to northern Norway and Hudson Bay, while also funding a 39-year-old naval lieutenant, one James Cook (1728–79), to observe the transit from Tahiti.

Cook left England aboard the ninety-eight-foot *Endeavour* – 'a cross between a Dutch clog and a coffin,' as one historian has described it[18] – on 26 August 1768, taking with him a professional astronomer, Charles Green, crates full of telescopes, clocks and meteorological equipment as well as seventeen sheep, several dozen ducks and chickens, and a milk goat that had circled the globe once before. The French Ministry of Marine ordered all its commanders to refrain from interfering with Cook and immediately release him should they have detained him, since he 'was doing important work for humanity'. Cook duly arrived in Tahiti, without French interference.

With six weeks to the transit, he issued strict orders against the trading of metal objects with the island's females, who adorned their thighs with intricate tattoos of arrows and stars and who would happily exchange their favours for a nail or two (at first, it was 'one ship's nail for one ordinary fuck',[19] but hyperinflation soon set in). In their enthusiasm the crew of an earlier ship, the *Dolphin*, had pried so many nails from their craft as to render it almost unseaworthy. Despite Cook's best endeavours, every kind of metal object – cutlery, cleats, cooking utensils – kept vanishing.

The *Endeavour* returned to England on 17 July 1771. Cook's observations, along with everyone else's, were sent to Paris, where they were to be synthesized by the leading French astronomer of the day, Joseph Lalande. As he analysed the results, it became clear that the host of observers had been unable to establish the exact moment when the edges of planet and star appeared to touch. As the black dot met the rim of the disc, it appeared to ooze into the Sun (whose edge has been theatrically designated 'the Terminator'), a phenomenon that had caused the watchers to disagree by several seconds as to the moment of intersection.* Despite these variations (and the several problems caused by international time differences), Lalande managed to calculate the distance at 95 million miles, just 2 million miles off our present figure. In 1894 the American astronomer William Harkness would come up with the figure

---

*This problem, known as 'the Black Drop effect', was probably due to refraction.

of 92,797,000 miles, at most only 158,000 miles below the modern value which is variously estimated at between 92,955,887.6 and 92,750,600.02.

The most recent transit occurred on 8 June 2004. I witnessed its last hour, atop a friend's high-rise in mid-Manhattan. By then, as one observer was to put it, 'the Sun's beauty spot had . . . shifted to the other cheek'. The rim of Venus seemed uneven, due to the heat waves surging off the city. I recall having little sense of the planet, which seemed to be *within* the great star, an ant inching its way into a translucent orange. When it reached Terminator Point, it seemed as if a hole had been drilled in the Sun's bottom-right-hand quadrant, allowing its tiny visitor to escape. Today scientists can radar-range the planets, and deep-space probes have relegated transit campaigns to the history books. When the next transit takes place, on 6 June 2012, how many will realize the dramatic history of its tracking?

☼

Venus may have preoccupied astronomers and governments, but the general public had moved on. New ways of playing with light – in camera obscura and camera lucida, magic box and trick lantern – were all the rage from Samuel Pepys's time on, while telescopes gained a particular hold on people's imagination. With the advent of yet more powerful models, the skies were becoming an open map. The American poet laureate Ted Kooser voices this new sense of advancing discovery:

> *This is the pipe that pierces the dam*
> *that holds back the universe,*
>
> *that takes off some of the pressure,*
> *keeping the weight of the unknown*
>
> *from breaking through*[20]

Galileo's device had been used interchangeably for terrestrial and celestial purposes, a typical model being five or six feet in length. Magnification could be increased simply by extending the barrels; by the 1670s these were as long as 140 inches. A 'telescope race' was on.

However, this method of magnification had reached its limit; no further increases of power were possible until the 1730s, when James Hadley and others solved some of the practical difficulties that had plagued Newton's prototypes, and soon reflecting telescopes with primary mirrors up to six inches in diameter were being made. Sky-searchers now had an arsenal of instruments: the achromatic lens (to counter defects in light dispersal), the optical glass, the micrometer (which measures small angles) and cross hairs (for fine alignment). By 1780, in the hands of William Herschel (1738–1822), the reflecting telescope with parabolic ground mirrors enabled solar science to emerge as a specialized area of research.

Building the most powerful telescopes of his time was only one of Herschel's accomplishments. The third child of an oboist in the Hanoverian footguards, he had joined his father's band at fourteen and stayed with it until 1757, when his battalion became so heavily engaged in the Seven Years War that music was put aside. On his father's advice, he absconded to England ('Nobody seemed to mind,' he noted, although the Elector of Hanover was also the British sovereign) and earned his living as a musical copyist, conductor, composer, music teacher, violinist and organist in the fashionable spa of Bath. He soon became preoccupied with what he called 'the construction of the heavens', fashioning hardwood telescope tubes 'as elegant as cellos' which he topped off with magnifying eyepieces made of cocus, the wood used in oboes, all to maximize what he called 'space-penetrating power'.

In 1781 he became the first person to discover a planet when he identified Uranus, lying further out than had ever been imagined: at a stroke its detection doubled the radius of the known solar system – and more than doubled Herschel's finances, allowing him to dedicate his life to astronomy. Summoned to an audience with George III, he wrote to his sister Caroline: 'I will make such telescopes & see such things . . . that is, I will endeavour to do so.' By 1783 this 'middle-aged German musician residing in an English resort town',[21] spending night after night on his observation platform, rubbing his face and hands with raw onions to keep out the cold, had established the general direction and speed of the Sun's movement in space, and analysed the motions of seven bright stars to prove that part of their movement was due to the Sun's pull.

Two years later, with his sister as his long-suffering assistant, he used star counts to map the Milky Way, the results supporting Thomas Wright's proposition of 1750 that that galaxy was one vast spinning disc, with the Sun situated within it. To ancient astronomers, the faint luminous band visible across the sky on clear nights was like a flow of milk yielded by some heavenly cow. The Greeks called it 'Kyklos Galaktikos' or 'the Milky Circle', the road-building Romans 'Via Lactea'. Although Democritus had been partially right in suggesting that the Milky Way consists of crowds of stars, Herschel showed that it was actually a huge aggregate of stars, nebulae, gas and dust. 'He has discovered fifteen hundred universes!' exclaimed the novelist Fanny Burney after visiting his Windsor observatory in 1786. 'How many more he may find who can conjecture?'[22]

By 1800 Herschel had mapped some 2,500 diffuse cloud-like structures that he dubbed 'nebulae' (the Latin word for 'clouds'), classifying them and fixing their positions in the sky. He called the Orion nebula, a knot of congealing gas 1,600 light years* from Earth, 'the chaotic material of future Suns' – which is precisely right. The Sun's core he described as 'a solid globe of un-ignited Matter'.

In that same year – which also saw Beethoven's First Symphony and Wordsworth and Coleridge's *Lyrical Ballads* – he extended Newton's experiments on light by demonstrating that invisible 'rays', as he was to name them, were to be found beyond the red end of the solar spectrum. One day he was writing up his experiments to quantify the relative heating power of different colours by letting a spectrum of light fall on a set of thermometers. To his surprise, he discovered that the greatest heat was generated beyond the red end of the spectrum where nothing was visible. 'Radiant heat,' he reported, 'will at least partly consist, if I may be permitted the expression, of invisible light.' Building on Hooke's researches of a hundred years before, he had discovered infrared radiation – essentially the transmission of heat – which showed that the Sun's heat was mostly carried as invisible rays that behaved like light but could not be discerned by the eye.[23] Herschel's discovery would allow scientists to gauge from starlight how far off a star was, and how large.

Herschel also continued to be interested in the hardware of his science. While he and his colleagues had more than adequate tools to

_____

*A light year, not a measure of time but a unit of length, is the distance that light can travel in one year – c. 5.900 trillion miles, or about 63,000 times the distance between Earth and Sun.

investigate the Sun's apparent motions, its distance from the Earth and the other planets and the various oscillations of the Earth on its axis, they lacked the wherewithal for a really detailed study. When William's son John (1792–1871) followed him into astronomy, the two collaborated in creating a 20-foot-focal-length, 18.25-inch aperture reflector, the first in a line of literally world-changing instruments.

One of William's biographers has characterized his achievement in transforming 'the starry heavens from a static backdrop against which to measure planetary positions, into a vast dynamic region in which stars evolved from clouds of nebulous material'.[24] Here Herschel was building on the reflections of Immanuel Kant (1724–1804), whose *A General History of Nature and Theory of the Heavens* (1755) had proposed that the Sun and planets were formed by the condensation of a rotating disc of interstellar matter. Kant (himself a skilful lens-grinder) added a 'nebular hypothesis' of planetary formation, reasoning that diffuse nebulae – dim clouds of dust and gas that were only first being well observed in his lifetime – must collapse under the force of gravity. As they did so, they would spin out into a disc, which in time would form itself into stars and planets.

Kant, however, was no mathematician, and his ideas did not receive a convincing scientific basis until the great Pierre-Simon Laplace (1749– 1827), who argued that the Sun and its system were formed by the gravitational collapse of an initially rotating diffuse gas cloud.

Throughout the nineteenth century, Laplace was cited as having pro- vided the mathematical proof (not offered by Newton) that the solar system was built like a clock.[25] God had no place in such a universe. 'Citizen First Consul, I have no need of that hypothesis,' Laplace is said to have told Napoleon disdainfully on being asked where the Creator fitted into his equations. By the time Laplace died in 1827, the battle lines were drawn for some of the greatest arguments of the nineteenth century. Astronomers had lost some of their awe of the Sun, and no longer worried about its place in the divine scheme. They had set about studying it as a star, and were won- dering what it was composed of, how it acted on the Earth, what it could tell them about the rest of the universe. Until Newton's time it had been a great *pushing* object, imposing itself on mankind with its heat and blinding light. Newton had shown that the Sun actually *pulls* more – and has an impal- pable power to organize everything around it.

11

ECLIPSES AND ENLIGHTENMENT

He ... held out his right hand at arm's length towards the
sun. Wanted to try that often. Yes: completely. The tip of his
little finger blotted out the sun's disc. Must be the focus where
the rays cross. If I had black glasses. Interesting. There was a
lot of talk about those sunspots when we were in Lombard
street west ... Terrific explosions they are. There will be a
total eclipse this year; autumn some time.

James Joyce, *Ulysses*[1]

When I was a schoolboy of perhaps ten or eleven, one of my masters
would end his class by making us keep silent for its final minute. It was
his way of showing us how long time could feel. And so it did: those sixty
seconds seemed to go on for ever as we waited for the end-of-lesson bell,
when we would burst out, like swimmers shooting to the surface.

Some forty years later, on 23 November 2003, I waited as the Moon
moved across the Sun high in the heavens. I was standing with about sixty
fellow eclipse-watchers on Antarctic ice three and a half miles deep, some
way outside the Russian base of Novolazarevskaya – in Queen Maud Land,
positioned at 70° 28' W and 11° 30' E. Ice surrounded us, its whole edge
lit to a searing, fluorescent blue. The Sun's masking was to last one minute
and forty-eight seconds, a brief enough period but one that would seem,
again, to stretch for an eternity. Only this time I wouldn't want it to end.

My journey had started with a flight from New York City to Cape Town,
where a decommissioned Ilyushin-76 military cargo jet waited to take us

Solar eclipse, November 2003, Antarctica:
'Something about the Way You Look Tonight' – which had a special appropriateness

on the six-and-a-half-hour flight to the ice strip near Novo. The night
before take-off, our group, ranging from an Arizona professor of astronomy to a poledancer from Chicago realizing her life ambition, gathered
for a briefing on conditions at base camp. There was the danger of being
misled by mock Suns, or sundogs, we were told – mirages created by light
refracted through ice crystals; of 'white-outs', like being inside a ping-pong
ball; and freakish 'whirlies', compact violent gusts that would scoop up any
object and hurl it away. These, however, paled before the current weather
conditions: roaring winds of 120 kph and temperatures of –28°C were
making it too hazardous to travel. Continuous storms over the last twelve
days had knocked down or blown away almost everything that had been
set up for us: equipment left by a previous group was totally lost beneath
the snow; even the two small transport planes were buried up to their
wingtips. The site team had been unable to leave their tents for the last
twenty-four hours, and the latest reports forecast a second, even fiercer
front in two days' time. An eighteen- to thirty-hour weather break might
give us just enough of a window to fly in, witness the eclipse and lift off
ahead of the second storm, but the official advice was plainly to abandon
the venture. Our chain-smoking Russian pilot said we might have a

chance. Next to me, someone whispered that no civil aviation plane had ever survived such an attempt. Our pilot stubbed out one more cigarette. 'Well, I am ready if you are.' And so we went.

We touched down in bright afternoon sunshine with a few hours to spare. Back in Cape Town, I had asked one of our group, Carol, a violinist with the Scottish Symphony Orchestra, if she would play for us if we smuggled in a violin. She had looked doubtful, but nodded. And that was how, within an hour of landing, the snowfields rang to Mozart's G Major Concerto, the 'Blue Danube Waltz', Elton John's 'Something about the Way You Look Tonight' and Massenet's 'Meditation' from *Thaïs*. A couple I had christened 'the Lovers' – a New York-based British businessman and a svelte, dark-haired Parisian lawyer – suddenly emerged from our one tent, she in just a purple slip and ski boots, and tangoed in a tight embrace, their movements surprisingly sensuous in the cold. Behind them brooded the Ilyushin-76, and beyond that nothing but millions of miles of frozen continent.

Although little wind was blowing by now, it was numbingly cold – about –22°C: we assembled about 10.15 that evening and would remain in place, thickly muffled, until after 11.10 p.m. Through my protective shades, I watched the Moon devouring the Sun: this phase took about three-quarters of an hour. Then, about ten minutes before totality, great irregular alternating bands of shadow and vivid light began to pulse towards us from north-west of the star, like smoke rising from the snow: patterns made by rays from the last points of the solar crescent interacting with the Earth's upper atmosphere.

Just before total eclipse, we saw great red flares bursting off the Sun's surface, at about one, five and eight o'clock. Our shadows grew darker and darker, until the Moon disc had finally put out the entire Sun and all that remained was a black circle hanging in the sky, set against a glowing halo like a giant sunflower. The horizon turned a deep red, contrasting with an upper crest of brilliant green. I took off my protective glasses – the Sun was almost below the horizon now, and we watched its spectacular dance with the Moon and the edge of the Earth. Slowly, satellite parted from star and a crescent reappeared, overwhelming the corona and forming a beautiful diamond ring. We were the first human beings in history to witness a total solar eclipse from the Antarctic, but

A partial eclipse, seen against a crescent on Islamabad's grand
Faisal mosque, Pakistan, 1 August 2008

only the latest witnesses of what remains one of the great marvels of
nature. And, while we looked on in awe, we no longer did so possessed
by superstition or fear. Science has not only enlightened us, it has
learned to use eclipses for its advancement – a revolution indeed.

☼

The tradition of pursuing eclipses in far-off lands is part of that revolution, one that began in the eighteenth century, its first great practitioner
being Captain Cook, of *Endeavour* fame. 'A grave, steady man,' as Boswell
would describe him, he had become intrigued by eclipses while reading
Charles Leadbetter's *Compleat System of Astronomy*, which declared that a

> Person being well skill'd in Astronomy . . . may, by the knowledge of
> Eclipses . . . determine the true Difference of Meridians between *London*,
> and the Meridian where the ship then is; which reduced to Degrees and
> Minutes of the Equator, is the true Longitude found at Sea.[2]

In August 1766 Cook experienced his first solar eclipse – off the
Burgeo Islands along the Newfoundland coast. He was lucky to see it at

all, as a fog lifted only at the last moment, but he seized the opportunity to work out the longitude of Newfoundland. His calculations were accurate enough that his reckoning was only three miles off across the whole width of the Atlantic; his paper to the Royal Society, delivered in April 1767, was so acclaimed that it secured him the official backing needed for his great voyages of discovery.

Over the next eleven years he would see the Sun go out four more times – each time totally: on 10 May 1771 off Ascension Island in the South Atlantic; on 6 September 1774, off New Caledonia; on 5 July 1776, from Tonga (being twelve hours ahead, the occlusion coincided with the eclipse of British power in North America: the Sun must have darkened almost as the Declaration of Independence was ratified); and his last on 30 December 1777, from a place he christened Eclipse Island, off Christmas Island (the one in the central Pacific, about 4,000 miles north-east of Australia). He knew from Tobias Meyer's *Nautical Almanac*, published in 1767, that wherever the Sun and Moon had identical coordinates, there he would find an eclipse. By the time he was murdered on Hawaii in February 1779, he had observed more solar eclipses than anyone else in history.

The average eclipse lasts just under three minutes, the longest ever recorded being over the Indian Ocean in 1955, running 7.08 minutes. Because the Earth and Moon have to be precisely aligned for a total eclipse to take effect over any given location, such events are relatively rare. A maximum of seven per year is possible, consisting of five solar and two lunar, or four and three; a minimum, consisting of two solar eclipses only. Within the whole solar system, the Earth is the only planet to experience such pronounced shadowings. The likelihood of witnessing a total eclipse from any arbitrarily selected point is once every 375 years, an annular eclipse every 224. To view an eclipse at all, one must be in a narrow band traced out by the Moon's shadow, at its widest no more than 112 miles. The shadow moves at 1,060 to 2,100 miles an hour and completes its journey in three to four hours.*

*Every part of Britain and Ireland has witnessed a total eclipse at some time over the last 2,000 years, except for a small tract at Dingle on Ireland's west coast. Yet the obscure island of Blanquilla, off northern Venezuela, was darkened by two occlusions within 3.5 years, and a spot in the Java Sea saw four eclipses within 13.7 years; a region near 29° E and 23° 30' N in the desert of southern Egypt will see five such eclipses over the course of 31.8 years – 14 February 2325; 20 June 2327; 5 February 2334; 31 July 2353; and 23 November 2356.[3]

As the scientific view of the universe deepened and grew in influence, observers began to speculate on a range of associated phenomena. On 15 May 1836 an annular eclipse passed over northern Britain. A retired stockbroker who had helped found the Royal Astronomical Society, one Francis Baily of Inch Bonney, near Jedburgh, described what have come to be known as 'Baily's Beads'. Just before the Moon started eating into the Sun, he wrote, a row of lucid points, like a bright necklace, irregular in size and spacing, formed around that part of the Moon that was about to enter the Sun's disc. What he was seeing were actually the last rays of sunlight streaming along those of the Moon's innumerable valleys that present themselves head-on to the Earth at the moment of eclipse; but it was the suddenness that impressed:

> Its formation, indeed, was so rapid that it presented the appearance of having been caused by the ignition of a fine train of gunpowder. Finally, as the Moon pursued her course, the dark intervening spaces . . . were stretched out into long, black, thick, parallel lines, joining the limbs of the Sun and Moon; when all at once they *suddenly* gave way, and left the circumference of the Sun and Moon in those points . . . comparatively smooth and circular, and the Moon perceptibly advanced on the face of the Sun.[4]

This same phenomenon had been described fifty-six years before, at a total eclipse over Penobscot, Maine, on 27 October 1780; but it was Baily's vivid description that secured his place in the poetry of astronomy.

Next to come under scrutiny were two happenings that we had seen in Antarctica: solar prominences, which look like reddish filaments, or blobs, in the lower corona but which are actually large accumulations of relatively cold gas suspended high in the Sun's atmosphere by its own magnetic field; and coronas themselves, the haloes of white light that appear around the Moon at totality – 'Heaven's awful rainbow', in Keats's phrase. Coronas were described as early as AD 96, the term first being applied in 1563. At the eclipse of 9 April 1567, the great Jesuit astronomer Christopher Clavius reckoned them as the Sun's uncovered rim. Kepler disproved this, but only with the even less accurate explanation that the halo-like effect emanated from the Moon itself; it would take another 300 years to discover that the light came from the ionized gases that form the Sun's turbulent outer atmosphere. They have three components: polar rays, which extend into space for at least several times the Sun's diameter; and the inner and outer equatorial coronas, the long elliptical streamers that, concluded Mrs D. P. Todd, the foremost amateur astronomer of her day, 'delicately curving and interlacing, may tell the whole story of the Sun's radiant energy'.[5] If only she had known how insightful her prediction would prove to be. The corona's pearly light is less than a millionth as intense as the central Sun's, and so can be seen only during total eclipses, but such observations tell us how other stars behave, whatever their distance from Earth.

☼

Over the last 500 years scientifically prepared man has used eclipses to take advantage of the superstitions and fears of less informed peoples. Examples abound both in real life and in fiction. During his voyage of 1504, Christopher Columbus was stranded while reprovisioning in Jamaica. Some of his officers mutinied, and Columbus and the men loyal to him found themselves at the mercy of the island's Arawak inhabitants, who refused to bring food, leaving them to eat rats. From his astrological charts, he knew that a lunar eclipse was imminent, so he waited for the fateful day and sent a message to the Arawak that in his anger he would make the Moon rise 'inflamed with wrath'; on cue, it began to disappear. The stunned islanders beseeched his mercy, whereon he retreated to his cabin, used a half-hour sandglass to time how long the eclipse would last, then emerged to declare that God had granted his prayer. The Moon was restored, the ship restocked.

H. Rider Haggard (1856–1925) put the same idea to use in his 1885 classic, *King Solomon's Mines*. The first popular novel in English to be set in Africa, it recounts the adventures of three British gentlemen – the mighty hunter Allan Quatermain and his companions, Sir Henry Curtis and Captain Good, RN, together with their formidable attendant, Umbopa – searching for the fabled mines of the title in the kingdom of Kukuanaland, where they soon find themselves at the mercy of the great King Twala. Conferring with his companions, the blunt-spoken Captain Good pulls out an almanac: 'Now, look here, you fellows, isn't tomorrow the fourth of June?'

Sure enough, it records that a total solar eclipse will begin at 11.15 Greenwich Mean Time. 'Visible in these Islands – Africa, etc. . . . There's a sign for you. Tell them you will darken the Sun tomorrow.'

The intrepid quartet attempt to bluff their way through, but when Quatermain rhetorically enquires of King Twala whether any man can put out the Sun, the great chief laughs out loud. 'That no man can do. The Sun is stronger than man who looks on him.' Quatermain announces that this is precisely what his company will bring about, and next day at the appointed moment the Sun starts to disappear. Haggard was more sympathetic than most of his contemporaries to African culture, and has the anciently evil witch doctress, Gagool, cry to the trembling multitudes that all will be well: 'I have seen the like before; no man can put out the Sun; lose not heart; sit still – the shadow will pass.' But the dying Sun sweeps her power away. 'The dark ring crept on. Strange and unholy shadows encroached upon the sunlight, an ominous quiet filled the place, the birds chirped out frightened notes, and then were still; only the cocks began to crow.' In the gathering stillness, a boy cries out, 'The wizards have killed the Sun. We shall all die in the dark.' The Englishmen's power seems absolute, and a chastened king accedes to their demands.*

Haggard may well have inspired Mark Twain's *A Connecticut Yankee in King Arthur's Court*, in which a modern man, Hank Morgan, struck on the head by a crowbar, is precipitated into Arthurian England. It is 19 June 528, and the locals are again hostile, threatening to burn the hapless Morgan until he too learns that a solar eclipse is about to occur. Twain's satire appeared in 1889, less than four years after Haggard's novel was published in the US, but it is also possible that Twain pinched the idea straight from history. He has his hero explain, 'It came into my mind, in the nick of time, how Columbus, or Cortés, or one of those people, played an eclipse as a saving trump once, on some savages, and I saw my chance. I would play it myself, now; and it

---

*This episode would become infamous, as it was fraught with scientific mistakes: the path of totality could never cover both 'these [British] Islands' and southern Africa; there was a full Moon the night before the eclipse, although solar eclipses can only come about when the Moon is new, i.e. sunward of the Earth; and totality is described as lasting 'an hour or more'. In later editions, the chapter was rewritten, and the solar eclipse became lunar, with a consequent dimming of dramatic power.

Hergé's Tintin uses an eclipse to outwit his Inca captors

wouldn't be any plagiarism, either, because I should get it in nearly a thousand years ahead.'*

During the nineteenth century superstition was still strong enough that, on 7 September 1820, the House of Lords suspended its divorce trial of Queen Caroline for the duration of an eclipse. But it was usually among the less educated that such atavistic fears persisted – giving rise to visions and religious inspiration. In the summer of 1831 a God-fearing Virginia slave acted on a vision vouchsafed him during the previous February's darkening: on the night of 21 August Nat Turner and seven companions massacred their master's household, then ravaged several other slaveowners' houses nearby, killing more than fifty men, women

---

*The eclipse trick is arguably one of the central tropes of the clash of civilizations. In 1949 Hergé published *The Prisoners of the Sun*, in which Tintin, Captain Haddock and Professor Calculus fall captive to an Inca kingdom which has survived into the modern world. The presiding chieftain commands that the trespassers be burned alive on a pyre lit by the Sun, but Tintin realizes that an eclipse is imminent, and at the crucial moment shouts out, 'O God of the Sun, sublime Pachacamac, display thy power, I implore thee! If this sacrifice is not thy will, hide thy shining face from us!' The Sun goes out and the Indians run about in terror.

and children. After remaining at large for nine days, Turner was captured, tried and hanged, but not before he had made his 'confession'
about the effect the eclipse had had on him:

> By signs in the heavens that it would make known to me when I should
> commence the great work – and until the first sign appeared, I should
> conceal it from the knowledge of men – and on appearance of the sign
> I should rise and prepare myself, and slay my enemies with their own
> weapons . . .[6]

William Styron imagined what Turner might have experienced in his
1967 novel, *The Confessions of Nat Turner*.[7] 'I looked up then to see the Sun
wink slowly out as it devoured the black shape of the Moon. There was no
surprise in my heart, no fear, only revelation, a sense of final surrender.'*

The same year that Twain published *A Connecticut Yankee*, a total
eclipse ignited a major new religion. It originated with an orphaned
Paiute Indian, born in Nevada around 1858, adopted by a local farmer
and given the unassuming name of Jack Wilson. Like Turner, he grew up
a devout child, and by his early twenties had become a local holy man,
impressing his neighbours with the wisdom of his utterances – and
taking a new name, Wovoka (which translates as 'Cutter' – he had probably been a woodsman in his teens). In late 1888 he fell critically ill with
scarlet fever. An eclipse fell at the New Year, and, when the Sun
reappeared, several Paiutes burst into the prophet's cabin to find him in
a deathly coma. 'The very hour of his translation,' opines his principal
biographer, Paul Bailey, 'had been a climactic one. To be ushered out of
mortality in the awful moments of the Sun's death had all the portents
of divine timing and celestial intervention. And the credulous Paiutes,
after ascertaining with knife and fire on Wovoka's seemingly lifeless flesh

---

*In May 1858 an eclipse passed over Peru, Lieutenant Gilliss of the US navy recording: 'Two
citizens of Olmos stood within a few feet of me, watching in silence, and with anxious coun-
tenances, the rapid and fearful decrease of light. They were wholly ignorant that any
sudden effect would follow the total obscuration of the Sun. At that instant one exclaimed
in terror, "*La Gloria*," and both . . . fell to their knees, filled with awe. They appreciated the
resemblance of the Corona to the halos with which the old masters have encircled their
ideals to the heads of Our Savior and the Madonna, and devoutly regarded this as a manifestation of the Divine Presence.'[8]

that it was no ordinary sleep, were ready enough to ascribe the Sun's escape from the black monster to the high assistance of their heaven-borne prophet.'[9]

When Wovoka miraculously revived, he told his followers that he had been carried away to a green land where the dead lived again, and pre-scribed the special dance that would summon up their spirits. 'Pretty soon now, the Earth will die; but Indians must not be afraid, because Earth will become live again, just like Sun died and came live again.' He promised that the white race would be swept away by a torrent of mud and water; his people would become young and free of disease and pain; the land would be fertile, game would return in abundance and an Indian paradise repossess the country.

Such a vision was difficult to resist. In a matter of months the Washoe, the Bannock and the Pit River tribes were treading the slow measures of what white men called the 'ghost dance'. Apostles from the Paiute car-ried his message to the Walapai, Cohonino, Mohave and even the far-off Navajo. 'No religious movement in history,' asserts Bailey, 'ever swept with greater speed or more dramatic impact across the face of the land.' Soon Mormon settlers were shuffling alongside their native brethren as the new religion swept south, east and north. The Bureau of Indian Affairs and the Interior Department became alarmed, but they need not have worried. Wovoka unwisely fixed a precise date – the year 1900 – for this transformation. When nothing happened, the great movement that had been set alight by the eclipse of '89 quickly collapsed. By the time he died in 1932, Wovoka was little more than a faded footnote to history.

☼

Such stories notwithstanding, worldwide there has been an inexorable change in attitude; new mythologies have replaced older ones, and eclipses not only no longer arouse the old awe, but now we use them for our sport. In the eighteenth century, cheating by moving a die with one's little finger was 'to eclipse'. On the Mississippi during the nineteenth century it became a tradition so to name the fastest boat. A partial eclipse on 28 May 1900 hap-pened to fall in a US election year, and followers of the two candidates – the Democrat William Jennings Bryan and the Republican William McKinley – gleefully handed out campaign buttons that read 'Partial eclipse will be

Parisian women crowd together to view the eclipse of 26 August 1892.
Some are using special filters to avoid injury

total in November' or 'Imaginary eclipse visible only at Kansas City July 4' –
a reference to the city where Bryan was nominated.*

In modern times the word 'eclipse' has been appropriated almost at will,
with little regard for context. The word, which once augured world-ending
catastrophe, is currently the brand name for a popular chewing gum.

---

*A quarter of a century passed before Bryan renewed his involvement with eclipses – again
unhappily – during the famous Scopes trial of 1925 in Tennessee. John T. Scopes, a young
high school teacher, was charged with teaching Darwin's theories of evolution in violation
of a state law that forbade 'any theory that denies the story of the Divine Creation of man
as taught in the Bible'. Bryan had campaigned for the law and offered to give evidence for
the prosecution. Ranged against him was Clarence Darrow, possibly the best-known
defence lawyer in American history. In his withering cross-examination of Bryan, Darrow
quoted the seven days of Creation in Genesis 1: 14–19, showing that God makes the Sun
on the fourth day of Creation, yet there is day and night for the previous three. How can
this be? Further, the same passage makes it clear that the Sun rules the day, the Moon the
night – so how did Bryan explain eclipses? The strain proved too much for the old cam-
paigner: Bryan died five days after the trial ended.

☼

However, nothing can get in the way of the sense of wonder we still feel when under the actual shadow of an eclipse. Novelists from Fenimore Cooper to Dorothy L. Sayers have recorded such scenes, either in their fiction or in their private notebooks. One of the finest descriptions, capturing that sense of momentary contact with the full mystery of created things, comes from Virginia Woolf, in Yorkshire to witness the 1927 eclipse, the first totality viewable in England for more than 200 years. The next day, 30 June, she wrote in her diary:

> Now I must sketch out the Eclipse . . . Rapidly, very very quickly, all the colours faded; it became darker and darker as at the beginning of a violent storm; the light sank and sank; we kept saying this is the shadow; and we thought now it is over – this is the shadow; when suddenly the light went out. We had fallen. It was extinct. There was no colour. The Earth was dead. That was the astonishing moment; and the next when as if a ball had rebounded the cloud took colour on itself again, only a sparky ethereal colour and so the light came back. I had very strongly the feeling as the light went out of some vast obeisance; something kneeling down and suddenly raised up when colours came. They came back astonishingly lightly and quickly beautifully in the valley and over the hills . . . It was like recovery. We had been much worse than we had expected. We had seen the world dead. This was within the power of nature.[10]

I know of no other record so moving – yet the eclipse lasted only twenty-three seconds.* But then length of time doesn't seem to matter –

---

*Woolf would use her experience again in her essay 'The Sun and the Fish' and once more, in 1931, when she re-creates her moment of rapture in *The Waves*, in the visit to Hampton Court that marks the novel's climax: 'The scene beneath me withered. It was like the eclipse when the sun went out and left the earth, flourishing in full summer foliage, withered, brittle, false . . . How then does light return to the world after the eclipse of the sun? Miraculously. Frailly. In thin stripes. It hangs like a glass cage. It is a hoop to be fractured by a tiny jar. There is a spark there. Next moment a flush of dun. Then a vapour as if earth were breathing in and out, once, twice, for the first time. Then under the dullness someone walks with a green light. Then off twists a white wraith. The woods throb blue and green, and gradually the fields drink in red, gold, brown. Suddenly a river snatches a blue light. The earth absorbs colour like a sponge slowly drinking water. It puts on weight; rounds itself; hangs pendant; settles and swings beneath our feet. So the landscape returned to me.'[11]

the experience is generally memorable. Isaac Asimov relates how, on 30 June 1973, he came to witness a total eclipse off the coast of West Africa aboard the cruise ship *Canberra*. 'What impressed me most was its ending. A tiny dot of bright light appeared, and suddenly spread out in half a second to become too bright to look at without filters. It was the sun coming back with a roar – and that was the most magnificent astronomical spectacle I ever saw.'[12] The palaeontologist Stephen Jay Gould also, like Woolf and Francis Bailey, stresses the quality of *suddenness*: 'The sky turns off as if a celestial janitor threw a switch. For the sun is powerful, and a fraction of one percent of sunlight is daytime, while totality is night-time – and the transition is a moment, a twinkling of an eye.'[13] 'In ordinary conditions,' he adds, 'it takes several hours for a person's color discrimination to disappear. In an eclipse it happens in a flash – one moment there is color, then it dies out in a rush.'

For the eclipse of 2001, Gould, having ordered all his students to take time off to observe it, then journeyed from his Boston home (where the skies were shrouded by rain) to New York City to witness it in full. Later he describes 'this remarkable event':

> In this age of artificially induced full-body shake-me-ups, . . . we hardly think that anything so subtle, albeit pervasive, as the character of surrounding sunlight could move our passions, or even invite our notice . . . The sky did not darken precipitously over New York on May 10. But we are exquisitely sensitive to the usual character of light, even though we may not explicitly credit our awareness, and may not be able even to state what feels so odd . . . the day turned eerily somber, while sunlight continued to reign – and people noticed, and trembled ever so slightly . . . noticed and stood still as a sky full of daylight darkened to the level of a clearly nonexistent thunderstorm. A woman said to her friend, 'holy shit, either the world is about to end, or it's going to rain – and it sure as hell ain't going to rain.'

If these are unusual sentiments for a scientist, they do demonstrate that there is still something primal in our response. In John Updike's story 'Eclipse', a man wanders into his backyard one afternoon with his two-year-old daughter to watch a near-total eclipse:

Sun and Moon dance together on the horizon, Antarctica, 2003

I looked up, a split-second Prometheus, and looked away. The bitten silhouette of the Sun lingered redly on my retinas. The day was half cloudy, and my impression had been of the Sun struggling, amid a furious knotted huddle of black-and-silver clouds, with an enemy too dreadful to be seen, with an eater as ghostly and hungry as time.

He calls out to his elderly neighbour sitting on her porch how she likes the eclipse. 'I don't,' she answers, adding, 'They say you shouldn't go out in it. There's something in the rays.' At last the darkness passes:

Superstition, I thought, walking back through my yard, clutching my child's hand as tightly as a good-luck token. There was no question in her touch. Day, night, twilight, noon, were all wonders to her, unscheduled, free from all bondage of prediction. The Sun was being restored to itself and soon would radiate its influence as brazenly as ever – and in this sense my daughter's blind trust was vindicated. Nevertheless, I was glad the eclipse had passed, as it were, over her head; for in my own life I felt a certain assurance evaporate forever under the reality of the Sun's disgrace.[14]

☼

Modern technology has helped us to understand parts of our own world through eclipses. In 1872 the great scientist Pierre-Jules-César Janssen (1824–1907)* observed an eclipse at Guntur in north-east India and just before and after totality discerned a yellow spectral line in the solar prominences, so bright as to indicate some previously unknown chemical element. The English astronomer Norman Lockyer independently arrived at the same conclusion, naming the new presence 'helium' in the belief that it was to be found only in the Sun. By 1895 it had been isolated and identified as a substance found on Earth as well. Further, studies of past eclipses have revealed repeated slight changes in the Earth's rotation, and variations in its spin rate.[15]

The pre-eminent scientific employment of eclipses, however, was made at the height of the First World War. Einstein had just published his general theory of relativity, part of which stated that gravity would affect light, so that light passing a major source of gravitation would be bent out of straightness.[16] Although this was a radical modification from the ideas of Newton, it explained a phenomenon that had long troubled astronomers: the discovery that the perihelion – the closest orbital point to the Sun – of the planet Mercury was precessing slightly faster eastwards each year than the gravitational effects of the other planets could account for. Einstein's argument that gravity was a field, not a force, explained the discrepancy. On making his discovery, he wrote, 'for three days I was beside myself with joyous excitement'.[17] But it remained to be proved.

In 1917 a professor of astronomy in the neutral Netherlands sent a copy of Einstein's paper to the Royal Astronomical Society, where it was read by both the leading British astrophysicist of the day, Arthur Eddington (unpreoccupied by wartime duties because a strict Quaker),

---

*Trapped in Paris in 1870 by the Franco-Prussian War, Janssen had memorably hopped aboard a hot-air balloon and sailed out over the astonished besiegers, south, ever south, before finally coming down in Oran in North Africa and observing the eclipse there, only to be frustrated by cloud. His exploit may have fired the imagination of Arnold Bennett, who in *The Old Wives' Tale* (1908) has Sophia, one of his sister heroines, stranded in the same siege. Her husband Gerald, like Janssen, escapes by balloon, but is never seen again – an exit unique in fiction.

and by the Astronomer Royal, Sir Frank Dyson. The latter, a specialist in solar eclipses, realized that light's relationship to gravity could be tested by studying the stars visible in the immediate neighbourhood of an eclipsed Sun. Were Einstein right, these would have their light bent by the Sun's gravitational field, and would show up on photographs as slightly displaced from their positions in photographs taken when the Sun was elsewhere in the sky.[18]

A solar eclipse was due on 29 May 1919, and Dyson, studiously ignoring the fact that he was trying to prove a theory that emanated from a scientist in perfidious Germany (feelings against which ran so high that less than two months after the eclipse, when the International Astronomical Union was formed, German and Austrian scientists were specifically excluded), won government funding for a two-pronged expedition, which sailed at the beginning of March, parting company at Lisbon. Eddington's party travelled to the Portuguese island of Principe, off West Africa; the other set off for Sobral, in north-east Brazil, further along the path of totality.

By how much would gravity bend light? Einstein's equation predicted that a ray just grazing the surface of the Sun would be pulled by 1.745 seconds of arc – twice the deflection assumed by classical Newtonian theory. (An arc second is 1/3,600 of a degree, so one is dealing with minute differences: hence the difficulty of measuring them.) Before he left London, Eddington, his principal assistant, E. T. Cottingham, and Dyson talked late into the night, and Cottingham asked what would happen if the eclipse photographs revealed not Newton's, not Einstein's, but *twice* Einstein's deflection. 'Then Eddington will go mad,' replied Dyson breezily, 'and you will have to come home alone.'

By mid-May Eddington and his party were in place. When the pinpoints of five stars came into view, Eddington had eight minutes to take his pictures: 'I did not see the eclipse, being too busy changing plates . . . We took sixteen photographs.' Over the next six nights he developed the plates, comparing them with photographs of the same stars taken when they had not been close to the Sun. In the first ten shots, thin cloud, while not enough to obscure the eclipse, was sufficient to mask the crucial stars; but the next two plates were usable, and quite enough to supply the proof he was looking for. On the evening of 3 June Eddington turned to Cottingham. 'You won't have to go home alone.' The two sets of photographs differed by almost the exact amount predicted by Einstein.

The expedition won headlines everywhere, *The Times* declaring, 'REVOLUTION IN SCIENCE/NEW THEORY OF THE UNIVERSE/NEWTONIAN IDEAS OVERTHROWN', and judged the finding 'one of the most momentous, if not the most momentous, pronouncements of human thought.' *The New York Times* added jocularly, 'stars not where they seem or were calculated to be, but nobody need worry', for Einstein knew where they were. It made the forty-year-old physicist a global hero overnight, and was also the first major instance of restored international scientific collaboration before peace had even been signed.*

Eddington (who years later would reconfirm general relativity with his own magisterial work on the behaviour of light under the intense gravity of white dwarfs) celebrated by producing a pastiche of Fitzgerald's translation of the *Rubáiyát of Omar Khayyám*, two of whose verses run:

> *And this I know; whether Einstein is right*
> *Or all his theories are exploded quite,*
> *One glimpse of stars amid the Darkness caught*
> *Better than hours of toil by candle-light . . .*

> *Oh leave the Wise our measures to collate.*
> *One thing at least is certain, LIGHT has WEIGHT*
> *One thing is certain, and the rest debate –*
> *Light-rays, when near the Sun, DO NOT GO STRAIGHT.*[20]

The rest of the scientific world also rejoiced, if not so impishly. 'It is doubtful,' reflects Thomas Crump in his history of eclipses, 'whether eclipse astronomy will ever again achieve a result of such cosmic importance.'[21] It has yet to do so.

---

*The episode is memorably re-created in the opening chapter of Nicholas Mosley's novel *Hopeful Monsters*, which won the Whitbread Book of the Year Award in 1990.[19]

12

THE SUN DETHRONED

GRAND MASTER: Oh, if you knew what our astrologers say of the
coming age, and of our age, that has in it more history within
100 years than all the world had in 4,000 years before! . . .

Tommaso Campanella (1568–1639), *The City of the Sun*[1]

We must make use of physics for warfare.

Nazi slogan, to be reworded by the German scientist Werner
Heisenberg as: 'We must make use of warfare for physics.'[2]

The Sun does not stand still, or even move in a simple pattern. At different times, and under different examinations, it may appear a solid
orb, a ball of fire or a constantly discharging source of winds, flares, spi-
rals and radioactive particles. Its latitudes rotate at different speeds, and
its whole surface surges up and down about four kilometres every hundred and sixty minutes – although even to talk about the Sun's 'surface'
is misleading since, being largely a collection of gases, it has none. Seen
from other planets, it is larger or smaller than when viewed from the
Earth; and its atmospheric effects are different on Mercury, say, or than
on the torrid mists of Venus, and different again on our planet. By the
end of the eighteenth century, astronomers had recognized that it was
simply one star among many – just the one that happened to be nearest
to us – and had estimated its distance, size, mass, rate of rotation and
movements in space to within ten per cent of today's values. But they
continued to gnaw away at other problems.

What was going on inside it? What caused it to shine? How old was it, and how did it relate to other heavenly bodies? Only now they were finding many of the answers. Between 1800 and 1950, aided and abetted by advances in technology, the Industrial Revolution and a new enthusiasm for scientific enquiry, hardly a year went by without astronomers, physicists, chemists and geologists finding something that added to our understanding.

In just the first half of the nineteenth century, for example, absorption lines in the solar spectrum were mapped, and electromagnetic induction and electromagnetic balance were being discovered, both of which increased our knowledge of how the Sun draws us to it. During the same period, the Sun's energy output – the 'solar constant' – was measured, leading to a greater appreciation of both its temperature and its influence on climate. In 1860 the Vatican's chief astronomer photographed the eclipse of 18 July, showing that the corona and prominences were real features, not optical illusions or deceivingly illuminated mountains on the eclipsing Moon.

The list of specific discoveries runs close to 200. In just the single decade 1871–80, for instance, came electromagnetic radiation, the solar distillation of water (the Sun as purifier), and the first radical re-estimate of the Sun's age at 20 million years and of the length of time it would have taken to contract to its current size – both of which were to touch off a major debate between scientists and literal interpreters of the Bible. Other discoveries were less contentious: the Sun's surface temperature was estimated at 5,430°C, 9,806°F, while its core was reckoned to be gaseous, with temperatures steadily decreasing from its centre to the surface. Several new instruments (the heliospectroscope, star spectroscope, telespectroscope) were invented; and all fixed stars were found to display only a few combinations of spectra, depending on their physical-chemical nature, 'an achievement of as great significance as Newton's law of gravitation'.[3]

In recent years writers such as Dava Sobel, Timothy Ferris and Bill Bryson, as well as popularizing scientists like Carl Sagan and Stephen Hawking, have demystified such esoteric subjects as dark matter (invisible but still exercising attraction, whose presence we infer from its gravitational pull, though we don't know what it is) and black holes (familiar, but still difficult fully to imagine). But what about deuterium, the ionosphere, even the Coriolis effect? Such words and phrases are part of a discourse that can seem closed to most of us. But two or three of the main strands of research can be braided together to show how, between 1800 and 1953, our knowledge of – and our feelings about – the Sun were transformed.

☼

Towards the middle of the nineteenth century, the French philosopher Auguste Comte was asked what would be for ever impossible. He looked up at the stars: 'We may determine their forms, their distances, their bulk, their motions – but we can never know anything of their chemical or mineralogical structure; and, much less, that of organized beings living on their surface.' Within a surprisingly few years mankind would have grasped those very things. To put into context how quickly research expanded and focus shifted: in 1800 only one stellar 'catalogue of precision' was available; in 1801 J. J. Lalande published one describing 47,390 stars; and in 1814 Giuseppe Piazzi added another 7,600. Others followed; between 1852 and 1859 alone some 324,000 were supplemented. Photographic charting began in 1885, and by 1900 the third and concluding volume of a work appeared that located a massive 450,000 stars, the fruit of collaboration between scientists in Groningen and John Herschel in Cape Town. It became abundantly clear that the Sun did not hang as the supreme object in the sky but was just one more star, an unspectacular one at that. Scientists could now concern themselves with this multitude of other suns, many far larger and heavier than ours.*

---

*In the first quarter of the nineteenth century a German astronomer formulated Olbers' Paradox: were there to be an infinite number of stars in the universe there would be no night, the sky would be one huge glare and we would be drowned in – indeed, could not have come into existence under – such limitless stellar energy.[4] A good explanation is given by fifteen-year-old Christopher Boone, the autistic boy narrator of Mark Haddon's 2003 novel, *The Curious Incident of the Dog in the Night-Time*: 'I thought about how, for a long time, scientists were puzzled by the fact that the sky is dark at night, even though there are billions of stars in the universe and there must be stars in every direction you look, so that the sky should be full of starlight because there is very little in the way to stop the light reaching Earth. Then they worked out that the universe was expanding, that the stars were all rushing away from one another after the Big Bang, and the further the stars were away from us the faster they were moving, some of them nearly as fast as the speed of light, which was why their light never reached us.'[5] One should further add that no star shines for more than around 11 billion years.

Even this piece of knowledge was to be massively augmented: in the predawn hours of 6 October 1923 at the Mount Wilson Observatory, Edwin Hubble was examining a photograph of a fuzzy, spiral-shaped swarm of stars known as M31,* or Andromeda, which was assumed to be part of the Milky Way, when he noticed a star that waxed and waned with clockwork regularity, and the longer it took to vary, the greater its intrinsic brightness.

Hubble sat down and calculated that the star was over 900,000 light years away – three times the then-estimated diameter of the entire universe! As *National Geographic* reported the story: 'Clearly this clump of stars resided far beyond the confines of the Milky Way. But if Andromeda were a separate galaxy, then maybe many of the other nebulae in the sky were galaxies as well. The known universe suddenly ballooned in size.'[6]

Today we know that there are at least 100 billion galaxies, each one harbouring a similarly enormous number of stars. The Earth had already been demoted by the realization that it circled the Sun, and not the other way round; William Herschel and his son John, with their revolutionary telescopes, had revealed the riches beyond our solar system, and unwittingly demoted the Sun as well. Here was a third or even a fourth diminution: that not only were we orbiting what was just a minor star among a multitude in the Milky Way, but that the Milky Way itself was but one galaxy of an untold number.† For scientists, stellar astronomy was the future; minor stars need not apply. As A. E. Housman, poet and classicist and also a keen student of astronomy, tartly observed, 'we find ourselves in smaller patrimony'.[8]

☼

This displacement was occurring alongside the centuries-long and frequently vehement argument over the age of the Earth – and thus the age of both the Sun and the universe, an issue that provoked intense interest among astronomers, theologians, biologists and geologists. The date of

---

*Certain objects have the letter 'M' before their designated numbers in honour of Charles Messier, the French astronomer who in 1774 published a catalogue of forty-five deep-sky objects such as nebulae and star clusters, to distinguish them from comets. The final version of the catalogue was published in 1781, by which time the list had grown to 103.
†As today's solar scientist Sten Odenwald admits: 'Hardly any astronomer I know really enjoys space physics of the kind involved in studying the Sun–Earth system . . . The excitement of exploring how stars evolve, and the structure and contents of the universe, was a much more potent draw of attention and enthusiasm. Solar and space research was often seen as too local, and it was intellectually very messy physics, to boot.'[7]

Creation had been made particularly exact by the elderly Irish Protestant archbishop James Ussher (1581–1656), who took all the 'begats' in Genesis and the floating chronology of the Old Testament and – with a little helpful tweaking from Middle Eastern and Mediterranean histories, not least the Jewish calendar, which lays it down that Creation fell on Sunday 23 October 3760 BC – arrived at the conveniently round figure of 4,000 years between the Creation and the (most likely) birth date of Christ in 4 BC.

Were it not for the London bookseller Thomas Guy, Ussher's dating might have sunk into obscurity, as had those of the hundred or so biblical chronologers before him. By law, only certain publishers, such as the Cambridge and Oxford university presses, were allowed to print the Bible, but Guy acquired a sub-licence to do so, and in a moment of marketing inspiration printed Ussher's chronology in the margins of his books, along with engravings of bare-breasted women loosely linked to Bible stories. The edition earned Guy a fortune, enough to endow the great London hospital that bears his name. If this were not enough, in 1701 the Church of England authorized Ussher's chronology to appear in all official versions of the King James Version. The 'received chronology' soon became such an automatic presence that it was printed well into the twentieth century.

As the historian Martin Gorst puts it: 'The influence of his date was enormous. For nearly two hundred years, it was widely accepted as the true age of the world. It was printed in Bibles, copied into various almanacs and spread by missionaries to the four corners of the world. For generations it formed the cornerstone of the Bible-centered view of the universe that dominated Western thought until the time of Darwin. And even then, it lingered on.'[9] Gorst recalls coming across his grandmother's 1901 Bible and finding, set opposite the opening verse of Genesis, the date and time for when the world began: 6 p.m. on Saturday 22 October 4004 BC.*

Even in Ussher's time, however, freethinkers had begun to question

---

*As recently as 1999, a Gallup poll reported that forty-seven per cent of Americans believed God created the human race within the last 10,000 years. Stephen Hawking notes in *A Brief History of Time* that a date of 10,000 years fits curiously well with the end of the last Ice Age, 'which is when archaeologists tell us that civilization really began'.[10]

the accepted chronology. Travellers returned from far-off lands with reports of histories that went back much further than 4004 BC, and with the advent of new forms of enquiry based on scientific principles – *Nullius in Verba* ('Take Nobody's Word For It') runs the motto of Britain's Royal Society – it became the turn of natural philosophers to debate the age of the globe.

In 1681 a fellow don of Newton's at Cambridge, Thomas Burnet, published his best-selling *Sacred Theory of the Earth*, which argued that the world's great mountain ranges and vast oceans had been shaped by the Flood of Genesis; he got round the Old Testament statement that Creation had taken only six days by quoting St Peter's pronouncement that 'one day is with the Lord as a thousand years'. A shelf of books appeared offering similar explanations. Then, later that same decade, two British naturalists, John Ray and Edward Lhuyd, after examining shells in a Welsh valley and ammonites on the north-east English coast, concluded that both groups of fossils came from species that would have needed far more than 5,680-odd years in which to live and die out in such numbers. Their researches aroused a short-lived flurry of interest, followed by a period of neglect until the London physician John Woodward (1665–1728) posited that fossils were the 'spoils of once-living animals' that had perished in the Flood. But this theory too was met with ridicule: faith in Ussher's chronology endured.

It was Edmond Halley who, in 1715, became the first person to suggest that careful observation of the natural world (such as measuring the salinity of the oceans) would provide clues to the Earth's age. His ideas were eagerly adopted by Georges-Louis de Buffon (1707–88), who in his *Histoire naturelle* (1749) argued that the Earth was tens of thousands of years old, an idea that 'dazzled the French public with a timescale so vast they could barely comprehend its enormity'.[11] While Buffon did not sample the oceans, he did believe the Earth began not in the biblical single moment but as the result of a comet colliding with the Sun, the debris from this explosion coalescing to form the planets.*

Theologians at the Sorbonne erupted in fury, and Buffon was forced

---

*Buffon simply ignored the biblical account of Creation, possibly encouraged by the appearance, in 1770, of the first printed blunt denial of any divine purpose – *Système de la nature*, by the forty-seven-year-old French philosopher Paul Henri Dietrich d'Holbach.

to publish a retraction – privately declaring, 'Better to be humble than hanged.' But he would not be kept quiet for long. When another Frenchman, the mathematician Jean-Jacques Dortous de Mairan, demonstrated that the Earth did not get its heat from the Sun alone but contained an inner source, Buffon was galvanized into writing a history of the world from its inception. If the Earth were still cooling, then by finding the rate at which it lost heat one could calculate its age. Over the next six years he conducted a series of experiments, ultimately concluding that the Earth was 74,832 years old (a figure he believed to be highly conservative – his unofficial estimate was 100,000). His book was met by a chorus of disbelief. In April 1788 he died, and the following year the Revolution erupted. Buffon's crypt was broken into and the lead from his coffin requisitioned to make bullets.

Undaunted, the intellectually curious continued to search for explanations for all they found on Earth, and by the nineteenth century the division between the Church and science was greater than ever. Between 1800 and 1840 the words 'geology', 'biology' and 'scientist' were either coined or acquired specialized senses in the vernacular. Geology in particular was popular, and soon scientists and others were identifying and naming the successive layers of strata – in Byron's phrase, 'thrown topsy-turvy, twisted, crisped, and curled' – that formed the Earth's surface. A shy young British barrister named Charles Lyell (1797–1875), building on John Woodward's example, used fossils to measure Earth's age, finding them in lava remains in Sicily that he estimated to be 100,000 years old.

By 1840 the evidence that the Earth was exceedingly 'ancient' was overwhelming. Even if man had been around for only 6,000 years, prehistoric time had expanded beyond comprehension. Then in 1859 came *The Origin of Species*, and the great argument that linked man's evolution to that of the apes, undermining a literal reading of the Bible story. Dr Ussher's reckonings were at last no longer feasible. Cambridge University Press removed his chronology from their Bibles in 1900; Oxford followed suit in 1910. By then Mark Twain had already imagined the Eiffel Tower as representing the age of the world, and likened our share of that age to the skin of paint on the knob at its pinnacle.

Mighty doctrines, when overthrown, open up amazing perspectives. Once science had extended the age of the Earth to over 100,000 years, an array of new ideas emerged. What age to give now to the Sun? After Darwin, scientists had to consider that it had been releasing its energy over *millions* of years. Natural selection required that the solar system be of a previously inconceivable age. It is only because our Sun is not particularly powerful and burns relatively slowly, being tolerant to the sensitive carbon-based compounds that are the basis of Earthly life, that evolution (not that Darwin ever liked that word) could occur.

Although the argument about Earth's age rumbled on through the latter decades of the nineteenth century, questions about its star were drawing more attention from scientists. As Timothy Ferris puts it: 'The titans of physics chose to focus less on the Earth than on that suitably grander and more luminous body.'[12] The noted German physicist Hermann von Helmholtz (1821–94) tried to explain what the Sun drew on for its power and in 1854 came up with the idea that, just like all other stars, it must be steadily contracting under its own gravity, so creating an energy by that very compression. However, although Helmholtz's theory might apply to a degree, such a method would only enable the Sun to shine at its present rate for a few million years; consequently, it had to be powered another way.

In the summer of 1899 a professor of geology at the University of Chicago, Thomas Chamberlin (1843–1928), published a paper on how the Sun was fuelled that challenged one of the then basic assumptions of astrophysics:

Is present knowledge relative to the behavior of matter under such extraordinary conditions as obtain in the interior of the Sun sufficiently exhaustive to warrant the assertion that no unrecognized sources of heat reside there? What the internal constitution of the atoms may be is as yet open to question. It is not improbable that they are complex organizations and seats of enormous energies.[13]

Within five years the central principles of physics, and consequently the basic assumptions about the functioning of the Sun, had been remade.

☼

In the spring of 1896 Henri Becquerel, a Paris physicist, inadvertently stored some unexposed photographic plates wrapped in black paper underneath a lump of uranium ore he had been using in an experiment. (Uranium – named after the planet in 1789 – is the most complex atom regularly found in nature, composed primarily of uranium-238 – as would be discovered, composed of 92 protons and 146 neutrons – plus a small percentage of the less stable uranium-235, three neutrons the less.) When after a few weeks he developed the plates, he found that it was as if they had been exposed to light, for they had been imprinted with an image of the silvery-white mineral, which, he thought, must be emitting a 'type of invisible phosphorescence'. Marie and Pierre Curie, following up on this lucky discovery, would later name the phosphorescence 'radioactivity'. The most extraordinary aspect was that uranium radiated heat without cooling down to the temperature of its surroundings; but the metal was not easy to obtain, nor was the radiation it produced particularly impressive, so the Curies' findings were generally ignored. Then in 1898 they discovered that pitchblende, another uranium ore, emitted much greater amounts of radioactivity than other sources of the metal. Was it possible that there were elements on Earth that were releasing tremendous amounts of unseen energy, waiting to be tapped?

Soon the further, rarer element (named 'radium') that the Curies had extracted from pitchblende was being written up in mainstream papers as the miracle metal, the most valuable element known to man, capable of curing blindness, revealing the sex of a fetus, even turning the skin of a black man white. 'A single gram would lift five hundred tons a mile in height, and an ounce could drive a car around the world.'[14] Advertisements promoted its use in 'radioactive drinking water' for the treatment of gout, rheumatism, arthritis, diabetes and a range of other afflictions. People demonstrated radium at parties and went to radium dances, while the word became a fashionable brand name: Radium beer, Radium butter, Radium chocolate, Radium condoms, Radium suppositories and even Radium contraceptive jelly. By 1904 scientists were also claiming radium to be a source of energy for the Sun.

The question was whether the energy was coming from within the atoms of radium or from outside. The great physicist Ernest Rutherford (1871–1937), a New Zealander who had worked on the nature of atoms

An advertisement for Tho-Radia, a cream based on radium and thorium,
marketed in the 1930s as a breakthrough in beauty preparations

both in England and Canada, began to explore atomic nuclei, combining
forces with a Canada-based English chemist, Frederick Soddy (1877–
1956). Radium, Rutherford had established, generates enough heat to
melt its weight in ice every hour, and will continue to do so for a millennium or more: the Earth stays warm not least because it is heated by
radioactive decay in the rocks and the molten core that lies at its centre.*

The two men were soon experimenting on thorium, a radioactive element not unlike radium, and found that it independently yielded a
radioactive gas – one element transmuting into another. This result was so

---

*Most of the planet's internal heat is generated by four long-lived radioisotopes – potassium-
40, thorium-232 and uranium-235 and -238 – that release energy over billions of years as
they decay into stable isotopes of other elements. The Earth's core is a hot, dense, spinning solid sphere, composed primarily of iron, with some nickel. Its diameter is c. 2,400
kilometres, some 19 per cent of the Earth's (12,740 kilometres). Surrounding it is a liquid
outer layer 2,200–2,300 kilometres thick, 85 per cent iron, whose heat-induced roiling creates the Earth's magnetic field. The two cores together comprise an eighth of the Earth's
volume but a third of its mass. Over a period of 700–1,200 years the inner core will make
one full spin more than the rest of the planet, intensifying its magnetic field.

startling that when Soddy told Rutherford about his finding, the New Zealander shouted back across the lab floor: 'Don't call it *transmutation* – they'll have our heads off as alchemists!' They proceeded to prove that the heavy atoms of thorium, radium and other elements they discovered to be radioactive broke down into atoms of lighter elements (in the form of gases), and as they did so they threw off minute particles – which, dubbed alpha and beta rays, were actually the elements' main output of energy.

Further experiments by Rutherford showed that most of an atom's mass lay at its core, surrounded by a crackling web of electrons. He and Soddy speculated that radioactivity of the same kind could be powering the Sun, but while their work was considered worthy of further investigation, it was not seen as astrophysically revolutionary. Some forty years later, Robert Jungk would write in his groundbreaking account, *Brighter than a Thousand Suns*: 'Professor Rutherford's alpha particles ought really, at that time, to have upset not only atoms of nitrogen but also the peace of mind of humanity. They ought to have revived the dread of an end of the world, forgotten for many centuries. But in those days all such discoveries seemed to have little to do with the realities of everyday life.'*

Soddy did his best to explain the discovery, first in *The Interpretation of Radium* (1912) then in *The Interpretation of the Atom* (1932), arguing that, before atomic decay was known, the only explanations for the Sun's energy output had been chemical, and thus short-term and puny; but once we had learned about radioactivity, the storm of subatomic reactions that revealed themselves were on something like the right scale to power the Sun.

Nearly a decade passed before anyone built on Soddy and Rutherford's discoveries. Eddington, fresh from his successes off the coast of Africa, now conducted a wide-ranging study of the energy and pressure equilibrium of stars and went so far as to construct mathematical models of their temperature and density ('What is possible in the Cavendish laboratory,' he famously commented, 'may not be too difficult in the Sun'). He estimated that the Sun had a central temperature of 40 million degrees, and argued that there must be a simple relationship between the total rate of

---

*During the last year of the First World War, Rutherford failed to attend a meeting of experts who were to advise the high command on new systems of defence against enemy submarines. When censured, he retorted: 'I have been engaged in experiments which suggest that the atom can be artificially disintegrated. If it is true, it is of far greater importance than a war.'

energy loss from a star (its 'luminosity') and its mass. Once he knew the Sun's mass, he reckoned he could predict its luminosity.

To the ancients it had seemed obvious that the Sun was on fire, but to late-nineteenth- and early-twentieth-century physicists that was unacceptable: it was just too hot to be chemically burning. So the question remained. As John Herschel put it, the 'great mystery' was

> To conceive how so enormous a conflagration (if such it be) can be kept up. Every discovery in chemical science here leaves us completely at a loss, or rather, seems to remove farther the prospect of probable explanation. If conjecture might be hazarded, we should look rather to the known possibility of an indefinite generation of heat by friction, or to its excitement by the electric discharge . . . for the origin of the solar radiation.[15]

Since research into the Earth's age was now suggesting it was more than 2 million years old, the Sun must have been shining for at least that length of time. What process could possibly account for such a startling output of energy? As the Ukrainian-born American physicist George Gamow (1904–68) put it, 'If the Sun were made of pure coal and had been set afire at the time of the first Pharaohs of Egypt, it would by now have completely burned to ashes. The same inadequacy applies to any other kind of chemical transformation that might be offered in explanation . . . none of them could account for even a hundred-thousandth part of the Sun's life.'[16]

Cosmologists looked to Gamow's fellow astrophysicists for an answer, and Eddington offered two: the first was that electrons and protons, having opposite electric charges, annihilated each other within the Sun's core, with the concomitant conversion of matter to energy. About a year later he advanced his second, and correct, solution: the Sun, by fusing protons, created heavier atoms, in the process converting mass to energy. But how could such fusion take place in the Sun's consuming heat?

A feature of these years was that many vital discoveries were made by outsiders – scientists whom no one regarded as solar physicists before their contribution to the field.[17] Eddington had a young disciple. Cecilia Payne (1900–80) had been five years old when she saw a meteor and decided to become an astronomer. Following university, she was introduced to

Eddington, who advised her to continue her studies in America; and she became the first student, male or female, to earn a doctorate from the Harvard College Observatory. The examiners judged her 1925 thesis, a photographic study of variable stars, the best ever written on astronomy.

Her solution to the temperature problem was to use Rutherford's discoveries of atomic structure to show that all stars had the same chemical make-up: where their spectra differed, it was from physical conditions, not from their innate composition. Hydrogen and helium were by far the most abundant of the Sun's fifty-seven known elements, as they were in other stars.[18] Despite this conclusion, in her final list of chemical elements in the Sun she omitted hydrogen and helium, branding her own argument as 'spurious'.

It later transpired that her supervisor, the renowned Princeton astronomer Henry Norris Russell, had tried to talk her out of her theory. 'It is clearly impossible that hydrogen should be a million times more abundant than the metals,' he wrote to her, reiterating the conventional wisdom.[19] But Payne's arguments nagged at him; he re-analysed the Sun's absorption spectrum and finally accepted that she was right: giant stars indeed had an outer atmosphere of nearly pure hydrogen, 'with hardly more than a smell of metallic vapors' in them. Stars fuse hydrogen into helium, releasing a continuous blast of energy. And when the Sun's great energy was generated by the transformation of chemical elements taking place in its interior – as Gamow impishly noted – what was happening was precisely that 'transformation of elements' that had been so unsuccessfully pursued by the alchemists of old.[20] In principle, the burning of hydrogen into helium would allow the Sun to shine for about 100 billion years.

The next step was to understand nuclear fusion. The late 1920s and early 1930s witnessed a shift to research on the atomic nucleus,[21] one particular centre being the Institute for Theoretical Physics at Copenhagen University, directed by the estimable Niels Bohr (1885–1962), 'who dressed like a banker and mumbled like an oracle'.[22] By the 1920s Bohr had attained international stature, able to attract many of the greatest physicists of the day, among them George Gamow. This remarkable Ukrainian had a reputation both for scientific innovation and for playfulness (he would illustrate his scientific papers with drawings of a skull and crossbones to indicate the danger of accepting hypotheses about fundamental particles at face value). In 1928 he showed how a positively charged helium nucleus (an 'alpha particle', of the kind emitted in such great numbers by the Sun) could escape from a nucleus of a particular metal found on Earth – uranium – despite the binding electrical forces within the metal.*

Gamow not only showed how alpha particles broke *out* of nuclei but he went on to show how they could break *in*. At Cambridge, two of his colleagues, John Cockcroft and Ernest Walton, set about applying Gamow's theory, testing whether very high voltages would propel the particles through the outer walls of the nuclei. In 1932 they succeeded: for the first time a nucleus from one element had been broken down into another element by artificial means, a feat that became known as 'splitting the atom'. In that same 'year of miracles', another Cambridge man, James Chadwick, discovered the neutron, the most common form of particle lurking inside virtually all nuclei. Suddenly, a whole variety of powerful reactions could be detected and even induced in the subatomic world. What was now obvious was that such discoveries bore out Payne's view that these were the reactions powering the Sun.[24]

The high tide of crucial new work roared on. In 1934 the French physicist Frédéric Joliot and his wife Irène Curie (daughter of Pierre and Marie) proved that by bombarding stable elements with alpha particles, 'a new kind of radioactivity' could be effected. A few weeks later, the Italian physicist Enrico Fermi bombarded uranium with neutrons and reported similar results.

In 1938–9 Hans Bethe (1906–2005), the great German-American atomic physicist, originally from Strasbourg but by this time at Cornell, wrote a series of articles culminating in 'Energy Production in Stars', which explained how stars – including the Sun – managed to burn for billions of years. He had been cataloguing the subatomic reactions that had been identified to date, but up until 1932 these had been only a

---

*The general spirit of the Copenhagen hothouse can be gleaned from a trip several of the scientists made to the cinema in 1928. After seeing a Western, Bohr contended that he knew why the hero always won the gunfights provoked by the villain. Reaching a decision by free will always takes longer than reacting instinctively, so the villain who sought to kill in cold blood acted more slowly than the hero, who reacted spontaneously. To test this 'in a scientific manner', Bohr and his band of fellow researchers sought out the nearest toyshop and bought two guns, with which a showdown was duly enacted, Bohr taking on the role of hero and the six-foot-four Gamow that of villain. Bohr's theory was adjudged correct.[23]

fistful. Suddenly, there came a landslide, and he could now see which particular ones explained the Sun's workings. He posited that its abundant energy was the result of a sequence of six nuclear responses, and it was this process that lit all the stars. Simply put: the Sun was what would become known as a nuclear reactor.[25]

That same year the German physicists Otto Hahn and Fritz Strassmann demonstrated that what Fermi had witnessed in 1934 was actually the bursting of the uranium nucleus. Two of their colleagues, Lise Meitner and Otto Robert Frisch, went even further, showing that when a uranium atom is split, enormous amounts of energy are released. (Frisch asked a biologist colleague for the word that described one bacterium dividing into two, and was told 'fission', and thus the splitting of atoms acquired its name.)

The Hungarian Leó Szilárd was the next to make his mark. Although even Einstein believed that an atomic bomb was 'impossible', as the energy released by a single nucleus was so insignificant, Szilárd, spurred on by memories of reading H. G. Wells's 1914 novel *The World Set Free*, which predicted just such a weapon, confirmed that neutrons were also given off in each 'splitting', creating the possibility of chain reactions whereby each fission set off further fissions (an idea Szilárd conceived one dull September morning in 1933, waiting at a red light in London's Bloomsbury), so that the energy of splitting a single uranium nucleus could be multiplied by many trillions, releasing energy exponentially.[26] And not only was fission possible; it could be produced at will. Opined the Cambridge scientist C. P. Snow: 'With the discovery of fission . . . physicists became, almost overnight, the most important military resource a nation-state could call upon.'[27]

By the advent of the Second World War, both Allied and Axis scientists were keenly aware that nuclear fission might be converted into some kind of weapon, although no one was sure how.* Still highly sceptical, Einstein declared that the challenge of inducing a chain reaction and

---

*With the war under way, one of the routine duties of R. V. Jones, the scientific adviser to MI6, was to read through a monthly summary of German scientific publications. Early in 1942 he perused the latest report, and the next minute was running like a lamplighter down the corridor. 'The German nuclear physicists have stopped publishing!': the War Cabinet met within the hour.

Albert Einstein (1879–1955) with the US theoretical physicist
Robert Oppenheimer (1904–67) during the time they worked together in the days
before the Manhattan Project

making a fission bomb was akin to going out to shoot birds at night in a land where there were very few birds; but secretly, in a letter to President Roosevelt dated 2 August 1939 (drafted along with his friend Leó Szilárd), he urged that funds be devoted to developing fission-based weapons: 'A single bomb of this type, carried by boat and exploded in a port, might very well destroy the whole port together with some of the surrounding territory.'[28]

Much alarmed by this letter, that autumn Roosevelt earmarked a small amount of money for the study of fission: small because it was assumed that any bomb would require so many tons of uranium that it was theoretically possible but probably not a practical option. However, in early 1940 two German refugees in Britain, Otto Frisch and Rudolf Peierls, calculated that only a few pounds of the 235 isotope were required. Other scientists in Britain conceived a gaseous diffusion technique, and, put together, these two advances galvanized the government to lobby for the US-based research to be shifted into more professional hands and be more effectively financed.

The Americans listened. In 1942 Brigadier General Leslie R. Groves Jr. took over what became known as the Manhattan Project. Guided by the solar physicist J. Robert Oppenheimer (1904–67), Groves brought together the top nuclear scientists of the day, supported by unprecedented funding and manpower.[29] The project took over some thirty sites across the US and Canada – Oak Ridge, Tennessee; several in central Manhattan; Chalk River, Ontario; Richland, Washington; and its headquarters at Los Alamos ('The Cottonwoods'), a small school on a ranch near Santa Fe, New Mexico. At its peak the project employed more than 130,000 people (most of whom had no idea what they were working on). But it was still far from inevitable that the United States would become the first country to build the bomb. On the eve of the Second World War, Germany was at least as far advanced in nuclear physics, and it had plentiful supplies of uranium.[30]

When a nucleus of uranium-235 absorbs a neutron, it splits into atoms of strontium and xenon, releasing energy plus twenty-five neutrons for every ten atoms. Uranium-238, on the other hand, absorbs neutrons and does not divide, so there is no such reaction. Raw uranium dug from the ground contains less than one per cent uranium-235. A bomb needs to be eighty per cent pure uranium-235; otherwise the uranium-238 blocks the chain reaction. For the scientists on the Manhattan Project, the problem was how to separate the two. The breakthrough came at Oak Ridge: a series of 'racetracks' – really huge belts of silver magnets – were devised to pull gaseous uranium through vacuum chambers, separating the weapons-grade U-235 from its heavier, benign sibling. 'You used to walk on the wooden walkways and feel the magnets pulling at the nails in your boots,' a scientist there recalled. On one occasion, a man carrying a sheet of metal walked too close to the racetrack and found himself stuck to the wall. Everyone yelled for the machines to be shut down, but since it would have taken a day or more to restart everything, the engineer in charge refused. Instead, the unfortunate had to be prised off with two-by-fours.*

---

*The early labs were hardly sophisticated places: stray radiation induced radioactivity in everything from gold teeth to the zippers on trouser flies. A large proportion of those who worked on the Manhattan Project died at suspiciously early ages.

When the Third Reich surrendered on 8 May 1945, the Manhattan Project was months away from a weapon. To hasten victory in the Pacific, Oppenheimer decided to conduct a test, and on 16 July, in the desert north of Alamogordo, New Mexico, released the equivalent of nineteen kilotons of TNT, far mightier than any previous man-made explosion. The news was rushed to Truman, who tried to use it as leverage on Stalin at the Potsdam Conference, to no avail. After listening to the arguments from his scientific and military advisers, and hoping to avoid an invasion calculated to exact a million and a quarter Allied dead – nearly double the British and United States fatalities in the whole war up to that point – Truman ordered that the weapons be used against Japanese cities, and on 6 August a uranium-based bomb, 'Little Boy', was dropped on Hiroshima.* Three days later, 'Fat Man' was loosed on Nagasaki. The pilot on the latter mission recalled: 'This bright light hit us and the top of that mushroom cloud was the most terrifying but also the most beautiful thing you've ever seen in your life. Every color in the rainbow seemed to be coming out of it.' The bombs killed at least 100,000 people outright, another 180,000 dying later of burns, radiation sickness and related cancers. The Atomic Age had begun. Hans Bethe, aghast, devoted the rest of his life to checking the weapon's 'own impulse', as he put it: 'Like others who had worked on the atomic bomb, I was exhilarated by our success – and terrified by the event.' Winston Churchill told the House of Commons that he wondered whether giving such power to man was a sign that God had wearied of His creation. 'I feel we have blood on our hands,' Oppenheimer told Truman, who replied: 'Never mind. It'll all come out in the wash.'

On 23 September 1949 the Soviet Union tested its first nuclear device, also a fission bomb. The amount of energy practically releasable by fission bombs (i.e. atomic bombs – or A-bombs – which produce their destructive force through fission alone) ranges between the equivalent of less than a ton of TNT to around 500,000 tons (500 kilotons). The other, vastly more powerful category of bomb obtains its energy by fusing

---

*In May 1946 *The New Yorker* sent the Japanese-speaking journalist John Hersey to write an extended account of what had befallen. His 30,000-word report took up the whole of the 31 August issue (no cartoons, verse or shopping notes), which sold out within hours. One haunted subscriber had pre-ordered a thousand copies: Albert Einstein.[31]

XX-28 George, a 225-kiloton atomic bomb exploding
on 8 May 1951

light elements (not necessarily hydrogen) into slightly heavier ones – the same process that fires the stars. Duplicate this, and one can make weapons of almost limitless power. Known variously as the hydrogen bomb, H-bomb, thermonuclear bomb or fusion bomb, these work by detonating a fission bomb in a specially manufactured compartment adjacent to a fusion fuel. The gamma rays and X-rays emitted by the explosion compress and heat a capsule of tritium, deuterium or lithium deuteride, initiating a fusion reaction.

In 1952 'a blinding flash of light' from an explosion that the US set off on a small island in the Pacific signalled the detonation of the first hydrogen bomb, and for a split second an energy that had existed only at the centre of the Sun was unleashed by man on Earth. In the intense debate that followed, some scientists, such as the ferociously anti-Communist Edward Teller, who had originally posited such a bomb, argued that nuclear power was a good thing and that research be continued; others, Oppenheimer, Einstein and Bethe among them, felt that their dreams for what physics might accomplish had turned into darkness and blood. What went almost unnoticed was that splitting the atom heralded yet one more demotion for the Sun: its other-worldly energy was no longer unique. As Bohr exclaims to Werner Heisenberg in Michael Frayn's play *Copenhagen*: 'You see what we did? We put man back at the centre of the universe.'

PART THREE
THE SUN ON EARTH

KEY ⎰ //// Rich honey bronze
    ⎱ \\\\ Beetroot
        ⁘ Slug

In the 1950s, the English author Stephen Potter made fun of the 'new' craze for sunbathing in his best-selling book *Lifemanship*, laying out three possibilities of how to tan

13

SUNSPOTS

Some have thought that they were great masses of *scoria*, or cinders, or dross, floating on a sea of molten rocks; others have thought that they were the tops of mountains appearing above a great sea of fire; others still have thought that they were clouds of black smoke, floating over the face of the Sun; but now it is quite generally agreed, among men of science, that they are vast openings through the envelope of the Sun, allowing us to see away down, down, down into an immense abyss . . .

William Urmy, 1874[1]

Castel Gandolfo, the papal summer palace, lies some thirteen miles south-east of Rome, overlooking Lake Albano. One sun-drenched afternoon in late October 2003, I climbed the steep hill to what is as much stronghold as spiritual residence. The estate extends all the way down to the lake – 136 acres embracing a working farm, an array of monuments, a garden designed by Bernini, an imposing baroque fountain and the ruins of a villa of Domitian's, acquired by Clement VIII at the beginning of the seventeenth century. The whole site is spectacular.

I was there because the papal residence also houses a fully operational observatory and an extensive astronomical library. One of Clement's sixteenth-century predecessors, Gregory XIII, commissioned a study of the calendar, an initiative that inspired what became three observatories, two just outside Rome and the third within the Vatican itself. In 1891 Leo XIII moved the Vatican observatory to a spur of the Seven Hills behind St Peter's, where for more than four decades the Church's scientists worked; but eventually the growth of Rome so brightened the night sky that observation became almost impossible, and in 1933 the centre relocated to Gandolfo. Two new telescopes were built and an astrophysical laboratory installed. The library burgeoned: *The Da Vinci Code* avers that more than 25,000 books on astronomy are housed there, although its long-time librarian, Juan Casanovas of the Society of Jesus doubts that number.

Father Casanovas was at the castle entrance to welcome me. An imposing figure, over six feet tall and with completely white hair, he seemed happy to help a fellow enthusiast. I had written asking if I could see some of his prize volumes, look over the rooftop observatory and talk to him about sunspots, on which he is an authority. We walked together up a narrow side staircase. The whole place seemed deserted and full of echoes, the Pope having long since taken his retinue back to Rome.

As we reached the library, Father Casanovas told me of its first editions of Copernicus, Newton, Kepler and Brahe, but what he led me to was an outsize volume bound in weathered brown leather: Galileo's personal register, in which he recorded his first sightings of blemishes on the Sun. I turned the leaves gingerly, taking in his original diagrams, evidence that the great daytime star was not after all immaculate: a revelation sufficiently dangerous in its theological implications that for over two years Galileo told no one. Then came Fabricius, and Harriot and finally Scheiner, and the Pisan's competitive instincts took over.

Father Casanovas moved away, leaving me to my musings. What had it been like for those even before Galileo, such as Aristotle's best pupil Theophrastus, who saw spots around 325 BC? Why hadn't his eyes been burned out of his head? Likely enough, instead of looking at the Sun directly he had examined its reflection, or gazed at it through some translucent mineral, as had the Chinese, observing it through wafers of jade. The sun-watchers of Xi'an and elsewhere evidently had no idea what sunspots were; neither had Galileo, but as he examined these impurities stealing across the solar surface he at least deduced that the Sun

The drawings that Galileo published in 1613 in his *History and Demonstrations Concerning Sunspots and their Properties.* When his thirty-five suns are made into a flip-through book, the motions of the spots are easy to track, as the drawings were made close together in time

rotated, and used the very drawings in front of me to estimate how fast it did so . . .

My thoughts were interrupted by Father Casanovas, who appeared suddenly at my shoulder to indicate that I should follow him, and strode off down one of the long corridors so that I had to hurry to keep up. Several turnings later we arrived at his office: everything in place, files neatly stacked, working papers carefully squared off. A computer was humming, its screen a kaleidoscope of red, orange and yellow, the latest images from SOLO, the Solar and Heliospheric Observatory, the joint project of NASA and the European Space Agency to photograph the Sun. I had never seen pictures so dramatic. 'This month is a high point for sunspots,' Father Casanovas explained. 'Huge storms are coming because of them.' He rummaged through a shelf of books and, with a grunt of satisfaction, handed me a slim volume, in English, on the history of sunspots. The cover proclaimed its author: J. Casanovas of the *Specola Vaticana.*

That evening, back in my hotel room near Rome's great central station, I settled down to read. Bit by bit I came to see that the discovery and understanding of sunspots is one of the most exciting stories in science, and also hugely intricate, stretching over continents and centuries. The two earliest reports of sightings occur in that classic Chinese text, the *I Ching* (*The Book of Changes*, which dates back anywhere from 5000 to 8000 BC), and tell of a *dou* and a *mei* on the Sun (both meaning a darkening or obscuration).[2] Over the ages, Chinese and Korean astronomers noted about 150 spots, likening them to the eggs of hens, swallows, ravens and other birds. Virgil wrote about the Sun being 'chequered with spots' in one of his pastorals; Gregory of Tours, in the late sixth century, described 'blood-red' clouds on the Sun; and Einhard in his *Life of Charlemagne* (c. 807) recounted a 'black-coloured spot [that] was to be seen for seven days at a stretch'. On 8 December 1128 John of Worcester, a monk who contributed to the Anglo-Saxon Chronicle that detailed English life (including astronomical events) from the birth of Christ to the accession of Henry II in 1154, drew the Sun marked by two large dark spots – a drawing that modern astronomers believe to be accurate. He lacked a telescope, so for him to detect not only the spots themselves but also their penumbras (outer shadows) implied that these maculations were large indeed. Even so, not even this discovery excited comment. Centuries before, Aristotle had declared the heavens incorruptible, and the Church in due course concurred: so, in Europe at least, any such sightings were either ignored or ascribed to transits of Mercury or Venus.

The introduction of the telescope would change all this. In his history, Father Casanovas equates the impact of Galileo's description of his sightings, as published in *The Starry Messenger* in 1610, with that of man's first landing on the Moon.[3] However, early telescopes were of poor quality, so most observers continued to believe that sunspots were somehow not actual parts of the Sun, and whatever interest Galileo had stirred soon faded away (Cyrano de Bergerac, a close friend of one of Galileo's pupils, being one of the few who continued to be fascinated by them.)*

The next advance came about by accident. As telescopes increased in number, generations of amateur astronomers strove to discover the planet that was believed to orbit between the Sun and Mercury. Samuel Heinrich Schwabe (1789–1875), a Dessau pharmacist-turned-astronomer, knew his best chance of detecting this hypothetical body was during its passage in front of the Sun; but he also realized that he might confuse such an object with a sunspot. So, from 30 October 1825 on, he meticulously recorded *everything* he saw in the heavens. After nearly two decades, he had still not found his planet, but he had tripped over something far more important.

'From my earlier observations,' he wrote in an article that appeared in 1843, 'it appears that there is a certain periodicity in the appearance of sunspots.'[5] His accompanying table gave powerful evidence of a cycle, with spots appearing in clusters. Once a circuit was completed, the solar disc might be clear for weeks, but the existence of the cycle itself seemed indisputable. A summary of his observations shows, for example, that the low point of one cycle was in 1833, with the fewest clusters and the highest number of days without visible spots; the next low point fell about ten years later. The high points can be seen in 1828, and again in 1837:

---

*Curiously, one possible exception to this lack of interest lay in France. In *Sunspots and the Sun King*, Ellen McClure, a professor of French history, points out that Henri IV was assassinated in the same year that Galileo made his first observations, and suggests that both events represented a challenge to order and hierarchy. Henri's murder renewed questions about his legitimacy, Galileo's discovery brought the Sun into the realm of terrestrial corruption. It became urgent to restore faith in an order grounded in asserted permanence and transcendence. Efforts to achieve this, she writes, 'resulted in the creation of the Sun King, whose authority and monarchical identity were constructed at least in part to counter the destructive implications of the sunspots.'[4] Two hundred years on, during the Congress of Vienna, Napoleon was lampooned in the British press in *Napoleon and the Spots in the Sun*, which suggested that the defeated Boney had been packed off to the Sun because he had nowhere else to go, and further that he, being himself but a cluster of sunspots, blocked sunshine from the Earth, causing that year's dismal weather.

Year	No. of clusters	Days when no spots were observed	Observation days
1826	118	22	277
1827	161	2	273
1828	225	0	282
1829	199	0	244
1830	190	1	217
1831	149	3	239
1832	84	49	270
1833	33	139	267
1834	51	120	273
1835	173	18	244
1836	272	0	200
1837	333	0	168
1838	282	0	202
1839	162	0	205
1840	152	3	263
1841	102	15	283
1842	68	64	307
1843	34	149	324

At first, Schwabe's analysis attracted little attention, but when the Irish astronomer and explorer Edward Sabine (1788–1883) saw the article he realized that the cycle Schwabe had uncovered correlated with the fluctuations he had observed in the Earth's magnetic field. Then a Swiss observer, Rudolf Wolf (1816–93), came up with a way to compute the average number of spots over even longer periods, refining Schwabe's estimate to 11.1 years.

Over the next twenty years Wolf put together a census reaching as far back as 1745. As he continued to reconstruct yet earlier activity, he realized how few spots had been observed between 1645 and 1715. This hiatus coincided with the coldest part of the so-called Little Ice Age in Europe and North America alike (1645–1714), when tidal waters like the Thames and even the canals of Venice iced over. But because of the lack of interest in sunspots, it would be over two centuries before anyone made a connection.

However, it would take the scientific community a while to accept both Schwabe's correlations and Wolf's 'measure of solar acne'.[6] Enter that energetic Prussian, Baron Alexander von Humboldt (1769–1859), who has been described as 'a combination of the measured studiousness of astronomer Carl Sagan and the down-and-dirty enthusiasm of *Titanic* discoverer Jim Ballard'.[7] Humboldt was intrigued by Schwabe's researches, having in his youth spent five years travelling throughout South and Central America (European newspapers reported him dead on three separate occasions), where besides studying its plants, animals, rivers and volcanoes he had taken magnetic measurements wherever he went, and found that field strength varied widely. Sunspots seemed the likely cause. His enthusiasm would make the study of the relationship between sunspots and magnetic fields an acceptable scientific activity (so much so that now over 200 terrestrial magnetic observatories are dispersed over the globe), and in 1851 he included Schwabe's updated table in his encyclopedic five-volume compilation of the natural sciences, *Kosmos*. His endorsement made scientists all over the world take Schwabe seriously.[8]

Schwabe also influenced the Englishman Richard Carrington (1826–75), a gentleman brewer and experienced astronomer who distilled over a decade of research into *Observations of the Spots of the Sun* (1863). On

Christoph Scheiner (1573–1650) observing sunspots with the help of an assistant

1 September 1859 Carrington had been tracking a group of spots from his observatory when all at once, as he wrote, 'two patches of intensely bright and white light broke out'. As he stared, incredulous, the two spots intensified and became kidney-shaped. He dashed outside hoping to find another witness, but the flare was brief, just five minutes, the result – as we now know – of magnetic currents criss-crossing and short-circuiting one another, at 420,000 miles an hour. Less than seventeen hours later (the light and X-rays from a flare take only eight minutes to reach Earth, but the heavier particles from each eruption need between eighteen and forty-eight hours), a great magnetic storm erupted over the globe, its auroras staining the skies as far south as Cuba. Was there a causal relationship between flare and storm? Carrington thought so, but cautioned, 'One swallow does not make a summer', and never took his hunch further.

Yet 'Carrington's flare,' writes Stuart Clark, 'was a tipping point in astronomy. The sudden demonstration of the Sun's ability to disrupt life on Earth catapulted astronomers into a headlong race to understand the nature of the Sun.'[9] By the 1890s the prevailing wisdom was that sunspots were made up of strong cyclones, which prompted the great American astronomer George Ellery Hale (1868–1938) to examine them for magnetic activity. Employing a spectroheliograph (a combination of spectroscope and something similar to a film camera), he took photographs of the Sun that showed large blotches of hydrogen being swept into the centre of a sunspot as if caught in a whirlpool. He observed two large flares, each of which was followed by a huge magnetic storm on Earth, 19.5 and 30 hours later respectively. By 1908 he had demonstrated that the spots were indeed gigantic cyclones in the solar atmosphere, similar in formation to the hurricanes and whirlwinds that build up in the West Indies and ravage the Gulf Coast.* Sunspot activity and the Earth's climate seemed not only connected; they seemed to be so *magnetically*.

Hale's conclusion that the spots were intensely magnetic was consistent with observations of the streaks visible above sunspot areas during total eclipses, which resembled the field lines of a bar magnet. Looping out from the Sun's surface, these streaks speckle it with positive and

*The first photograph of the Sun, a daguerreotype, was taken in 1845 by the French physicists Armand-Hippolyte-Louis Fizeau (1819–96) and Léon Foucault (1819–68) – the latter best known for the Foucault pendulum, which demonstrates the Earth's rotation.

negative lines of magnetic force, and channel the storms into eruptions that extend thousands of miles beyond the surface before being dragged back.[10] Hale's reasoning let observers gauge sunspot radiation, which originates in layers deep within the Sun: it also suggested what the spots were and how they affected the climate of Earth.

Sunspots could be said to be a bit like snowflakes, in that each one is unique but of similar structure. All are roughly circular, averaging from 3,000 to 30,000 kilometres in diameter – although Hale saw one of 130,000 kilometres, ten times Earth's diameter. At each spot's roiling centre lies what is called an umbra ('shadow'), which appears dark only because it contrasts with the more luminous surface around it, but if isolated would be as bright as a full Moon set against a black sky. An umbra typically has a temperature of about 4,300 Kelvin (a centigrade measure of heat, where –273°C is absolute zero), roughly 2,100 K cooler than the surrounding photosphere and embedded about 450 miles deeper. The umbra, comprising nearly a fifth of the area of an average spot, is surrounded by a fibrous grey filamentary *penumbra*, which resembles the petals of a flower. Taking up the remaining eighty per cent, it is about three-quarters as hot as the photosphere (which constitutes the surface of the spot). In summary, a sunspot appears in three different forms, depending on depth: the photosphere, the penumbra and, at its deepest, the umbra.[11]

It is still a mystery why a spot's core should be less hot than either its penumbra or the solar surface: all we know for sure is that within each of these great funnel-shaped vortices in the outer layers of the Sun a giant cooling process is at work, and that the most powerful focus of its magnetic field lies at its darkest, or coldest, point, the umbra.[12]

Some years before Hale's researches, the superintendent of the Royal Observatory at Greenwich, Edward Walter Maunder (1851–1928), had been making some important discoveries of his own. Keen on sunspots since the age of fourteen, by the time he was twenty-six he was measuring how large they could get. Over the next thirty years he embarked on a massive photographic project, accumulating several thousand pictures of 5,000 groups of spots, coming to think romantically of this task as capturing the Sun's portrait, and once writing that taking the spectrum of a sunspot was like seeing into its soul.[13]

He found that the magnetic flux increased rapidly during a spot's early life, before sinking into a slow decline over a matter of days. His photographs showed that when the latitude of sunspots was plotted against a series of eleven-year cycles, their positions formed a pattern, that (with a little imagination) resembled three butterflies, travelling west: the spot furthest ahead in its migration was designated the 'leader' and had one magnetic pole, while its followers showed an opposite polarity. Just as Galileo had demonstrated, all spots crossed the solar disc in straight lines, and hunted in pairs. They would start out close together, then move apart as the group travelled, sometimes at distances that were up to twenty degrees of the Sun's circumference, while always remaining parallel to the equator. At the end of every cycle the polarity reversed, so that the northern hemisphere 'leaders' took on negative polarity, and the southern positive. A cycle would have to run through two successive eleven-year sequences – twenty-two years in all, give or take a few months – to complete itself. Surprisingly, at the height of the cycle, when sunspots were most numerous on the solar disc, the Sun shone with a greater intensity than it did when its spots were fewer. A cycle is part of an overall pattern of solar activity, involving not just the movement of spots but also of prominences, flares, showers of solar wind particles, cosmic rays and energetic protons, a minute but potent proportion of which plunge on to Earth, some within fifteen minutes of leaving the solar surface.

To put it another way: as the Sun throws its tantrums, the magnetic fields erupting from its interior discharge high-energy cosmic rays. Maunder speculated that sometimes these rays are deflected past Earth, which cools as a result so that, for example, in 1536 Henry VIII and his court were able to sleigh over the Thames all the way from central London to Greenwich, while during the terrible winter of 1709 wine froze in the glasses before the Sun King at his banqueting table.

☼

The next step forward came from an unlikely place. Shortly after the First World War, Andrew Douglass (1867–1962), an academic at the University of Arizona, pioneered a new science called dendrochronology – the study of tree rings, whose successive widths (broad in years that were good for growing, narrow in bad years) provide a record of climatic variation over a tree's lifetime. He assumed that trees, the longest-living organisms on Earth, were the only objects of the plant kingdom that could offer reliable records, since everything else died into the soil; and noted that years of rapid growth were followed by periods of retarded development; then rapid growth resumed. On average, ten to twelve rings separated any two intervals of development.

One morning in 1922, out of the blue, he received a letter from Maunder outlining his theory about the effects of the absence of sunspots between 1645 and 1715 and suggesting that 'you should see this in the tree rings'.[14] His curiosity piqued, Douglass began studying beams from old buildings as well as very old trees, for instance Arizona pines and Californian redwoods; and sure enough their rings revealed a pattern of slow growth for the same period that sunspots vanished and Earth suffered its long cold snap. This was still not enough for general acceptance,

Section of a Scotch pine from a forest in Prussia, planted in about 1820 and cut in 1912. Arrows, placed by Douglass, mark the years of maximum sunspot activity, showing an apparent correlation with maximum growth

*In 1937 the memorably named scientist H. True Stetson, of MIT and Harvard, came up with new evidence – rabbit skins. The pelt records of the Hudson Bay Company showed remarkable variations in the number of fox, lynx and rabbit taken over periods of ten to eleven years. Nearly every peak corresponded 'quite closely' with the scarcity of spots. 'If sunspots

however, and Maunder died in 1928, his theories unrecognized.*

Maunder had not been alone in his theories about the effects of sunspots – or in having to endure the scepticism with which such theories were met. From the time spots were first observed, some people have speculated about how they affect life on Earth, while others have mocked them: as early as December 1795 William Herschel delivered the first of what was intended to be a series of lectures on the Sun and its effects on Earth to the elite members of the Royal Society, in which he said that he had uncovered five periods of low sunspot activity during which the price of wheat had been high, a fact he connected to unusually long spells of drought. Most of his audience laughed at him (so much so that he felt forced to cancel the remaining talks) – but then he was also airing the theory that the Sun was cool at its centre and inhabited, a belief that in Herschel's day was enough to have one declared a lunatic (or perhaps a solatic).*

To ascribe happenings on Earth to spots on the Sun seemed destined to remain the stuff of fiction, and, sure enough, in 1892 Mark Twain published a novel, *The American Claimant*, in which the idea of solar blemishes controlling our climate is carried to its logical extreme – sunspots as big business. In an explanatory note, Twain's tongue is firmly in cheek:

> This is an attempt to pull a book through without weather. It being the first attempt of the kind in fictitious literature, it may prove a failure, but it seemed worth the while of some daredevil person to try it, and the author was in just the mood . . .[17]

By the novel's end, the scheming Colonel Mulberry Sellers comes

---

have anything to do with rabbit population,' Dr Stetson continued, 'and sunspot years have been favorable for tree growth, one might wonder why rabbits appear most numerous near sunspot minima and most scarce near sunspot maxima. Perhaps the trappers, stimulated by sunspots, have been more energetic in depopulating the rabbit world during years of sunspot maxima, or perchance other animals who are natural foes of the little four-footed creatures thrive best at sunspot maxima.'[15] This argument too failed to sway the doubters.
*In 1787 a certain Dr Elliot had fired a pistol at a couple out walking and was only acquitted of attempted murder because no bullets were found at the scene. His counsel had pleaded insanity on the grounds that Elliot held the belief that the Sun's light 'proceeds from a dense and universal aurora, which may afford ample light to the inhabitants of the surface [of the Sun] . . . It may easily be conceived to be far the most blissful habitation of the whole system.'[16]

up with his grandest plan yet for making a fortune: he will reorganize the Earth's climates, furnishing them to order and taking old climates in part payment. How? By utilizing 'the spots on the Sun – get control of them, you understand, and apply the stupendous energies which they wield to beneficent purposes in the reorganizing of our climates'.[18]*

Twain, of course, was making fun of crazy business schemes as much as he was satirizing the wayward theories of scientists. But Maunder's conviction that a connection *did* exist between solar activity and fluctuations in the Earth's magnetic field would one day find widespread acceptance. It would take a while, however, because other scientists continued to be either uninterested or sceptical. After the Second World War, physicists intensified their monitoring of solar events. Yet for all their prodding and probing, evidence of the effects of sunspots remained sporadic: as late as the 1960s a young researcher entertaining any correlation between solar phenomena and patterns in the terrestrial climate branded himself a crank. Then came Eddy.

John A. Eddy (1931–2009) was an astronomer based at the University of Colorado, which has a High Altitude Observatory where this versatile scientist was able to study Jupiter's atmosphere, the solar corona, the history of solar physics, even American Indian astronomy. Intrigued by Maunder's theories, which he later likened to 'deciphering the Dead Sea scrolls of solar physics', he couldn't understand why they had been written off. In the early 1970s he went to the Laboratory of Tree-Ring Research in Tucson, but he was unable to replicate Douglass's match between tree rings and climatic variations. Undaunted, he next tracked the history of auroras – which, after all, owe their life to sunspots – and established how few there were during the era that he titled the 'Maunder Minimum'. When the Sun was magnetically active, he discovered, it grew many spots, whose increased

---

*All this recalls an even earlier satire, Samuel Johnson's 1759 philosophical novella, *Rasselas*. In the chapter 'The Astronomer Discovers the Cause of His Unease', the savant in question tells Imlac, the story's narrator, 'I have possessed for five years the regulation of weather, and the distribution of the seasons . . . The Sun has listened to my dictates, and passed from tropic to tropic by my direction; the clouds, at my call, have poured their waters, and the Nile has overflowed at my command.'[19] Imlac quickly takes his leave.

aggregate magnetism reduced the Earth's exposure to radiation. What does not get through collides with other particles in the atmosphere, amalgamating into carbon-14 (the element's main radioisotope), which shows up in tree rings. And, said Eddy, coming to a triumphant point, the rings displayed a prolonged increase in carbon-14 between 1650 and 1715.*

Judging he might have found the final piece of the puzzle, Eddy now turned to the ideas of the Serbian engineer Milutin Milanković (1879–1958), who believed that ice ages on Earth were caused by small variations in the amount of sunlight, the result of gradual cyclic changes in the shape of the Earth's orbit.[21] He argued that there are three types of such orbital variation (sometimes referred to as 'stretch, wobble and roll'), in cycles operating over periods of approximately 100,000, 22,000 and 40,000 years, and that these changes affect the amount and incidence of solar radiation reaching the Earth.[22]

The first cycle, he argued, is dictated by the shape of the Earth's path around the Sun, and the degree to which it departs from a circle (its 'eccentricity') to become more elliptical. As the circle 'stretches', the distance between Earth and star alters, so affecting the amount of radiation we receive.[23] The second cycle arises from the wobble in the Earth's axis of rotation, which affects when the seasons fall, slowly swinging the northern hemisphere – and reciprocally the southern – closer to, then further from, the Sun. The northern hemisphere contains much more land than the southern, and land reacts more rapidly to temperature changes than do oceans. The resulting swings in heating modify weather patterns. The third cycle is created by a small oscillation in the tilt of the Earth's axis of rotation. Over the course of about 40,000 years, this tilt varies between

---

*Since Eddy, scientists have looked again at the annual layers of silt brought down by rivers and found in lake sediment cores, and also studied cave minerals and deposits, pollen counts, geological boreholes, coral and mountain glacier deposits – all of which bear witness to the chill presence of the Maunder Minimum. Recent research has also helped explain one of the great puzzles of the Little Ice Age – that it appears to have been a peculiarly European phenomenon. Peaks in ultraviolet radiation emitted by the Sun have been shown to boost ozone formation in the stratosphere – the layer of the atmosphere that lies 20–50 kilometres above our heads – which in turn absorbs more ultraviolet and heats up. Stratospheric winds affect the weather we experience, and Europe is particularly susceptible because it sits under the northern jet stream at a longitude that makes it vulnerable.[20]

21.5° and 24.5°, and when it is at its smallest (as at present, remaining so for about 9,800 years from now) differences between summer and winter are reduced.*

By employing Milanković's theories about cycles, Eddy was able to give more credibility to Maunder's claims. He buttressed his case with evidence of temperature changes stretching back nearly 300,000 years, using data from mud and clay dredged up from the seabed: such temperatures rose and fell over the centuries with the amounts of sunlight received – and fitted in well with Milanković's calculations.

In a landmark paper published in *Science* in 1976, Eddy concluded that the Earth has undergone a total of eighteen periods of minimal sunspot activity in the last 8,000 years – one of which was the Little Ice Age. Maunder was finally vindicated, and his conclusions about what – thanks to Eddy – became known as the Maunder Minimum would be acclaimed as 'the most significant event in the history of solar observation'.

☼

Within our solar system, only Venus, Earth and Mars have solar-influenced atmospheres. Venus is a boiling mess, Mars a frozen desert, while the Earth, though subject to far less extreme temperatures, is under a constant flux, its climate overwhelmingly driven by the Sun. But how much of what happens to us derives specifically from sunspots, which are, after all, just the visible manifestations of magnetically active regions on the Sun?

Generally, spots are blamed (rarely thanked) for many events, from the drop in cast-iron production to the incidence of suicide in northern climates; time taken to complete sea voyages; shipwrecks; disappearing radio programmes (especially in the short-wave, high-frequency bands); carrier or racing pigeons losing their way; enormous increases in car production; myocardial infarctions; convulsive seizures and hallucinations;

---

*The traditional objection to Milanković's theory about ice ages is that, if their timing were set by variations in the sunlight falling on a given hemisphere, why didn't the southern hemisphere get warmer as the northern hemisphere cooled, and vice versa? The answer is that changes in atmospheric CO_2 and methane link the two hemispheres, warming or cooling the planet as a whole; thus global warming and the natural emission of greenhouse gases have reinforced one another.[24]

even the outbreak of epidemics, wars and revolutions – it being argued that the French Revolution began in 1789 rather than four or five years later because the winter of 1788 was extremely cold.*

Most physicists take it as given that sunspots can disrupt compasses and various electrical transmissions, yet it took a while for word to reach beyond the scientific community. In 1953, for instance, during the proceedings of the House Un-American Activities Committee, Roy Cohn was interrogating Raymond Kaplan, a supposed Communist sympathizer and chief radio engineer for the Voice of America. The VOA (as it was known) had been set up in February 1942 to promote favourable views of the US abroad, and Cohn had tagged it as one of seventy-eight suspect organizations intent on subverting American interests. He pointed out that some programmes were not being broadcast as widely as they should have been. Disloyal people were responsible: could Kaplan name names?

COHN: There are VOA towers where radio messages could not get out to countries that we wanted to reach –
KAPLAN: The transmission towers couldn't reach certain countries –
COHN (raising voice): Disloyal people working for –
KAPLAN (interrupting): It's more complicated than that. For example, sunspots: they affect the transmission, it's not entirely clear how –
COHN (interrupting): Sunspots? [General laughter] (Softly, but derisively) Sunspots . . . Or perhaps disloyal Americans selling out our country.

Within three weeks of that hearing, Kaplan threw himself in front of an oncoming truck, dying at the age of forty-two.[26]

Cohn's dismissal of sunspots was echoed by an unlikely source – the James Bond books, the first of which appeared in April 1953. In *Moonraker* (1955) Ian Fleming introduces a 'hand-picked team' of MI5 boffins 'whose only interest was the world of microwaves [and] sunspots'.

---

*Both the Russian scientist Valerii Orloff and the American historian William James Sidis have connected revolutions and sunspots.[25] The American Revolution, the French Revolution, the Paris Commune and both Russian Revolutions of 1905 and 1917 all fell fairly close to the years of maximum solar activity.

However, his thinking would evolve, and in both *Goldfinger* (1959) and *Thunderball* (1961) he depicts sunspots as seriously interfering with radio messages.*

There is now no doubt that sunspots can violently disrupt events on Earth. Following the invention of electric telegraphy and the telephone, telegraph and telephone lines were strung across the world. During solar storms, some operators were almost electrocuted by sunspot-induced surges. 'At one place the operator received seven shocks,' reads one account. 'At another place the telegraphic apparatus was set on fire; and in the city of Boston a flame of fire followed the pen of a telegraphic instrument.'[28] On 12 March 1989 electric currents streaming up to sixty miles above the ground hit Quebec's power grid and in less than sixty seconds incapacitated half its generating system: the shutdown lasted more than nine hours, left 7 million people without electricity and cost the province between 3 and 6 billion dollars. Other disruptions, not quite on that scale, took place all across North America.[29] As the world becomes more reliant on electronics, satellites and similar technologies, we can expect even greater dislocations, especially since the erratic nature of sunspot flares can leave scientists as little as thirty minutes to give warning. And the immense magnetic storm of 1989 was only a *third* as intense as that which Richard Carrington had witnessed 130 years before, when some 20 billion protons sandblasted every square inch of our planet's atmosphere. The next solar maximum is going to be the most active for fifty, or possibly the last hundred years, and has been forecast to start some time between 2010 and 2012.

It was time to return to Eddy and his tree-ring evidence. In January 2005 I visited Mike Baillie, Professor Emeritus in the School of Archaeology and Palaeoecology at Queen's University, Belfast, who specializes in dendrochronology. Despite the fact that he has constructed a table of tree rings going back 7,000 years, he turned out to display a bracing scepticism (and tart humour) towards his chosen field. 'We really don't know what

---

*Long-distance radio signal reception can be seriously compromised by varying sunspot activity.[27] During the preparations for D-Day, precautions were taken to ensure that the Allied landings did not coincide with a shortwave blackout due to an errant solar flare. Daily 'top-secret' reports on the Sun's state would be filed from an observatory high above the Fremont Pass in Colorado, which strategists thousands of miles away studied carefully before planning the attack.

the relationship is between sunspots and tree rings,' he admitted as we pored over various papers in his overflowing office, about the same size as Father Casanovas's but on the opposite end of the tidiness scale. 'Eddy supposes that there were no "global events" before tree chronologies began; then one discovers that there were. And one must be careful: a 22.2-year sunspot cycle doesn't just translate into two 11-year cycles.'*

At this point he got up from his chair – dislodging two small bits of stone and a stack of student papers – to hunt down a thin magazine: *The Tree-Ring Bulletin*, containing a long article by two researchers from the University of Arizona.[32] The authors argue that while it is conceivable that general trends in sunspot activity might be reflected in tree-ring growth, they couldn't find any convincing evidence of a consistent relationship. 'This came out in 1972,' said Baillie, standing over me as I read. 'There's been nothing better since.' Eddy produced his groundbreaking essay a full four years later – without ever mentioning this awkward argument. My host turned back to his chair. 'Of course sunspots *do* probably have an effect on climate, but so do other factors . . .' For a moment he seemed lost in his own thoughts. 'We don't know what effect volcanoes have. We don't have good returns for past tsunamis, or how often the Earth has been struck from space; but people think I'm a little crazy worrying about all this.'[33]

He didn't seem crazy at all. Baillie was talking several years before the eruption of Eyjafjallajökull in April 2010, but in April 1815 Indonesia's Mount Tambora blew up with even greater force – so much so that a

---

*By the end of the twentieth century, sunspots became among economists a shorthand for any extrinsic uncertainty that sets a crisis in motion – an accidental random shock that, by altering expectations, leads to an economic downturn. Roger Guesnerie is an authority on sunspots as a branch of chaos theory: 'The idea of sunspot equilibrium,' he writes, 'has been one of the most important ones . . . in economic theory in general over the last twenty years. It has made us reassess our understanding of economic fluctuations, of dynamic economies, and of the rational expectations hypothesis.' Thus, 'sunspot theory' began as an insult and has developed into a useful instrument of descriptive economics.[30]

Meanwhile, E. A. Wrigley of Cambridge University argued in a seminal lecture in November 2002 that England's Industrial Revolution was mainly due to the Sun. The Dutch Republic enjoyed its golden age in the sixteenth and seventeenth centuries because – as argued by Adam Smith in *The Wealth of Nations* – it had the ideal conditions for economic growth. For Wrigley, England's revolution was not so much economic as physical, chemical and biological – reliant on the photosynthetic products of millions of years of plant growth being 'frozen' in geological strata. England's boom from the sixteenth century on tapped into the energy capital locked up in coal deposits 300 million years before.[31]

third of it vanished in rocks and dust, possibly the biggest eruption in recorded history: 10,000 people were killed outright, with ash and sulphuric acid blasting 27 miles into the atmosphere. Global temperatures dropped, food prices soared, there were riots across Europe, famine and epidemics.[34] Byron was so depressed that he wrote a poem about the Sun itself being destroyed: 'I had a dream. Which was not all dream/ The bright sun was extinguish'd, and the stars/ Did wander darkling in the eternal space/ Rayless, and pathless, and the icy earth/ Swung blind and blackening in the moonless air . . .'

For all that, it does not help when other scientists propose so many cycles, and correlate them with so many different events: besides the 11-year one, the Sun is supposed to exhibit 27- and 154-day cycles, the 22-year cycle of magnetic reversal found by Hale, a Gleissberg cycle of about 80 years and a 200-year cycle called the Seuss. Clearly, there are still questions to be answered. In June 2010, the respected science journalist Stuart Clark, writing in both *New Scientist* and the *Washington Post*, argued that the Sun's magnetic activity was diminishing and that the Sun may even be shrinking. 'There is evidence now available', he reported, that 'the Sun is inexorably losing its ability to produce sunspots. By 2015, they could be gone altogether, plunging us into a new Maunder minimum – and perhaps a new Little Ice Age.'[35]

Meanwhile new branches of research are growing up all the time. Data from SOHO have recently shown that gas flow affecting magnetic fields near the Sun's poles may have been responsible for the lack of sunspots, flares and other storms from 2008 through the first half of 2009, which have extended the usual lull at the end of the 11-year solar cycle for an extra 15 months, with sunspots virtually disappearing. But as of mid-April 2010, the Sun appears to be reawakening, and a huge coronal mass ejection shot out from its surface at about 1.1 million m.p.h., smashing into our upper atmosphere and setting off dazzling auroras around the Arctic and Antarctic – the most spectacular solar storm to hit Earth for three years.[36] We are back almost where we began: the Sun clearly creates climate; but just how much does it *vary* it? A vital question, tied as it is to the great debate on global warming.

14

THE QUALITIES OF LIGHT

For the rest of my life I will reflect on what light is.

Albert Einstein, 1917[1]

We say 'light', but we are thinking only of the Sun.

Graham Greene, *The Power and the Glory*[2]

Not surprisingly, light suffuses much of my account of the Sun; this chapter begins with mankind's attempts to track its speed and examines how light releases colour; then shifts to cultural history, following it from where it is brightest through how it affects those on playing field and battlefield, through its diminutions, until it finally gives way to night.

What *is* light? Pythagoras's contemporary, the philosopher and poet Empedocles, brilliantly intuited that it was a streaming substance, adding that, because it travels so fast, we are not conscious of its motion. But he also held, along with Plato, that there was a 'fire within the eye', as if we see through some kind of lantern. The more traditional belief was that an object emitted particles of light that bombarded the eye of the beholder, and that light was transmitted instantaneously. Not until 1638 was the idea advanced that light might have a measurable speed – the year that Galileo, in his polemic *Dialogue Concerning Two New Sciences*, put the notion into the mouth of the credulous Simplicio:

Everyday experience shows that the propagation of light is instantaneous; for when we see a piece of artillery fired at great distance, the flash reaches our eyes without lapse of time; but the sound reaches the ear only after a noticeable interval.[3]

Galileo was certain that light had a finite velocity, but his experiments were crude (he simply placed two lanterns on hilltops less than a mile apart), and he never came close to appreciating the tremendous speeds involved. The first serious estimate – when the Danish astronomer Ole Rømer (1644–1710), working from the Paris Observatory, calculated it to be 138,000 miles per second – was overshadowed by the debate between Huygens and Newton over whether light was made up of waves or particles. In 1728 the English scientist James Bradley, working on the premise that light travels at 10,000 times the Earth's speed in orbit, came up with the remarkably accurate figure of 185,000 mps, which means that sunlight reaches us a little over seven and a half minutes after leaving the Sun.

Light streaming into the concourse of New York's Grand Central Station. But what *was* light?

Nearly a century later, further experiment by the French rivals Hippolyte Fizeau and Léon Foucault, one employing revolving mirrors and the other rotating toothed wheels, yielded results of 190,000 mps.* Then in 1859 Gustav Kirchhoff (1824–87), a physicist from East Prussia, demonstrated that one could derive detailed information about celestial objects by studying the light they emitted: a transforming discovery that would lay the foundation for the new discipline of astrophysics. This was joined to a further breakthrough just a few years later – the equations that Einstein was to call 'the most profound and the most fruitful that physics has experienced since the time of Newton': the great Scottish scientist James Clerk Maxwell (1831–79) discovered that a particle or photon of light was actually a self-sustaining bundle of constantly oscillating electric and magnetic fields. As the CCNY (City College of New York) physicist Michio Kaku observes: 'Maxwell suddenly realized that everything from the brilliance of the sunrise, the blaze of the setting sun, the dazzling colors of the rainbow, to the firmament of the stars in the heavens could be described by the waves he was scribbling on a sheet of paper.'[5] Maxwell was even able to calculate the speed of light from the basic parameters of electricity and magnetism: almost a lifetime later, quantum physics would show that while light moves through space like a wave, it behaves like a particle when it encounters matter.

In 1887 Albert Michelson, a Polish-Jewish immigrant to the US, redesigned Foucault's experiment and came up with a speed for light of 186,355 mps – twenty times the Frenchmen's accuracy, and enough to make Michelson famous.[6] In the 1880s the general opinion was that light waves had to be transmitted through a 'luminiferous' – light-bearing – aether (as it used to be spelled), assumed to be invisible and all-pervasive.

---

*Fizeau's rotating wheel had regular small gaps in it. The light, passing through from a source on the further side of the wheel, was reflected by a mirror and, if the wheel were rotating fast enough, was returned through a different gap. Its speed could be calculated using the distance from mirror to wheel, the speed of rotation and the spacing between the gaps. Foucault altered Fizeau's method to include a rotating mirror rather than a spinning-wheel. A light source was shone on the mirror where it was reflected on to a distant second mirror, concave and fixed; then bounced off the fixed mirror back to the rotating one. Were the mirror rotating at high speed, the returning light hit it at a slightly different place, shifting the beam from its original path. The speed of light could then be measured by taking into account the speed of mirror rotation, the amount of shift and the distance between the two mirrors.[4]

Working with the chemist Edward W. Morley (1838–1923), Michelson argued that light required no such medium, and that its speed is constant whatever its situation: thus a beam from the headlamp of a speeding train will not move faster than that from a stationary beacon. In this he overreached, and Einstein would prove that its speed *does* vary: in deep space in a hard vacuum, far removed from solar and planetary masses, it will travel more slowly than when near the Earth or similar large masses: though not just size matters – through diamonds, it travels less than half as fast.

Until recently, the slowest speed recorded was just over 38 mph – slower than a racing cyclist – through sodium at –272°C. In 2000 a team at Harvard brought light to a standstill by shining it into a 'bec' (which stands for 'Bose-Einstein condensate', a medium for cooling atoms down to extremely low temperatures).[7] So light not only travels at different speeds; it can, amazingly, be halted altogether.

In 1802, two years after William Herschel discovered infrared radiation, the English physicist William Wollaston (1766–1828) found that if he shone light through a prism, in front of which he had placed a thin slit, the resulting spectrum displayed a series of parallel dark lines, like the cracks between piano keys; not until twelve years later did Josef Fraunhofer show that different wavelengths produce different colours, assigning a quantity to each colour, so making it possible to match the range of wavelengths to the visible 'spectrum' (Newton's word). The colour resulting from the shortest wavelengths was violet, red from the longest: anything beyond these two we cannot see.*

Fraunhofer next set about passing a sunbeam through a prism. Unlike his predecessors, he saw something unexpected – 'an almost countless number of strong or weak vertical lines which are darker than the rest

---

*Our eyes, each with some 125 million sensors, are good detectors but not great ones. The retina or screen at the back of the eye on which the lens casts its image contains two different elements. Rods register form; cones (so named because of the shape of the receptive part of the cells) colour. Animals that are active mostly during daylight use a mixture of rods and cones, but the retinas of nocturnal creatures are packed almost entirely with rods. Owls, for instance, perceive the world largely in monochrome, but are able to do so in light so dim that other creatures cannot see at all.[8] Most vertebrates have at least two classes of cone; many birds, turtles and fish four or five. Humans get their sense of colour from three different types of receptor (or cone) in the eye's retina. As they are stimulated in different proportions, our visual system constructs the colours we see.

of the colored image'. Surprised by this discovery – some lines appeared almost a perfect black – he knew this was no optical illusion; rather, these were 'gaps' in the solar spectrum, registering the absence of certain chemical elements in the Sun itself. As no two elements share the same lines, he could examine each set of colours and identify the element that caused it.

In 1854 an American, David Alter, ratcheted the argument up a notch by suggesting that each element had its own pattern of coloured lines – a unique signature. The science writer Stuart Clark explains the significance of this finding: 'If astronomers could determine which lines were produced by which vapours, they would gain a power beyond belief: the ability to deduce the chemical composition of celestial objects.'[9] A few years on from Alter, Gustav Kirchhoff and Robert Bunsen (for ever celebrated for the Bunsen burner – not his invention but named after him)* combined prism and telescope into something they called a spectroscope which one evening, on a whim, they turned on a fire raging in the nearby city of Mannheim and detected the presence of barium and strontium. If they could analyse the make-up of a local conflagration, why not the surface of the Sun? And so, several experiments later, they did. They argued that Fraunhofer's dark lines derived from the absorption of wavelengths in the Sun's atmosphere, and that the Sun's surface consisted of a hot radiant liquid, the studying of which would reveal all its parts.[10] They had done what Comte believed impossible: investigated the chemical composition of an object without actually having chipped off a piece for analysis.[11]

At the same time that Kirchhoff and Bunsen were building on Fraunhofer's ideas, another major figure of nineteenth-century physics, John Tyndall (1820–93), discovered that when a beam passes through a clear fluid holding small particles in suspension, the shorter blue

---

*Bunsen was one of history's great experimenters. While still a young man, he was working on a sample of poisonous cocodyl cyanide when it exploded, shattering the protective mask he was wearing, destroying the sight in his right eye and almost destroying him. Undeterred, by 1861 he had identified sodium, calcium, magnesium, iron, chromium, nickel, barium, copper and zinc in the Sun. He claimed that he never found time to marry, although a better reason may be that the chemicals he worked with made him stink. The wife of one of his colleagues said of him: 'First I would like to wash Bunsen and then I would like to kiss him, because he is such a charming man.'

wavelengths scatter more than the longer red ones: a clear cloudless day-time sky shows as blue because particles in the air scatter blue light more than red.*

☼

The Sun is reckoned to be brighter than eighty-five per cent of the stars in the Milky Way (most of which are ageing red dwarves), but the brightest sustained light on the planet is not the Sun but the 'Sky Beam' at the Luxor Resort and Casino in Las Vegas: aimed straight into space, it derives from thirty-nine xenon lamps, each of seven thousand watts, every one about the size of a washing machine. The resort's technical manager has explained that every night, before the Beam is switched on, strobe lights flash for thirty seconds: 'We don't want to surprise any pilots.'[15]† On the other hand, in 1911 Rupert Brooke was staying with Virginia and Leonard Woolf when, mid-composition, he asked, 'Virginia, what is the brightest thing you can think of?' 'A leaf with the light on it,' she replied, and that was the phrase that completed his poem.[18] But then Mrs Woolf never visited Vegas.

*It may be no coincidence that our vision is adjusted to see the sky as a pure hue: we have evolved to fit in with our environment, and our ability to separate natural colours clearly is likely a survival advantage. Yet 'blue' went unalluded to until quite recent times: in the many hundred allusions to the sky in the Rig-Veda, the Greek epics, even the Bible, there is no mention of the colour.[12] In a 1968 study, Elizabeth Wood writes: 'Physicists have found that light-scattering by the air molecules themselves accounts for the hue and density of the blue of the sky, whereas scattering by particles larger than molecules detracts from the blue color and gives the sky a milky appearance.'[13] However, not all the blue is caused by scattering: light under the sea's surface is blue because water absorbs its longer wavelengths through depths up to about twenty metres; if you cut yourself below a certain level your blood will not appear red.[14]

†Las Vegas and the Sun have an unusual quality in common: an abundance of neon glow. In 2005 scientists discovered that this element, the fifth most common in the cosmos, is some three times more abundant in the Sun than previously thought.[16] Las Vegas is per-haps uniquely a place where neither light nor time is seen as a welcome necessity. 'The absorptions of casino life make mock of the futile cycles of Sun and Moon,' writes Anthony Holden in his poker classic *Big Deal*. 'Daylight, in Vegas, is the pushy, intrusive irritant delay-ing the beauties of neon-clad light – brighter, more beautiful and far more exciting. There are no clocks on view on any Las Vegas casino floor, another deliberate ploy by the man-agement to speed your passage further and further from reality. The only time you keep in Glitter Gulch is that set by your body clock . . . For reasons no one can satisfactorily explain, poker is essentially an act of darkness. By which I mean that it feels odd to play poker in daylight.' The word 'poker' comes from the German *pochen*, meaning to boast or bluff – neither activity associated with clarity or openness.[17] Las Vegas is not so much a city of darkness, however, as one of sinister light.

Sunlight still has its special power, and throughout history soldiers in particular have learned both how to use that power and to guard against it. When warfare still allowed for some mutual consideration, soldiers would put a reflecting object next to the wounded, so the enemy would hold their fire. But more often than not sunlight has been employed offensively. In 1805 during the Battle of Austerlitz, Napoleon ordered his men to leave their commanding position on top of a small hill, ceding the ground to the Austro-Russian army. The next day the French troops were hidden by fog, but the enemy was exposed by the morning sunshine as a perfect target. Arriving at the battlefield of Borodino seven years later, Napoleon triumphantly alluded to his *soleil d'Austerlitz*. On occasion he would instruct his cavalry to put stockings of white fabric over their helmets; otherwise reflected sunlight would enable enemy artillery to find its range at a substantial distance. (The straightness of Roman roads is probably due to their legionaries' polishing their shields to provide bearing reflections to keep those roadways straight.)[19] Japan's oldest surviving book, the *Kojiki*, records that the legendary first emperor, Jimmu (c. 700–600 BC) once chose to fight facing the Sun; he was defeated, for it seemed he was trying to attack the star itself.

If this last instance seems fanciful, reflected light brought real danger enough, particularly in the days of colourful uniforms, which would disappear from the battlefield as field glasses became pervasive and the range, accuracy and capacity of weapons improved. In *The Wouldbegoods*, E. Nesbit has her child narrators chat with an artillery battery about to embark for the South African War, whose friendly commander tells them how, with the introduction of the rifle, every kind of visibility is an invitation to die: 'The guns will be painted mud-colour, and the men will wear mud-colour too.'[20] The term 'mud-colour' would become 'khaki' (Hindi for 'dirty' and soon to be synonymous with war's new loss of glamour). That great unorthodox Catholic, Charles Péguy, at the Battle of the Marne in September 1914, took out his field glasses to scan an enemy position and was instantly shot through the head. For much of that war the French continued to wear their blue coats and scarlet trousers – a machine-gunner's dream. But that conflagration was a hinge-point: by 1914 countries like Britain already had their men in khaki; the rest would soon follow.

Sunlight can reveal but it can hide too. In air combat, pilots seek to position themselves 'up-sun' – i.e. with its glare behind them – before they swoop down on the enemy.[21] The same principle applies at sea: on 1 November 1914, off the coast of central Chile, a Royal Navy squadron of four ships commanded by Rear Admiral Cradock took on eight German ships led by his friend, Vice Admiral von Spee, each manoeuvring to get the afternoon sun into the others' eyes, a serious disadvantage in the days before radar, when the naked-eye calculations of the gunlayers and fire controllers were crucial. A historian of the campaign describes what followed:

> Cradock attempted to close the fire battle quickly, to find the sun behind him blinding the German gunners, but von Spee, with speed-ier ships, kept out of the range of the British guns and delayed the battle until the sun had set.[22]

'We were silhouetted against the afterglow,' recalled an officer on one of the surviving British ships, 'with a clear horizon behind us to show up the splashes from falling shells while the [enemy] ships were smudged into low black shapes scarcely discernible against the background of gathering darkness.' The Germans closed in for the kill: in all, 1,654 offi-cers and men lost their lives, while just three German sailors were wounded in the first British naval defeat since 1812.*

Sunlight can also dictate fortunes in the sporting arena. In NASCAR racing, a strong Sun helps the cars' tyres to grip better. But the cheaper stands at American football and baseball games are called 'bleachers' because they are whitened by exposure to the Sun, and baseball and cricket are filled with examples of catches missed under its glare.[24] Not only catches. Visiting Lord's, the cricket-loving G. H. Hardy, a world authority on number theory, saw a batsman unsighted by a reflection from an unknown source. It was eventually tracked down to 'a large

---

*The Sun's rays can also take their toll in times of peace. A young Bob Dylan crashed his Triumph 500 motorcycle at the crest of a hill near Woodstock. As he told the playwright Sam Shepard, 'I was driving right straight into the sun, and I looked up into it even though I remember someone telling me a long time ago when I was a kid never to look straight at the Sun 'cause you'd get blinded . . .'[23]

pectoral cross reposing on the middle of an enormous clergyman. Politely the umpire asked him to take it off' – to the delight of the gleefully anticlerical Hardy.[25]

'The Fight', William Hazlitt's famous essay of 1822, describes a bare-knuckle boxing match – outlawed since 1750 – held one December in the countryside west of London, '. . . the ring surrounded by covered carts, gigs and carriages'. The contenders are the silky-skinned, panther-like Hickman ('the Gas-Man') and the massive, knock-kneed Bill Neate. The two toss for who will begin with their back to the Sun, and the Gas-Man wins. 'Neate seemed, not cowed, but particularly cautious. I saw his teeth clenched together and his brows knit close against the Sun.' However, part-way through the bout the men swap positions, and soon it is Hickman who has the early-morning light in his eyes, and the fight ends in an emphatic victory for 'heavy, clumsy, long-armed Bill Neate'.[26]*

The great tennis player and coach Harry Hopman admitted that he went out of his way 'to get all the help I could from the Sun', routinely practising for up to half an hour before an important match. 'The purpose of this practice session was to find out where the Sun would be when my match began.'[27] In late-nineteenth-century Britain, soccer was played on the half-Saturday that workers were allowed off. During the winter, the Sun's early setting made it difficult to finish games in decent light. Floodlights were introduced in the 1880s, but for technical reasons were not widely used until the 1950s; thus, for decades Britain's national pastime was played far into the gloaming: the Sun as spoilsport.

Solar interference can also be used to rationalize bad decisions, something I have experienced first hand. In 1994 the Commonwealth Fencing Championships were held in a large ski chalet perched above a resort seventy-eight miles north of Vancouver. It was an exhilarating venue, with snow-covered mountains capping the all-white-clad fencers.

---

*The duel in the Sun has become a regular narrative trope. Thus, in the film *Nevada Smith* (1966) Steve McQueen complains during a shooting match, 'That's not fair! The Sun was in my eyes, and I wasn't expecting it!' His opponent-cum-teacher, who has just tossed a plate in the air for his rookie pupil to shoot at, replies, 'Do you expect a man to walk out in the open for you, when the Sun's at your back?'

I was a member of the Northern Irish team, and by late afternoon the sabre event was down to its last eight competitors. As I squared off against my Canadian opponent, I knew that whichever of us won would receive at least a bronze medal. The bout swung first one way, then the other, until we reached 14-all, with one hit to go. My opponent attacked, I parried and riposted. As he made no attempt to defend himself, but continued with a second attack, I knew the hit was properly mine. However, I hadn't reckoned on the referee. In international fencing, an official's nationality is not always scrutinized, and here ours was – Canadian. He said he couldn't award the hit, as the Sun had made it impossible to see whether I had parried the first attack effectively. As the setting Sun was *behind* him, and the chalet lighting was good, this was a difficult decision to accept, and some of my teammates gave loud vent to their feelings. The fight eventually restarted, and this time my opponent made a successful lunge, to win the bout and eventually the gold.

☼

'Love is sunshine, hate is shadow,' wrote Longfellow. The power of light is one thing, but its gradual diminution over the latter day, and indeed its momentary absence, are both part of the Sun's story. Shadows at their simplest are where light cannot reach, sometimes called 'light's negative traces', daytime darkness. Many ancient societies held men's shadows to be their souls. From their point of view, a shadow, having no substance, is a non-thing, an existential void. The East African Wanika people are even afraid of their shadows, perhaps believing that these inseparable emanations can watch all their actions and bear witness against them. Zulus say that a corpse throws no shadow. Peter Pan's snaps off when he leaps out of the Darlings' window (to be put in a drawer and later sewn back on by Wendy), Barrie perhaps subconsciously responding to the idea of a shadow being the externalization of one's self, and Wendy and her siblings are being asked to leave their real selves behind. 'Who is it that can tell me who I am?' pleads King Lear, to which loyal Kent replies, 'Lear's shadow.'

At the other end of the cultural scale comes the ritual of Simloki, or Soldier Mountain, a five-thousand-foot extinct volcano near the California–Oregon border. As the Sun retreats in late afternoon,

Simloki's shadow will stretch until it falls on the far side of a valley some twelve miles away about an hour and a half later. Simloki is a holy place to the nearby Ajumawi people, its shadow venerated as a spirit of great power. According to legend, two of the tribe's outstanding mythic forebears agreed to race against the shadows of large trees or mountains as these swept across the valley. From this tradition came an earthly contest. Anyone who outraces Simloki – a possible outcome, though rare – is believed to receive supernatural graces. Since the shadow is itself a spirit being, with commensurate powers, it is a unique opponent, and must be treated with proper respect. Thus, should a runner glance over his shoulder in mid-race, he will be instantly struck down. Ajumawi braves make sure never to look back.

Mountaineers are used to seeing a phenomenon known as 'The Brocken', a ghostly shadow that makes it look as if the climber's shadow has become a holy figure with a halo. In fact, it is the result of shadow and diffraction effect – a shadow plus an image.[28] Webster's Dictionary lists twenty-three meanings for 'shadow', including a parasite, a spy/detective and a stand-in for the devil, while 'to cast a shadow' is to threaten happiness, friendship, fame. In the Middle East, those who bring bad luck 'cast a yellow shadow'; and in *The Lord of the Rings* Tolkien creates the forbidding land of Mordor 'where the shadows lie'. In ninja fiction, the cult stories based on the fourteenth-century Japanese assassins, these deadly figures can morph into 'living shadows'. More innocently, Japanese culture arranges flowers into *In* (shade) and *Yō* (Sun). Shadows have even played their part in murder trials. In the 1920s a famous advocate nailed his man by showing a photograph of the defendant coming out of a church, where the shadows indicated a different time of day than the man claimed.[29]

The equation of darkness with evil acts is natural enough; yet one category of people specifically works at night to avoid the day: the sub-untouchables of India. To set eyes on such a person, even at a great distance, was deemed to pollute the viewer, and these wretches, if they ventured out in daytime, were usually murdered.[30]* India was, and is, full

---

*On the other hand, in many societies it was at night that servants frolicked, their masters long asleep. Slaves' spirits rose as the Sun fell. A book on the subject of night quoted a North Carolina planter's judgement that for slaves '"Night" was "their day".'[31]

A peasant uses a sunshade, Cambodia, 1952

of such dark laws. For instance, through the centuries it has been regarded as a profound offence to stand in a king's shadow or, worse, to let one's own fall across one's sovereign. Pollution by shadow – getting between the Sun and a person of higher caste – was a deep outrage, while even a man of much lower birth might go out of his way to avoid having his shadow trodden on: a French politician, visiting in 1957, recalls villagers spending their daylight hours in trees to avoid this degradation.[32]

Darkness was not always to be shunned, and in hot climates shade was often coveted, even seen as a symbol of majesty. As potentates were frequently on the move, the twin demands of comfort and symbolism led to the invention of the umbrella (named from the Latin *umbra*, shade or shadow) as a portable canopy. The sixth-century Sanskrit epic *S'akuntala* likens a ruler thus protected to a sunshade for his subjects, in that 'the sovereign, like a branching tree, bears on his head the scorching sunbeams, while the broad shade allays the fever of those who seek shelter under him'.[33]

We now use this almost horizontal light to discover the site of lost villages. The means is low-level oblique photography, a child of necessity born of the First World War. Air photographs taken at an angle had been produced even before the invention of the aeroplane, but after 1918 the skills acquired by air observers in the Great War were developed for peacetime surveys, and the Second World War only refined photographers' ability to discover what was lying beneath what was obvious on the ground. Once a land surface has been disturbed by human or natural agency, the imprint of that disturbance remains indefinitely, and where features exist in relief they can be effectively photographed from above by taking advantage of shadows.

Sometimes such photos not only reveal new features but suggest an order of succession – an introduction to not one but several older worlds. Skilful attention to such 'shadow-sites' emphasizes outstanding features of form and shape better than ground surveys, and can convey the variations in tone of soil or surface growth that reveal the existence of buried structures. The discovery in the 1920s of Woodhenge in Wiltshire, the timber counterpart of Stonehenge, is one example: literally thousands of lost villages and Roman camps have been revealed by this method.[34]

In Hindu mythology the Lord Vishnu visits the underworld carrying an umbrella, while in ancient Egypt and other Arab lands umbrellas (often made from palm fronds) were carried over princes of the blood as one of their privileges. But soon that amenity was democratized, and an 1871 health manual for Anglo-Indians states authoritatively that 'during the hot season it is unsafe for any European to walk out in the Sun, on the plains, between eight o'clock in the morning and four in the afternoon ... without at least the protection of an umbrella'.[35] Robinson Crusoe constructed one for himself in imitation of those he had seen in Brazil, 'where they were very useful in the great heats there ... It ... kept off the Sun so effectively that I could walk out in the hottest of the weather.' Early versions in both Britain and France were consequently known as 'Robinsons'.[36*]

Light modulates continuously over the day, each stage with its own character: dawn, sunrise, noon, twilight, dusk, gloaming, sunset. In Robert Louis Stevenson's *The Strange Case of Doctor Jekyll and Mr Hyde* (1886), Mr Utterson beholds 'a marvellous number of degrees and hues of twilight'.[37] The battle-weary subaltern in R. C. Sherriff's *Journey's End*, as dawn rises over the wire and trenches of the ravaged Western Front, exclaims: 'I never knew the Sun could rise in so many ways till I came out here – green, and pink, and red, and blue, and grey. Extraordinary, isn't it? ...'[38] When the Sun is rising or setting – shining flat across the ground – every feature on Earth is caught up and displayed. Twilight inspired Hegel famously to write: 'Only when the dusk starts to fall does the owl of Minerva [the Roman symbol of wisdom] spread its wings and fly' – his way of saying that philosophy comes to understand a historical condition just as it is on the point of disappearing.[39]

Just before the Sun sets there comes 'that brief time between day and night when great plans are hatched and all things are possible'.[40] Nabokov in his memoirs makes a point of mentioning 'the lovely Russian word for dusk' – *soomerki*.[41] It is at twilight, but only then, that an Orthodox Jew may study secular matters.[42]

Writers have risen to this moment. One summer evening, as William Faulkner struggled for a title for his new novel, his wife remarked on the distinctive hue of the southern sky. Faulkner was duly inspired and the book had its title: *Light in August*. To Rat, at dusk in *The Wind in the Willows*, 'the light seemed to be draining away like floodwater'.

---

*In Europe, carrying an umbrella was long regarded as effeminate. The first Englishman who, from 1750 on, regularly sported one (made from oiled silk) is said to have been the sickly but dandified Jonas Hanway, founder of the Magdalen Hospital. Women may have carried them at an earlier date, and Beaumont and Fletcher allude to the new machine in their 1624 comedy *Rule a Wife and Have a Wife*:

Now are you glad, now is your mind at ease; /Now you have got a shadow, an umbrella, To keep the scorching world's opinion /From your fair credit.

The great ball of Krishna, Mahabalipuram, on the south-east coast of India, 1971:
a wonderful photograph, and a cooling shadow

'Real dusk,' writes Henry James in *The Portrait of a Lady*, resorting to a
parallel image, 'would not arrive for many hours; but the flood of
summer light had begun to ebb, the air had grown mellow, the shadows
were long upon the smooth, dense turf.' Others have called twilight
'blind man's holiday', 'l'heure verte', even 'entre le chien et le loup', the
time when the light is such that it is impossible to distinguish between
dog and wolf.[43]

It is said that pirates, who often attacked at night, would wear an eye-
patch, which they would remove before boarding. Having one eye not
geared to daylight was meant to help as they rushed below decks in
search of prey or booty; but in fact eyepatches decrease one's field of
vision and depth perception – not a good thing when leaping from a
ship. Removing a patch could also inflict momentary blindness, as the
pirates' eyes adapted. Modern snipers do practise something similar,
though: US army patrols have a stand-down period while being inserted
for night action, so their eyes may adjust.

The fullest moonlight is just too weak to awaken colour discrimination
in the human eye, but, because the Earth is round, most clouds catch the

Sun's rays long after it has set beneath them, forming at such a height that they reflect the sunlight just over the horizon and glow with a ghostly hue – a phenomenon called 'noctilucence'. The distinctive light cast by the dying Sun on the Earth, seen as a blue-grey front of light that rises in the East after a clear sunset and often tinged with pink, is 'twilight edge'. Moments later comes 'civil twilight', the minutes after sunset until the Sun is six degrees below the horizon; 'nautical twilight' covers between six and twelve degrees below, and between twelve and eighteen is 'astronomical twilight', an acknowledgement that it is finally getting dark enough for skywatchers to begin work. Once the horizon has lifted 18° above the Sun, true night begins.

Reflecting on his travels in Africa, Carl Jung wrote: 'When the great night comes, everything takes on a note of deep dejection, and every soul is seized by an inexpressible longing for light . . . It is a maternal mystery, this primordial darkness. That is why the Sun's birth in the morning strikes the natives [of Africa] as so overwhelmingly meaningful. The moment in which light comes is God. That moment brings redemption, release . . . The longing for light is the longing for consciousness.'[44]

15

BENEATH THE BEATING SUN

The heavens are red-hot iron and the earth is burning brass,
And the river glares in the Sun like a torrent of molten glass,
And the quivering heat haze rises, the pitiless sunlight glows
Till my cart reins blister my fingers as my spectacles blister my nose.
Heat, like a baker's oven that sweats one down to the bone,
Never such heat, and such health, has your parboiled nephew known.

Rudyard Kipling, 'Dear Auntie, Your Parboiled Nephew'[1]

I am a typically fair-skinned Englishman, ginger-haired and freckled: even a few minutes under the Sun and any areas I have left uncovered will turn an unsightly red, painful reminder that its rays produce not only vitamins: they also scorch. From wretched childhood afternoons on, I have been aware of what the Sun can do to the human body, even though my experiences have generally been of British summers, the mildest broiling possible. In recent years I have stood in the middle of the Arizona desert, where temperatures climb over 120°F (49°C), and pondered what it must be like to be burned alive. The force of the south-western sun bleeds colour out of fabrics and wood faster than anywhere else on the continent, but even that heat does not come close to matching the hottest surface temperature ever recorded – 136°F (58°C) on 13 September 1922, in the Libyan Sahara. In the US, Greenland Ranch, California, seared by 134°F on 10 July 1913, holds the official record, although the true all-time high was most likely endured during forty-three consecutive days between 6 July and 17 August 1917 in Death

Valley, California – so named in 1849 by one of eighteen survivors of a party of thirty that attempted a short cut to the goldfields.

In his book *Equator*, Thurston Clarke describes a journey he made in 1986, stopping off at points along the zero-degree latitude, easterly from French Martinique in the Caribbean to Mount Cayambe, in Ecuador, whose name is Spanish for 'equator'. 'I have become a connoisseur of heat,' he writes at the outset, and indeed he creates a convincing taxonomy of his subject: 'There is the heat that reflects off coral and scorches and softens the face like a tomato held over a fire. There is the greasy heat of a tropical city, a milky heat that steams a jungle river like a pan of nearly boiled water, a blinding heat that explodes off tin roofs like *paparazzi*'s flashbulbs, and a heat so lazy and intoxicating that all day you feel as though you are waking from a wine-drugged nap.'[2]

When he reaches Lambaréné, the island where Albert Schweitzer established his mission amid the swamps of the Ogooué, principal river of Gabon in West Central Africa, he says he feels as if 'wrapped in hot towels', and recounts the time that Le Grand Docteur, making his sweat-soaked rounds, was asked whether such heat bothered him. He never permitted himself to think about it, he said, then added, 'you must have great spiritual activity to fight the heat'. A handful of great men throughout history – George Washington and General MacArthur among them – have added to their aura of majesty by apparently not sweating at all. Barack Obama is said to be of their number.

Heat has long been the enemy of the unprepared traveller, particularly of soldiers on the move. In June 1578 Sebastian I of Portugal took a large force to North-West Africa, where his 'armour grew so hot that he had to have water poured over his body under the metal plates, and the suffering of his troops, who could not afford this luxury, must have been extraordinary'.[3] The army was annihilated, the prelude to a run of disasters: within three years Portugal had fallen to the Spanish Hapsburgs. In China at around the same period, a different kind of army under a hot sun – coolies digging canals – ended each day with its backs 'chapped like the scales of a fish'. Some historians still perpetuate the legend that when, on 24 June 1812, Napoleon and his Grande Armée of 691,501 (a suspiciously exact figure, but still an acceptable approximation for the largest force in European history since Xerxes invaded Greece) crossed the River Niemen and headed towards Moscow, they were destroyed by the cold. By the time the army reached Smolensk in mid-August, a third had indeed been lost – but to a weeks-long heatwave. As Pierre Bezukhov comments in *War and Peace*, 'Russia and hot weather don't go together'.[4]

The Sun imposes a steep threshold of effort in a very short time, and the wise general seeks to protect his troops from the heat of day; but sometimes military tradition or plain ignorance trumps common sense – Scottish Highland regiments, for instance, got up in kilts, found the backs of their legs got charred (especially after being in skirmishing line on their bellies), compromising their ability to march. For centuries, British troops died routinely of heatstroke during battle, because there was no drill for taking off one's coat – a deadly tradition that continued well into the twentieth century. In the Western Desert of Egypt and Libya during the Second World War, British soldiers in the Eighth Army wore open-necked shirts and shorts, their quartermasters ignoring the fact that the best way to keep off the Sun would have been with loose-fitting, lightweight clothes that covered more skin.

Certainly Indians have long understood this and dressed accordingly – with good reason. Their country seems to have been the most systematically awful place for heat, a place where things *teem* most, where unfriendly micro-organisms and long-lasting infections conspire against you. Harry, Olivia's lover in Ruth Prawer Jhabvala's *Heat and Dust*, exclaims: 'I stayed in my room all day yesterday and this morning. What else can you do in this hideous terrible heat? Have you looked outside? Have you seen what it's like? . . . No wonder everyone goes mad.'[5] A junior officer under the Raj who failed to wear his topi (sun helmet) outdoors would be confined to barracks for fourteen days. For the British, going out at midday became almost an act of impiety – certainly of self-destruction: Carnahan, the broken protagonist of Kipling's *The Man Who Would Be King*, dies after wandering about bare-headed at noon for a mere half-hour.

'Mad dogs and Englishmen . . .' Noël Coward's famous ditty begins:

In tropical climes there are certain times of day
When all the citizens retire
To tear their clothes off and perspire.
It's one of those rules that the greatest fools obey,
Because the Sun is much too sultry
And one must avoid its ultry-violet ray . . .[6]

The song had its origins in Coward's disdain for British colonial society, whose refusal to adapt to foreign habits he observed in Malaya, which he visited in 1930. During a later car ride from Haiphong to Saigon, he composed the whole piece entirely in his head. He may have been inspired by Kipling, who wrote in *Kim* that 'only the devils and the English walk to and fro without reason', and later: 'and we walk as though we were mad – or English', but the phrase is Coward's.[7]

Outposts of the British Empire seldom provide a welcoming climate, and the lack of knowledge about how to deal with weather extremes accounted for much suffering, even loss of life. Charles Sturt, leader of the Royal Geographical Society's 1844 expedition into the Australian desert in search of a legendary inland sea, recorded: 'So great is the heat that . . . our hair has ceased to grow, our nails have become as brittle as glass. The scurvy shows itself upon us all. We are attacked by violent headaches, pains in the limbs, swollen and ulcerated gums . . .'[8] In British East Africa (modern Kenya) in the early years of the twentieth century, everyone buttoned on spine pads of quilted flannel, as it was believed that the Sun affected the backbone. Men wore cummerbunds under their revolvers to protect their spleens, while people with metal-roofed houses at high altitude wore hats indoors, as they thought the Sun penetrated iron. The first permanent doctor in Nairobi, setting up practice in 1913, held that people with blue eyes should wear dark glasses, and everything should be lined with red – a colour believed to modify solar rays.[9]

Try as they might, the colonial British were trapped. In his fine study of the Indian Mutiny, Richard Collier writes:

Dawn, 14 June 1857. Implacably the Sun rose above Cawnpore . . .
Even in other years, granted the gift of shade, such heat had driven

men to lock themselves in darkened rooms and blow out their brains. On this day five, perhaps six, people within Sir Hugh Wheeler's entrenchment would die of sunstroke. A band of steel seemed to encase their temples followed by a great drowsiness; they sank vomiting to their knees, faces blackening as they died. Today some muskets would split apart like firecrackers, touched off by the Sun. From the mud walls, men would see, across the arid sandy plain, a strange chimera of forest glades and blue sparkling water. Against their wills they would suck in the unbelievable, unbreathable smell of the entrenchment, like the contents of a thousand privies rotting beneath the Sun. Small children pleading for a drink of water would be told to suck on a leather strap to quench their thirst.[10]

Even before the uprising, life had been barely tolerable. Families sat in semi-darkness, officers in undress summer uniforms, their ladies in loose white muslin. Church services, so much a part of empire, were kept brief, but because of the heat few would kneel in worship. In one Calcutta station,

> The British did not know it but they were really fighting a losing battle against the Sun, a Sun that caused tempers to flare murderously, that made social standards at times of rigid importance, at other times a burden not worth the shouldering. For eight months of the year the Sun was a monstrous weight pressing down on all their lives, curtailing movement, hedging them in darkened rooms. The Sun sent pianos out of tune, dried ink as it touched paper, melted peaks of officers' bell-shaped shakos [tall, cylindrical caps] to a sticky black glue . . . By common consent the greatest beauty had the palest cheeks in the station . . . Houses were designed with one purpose: to cheat the Sun of victory.

Intense heat of this kind can bring illnesses peculiar to equatorial countries, such as miliaria, where sweat collects beneath the skin – one of the commonest diseases for soldiers fighting in the tropics during the Second World War; tropical acne; cataracts; cholinergic urticaria, an eruption of one- to three-millimetre wheals, mainly on the torso;

phototoxicity and photoallergy – eruptions caused by the interactions between drugs and sunlight; solar eczema; and homopolar ringworm, in which the worm raises bumps on the skin where hair doesn't grow. Such diseases accounted for twelve per cent of all outpatient visits in the Vietnam War.

Great heat has ever been wrapped in fear and superstition. In ancient Greece, high noon was when demons came out, and it is no accident that Pan, a distinctly sinister god, is Lord of Noon. Of course, all regions are not created equal. The incidence of sunshine over the Earth varies from over four thousand hours per year (more than ninety per cent of the longest period possible) to fewer than two thousand. The Sahara gets the most, areas around Iceland and Scotland the least.

Until the current era of extreme population mobility, a person's skin colour was likely to be influenced by where he or she lived. Around the equator dark-skinned people would prevail, because dark skin contains above-average amounts of melanin – a pigment that is also a sunblock – which helps them adapt in areas saturated with light; light-skinned people would be more common towards the Poles.

Skin colour has come to have symbolic and even moral values divorced from its origins, so that throughout human history it has been used to assess the relative superiority of races as well as professions and classes. As the science writer Jonathan Weiner has noted, 'Skin color, the biggest visible difference between the populations of human beings on this planet, around which so many bitter and divisive myths have been spun, is really nothing more than a schematic diagram of planetary levels of ultraviolet radiation.'[11] Some dark-skinned people hold that, as their colour shows that they are nearer to the Sun, it also implies greater closeness to God; but in general there has been and continues to be a prejudice against darker skins in much of the world. Common definitions for blackness listed in the OED, some dating back to the 1500s, include 'foul', 'dirty', 'wicked' and 'horrible'. Barack Obama tells in *Dreams from My Father* of flicking through an old copy of *Life* magazine when he was a child and coming across a picture of a man who had been chemically treated to lighten his complexion. His skin, writes the President, had a 'ghostly hue', as if 'blood had been drawn from the flesh'.[12] The man 'expressed some regret about trying to pass himself off as a white man, was sorry about how badly things had turned out. But the results were irreversible. There were thousands of people like him, black men and women back in America who'd undergone the same treatment in response to advertisements that promised happiness as a white person.'

Even when it didn't have racial – and racist – implications, the prejudice favouring lighter skin has been consistent through the ages, partly because skin colour has served as a signifier of social, economic and religious standing, since those who performed manual labour outdoors were much more likely to be tanned than those whose professions were better paid and more highly esteemed (or those who were wealthy enough that they did not have to work at all). In Minoan Crete (2600 BC on), women kept out of the light to maintain an ethereal appearance.* From the mid-tenth century, European women were whitening their skins with potions every bit as destructive as the one used by the man in *Life*. Arsenic was the skin-whitener of choice, even though its deadly properties were well known – as shown during the Italian Renaissance, when one Giulia Toffana created an arsenic-based powder designed for wealthy women to kill their husbands. The women applied the powder to both their pudenda and their cheeks: husbands who kissed too passionately rarely survived. Some 600 widowhoods later, Toffana was detected and put to death, but still women whitened their faces with arsenic. Not only their faces: in 1772 British ladies started bleaching their hands with the poison to give them a porcelain tone, prompting the great potter Josiah Wedgwood to advertise black teapots, against which background the hands of the hostess looked even whiter.

Shakespeare seems not to have shared this preference. Three of his most delightful heroines are described as sun-browned – Olivia in *Twelfth Night*, Julia in *The Two Gentlemen of Verona* and Beatrice in *Much Ado*. He

---

*As against this, there is the odd fact of ancient Greek statuary: it is said that in Rhodes, Athens and Olympia alone, the extraordinary figure of 73,000 statues could be seen. If we disbelieve this estimate, Pliny the Elder (c. AD 23–79) more reliably informs us that the statues per city still numbered in their thousands, and that one could go nowhere in the Greek world without encountering life-size bronze simulacra. Many statues were of painted stone, or polychrome, or gilded wood; but more still suggested that to be bronzed (although we do not know exactly what kind of colour this would have been) was to be perfect, or at least envied. What must it have been like to have lived among such figures? Would it have suggested that to be suntanned was to emulate the gods?[13]

himself, according to Sonnet 62, was 'Beated and chopp'd with tanned antiquity',[14] as he realizes when looking in a mirror. He was certainly well aware of the prejudice of the day. Sonnet 127 asserts that, 'In the old age black was not counted fair, /Or, if it were, it bore not beauty's name . . .'

During the reign of Queen Elizabeth I some women were so keen to appear pale-skinned that they used whitening agents that contained lead oxide, which, accumulating in the body, brought on grave pathologies, even paralysis and death. Others spread egg whites over their faces to obtain a glazed look, and painted thin blue lines on their foreheads to give off a translucent glow.* Elizabeth herself, traversing England on horseback to urge loyalty on her people, did not have the luxury of remaining indoors; but back at court she hid her weather-beaten features behind a lotion called 'ceruse', a lead-based skin whitener used in ancient Rome and revived in the Renaissance. In seventeenth-century journeyman portraits, women's faces are customarily rendered as corpse-like white ovals. Later, the development of coach travel reinforced the fashion: being pale showed you could afford covered transport. (As the Brazilian proverb has it, then as now, 'money bleaches'.)

Under Charles II, heavy make-up was universal at court. The most perilous of the beauty aids employed were white lead, arsenic and mercury, all of which not only ruined the skin over time but also made the hair fall out, corroded the stomach, brought on the shakes and often killed the aspirant beauty outright. The danger in such cosmetics became widely known through the dramatic deaths of the courtesan Kitty Fisher and of Maria, Countess of Coventry, the elder of the two legendary Gunning sisters of George II's day. Kitty and Maria maintained a famous rivalry due to Kitty's affair with Maria's husband, George William. As for Maria

---

*It was even better to be 'blue-blooded', meaning 'high-born'. There are several possible origins for this term: it was the custom in late summer for the gentry to help gather in the harvest and thus expose their arms. Their skin would be pale, so their veins showed up blue, and the labourers believed their masters had blue blood. Another explanation comes from the execution of King Charles I in 1649, where he wore a white shirt. This would have been washed in a solution of calcium hypochlorite, which chemical, when it comes into contact with blood, provides a brilliant blue stain.[15] While both explanations may be true, the term is more likely the result of the outcry in Europe, especially in Spain, against aristocracies intermarrying with Jews and Muslims – one Pope called Philip II of Spain 'the spawn of infidels and Jews'. The less the ancestral intermarriage with Jews and Arabs, went the argument, the more the veins of white-skinned, generally peasant people showed up as blue.

Cleopatra Dissolving the Pearl, 1759 – Joshua Reynolds' portrait of the courtesan and actress Kitty Fisher. Lead-based potions taken to lighten her face killed her in 1767

(1733–1760), she caught the eye of the notorious playboy Augustus Henry Fitzroy, third Duke of Grafton and from 1767 to 1770 functionally Britain's Prime Minister, but the effort of retaining his affections came at great cost: she died at twenty-seven, after too many years of lead-based improvements. Catherine Maria ('Kitty') Fisher died in 1767, apparently from lead-based preparations, seven years after her rival.

As the eighteenth century gave way to the nineteenth, whiteness continued to be the primary prerequisite of beauty. There were occasional exceptions to this rule – throughout the centuries, for instance, the Anglo-Irish gentry, effectively a garrison community, cultivated the martial virtues in their women, whose ruddy features were evidence of their regular riding to hounds. But while a certain kind of lady (witness Bertie Wooster's Aunt Agatha) could be wind-burned – coarsening the complexion rather than tinting it – it was not done to go out in the sun.

The same kind of thinking found its way to America, where no Southern belle or Northern debutante dared go out without her parasol for fear of losing her lily-white colouring. Women in the South at the time of the Civil War took to chewing newspaper, as they had discovered that something in the ink whitened their faces. Freckles (the result of one's melanin granules being distributed unevenly: a freckle is just an activated high accumulation) were also to be avoided. Scarlett O'Hara's younger sister Suellen, finding the family carriage has momentarily stopped in the Sun, cries out in dismay: 'Oh Papa, can't we get going? I can feel the freckles popping out on my face.'

Even as late as the 1880s ladies in society were prepared to resort to extreme measures. The Parisian beauty Amélie Gautreau, the notorious 'Madame X' of Sargent's portrait, worked hard on her appearance: her 'skin was suspected by some of her rivals to be the work of an enameller and by others to have been achieved by judicious doses of arsenic'.[16] But by Madame X's time fashions were changing. In Victorian Britain, people had started to shun make-up, associating it with prostitutes and actresses (considered by many to be one and the same). Any hint of tampering with one's natural colour was frowned upon as vulgar – or worse. White skin as a mark of class may also have lost its claims as a result of urbanization, and because of the changing labour patterns brought on by the factories of the Industrial Revolution. With indoor work the norm for most working-class people, a pale skin was no longer a signifier of status. During 1914–18 few officers returned from the trenches with pale skins: it was a patriotic mark to have been out in the sun.

Besides, a new enthusiasm was in the wings. Seaside holidays, which had originally been recommended for reasons of health, were becoming a mark of the smart set. Those who had the money and the time to enjoy themselves wherever they wanted were choosing to go to the beach, and tanned skin came to be seen as a sign of membership in the propertied classes. As Paul Fussell recognized:

For the Mediterranean to be re-appropriated after the [First World] War, its most ubiquitous natural asset, the Sun, had to be redeemed from the social stigma it had borne in the nineteenth century. Then, the better sort of people had tended not to sit in it, believing that if its effects were indispensable to the welfare of flora, they were of very dubious value to persons . . .[17]

When in 1920 the couturier and society trendsetter Coco Chanel accidentally developed a tan while cruising the Mediterranean aboard a nobleman's yacht, her bronzed skin changed the course of fashion. Setting her imprimatur on the new look, in 1929 she declared: 'A girl simply has to be tanned.' The social prestige, or otherwise, of the colour achieved was keenly debated: how intense should one's tan be? A golden quality was best, said those in the know. As early as 1921, Daphne du Maurier recalled August afternoons where her family 'all lay out on the lawn like corpses to catch the Sun'.[18]*

Chief among the fashionable seaside venues was the French Riviera. Such consumptive luminaries as Robert Louis Stevenson and Aubrey Beardsley were early visitors. Then came Greta Garbo, Cole Porter, Dorothy Parker, Anita Loos, Noël Coward, and Zelda and Scott Fitzgerald ('[It] was a way of life, lived outdoors on the beach and under the sweltering Sun . . . ').[20] A year-round tan was especially envied.

But even as people were discovering the delights of beach life, there remained some ambivalence. In 1916 the first British *Vogue* advertised a Helena Rubinstein cream that allegedly removed 'impurities', including 'sunburn, freckles and discoloration'. ('Burn' and 'tan' were interchangeable, with 'tanning' having the unpleasant association of prepared leather.) And in the 1920s the open touring car, which allowed one's face to brown, went quickly out of fashion.

Hollywood was developing its own standards: black-and-white film tended to make tans look an unhealthy grey, so stars who began their careers in this period, such as Garbo, Ingrid Bergman and the two Hepburns, either kept themselves out of the Sun or used enough make-up so that their skins appeared luminously white on-screen. Another potential problem was identified quite clearly by Rudolph Valentino, who was careful to stay out of the Sun because he tanned so deeply that 'I become like a Negro'. He lightened his skin with a make-up prepared for him by Max Factor, while himself grinding up additional pigments to

*The novelist and poet James Hamilton-Paterson adds an intriguing observation from on board the Russian ship *Keldysh*, his home while he was writing his non-fiction study *Three Miles Down*: 'Why do Russians sunbathe standing up?'[19] But then generations of Russians must have been bemused by the lyrics of the Beatles song, 'I Am the Walrus': *Sitting in an English garden waiting for the Sun. /If the Sun don't come, you get a tan /From standing in the English rain.*

hasten the process, so that he might escape having to play bit parts as a swarthy villain.[21]

Conversely, the legendary Josephine Baker, having escaped black St Louis, where she was judged too light-skinned, became the toast of Paris as the star of 'La Revue Nègre', the black stripper in a banana skirt. She was soon promoting Bakerfix hair gel and Bakerskin darkening lotion, for Parisian women who wanted to imitate the Baker look (even though her own routine was devoted to skin-lightening milk baths and lemon rubs).[22]

While most white people certainly did not aspire to look black, the fashion in skin colour among Western Europeans and trend-conscious Americans was beginning to change. Cary Grant would work on his tan almost daily, spending his holidays from film-making on the nearest sunny beach, holding a mirror under his chin to reach nooks hidden from the light. Over the same period there was a corresponding change in lifestyles. Women left their houses to enjoy outdoor life – going on hikes, picnicking, playing lawn tennis and engaging in other acceptably 'feminine' activities. Throwing away centuries of tradition, on beaches throughout Europe and the US, they sunbathed, wore decorative Sun hats and shawls – not for protection but as fashion statements – and went swimming. Afterwards, they brushed brown- and beige-tinted powders and creams on the areas of the skin that the Sun had missed.

To cater to the new enthusiasm, Ambre Solaire suncream was invented in 1936, and Polaroid sunglasses were introduced the same year. By the end of the 1930s the fashion world was featuring clothes for women to show off their new tans: shoes were worn without stockings, and sleeveless dresses became stylish; bathing costumes that had covered women's legs now left them bare. Across the classes, the suntan had arrived. For those without the opportunity to be in the Sun or who didn't tan easily, aids based on the leg make-up that was women's answer to the scarcity of nylon stockings during the Second World War appeared on the market.

In France, access to the Sun – for both men and women – was finally democratized in June 1936, when the National Assembly introduced the two-week paid holiday, *les congés payés*, which 'brought a glow to workers' pale cheeks'.[23] Now at last working-class townspeople were learning how to picnic and to cycle, and getting their first amazed glimpse of the sea – but leisure time was so new to them that they had no holiday clothes, and

were photographed at play in suits and hats. Stephen Potter, author of a
series of humorous books on 'Lifemanship', brilliantly caught the pallid
Englishman's attitude to sunbathing at the start of the 1950s. The first
action at a summer weekend, he writes, of 'that confirmed hostess-nobbler,
P. de Sint' is to strip to the waist and sunbathe, and his skin soon turns 'a
rich honey bronze'. This is when his nemesis, Cogg-Willoughby, strolls by.

COGG-WILLOUGHBY: By Jove, you brown easily.

DE SINT: Do I?

COGG-WILLOUGHBY: Yes. You're one of the lucky ones.

DE SINT: Oh, I don't know.

COGG-WILLOUGHBY: They always say that the Southern types brown
    more easily.

DE SINT: Well, I don't know, I'm not particularly . . .

COGG-WILLOUGHBY: Oh, I don't know . . . Mediterranean . . .

Adds Potter: 'Cogg-Willoughby was able to speak this phrase with an
intonation which suggested that de Sint was of Italian blood at least, with,
quite probably, a touch of the tarbrush in his ancestry as well.'
Thereafter, de Sint spends the rest of the weekend 'trying to cover up
every exposed inch of his body'.[24]

But Potter, if he were ever a serious combatant, was fighting against
the tide. Within a decade of *Lifemanship* being published, the reversal in
skin-tone preferences was complete: bronzed skin on Caucasians signified both health and social standing, for reasons that were also a reversal
of those that had previously favoured paleness. President John F.
Kennedy was at least as obsessive as Cary Grant in his commitment to the
perfect tan. During the week that he was preparing his celebrated inaugural address,[25] he spent as much time tanning as he did working on his
speech.*

---

*According to a recent biographer:
   In 1961 a tan was considered evidence of well-being and a way to disprove the rumors
   about Kennedy's poor health. Hugh Sidey, who covered him for *Time*, remembers the
   whole family being obsessed with tanning. Jackie used a reflector, Bobby pursued the
   grail of a full body tan, and Kennedy himself had been traveling with a sun lamp since
   his 1946 congressional campaign. He worked tirelessly at keeping his tan dark enough
   to show up on black-and-white television.

The Sun viewed in extreme ultraviolet light by NASA's SOHO satellite. Bright spots mark the most active regions, where sunspots, flares, and coronal mass ejections form. The spectacularly varying colours for the Sun are explained by the fact that the EIT (Extreme ultraviolet Imaging Telescope) images the solar atmosphere at several wavelengths, and so shows solar material at different temperatures. In the images taken at 304 Angstrom the bright material is at 60,000 to 80,000 K; in those taken at 171 Angstrom (blue), at 1 million K. The 195 Angstrom (green) images correspond to about 1.5 million K, 284 Angstrom (yellow) to 2 million K; 304 Angstrom is orange

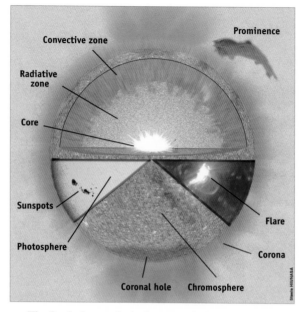

The Sun's three principal zones – the inner core, the transition zone (where particles begin their ascent towards the surface) and the convection zone, where energy circulates like porridge bubbling in a pot

A diamond ring during a total eclipse, photographed in 1999; it occurs as the Sun rises over the jagged limb of the Moon

These coronal mass ejections (CMEs) are like rubber bands that can be stretched and twisted until they break, releasing massive doses of hot air and energy

The coolest part on the Sun is the dark eye of a sunspot – 3,800°F, compared to 6,000°F for the rest of the surface. A sunspot this size could swallow the Earth with ease

The Sun changes from relative calm to a bright, tangled mass of magnetic loops in the years leading up to a solar maximum – the peak of the Sun's eleven-year cycle

Approx. size of Earth

A close-up of an erupting prominence with Earth inset at the approximate scale of the image. Taken on 1 July 2002

The two active regions (brighter areas) are almost diametrically apart. SOHO got a good view of the extensive areas above the Sun influenced by the powerful magnetic fields associated with the two regions

Progress of a Coronal Mass Ejection (CME) observed over an eight-hour period on 5–6 August 1999 by LASCO C3. The dark disc blocks the Sun so that the LASCO instrument can observe the structures of the corona in visible light. The white circle represents the size and position of the Sun

Like snowflakes, no two auroras are the same. In the fourth century BC Aristotle described 'glowing clouds' and a light like flames of burning gas – one of the first scientific accounts

A time-lapse photograph of the Sun changing in colour and brightness as it sets. The reddening is caused by aerosols and dust in the atmosphere, which scatters blue light more strongly than red, giving the Sun its distinctive tint

Sunrise seen from Mount Fuji, 4.37 a.m. on Midsummer Day, 21 June 2005

The seashore at Futamigaura, near the holy site of Ise: on New Year's Day homage is paid to the sun goddess Amaterasu. The 'Wedded Rocks' in the foreground are believed to shelter Izanagi and Izanami, the male and female couple from whom all the islands of Japan arise

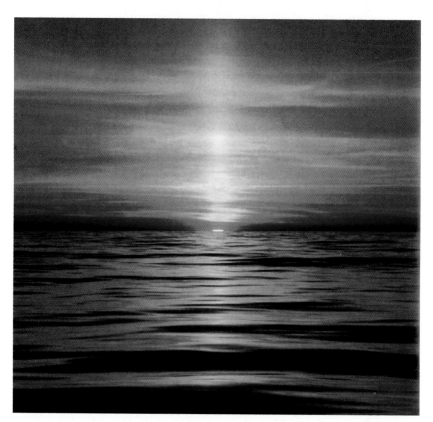

A pillar of light rises from the setting Sun in the North Atlantic, off the coast of Maine. Sun pillars will sprout from the top of the Sun when light reflects off the undersides of plate-like crystals in the air

John F. Kennedy deepens his tan at Palm Beach during the winter of 1944

The downside was that Kennedy's skin deteriorated rapidly. He might have looked youthful on television, in contrast to the sweating, sallow Nixon, but Gore Vidal, who shared a stepfather with Jackie, wrote that his stepbrother-in-law looked older than he did in his photographs: 'The outline is slender and youthful, but the face is heavily lined for his age. On the upper lip are those tiny vertical lines characteristic of a more advanced age.'[26] By 1961 the Sun was already taking its toll.

By the 1970s two entire generations had baked their bodies, and, by the 1990s, Disney was having trouble getting its female staff at Euro Disneyland to wear stockings – a good tan was worth flaunting. Conversely, skin whiteners and lighteners are still popular among many dark-skinned people worldwide. A label on a cosmetic manufactured in Paris but sold in South-East Asia asserts that its product is 'the first technology which regulates the different steps in the skin pigmentation process, to perfect the whitening efficiency'. In 2007 plastic surgeons in London announced that they had developed creams using plant extracts and concentrated vitamin C to lighten skin – adding that they hoped to drive backstreet suppliers of potentially dangerous preparations (often containing bleach) out of business.[27]*

From the Hawaiian prostitutes in James Jones's 1951 novel *From Here to Eternity* (who were kept out of the Sun, as their serviceman clients preferred their women as white as possible), to Jennifer Lopez, Michael Jackson and the heroines of modern Japanese novels, it is still true: light-skinned people want darker skins, and vice versa.†

☼

The whole fashion for the seaside holidays that started the trend for tans had begun with concerns about health – and eventually took on a distinctly erotic undertone. In a 1990s essay, the novelist John Fowles writes about the transformation of his hometown of Lyme Regis, in Dorset, as early as the eighteenth century, specifically because of the belief in the health benefits of sea and sun:

> How can something so nice have been ignored for so long? The change came, like most major human changes, from a conjunction of two factors . . . In this case the medical profession and the first Romantics spoke as one.[30]

---

*In David Davidar's engrossing first novel, *The House of Blue Mangoes*, set in India mostly in the years before the Great War, a village doctor concocts a special potion to lighten his patients' skins. Although the story is the product of Davidar's imagination, the author told me that his grandparents, who lived in just such a village, used talcum powder to pale their faces, while a cream called 'Fair and Lovely' is among several such products that still sell well in India today.[28] India's pursuit of a pale appearance is known as the 'Snow White Syndrome'; although in 2010 the Indian fashion industry promoted a campaign, 'the dawn of dusk', to combat prejudice against darker skin colour.

†Writing in *The Washington Post*, Jabari Asim makes the point that conventional wisdom in fashion- and lifestyle-magazine publishing has long held that dark faces on covers discourage 'mainstream' purchasers, which may explain why Beyoncé Knowles was the first black woman to adorn a *Vanity Fair* cover all by herself since Tina Turner in 1993. *Vanity Fair* is not unique in its timidity. When Halle Berry appeared on the December 2002 cover of *Cosmopolitan*, she became only the fifth black woman so featured, and the first since Naomi Campbell in 1990. When Italian *Vogue* of July 2008 featured black models exclusively, it was considered sufficiently newsworthy that *The New York Times* ran an article. News magazines show far better percentages. For example, Oprah Winfrey recently appeared on the cover of *Newsweek*, and this mostly passed without comment. Nor was her complexion tampered with; *Newsweek*'s rival, *Time*, had infamously proved the danger of that tactic when in 1994 it darkened a portrait of O. J. Simpson in order to make him appear more sinister.[29]

By 1780, Fowles asserts, the sea – its water, its air, the light and relief it gave landscapes – was all the rage. Between 1803 and 1804 Lyme played host to its most famous habituée, Jane Austen, and her family, and *Persuasion* shows her marked approval of what the town (if not its inhabitants) had to offer. As one of her characters comments: 'I am quite convinced that, with very few exceptions, the sea-air always does good. There can be no doubt of its having been of the greatest service to Dr Shirley, after his illness . . . He declares himself, that coming to Lyme for a month, did him more good than all the medicine he took . . .'[31]

The appeal of the French Riviera had similar origins. By the late eighteenth century it had become a fashionable destination for the British upper classes. Its popularity was solidified and much expanded in the 1830s, when Henry Peter Brougham (1778–1868), Lord Chancellor of England, travelling with an ailing sister, began visiting Nice at a time when it was little more than a fishing village on a picturesque coast, and built a house there. By the time of his death Nice and its neighbouring towns had become the sanatorium of Europe, and by 1874, shortly after the railway reached the Riviera, one observer calculated that 'between seven and eight thousand English invalids . . . annually spent the winter in the south'.[32]

The health benefits of being in the Sun got another boost in 1890, when the German bacteriologist Robert Koch showed that sunlight was lethal to the tuberculosis bacillus, and sanatorium designers were encouraged to exploit this discovery. However, most doctors were unconvinced, and the Sun's use as a medical aid in modern times is primarily due to Oskar Bernhard (1861–1939), who was surgeon-in-chief at a hospital in the Swiss valley of Upper Engadine when on 2 February 1902 an Italian was admitted with severe knife injuries. Eight days after surgery, the wound still gaped widely, and attempts to re-stitch it failed. The surrounding tissue was soft and spongy, and the whole discharged profusely, with no drying treatment having any effect. Dr Bernhard then took the unusual step of exposing his patient to the Sun:

> By the end of the first hour and a half there was a marked improvement noticeable, and the wound presented quite a different appearance. The granulations became visibly more normal and healthy, and the enormous wound skinned over quickly.[33]

English children enduring sun-lamp treatment. Such lamps were safeguards
against tuberculosis, which spread alarmingly both in Europe and the USA
after the First World War

Bernhard began to treat other patients with sunlight and found further healing effects: the discharge from foul-smelling wounds rapidly became odourless and, in addition to cleansing the lesions, sunlight had an analgesic effect. He decided to treat open tuberculous cavities the same way. At that time the disease was rampant, claiming a million lives in Europe every year. Bernhard's approach was soon widely copied, one expert declaring:

> . . . Every room occupied or visited by patients should be flooded by sunlight whenever possible, because of all disinfectants sunlight has been shown to be the most powerful.[34]

In 1903 the Dane Niels Finsen was awarded the Nobel Prize for his use of artificial sunlight to cure tuberculosis of the skin. He was the theorist; others like Bernhard were the main practitioners. That same year, another Swiss physician, Auguste Rollier (1874–1954), began moving his patients to 5,000 feet above sea level, the better to catch the ultraviolet light: the first patient he cured was his fiancée. Rollier pioneered a treatment in which very slow tanning in cool conditions was combined with rest and fresh air: a patient's feet would be uncovered first, followed gradually, over the course of fifteen days, by the lower legs, the thighs, the abdomen and the chest, until finally the entire body could be exposed. It seemed to work.

But treating tuberculosis with sunlight remained on the borderline of accepted medical practice. When in 1905 Rollier presented his first results to a Paris audience, it walked out. Yet at his peak he was operating thirty-six clinics, with more than a thousand beds. (The leading practitioner of heliotherapy in Britain, Sir Henry Gauvain [1878–1945], did not believe that sunlight alone could cure advanced cases and combined his patients' exposure to sunlight with sea bathing.)

Rollier's work attracted the notice of architects, and the number of designs for sunlit buildings increased. Le Corbusier may have been the standard-bearer in France, but between the wars numerous beach resorts were built in southern Britain to draw people out of their bad housing and – for a while, at least – into healthy, 'life-enhancing sunshine'. By 1933 sunlight was said to be a beneficial treatment for over 165 diseases. In the years after the Second World War, it was a standard medical precept that the more sun we got the better. Schoolroom windows were made so that ultraviolet rays could penetrate easily, and boarding-school matrons would line up their charges at bath-time to stand under sunlamps.[35]

The Sun's health benefits were first taken up in earnest in Germany, with the treatment of the wounded during the First World War, after which sunbathing was recommended for those recovering from the flu epidemic of 1918–19 (which killed 21 million worldwide) and for children with vitamin-deficiency ailments inflicted by the Allied blockade. Almost immediately, the 'naturalism' movement began to advocate nudity for the sake of one's general well-being: indeed, nudists were the first modern sunworshippers.

After the First World War, a generation of writers grew up basing their philosophies on various aspects of 'solar' thinking. The critic Martin

Green has labelled them *Sonnenkinder* – Children of the Sun – dividing them into the dandy, the rogue and the naïf, all set on bucking the traditional culture.[36] Many artists and writers were among these early disciples. As one historian wrote of Rupert Brooke and his intimates: 'These Children of the Sun were enthusiastic nudists.'[37] Before him, Walt Whitman had made naked sunbathing part of his nature cult, and in the early twentieth century Hermann Hesse followed suit, travelling to Italy to cure his headaches and gout.

For many years 'sunbather' was a euphemism for 'nudist', while one of the first associations in Britain devoted to nudity was called 'The Sun Bathing Society', and most naturists belonged to 'sun clubs'. Giving naked sunbathing (and by clear implication its erotic potential) the royal nod while revelling in the eastern Mediterranean sun in June 1936, Edward VIII stripped down to demonstrate his longings for 'freedom' on the yacht *Nahlin* with Mrs Simpson. Other European nations at first responded fitfully to the craze, but from around 1925 there set in 'the Solar Revolution', one of the most startling reversals in modern social history.[38] By the end of the decade the Sun had become the design motif of the era, appearing on everything from cufflinks to stained-glass suburban windows, garden gates to wireless sets.[39] But the enthusiasm remained greatest in Germany. Stephen Spender witnessed it first hand:

> Thousands of people went to the open-air swimming baths or lay down on the shores of the rivers and lakes, almost nude, and sometimes quite nude, and the boys who had turned the deepest mahogany walked among those people with paler skins, like kings among their courtiers. The Sun healed their bodies of the years of war, and made them conscious of the quivering, fluttering life of blood and muscles covering their exhausted spirits like the pelt of an animal: and their minds were filled with an abstraction of the Sun, a huge circle of fire, an intense whiteness blotting out the sharp outlines of all other forms of consciousness . . .[40]

Spender's description also hints at the erotic charge of baring oneself to these life-giving rays. As Fowles wrote of Lyme Regis:

For many decades sea bathing remained what it had been to Jane Austen, a medicinal activity . . . and those who still braved Neptune direct did it from wheeled cabins. But the Victorian spirit . . . began to see the siren plainly – that is, to sense the always implicit eroticism and sexuality of the beach.[41]

It was not just the bodily display: the Sun has always been an aphrodisiac – 'hot' had the meaning of 'lustful' well before 1500. Only under the Sun, wrote Lawrence Durrell, could 'the essential male and female relationship' flourish 'uncomplicated by mirages and falsities'. His fellow novelist Henry Green went further: 'the weather lies at the root of the way women and men behave', with the result that 'the English in their relations with each other are less frank than other nationalities to the extent to which their skies are less clear and so by the less amount of Sun they have'.[42] And indeed many English people looked with dismay at the new fashion and the skimpiness of attire that was part of its appeal. In Alan Herbert's *Misleading Cases*, he creates the story of a man arrested around 1920 for wearing bathing drawers, not a full bathing suit. A newspaper of 1925 records how in the south-coast town of Bournemouth attendants were empowered to prevent visitors from sitting on the beach in their swimming costume, and bathers were commanded to 'walk straight into the sea and straight back to their bathing tents'.[43] Sunbathing continued to be banned in Bournemouth until the early 1930s. And as late as 1941, in Agatha Christie's *Evil Under the Sun*, sunbathing is presented as close to exhibitionism.*

Albert Camus (1913–60), that great French-Algerian chronicler of philosophical dread and human absurdities, has provided the nearest to what one might call a philosophy of sunbathing. In 1939, in 'Summer in Algiers', a long essay on its pleasures, he wrote,[45] 'Algiers and a few other privileged coastal towns open into the sky like a mouth or a wound.' One fell in love first and foremost with the sea, but also 'a certain heaviness of the sunlight'. Following from that was not the sunbathing hedonism

---

*One of the twelve signs by which anti-terrorist agents are supposed to recognize male suicide bombers is that they will have paler skin on the lower half of their face – they will have shaved off their beards, to help them blend in and look less suspicious, and their chins will have had no recent exposure to the Sun.[44]

of the rich, but the delights of the poor and the dispossessed, whose 'pleasures have no remedies' and whose 'joys remain without hope'. In summer the young men go down to the sea to enjoy 'the water's gentle warmth and the women's brown bodies':

Not that they have read the boring sermons of our nudists, those protestants of the body . . . They just 'like being in the sun'. It would be hard to exaggerate the significance of this custom in our day. For the first time in two thousand years the body has been shown naked on the beaches . . . Swimming in the harbour in the summertime, you notice that everybody's skin changes at the same time from white to gold, then to brown, and at last to a tobacco hue, the final stage the body can attain in its quest for transformation.

More prosaically, in 1986 a leading dermatologist commented: 'Why does anyone insist on tanning their hide? Because it's sexy. It's a cultural characteristic; a sign of health.'[46]

16

SKIN DEEP

Some people get too much sun down here. You must be careful.

US Head of Special Operations, Rio de Janeiro, to Ingrid
Bergman, in Alfred Hitchcock's *Notorious* (1946)

Sunlight stimulates and enlivens; it is of help in almost all conditions. It is the greatest of all natural tonics – like good
champagne, it invigorates and stimulates; indulged in to excess
it intoxicates and poisons.

Sir Henry Gauvain, *The Times*, 11 May 1922

Sunshine causes two common skin cancers – basal cell carcinoma (from
which I have suffered) and squamous cell carcinoma – and possibly a
third, melanoma. More than a million cases of the first two are diagnosed
annually in the US, the mortality rate rising at about three per cent annually since 1980.[1] One American in six will develop skin cancer. In Britain
the affliction is twice as common in teenagers and children as twenty years
ago, and in the population as a whole has risen by twenty-five per cent in
five years.* The annual number of non-melanoma skin cancers diagnosed
in the UK is 100,000, a figure that could triple by 2035.[2] But take any

---

*The British remain largely unaware of the dangers, but such ignorance is not confined to
the UK. In the US annually, one person in 200 will be sufficiently injured by the Sun to
require medical attention, while one in 29,500 will die. The corresponding odds from, say,
the West Nile virus are one in 68,500 and one in a million, yet in 2006 major American
newspapers and wire services ran 2,240 articles on the virus.

By the early 1920s, 'heliotherapy' was widely practised – if frowned upon
by the medical establishment

country with hot summers and a passion for sunbathing, and the statistics point in the same direction.

Skin is the body's largest organ (laid out flat, the average adult man's
would cover about twenty-one square feet), and in many ways the most
vulnerable. Less than a millimetre thick in places, it is made up of several layers, principally the epidermis (the entirety of which our bodies
replace every thirty-five to forty-five days), the dermis (containing our
nerve endings, as well as our sweat glands and hair follicles, which
together regulate our temperature) and a layer of subcutaneous fat (containing the tissue that acts as energy source, cushion and insulator). The
epidermis has five sub-layers, the innermost of which, known as the
Malpighian, is what is affected by tanning. Within this layer are the cells
that produce melanin, the pigment that increases in response to sunlight
and causes our skins to darken.

Lack of melanin causes the serious condition known as albinism (from
the Latin *albus*, meaning 'white'), a term coined by Balthazar Tellez
(1595–1675), a Portuguese explorer who was surprised by how many of
the West Africans he encountered had light skin and blond hair. Albinism

is set off by a genetic disorder characterized by a lack of pigment in the eyes, hair and skin (more rarely in the eyes alone). The principal gene causing albinism prevents the system from creating the usual amounts of melanin. It affects males and females equally, and most species of animal – mammals, fish, birds, reptiles and amphibians. Nor are plants immune: the albino cactus, *Mammillaria albilanata*, protects itself from the Sun by an extremely dense profusion of stark white thorns. Because their lack of melanin means that they will burn under the slightest exposure, albino organisms must spend their lives in constant battle with the Sun. Among human beings, up to one in seventy-five carries the albino gene. It is a miserable inheritance: African albinos, for example, have a rate of skin cancer a thousand times greater than their pigmented brethren. No albino over the age of twenty is free of the disease, and in the sunniest parts of the continent only about one in ten lives past thirty.[3] Albinos are often legally blind or afflicted by severe vision problems, because their retinas do not develop and are unable to absorb enough incoming light. Since melanin also affects parts of the body that never see daylight, their hearing and nervous systems are also compromised.

Unsurprisingly, as with so much related to the Sun, albinism's outward stigmata have aroused myths and superstitions around the world, particularly in poorer countries. In Jamaica, albinos are viewed as lower forms of being, to be held apart as cursed. In Zimbabwe, a folk myth has it that having sex with an albino woman will cure a man of HIV; many such women have been raped (and infected) by HIV-positive men. Elsewhere in sub-Saharan Africa, albinos are deemed to possess magical powers, a belief that has led to the ghastly practice of their being hunted down – 45 of Tanzania's 7,000-strong albino population have been killed for body parts since 2007, fetching $2,000 a corpse. Fishermen on the shores of Lake Victoria weave albino hairs into their nets to increase their catch.[4]

Literature rarely portrays these unfortunates,* but there is a moment in *Moby Dick* that compares albinism with the unnerving whiteness of the great whale:

---

*What few portrayals exist are rarely positive. Evelyn Waugh, writing in his diary of some dinner guests, noted characteristically, 'Charles Drage with his wife. She is an albino with a mind like a damp biscuit.'[5] H. G. Wells's *The Time Machine* paints the evil and light-fearing Morlocks as cannibal hominids resembling albino apes; Elric of Melniboné, the albino

What is it that in the Albino man so peculiarly repels and often shocks the eye, as that sometimes he is loathed by his own kith and kin! It is that whiteness which invests him, a thing expressed by the thing he bears. The Albino is as well made as other men – has no substantive deformity – and yet this mere aspect of all-pervading whiteness makes him more strangely hideous than the ugliest abortion.[6]

Answering his own question, Melville argues that the whiteness of the albino conjures up the pallor of ghosts and phantoms: 'Whatever grand or gracious thing he will by whiteness, no man can deny that in its profoundest idealized significance it calls up a peculiar apparition of the dead.'

☼

Our bodies produce two different types of melanin, eumelanin (brown) and phaeomelanin (yellow and red). Red hair is an index of the body creating more of the latter, less of the former, which induces poor tanning. Skin changes colour because of the increased production of melanin under ultraviolet light. Laboratory tests on people of different skin types at the equivalent of noon on a summer day have shown that the most sensitive skins – those with the least amount of melanin – burn after fourteen minutes of exposure; the most resistant dark skin takes more than seven times longer. Ultraviolet light acts in three forms, A, B and C. It is UVA, 'black light', that tans the skin; at sea level, ninety-nine per cent of UV radiation is UVA. UVB is what reddens the skin in sunburn, and is considered the main cause of basal and squamous cell carcinomas, and possibly a significant one of melanoma. Although less likely than UVB to induce sunburn, UVA

protagonist of a series of fantasy novels by Michael Moorcock, is both hero and antihero; while the albino Arthur 'Boo' Radley, in *To Kill a Mockingbird*, is presented as 'simple'. There is also the stereotype of the cold-hearted albino evildoer – one thinks of Silas, the religious multi-murderer in *The Da Vinci Code*: the shortcomings stemming from his disease, besides the obvious one of conspicuousness, should have rendered him a particularly ineffectual killer for hire. According to the Apocryphal Book of Enoch, Noah was albino, as, more historically, were Seinei, twenty-second Emperor of Japan, Edward the Confessor (1042—1066), king of England, and the Rev. Doctor Spooner, the Oxford classicist whose tendency to errors of speech gave us the term 'spoonerisms'. The aberration has been related to his nystagmus, which jumbled his vocabulary along with his vision.

A mother and her albino child, photographed in 1905 during a famine in South Sudan.
Most forms of the condition result from recessive genes inherited from both parents

penetrates the skin more deeply, and is considered the chief culprit
behind wrinkling, leathering and other aspects of 'photoageing' – it
not only increases UVB's cancer-causing effects but also may directly
cause some cancers, including melanomas. Since UVC is filtered out by
the atmosphere and hardly ever reaches us, we don't have to worry
about it for now, but the depletion of the protective ozone layer may
change that in the future.

In 1985 the American Academy of Dermatology (AAD), alarmed at
the growing incidence of these cancers, became the first medical organization to launch a public health campaign. Three years later an AAD
conference concluded that 'there is no safe way to tan'. By the end of
that year a leading modelling agency director (Eileen Ford) added her
voice: 'The tanned look is dead.' Widely publicized as they were, these
views had minimal effect. In a 1997 *Seventeen* survey, two-thirds of the
teenagers responding said they 'look better with a tan and feel healthier,

Freewheelin' Frank Reynolds, secretary of the San Francisco branch of the Hell's Angels. At the Human Be-in at the Polo Fields at Golden Gate Park of 1967, which drew a crowd of 20,000 and was a major counter-culture event, Frank spent as long as he could gazing heavenwards. Larry Keenan, the photographer who recorded this self-imposed ordeal, wrote: 'When I saw Freewheelin' a few weeks later, he told me his eyes had gotten burnt staring into the Sun on acid.'

more sophisticated'; half added that they appeared more athletic. As *Women's Wear Daily* noted in 2000: 'It will be a cold day in hell before there's a shortage of bodies suntanning on the beach, and this summer a sun-roasted hide is more fashionable than ever.'

Today, between 20,000 and 24,000 tanning salons are listed in the US Yellow Pages, claiming 22 million-plus regular clients each year. The light emitted by the sunlamps in these establishments is about ninety-five per cent UVA, five per cent UVB. The industry turns over $5 billion a year – Sarah Palin while she was in office even installed a tanning bed in the Alaska governor's mansion – despite the fact that in recent years the World Health Organization, the American Medical Association and the AAD have all labelled such beds the health-peril equivalent of cigarettes. The UK is averaging a hundred deaths a year from sunbeds alone, and studies over the last four years have suggested that both outdoor and indoor tanning can be addictive – that is, UV light helps release feel-good chemicals that relieve pain and generate feelings of well-being, potentially leading to dependency. A British research project recently reported forty per cent of those interviewed (girls between fifteen and seventeen in Liverpool and Sunderland) saying they went to tanning salons – Britain has eight thousand – at least once a week. Understandably concerned, in November 2009 the Scottish Parliament forbade under-eighteens to use sunbeds, and similar legislation is being considered in other countries.*

☼

An oddity on any list of sun-related ailments is the long-held fascination with staring at the Sun. For normal unprotected eyes, it takes about thirty seconds of direct exposure to inflict damage (in most people, natural reflexes make them look away, contracting their pupils). When injury does result, it takes the form of a blind spot in the centre of one's field of vision ('solar retinopathy'): this can be temporary but more often proves permanent. Yet it is apparently so widespread an enthusiasm that the *Guinness Book of Records* now specifically refuses to consider entries for the longest time spent staring at the Sun.[7]

In 1840 Gustav Fechner (1801–87), the German philosopher who founded experimental psychology, was blinded for three years as a result of a Newton-like experiment, in his case peering at the Sun through coloured glasses to investigate the effects on the retina. His injury was the more serious because he was looking through a hole in the shutter of a darkened room, which meant his pupil expanded. Thomas Harriot, the co-rediscoverer of sunspots, made protracted observations of the noonday Sun, and reported that 'My sight was after dim for an houre'; while G. K. Chesterton (1874–1936) wrote a whole murder story, 'The Eye of Apollo', based on the deceiving notion that one could, with training, look the Sun in the face.[8]

Even more dangerous than high noon is sunset, because it makes sungazing easier, even though the strength of UV rays is barely diminished. A similar danger lurks in the dim light of a solar eclipse. In 1999, after an occlusion over England, one London eye hospital

---

*Sportswear companies have built extra sun protection into their clothes by washing them in sunblock, applying a coating to absorb rays or choosing tightly woven fabrics that keep out ultraviolet light. Most contact lenses now contain UV protection.

reported that ten per cent of its admissions that day experienced some permanent loss of vision. On examination, doctors could pinpoint the actual phase of the eclipse that a patient had watched by noting the 'sickle' on the patient's retina – the arc of swelling that corresponded to the crescent-shaped solar portion left uncovered at that moment by the Moon. Besides eclipse-watchers, the people most susceptible to solar injuries are those who gaze up at it in a drug-induced stupor.*

☼

Too little sunlight can be as much of a problem as too much: for example, the very same rays that cause cancers also provide essential vitamins. Before urbanization, most people spent hours each day in the Sun, so lack of exposure was not an issue; but for the last 200 years ever more people in the northern and western hemispheres have lived mostly indoors, so that many have grown up with insufficient sunlight, and thus insufficient vitamin D – the substance that is produced in our bodies by exposure to the Sun. Besides strengthening our immune systems, this vitamin is vital for the absorption of calcium, which protects against rickets (a condition that softens bones and consequently deforms skeletons), and also combats multiple sclerosis (people born in October – whose mothers have had a summer of sun – rarely get the disease; but this is maybe because UV light is doing something beyond making Vitamin D), rheumatoid arthritis, hypertension, premenstrual tension, diabetes, influenza and several cancers.[10] It has been documented that deaths from cardiovascular disease are higher in the winter, when less sun reduces levels of the vitamin.

The vitamin comes in two forms: D2, derived from plants, and D3, from ultraviolet B (UVB) rays and from animals (cod liver oil, salmon, mackerel, sardines and fortified dairy foods are particularly rich sources). According to Dr Michael Holick, the distinguished Boston

---

*It also appears that about twenty-five per cent of people sneeze when exposed to sunlight. Francis Bacon commented at length on the subject in *Sylva Sylvarum* (1635), so the malady has been known about for centuries. Dr Tom Wilson, a leading pathologist, writes: 'We do not know exactly why this happens, but it might reflect a "crossing" of pathways in the brain, between the normal reflex of the eye in response to light and the sneezing reflex. There is no apparent benefit from "sun-sneezing", and it probably is nothing more than an unimportant (but annoying) holdover of evolution.'[9]

The young girl in the centre of the photo, who suffers from the rare disease Xeroderma Pigmentosum (or XP), which affects one person in a million, must wear protective clothing for even the briefest of exposures to daylight, and goes through three bottles of sunscreen a week during the summer. On average, the rest of us can withstand sixty years of regular exposure to sun before skin cancers may develop; XP sufferers develop the same tumours by the age of ten

University professor of medicine, between ninety and ninety-five per cent of our intake comes from casual exposure to sunlight.[11] Both types are converted into an active form in the liver and kidneys. Between twenty and thirty minutes provide 10,000 units of the vitamin: an infant needs about 200 units every 24 hours, an elderly person 600, an adult about 1,000. The elderly are most at risk in part because they tend to stay indoors – in Europe as a whole, one old person in three lacks adequate vitamin D.[12] In Saudi Arabia, despite the ample sunshine, women get rickets and other vitamin D-deficiency-related conditions because their traditional clothing covers nearly all their skins.*

---

*This prompts the question, how long can one live without the Sun? Some people seem to exist almost entirely without sunshine. There have been communities that dwell in deep caves – the *Trogloditae* found in desert communities in Tunisia, Morocco and Libya, or the Miao who have dwelt for centuries in the mountain caverns of south-west China's Guizhou Province. In 1989 a young Italian, Stefania Follini, spent a hundred and thirty days in a New Mexico cave, sleeping ten hours at a time but often staying awake for twenty-four. She so lost track of time that

☼

I will limit the list of disorders brought on or aggravated by the Sun: they make for distressing reading. But one ailment with wide-ranging implications is porphyria cutanea tarda (from the Greek for 'red/purple'), a flaw in the body's ability to metabolise the red pigment in blood. While sunshine does not cause the disease, it exerts a profound effect on anyone who has it, which explains the fear of the Sun among sufferers.

Briefly: a component of haemoglobin, which helps transport oxygen around organs and tissues, is produced in eight stages, each catalysed by a separate enzyme. In porphyria, one step fails to take place, causing a backup. The body compensates by dumping the compounds, often into the skin, which forces pigments to accumulate there and in bones and teeth. Even at its mildest, the disease causes abdominal pain, nausea, vomiting, weakness, confusion, rapid heart rate and urinary problems.[14] At its worst, porphyrins, unthreatening in the dark, are transformed by sunlight into caustic, flesh-eating toxins, and if left untreated can eat away a victim's ears and nose and erode lips and gums to reveal red, disease-bared teeth.

The disease has become inextricably entwined with the vampire myth in general and Dracula in particular. Such myths go back thousands of years and appear in almost every culture (India boasts vampire-like supernatural beings who dissolve at the Sun's rising). The word 'vampire' came into the English language in 1732, possibly from the Slavic verb 'to drink', via best-selling books about a Serbian farmer who had reportedly been killed by a vampire and returned from the dead to feast on his neighbours. In 1897 the legend was reworked by an Irish writer named Abraham ('Bram') Stoker, who combined the unlikely callings of theatre manager and legal scholar. His inspiration was Henry Irving, the overbearing actor-manager of London's Lyceum Theatre who had

---

when Maurizio Montalbini, the sociologist conducting the experiment, informed her that her allotted four months' isolation was up, she thought she had been underground for only half that time. It would appear that the biological clock is not a perfect instrument, and will tend to run fast or slow if deprived of synchronizing clues.[13] But so far none of those who have experimented with periods underground have suffered long-term ill effects.

Max Schreck in the greatest of vampire films – *Nosferatu: A Symphony of Horror*,
directed by F.W. Murnau in 1922. The title is Slavic for 'plague-carrier'

already played the evil Hyde in *Dr Jekyll and Mr Hyde*, and whom Stoker
hoped would play the count in a stage version; and he gave his famous
figure many of the marks of porphyria – including a dread of sunlight.[15]
According to Stoker, vampires can attack only between sunset and sunrise, so while Dracula may walk about even under the brightest day, he
cannot exercise most of his powers except when the Sun is down.

The portrayal of vampires continued to flourish without much change
or critical commentary until, in 1985, the biochemist David Dolphin
suggested that it was porphyria that underlay the original legends,
pointing to the number of similarities between it and vampirism: porphyria victims are intensely sensitive to sunlight, and even mild
exposure can cause disfigurement (hence the myth of the vampire as
werewolf); to avoid sunlight, people with porphyria go out mainly at
night; today porphyria can be treated with injections of blood products
(centuries ago, sufferers might have treated themselves by drinking

blood);* porphyria is inherited, but the symptoms may remain dormant until triggered by stress; and garlic contains a chemical that makes the symptoms worse, turning a mild attack into an agonizing reaction.[17]

Professor Dolphin's argument was never widely accepted, and indeed any discussion of vampires and porphyria may seem an exercise in melodrama; but the long-standing connection does underscore our deep-seated belief that sunlight ensures health, along with its converse – our fear of those who, whether porphyric or albino, seem to shun the light.†

☼

Porphyria has a distant cousin: psoriasis. More than any other skin ailment, it is amenable to the healing effects of sunshine. Although the actual name (from the Greek *psora*, to itch) was not introduced until 1841, the condition was first discussed by Hippocrates (460–377 BC): some scholars believe it to have been included among the skin conditions called *tzaraat* in the Bible, and it has frequently been described as a variety of leprosy.‡ Unpredictable and infuriating, it is also the most baffling of disorders, characterized by skin cells that multiply up to ten times faster than normal: as they reach the outer skin and die, their sheer volume leaves dry red patches covered with white scale. Psoriasis typically breaks out on the

---

*At death, a great deal of blood generally collects in the lungs. As the body decomposes, internal gases build up. After about four days, the pressure causes a 'bloody purge', as the double lungful of blood is expelled through the mouth and nose, both of which remain full of blood, the face probably covered in it, and a large puddle collecting around the body. This may be the origin of the belief that some corpses were not 'properly dead' but continued to live and began to drink the blood of fellow humans. Ancient methods of dealing with the 'undead' may also have lent even more credence to the belief that they were still alive: when the bodies of suspected vampires were exhumed and a wooden stake driven through the chest, any pooled blood would spurt out, while the escaping gases would generate what sounded like a sigh. This was taken to be the release of the spirit of the undead from the curse of vampirism.[16]

†Vampire mythology is not the end of the fascination with porphyria: in 1966 the theory was advanced that the disease had set off the madness of King George III, and, since it is inherited, it has been traced back to Mary Queen of Scots and forward to Queen Victoria (who always felt uncomfortable in sunshine, as indeed reportedly other royals have been).[18] As of this date, the theory remains unproven, but it is by no means absurd.

‡The history of psoriasis is littered with treatments of dubious effectiveness but high toxicity: the application of cat faeces, for example, in ancient Egypt, onions, sea salt and urine, goose oil and semen, wasp droppings in sycamore milk, and soup made from vipers have all been applied. In Turkey, 'doctor fish' in the outdoor pools of spas are encouraged to feed on the skin of psoriatics, and cleverly consume only the affected areas. There is still no real cure.

knees, elbows and scalp but can affect the torso, palms and soles. Its prevalence in Western populations is around two to three per cent.[19]

Sufferers range from the late playwright Dennis Potter to the musicians Art Garfunkel and Elton John; and include three revolutionaries: Joseph Stalin, Abimael Guzman (the leader of the Shining Path movement in Peru) and Jean-Paul Marat (1743–93), the radical journalist of the French Revolution, famously murdered while in his bath. Marat spent most of his days in that tub, in water mixed with medications and with a vinegar-soaked cloth around his head like a turban, because he had developed psoriasis while hiding from the police in the Paris sewers. Had he ventured outdoors, under the healing rays of the Sun, he might twice over have saved his skin.

Another victim was John Updike, who wrote that the disease 'favors the fair, the dry-skinned, the pallid progeny of cloud-swaddled Holland [of whom Updike was one] and Ireland and Germany'. In his memoirs he recounted that he had been under its thrall from the age of six. 'Only the Sun, that living god, had real power over psoriasis,' he writes. 'A few weeks of summer erased the spots from all of my responsive young skin that could be exposed – chest, legs, and face.'[20] As an adult, he attended Massachusetts General Hospital for ultraviolet light treatment 'in a kind of glowing telephone booth'. For Updike, the relationship with the Sun became intense, almost personal:

> From April to November, my life was structured around giving my skin a dose of Sun. In the spring, at Crane's Beach, though the wind off the sea was cold, the Sun was bright and the hollows in the dunes were hot. I would go there alone, with a radio and a book, and without benefit of lotion inflict upon myself two hours of midday ultraviolet. I wanted to burn; my skin was my enemy, and the pain of sunburn meant that I had given it a blow. Overnight, the psoriasis would turn from raised pink spots on white skin to whitish spots on red skin; this signaled its retreat, and by June I could prance back and forth on the beach in a bathing suit without shame.

But then something odd happened. Though sunlight healed Updike's diseased skin, it also gave him headaches and made him nauseous, dislocating any attempts he made at consecutive thought – 'Writing,' he lamented, 'is a thoroughly shady affair.' Then, in his early forties, he found the psoriasis was fighting back: 'My skin, once so thirsty and grateful for solar rays, and so sensitive to them, had grown toughened and blasé . . . At forty-two, I had worn out the Sun.'

☼

The Sun (or rather its absence) is associated with one other very different illness: SAD, or seasonal affective disorder, so named in 1982 and defined as depression caused by the absence of sunlight during winter months. The official definition requires that a patient exhibit a lifetime of mostly winter depressions, including at least two straight years without any episodes out of season. Sufferers become anxious, irritable, incapable of concentrating or acting decisively, and avoid social activities. They tend to eat and sleep excessively, their sexual drive decreases, and they are beset by fatigue, feelings of worthlessness and guilt.[21] Women are three times as likely as men to be affected. SAD diagnoses account for about one case in three of clinically recognized depression, and only about ten per cent of its major manifestations. (A milder form, 'winter blues', is said to be at its worst from December through February, mainly afflicting those aged eighteen to thirty. There is also Non-Seasonal Affective Disorder: Non-SAD.)

SAD is believed to involve a disruption of the body's 'circadian rhythm' (from the Latin for 'around a day') – the roughly twenty-four-hour cycle that governs biological processes in both plants and animals and acts as a body clock, affecting sleep, alertness and hunger. The circadian 'clock' is located in the brain's suprachiasmatic nucleus (SCN), which is in the hypothalamus; it takes in information about the length of day and night from the retina and passes it on to the pineal gland, a pea-sized structure at the base of the brain – sometimes called 'the third eye' – which secretes the hormone melatonin. The amount of melatonin fluctuates cyclically, peaking at night (when it helps us to sleep) and ebbing during the day.[22]

Norman Rosenthal, the South African psychiatrist who gave a name to the condition, has himself been troubled by deep depressions since 1976, and now directs a company that makes 10,000-lux lamps (retailing at $250 each) to help the light-deprived; he and his students have produced many books and papers on SAD, reinforcing the impression of its ubiquity. Rosenthal has a website and runs clinics for sufferers – in the US alone estimated at 14 million 'severely affected' and a further 33 million with winter blues.[23] Forty-seven million Americans seasonally depressed? Common sense and experience, of course, tell us darkness *is* depressing – even in some of the sunniest parts of the world. For most people, dark days and long, cold nights *do* lower the spirits, but as one of the people I interviewed told me, 'Most people just get on with things.' Which raises the question: is SAD really a disease? There are studies that suggest that this is so, others that deny it, and yet others that argue that it exists but that the correlation between its occurrence and the absence of sunlight, or between it and the degree of latitude, is less direct than Rosenthal and others believe to be the case.

I became curious about what life was like in a place where the condition should be endemic – in the chilly polar wastes – and to find out set out for the island of Tromsø, 300 miles north of the Arctic Circle, some two hours' flight from Oslo. The capital of the Norwegian province of Troms, it is the largest permanent community so close to the Pole, with about 60,000 inhabitants. Even though it receives the same amount of sunshine in a year as the tropics, it does so in a single concentrated stretch during the summer. For ten weeks each year, from mid-November until the end of the third week in January, virtually no sunlight falls, a season that people call *mørketiden*, or 'the dark times', during which (at least according to many press reports) depression soars, as do the figures for mental and physical illness, divorce, arrests for brawling, and suicides.

I arrived expecting to be burdened by the oppressive gloom of a place in which, as *The New York Times* put it, 'The darkness and cold are enough to stiffen a soul nearly dead'.[24] To my surprise I found almost the opposite. For a start, there is no perpetual night. From about ten until noon most winter days the sky is silvery-grey, quite enough to see by, and the abundant snow reflects what light there is. I arrived late in the evening, and it was odd the next morning reading a local newspaper in which all the exterior photographs seemed to have been taken at night.

I had arranged to see several experts at Tromsø University – the noted astronomer Truls Lynne Hansen; a senior librarian; and various members of the psychology department. Four were Norwegian, the others being a Mexican and a researcher from Ohio (who said she had left her home state because she found it so depressing). I also read three doctoral dissertations on SAD. And found, to my further surprise, that conclusive evidence is definitely not yet in.

A Norwegian research paper of 1991, involving 128 participants, neatly introduces its conclusions with a quotation from Hippocrates: 'Whoever wishes to pursue the science of medicine in a direct manner must first investigate the seasons of the year and what occurs in them.' It mentions that from the late eighteenth century scientists regularly described patients with seasonal disorders, but ends by admitting that although SAD occurs at a time when people are exposed to less daylight, 'no definite proof of a causal relationship' can be found. Again, 'It is not yet clear whether what is identified as . . . SAD represents a distinctive affective syndrome, a subtype of recurrent affective disorder, or the most severe form of a widely distributed population trait.'[25] Another paper, completed in 1997, records that, of the tests carried out in the Oslo area, the examining psychiatrist was struck by the 'non-depressive' look of her patients, who 'generally scored low on the item "apparent sadness"'.[26]

Judith Perry, one of the Tromsø psychologists, has studied whether sensitivity to seasonal changes has led to eating problems, expecting to encounter SAD more prevalently the further north she went. But the results suggested the opposite: 20.7 per cent of those tested in Nashua, New Hampshire (42.30° N), for instance, exhibited symptoms of SAD, against a figure of 11.3 per cent in Iceland (62–67° N). 'The interpretation of this discrepancy remains obscure,' she admitted. 'Nevertheless, there is reason to question whether latitude per se is critical to the development of SAD.'[27] Another research initiative concluded: 'This is the second study to find the prevalence of SAD and sub-SAD to be lower among Icelanders or their descendants than among populations along the East Coast of the US.'[28]

Four other psychiatrists reported on a study of a hundred participants from Tromsø, whose abilities over a range of cognitive tasks (they were all prospective aeroplane pilots) were tested both in summer and winter: 'The conclusion is negative. Of five tasks with seasonal effects, four had disadvantages in summer . . . Although the ideas in the SAD literature and anecdotal evidence and the extreme latitude were all reasons to expect a winter deficit in cognition in this study there was little evidence for one: there was more support for a *summer* deficit.'[29] None of these doctors was denying the existence of SAD, just questioning whether it could be purely correlated with coldness and lack of light.

It was one of the Tromsø psychologists who put SAD into perspective. Her view was that it affected only those already suffering from clinical depression: it further depressed the already depressed. The researchers I talked to in Tromsø did not belittle the agonies depression can bring; but they believe that SAD has laid its grip on far fewer than is claimed by those who evangelize on the subject. On the other hand, there *are* those figures for crime and divorce, which may or may not be linked to depression . . . it seems to me that we are still in the early days of understanding how we react to the loss of light.

I left Tromsø recalling a research trip to Heidelberg just a month before, when I had escaped from a depressingly rainy day into a local tavern. I was served by a Spanish girl, who had symbols of Earth, Fire, Wind and Water, all framed within an image of the Sun, tattooed on the small of her back. I asked why she had chosen that design. 'It is because I love the Sun,' she replied. 'And while it may not be here in Heidelberg, it is always *here*' – and she vigorously patted her rump.

17

THE BREATH OF LIFE

Dear Proffesser, We are in the sixth grade. In our class we are
having an argument. The class took sides. We six are on one
side and 21 on the other . . . The argument is whether there
would be living things on earth if the sun burnt out . . . We
believe there would be . . . Will you tell us what you think?
Love and lollipops, Six Little Scientists

    Dear Children: The minority is sometimes right – but not
in your case. Without sunlight there is: no wheat, no bread, no
grass, no cattle, no meat, no milk, and everything would be
frozen. No LIFE.

Correspondence between schoolchildren
and Albert Einstein, 1951[1]

The Sun, with all those planets revolving around it and dependent on it, can still ripen a bunch of grapes as if it had nothing
else to do.

Galileo Galilei[2]

Several times a month, Bill Albers, head doorman of my New York apartment building, pushes under my door envelopes marked 'From the Man
at Re-Write', each containing some item from a magazine or newspaper,
which range from 'Ripley's Believe It or Not: Pistol Shrimp have claws
that fire shock waves to stun their prey, creating a wake of bubbles as hot
as the surface of the Sun!' to recondite articles on solar energy or the

latest theories about sunspots. One evening I told him I was writing about the Sun's effects on animals and plants, and, while I would be discussing animal behaviour, I had to begin with explaining photosynthesis. He asked how this was going. 'Slowly,' I told him. 'The whole process is so complex.' 'Complex!' he echoed with a laugh. 'It's simple – just repeat what you learned in primary school.'

Maybe he is right. I well remember being taught that our fuels and foods derive from plants, and in turn their energy comes from sunlight, with the main difference between us and plants being that we (and all other animals) absorb energy from the Sun indirectly, in the form of food, whereas plants receive it directly. The process by which plants do this is called photosynthesis ('putting together with light'), which occurs both in bacteria and in 'photoautotrophs' – i.e. any plant organisms that can synthesize their own food by using sunlight, converting energy from a physical activity into a chemical process. It is under that latter heading that most plants are classed.

By means of a compound called chlorophyll ('leaf green'), plants use the hydrogen of water to transform carbon dioxide into more complex chemicals based on carbon, including sugar molecules such as glucose; the water's oxygen is then liberated as a gas – a plant waste that, while precious to us, certainly has its drawbacks. As Bill Bryson observes in *A Short History of Nearly Everything*, oxygen, although critical to animal life on Earth, is more often than not toxic: 'It is what turns butter rancid and makes iron rust. Even we can tolerate it only up to a point. The oxygen level in our cells is only about a tenth the level found in the atmosphere.'[3]

Carbon is passed into leaves through the energetic effect of sunlight, and those leaves pass on the materials to form the wood of a tree or the petals of a flower, the chlorophyll within them capturing energy from the violet and red parts of the solar spectrum and transforming it by driving electrons around certain molecular circuits in a series of chemical reactions. The Sun engages over 25 million square miles of leaf surface every day. But just one to three per cent of the light falling on a green plant becomes biological energy; the rest is lost by transmission, reflection or ineffective absorption.[4]

Photosynthesis takes place in structures within cells called chloroplasts, typically only a few thousandths of a millimetre in width. They contain the chlorophyll and other chemicals, especially enzymes ('levellers': proteins that control specific responses), necessary for the various reactions. The scientists who work in the area don't yet entirely understand the complex biochemistry of photosynthesis, although it is the most important metabolic innovation in our planet's evolutionary history. Each day in midsummer, an average hectare of corn creates enough oxygen to meet the respiratory needs of about 325 people. The rates at which living beings die and consume one another are so high that we would all disappear within a human lifetime were it not for this process.

As far back as the 1640s, researchers suspected that plants required air and water to grow, and by the 1700s had begun to identify the individual gases involved in combustion, respiration and photosynthesis. Then a major step forward took place – thanks to a city brewery. In 1772 plans were well advanced for Cook's second voyage in search of the hypothetical Great Southern Land. Several scientists were given permission to join the expedition, and the Royal Society at first approved the inclusion of an astronomer/botanist named Joseph Priestley; but his notoriety as a religious and political freethinker led to his invitation being revoked. Priestley (1733–1804) instead took up the post of salaried literary companion to Lord Shelburne, a Whig grandee. While in Shelburne's service, he settled down to his experiments.

Priestley's previous job had been in Leeds, where his house abutted a brewery, and he had been drawn to experiment with the fumes given off by the fermenting beer, which he noticed remained in the vats to a depth of a foot or so, and did not mix with the air above it. This gas was carbon dioxide (or 'fixed air', as he called it), and his tests showed that lighted candles would go out when put into it. He knew that the air we breathe differs from carbon dioxide, but he wasn't sure how many other forms there might be. Soon after taking up his post with Shelburne, he demonstrated that air was a mixture of gases, not an element (as the ancient Greeks had propounded). He argued for the existence of a near-weightless substance called 'phlogiston', which he posited had to be present to create fire. On surer ground, he also identified nine separate gases, all of which he believed to be polluted versions of 'normal' air. These gases would later be named nitrous oxide ('laughing gas'), ammonia, sulphur dioxide, hydrogen sulphide, carbon monoxide, chlorine, silicon tetrafluoride, hydrogen chloride and something he called 'de-phlogisticated air' – an awkward term for what came to be called 'oxygen' (from the Greek, 'the acid-maker').

Priestley showed that a candle would burn more brightly when this gas was pure, and that even a mouse could survive in it (previous experiments had sent several mice, happily drunk on beer fumes, to an early grave*). The gas was 'between four and five times as good [i.e. for respirability] as common air', he said; given that we now know the air we breathe is twenty-one per cent oxygen, the rest being mainly nitrogen, this estimate was remarkably accurate. He showed that a candle in a closed container eventually spluttered out, but, after a sprig of mint was added, the air would invigorate a flame, not douse it: he had shown that plants release oxygen.

Priestley went on to demonstrate that oxygen is taken up by blood in the lungs, and that water is a compound of hydrogen ('water-maker') and oxygen, in the proportions of two to one, as measured by gas volume. But that was the end of his original work in this area. While he had demonstrated that plants revived the air after animals vitiated it, he failed to realize that they needed sunlight to do so – which seems odd, as he was well acquainted with the work of Stephen Hales (1677–1761), who had speculated that leaves were the 'lungs' of a plant, and asked: 'May not light also, by freely entering the expanded surfaces of leaves and flowers, contribute much to ennobling the principles of vegetables; for as Newton puts it in a characteristically probing query, "Are not gross bodies and light convertible to one another?"'.

Unfortunately, Priestley soon became less interested in science than in preaching against the divinity of Christ and campaigning against slavery. In 1791, a Church-and-King mob razed his meeting houses to the ground, moving on to his home, reducing it – laboratory and all – to

---

*In 1773 a poem by one Anna Barbauld described the experiment from the mouse's point of view, 'The Mouse's Petition to Dr Priestley, Found in the Trap where he had been Confined all Night' – perhaps the first animal-rights manifesto ever written. Commentators on Priestley have suggested that his interest in asphyxiation was spurred by reports of the Black Hole of Calcutta (1756), the horror in which forty-three out of sixty-nine people in hot, airless incarceration had died. Around the world, caves and similar confined spaces collect carbonic acid naturally, to deadly effect; twelve miles outside Naples, on a mountainside near Lake Agnano, is the Grotta del Cane (the Grotto of the Dog). Carbonic acid gas, being heavier than normal air, gathers at the bottom of the grotto, so while taller animals – such as humans – can breathe easily, smaller animals perish, or are thrown into convulsions. In the nineteenth century it became a loathsome local industry to push dogs into the grotto to regale tourists with the gas's effect.[5]

ashes. Eight rioters and one special constable were killed. The great scientist took himself off to America, and the baton passed to the Dutch botanist Jan Ingenhousz (1730–99), whose research revealed both that stale air could be revitalized only by the green parts of plants, and that this required the aid of sunlight.*

Ingenhousz also discovered it was the Sun's energy in the form of light, not warmth, that was essential for plant respiration. He demonstrated that, under the action of light, plants absorb carbon dioxide through minute pores in the surface of their green parts, giving off tiny bubbles of oxygen; in the dark, the bubbles eventually stop. 'It seems to be more than probable,' he wrote, 'that leaves are destined to more than one purpose':

> It is also probable, that the tree receives some advantage from the
> leaves absorbing moisture from the air, from rain, and from dew; for
> it has been found a considerable advantage to the growth of a tree, to
> water the stem and the leaves now and then . . . It will, perhaps, appear
> probable, that one of the great laboratories of nature for cleansing and
> purifying the air of our atmosphere is placed in the substance of the
> leaves, and put in action by the influence of the light.[7]

As the revolutions of science gathered speed, so too did the unravelling of the mystery of photosynthesis. In 1845 Julius Robert Mayer (1814–78) explained that plants converted light into chemical energy; just how this came about, however, was to preoccupy scientists for over a century. In the 1920s photosynthesis was confirmed to be a number of successive steps that combined two separate and opposite uses of sunlight: the photo-oxidization of water and the photo-reduction (i.e. deoxidation) of carbon dioxide.

During the 1950s and 1960s the American chemist Melvin Calvin

*A debate has raged through the centuries over what degree of credit Priestley deserves. Thomas Kuhn, in his study *The Structure of Scientific Revolutions*, favours either Antoine Lavoisier (1743–94) or Carl Wilhelm Scheele (1742–86) as the discoverer of oxygen, Jan Ingenhousz of photosynthesis; but his main point is that such breakthroughs are never as simple as legend or history suggests.[6] Priestley himself, writing to a friend about Ingenhousz's work, mentions the considerable difference between the day and night activities of plants as something 'he hit upon and I missed'. It was Lavoisier who measured another form of air called simply 'gaz' (a word not yet assimilated into English and meaning 'ghost, spirit'), and renamed it 'hydrogen'.

demonstrated that light reactions generating chlorophyll were not a protracted process but transformed the Sun's energy instantly.* Working
with green algal cells, he was able to identify at least ten intermediate
products formed within a few seconds. Over the past decade, researchers
have pieced together the story of how the photosynthesis necessary for
a flourishing biosphere first evolved on Earth.[10] Other researchers are
trying to work out how to duplicate the chemical processes involved, in
order to find new sources of energy. A pan-European collaboration
called Solar-H has been set up to find ways of extracting hydrogen by
means of sunlight, to help store energy as a fuel. A consortium in
Sweden is studying artificial analogues to photosynthesis. Other groups
are developing algae with small antennae to generate local, low-tech
biomass-energy. And so on.

Photosynthesis may at one level be easy to comprehend, but (as Bill
Albers and I came to agree) it is not simple. Indeed, studies of it are a
growth industry, and one's head spins at some of the questions that have
occupied recent years. How do plants respond to too *much* light? What are
photosynthetic bacteria doing inside grains of desert sand? If one reverses
the processes of photosynthesis, can water molecules be put back
together? How do the north and south faces of a vineyard differ in their
sugar production? Some of these questions, at least, can now be answered.

☼

The responses of the natural world to the Sun are so wondrous that they
can seem almost infinite. Several species of mollusc that register daylight
lay down a certain thickness of cells, which thickness – a sandwich of
daily imprints – directly correlates to the number of hours of daylight to

*The biochemist George Wald has argued that chlorophyll is so well suited to photosyn-
thesis that it is the only pigment that can function in the role, and that life on other planets
would have to evolve chlorophyll in order to photosynthesize. However, in April 2008
Scientific American conjectured what plant life might look like elsewhere, and reckoned that
photosynthesis that didn't necessarily involve cholorophyll was very likely on other planets, quoting H. G. Wells in *The War of the Worlds*: 'The vegetable kingdom in Mars, instead
of having green for a dominant colour, is of a vivid blood-red tint.' Mars, the magazine
argued, has no surface vegetation; but light of any colour from deep violet through the
near infrared could set off photosynthesis, which adapts to the spectrum of light that
reaches organisms, which in turn is the result of the parent star's radiant spectrum.[8] When
photosynthetic organisms first appeared on Earth, the atmosphere lacked oxygen, so they
could not have used chlorophyll; not even Earth was always green.[9]

which the mollusc was exposed, and one can gauge their age by the number of layers. Coral fossils found in Devon reveal an extraordinary periodicity in growth rings – about 400 in each annual set – by which evidence we can calculate that, some 370 million years ago, there were about 400 days in the year, each lasting about 22 hours.[11]

Many myths tell of the power of the Sun to affect nature. 'The Sun breeds maggots in a dead dog, being a god kissing carrion,' says Hamlet, reflecting one superstition. *Antony and Cleopatra* repeats another: that the Sun's rays engender serpents by spontaneous generation, an old mummy's tale disproved only in the seventeenth century. A hundred years on, the French inventor Joseph Nicéphore Niépce (1765–1833) coined the term 'actinism' to describe the Sun's ability to produce chemical effects in non-living things; some minerals, such as white marble, can become phosphorescent – light-emitting – by long exposure. Niépce noted that 'granite rocks, and stone structures, and statues of metal, are all alike destructively acted upon during the hours of sunshine'.[12] More, the Sun starts fires, sinks ships by warping their planks and in the world's hottest zones can 'roast [rocks] into disintegration'.[13] In 1814 the British scientist Humphry Davy subjected a diamond to intense heat using a huge solar lens: the gem eventually burst into flame, leaving a fine crust of charred carbon, proving that it was little more than a lump of coal.

Julius von Sachs (1832–97), one of the greatest of nineteenth-century German scientists, systematized what he called 'phototropism', from the Greek for 'light-turning' – the tracking of the Sun by organisms. (I saw this in action when in July 2006 I visited the most advanced tomato farm in southern Spain, where seedlings are turned on their sides twice daily, first one way then the other, so that their stems will always bend towards the Sun, making them stronger.)* Plants position

---

*Later in the day the seedlings are taken out again and stroked with small hairbrushes, building up the stems' resistance in order to strengthen them still further. A side effect of absorbing a great deal of sunlight is absorbing a great deal of heat: most plants' leaves are prone to heat stress – and if overexposed to ultraviolet rays shrivel and die. So the roofs of the greenhouses are now made of light-dispersing plastic and no longer of glass, which can lead to tomatoes getting sunburned. This is ironic, because recent research has suggested that eating tomatoes can protect against sunburn and premature ageing of the skin. Experts at Manchester and Newcastle Universities have found that the fruit improves the skin's ability to protect itself against ultraviolet light: the effect is thought to be caused by the pigment that makes tomatoes red.

themselves with extraordinary accuracy to catch as much sunlight as possible: gaze up at a forest canopy, and the leaves overhead will usually form a near-continuous ceiling, like the pieces of a jigsaw. They are not acting as some neighbourly cooperative but are instead fierce competitors for light. The more of it they can catch, the better their chance for survival, and some have evolved particularly ingenious mechanisms for doing so. Thus the giant edible aroid that flourishes in marshy parts of the Borneo rainforest not only has leaves ten feet across and a surface area of over thirty square feet; it coats the underside of its leaves with a purple pigment that catches the light after it has passed through the thickness of the leaf, giving the chlorophyll a second helping, as it were. Begonias found on the same forest floors have transparent cells in the upper surface of their leaves that act as tiny lenses gathering and focusing light on to the chlorophyll within.

Generally, competition drives plants to grow taller, but to do that they need structures to prevent them from toppling, so roots become thicker, either spreading laterally through the surrounding soil or plunging downwards. Trees offer a particularly efficient solution to the problems of photosynthesis, but we should be careful not to be earthbound in our thinking. As Oliver Morton writes:

> Think of a birch tree in winter, its leaves lost, its architecture revealed in dark lines against cold grey cloud. . . . Take away the tree's established 'common-sense' context by turning round, bending over and looking at it upside down through your legs. Its growth looks less like something pushed from the earth than it does something drawn from the sky . . . The bulk of the tree is not made from the soil beneath it – indeed, the soil is in large part made by the tree. . . . Trees are built from Sun and wind and rain. The land is just a place to stand.[14]

The Sun plays an even greater part in the life of those organisms that are constructed around a central core. This includes the largest family of flowering plants, the composites, especially sunflowers (translated from the Spanish as 'the Hearbe of the Sunne', then changing their name three times before settling on Helianthus, after the Sun god) and other flowers that are shaped like daisies (the word itself meaning 'day's eyes'). The sunflower, whose turning to follow the Sun was first scientifically described by Leonardo da Vinci in his botanical studies, was imported into Europe about 1510 from the Americas by the Spanish. The flower was sacred to the Aztecs and had been used as an emblem of the Sun god by the Inca; within a few decades of its arrival in Europe it had become an accepted symbol of loyalty because of the way in which it followed the Sun so closely. But not all sunflowers do that: the larger ones, with thick stalks and seeds, such as Evening Sun, Mammoth Grey Stripe and Sunspot, do; but smaller sunflowers do not, and may even turn away from it.*

David Attenborough begins *The Private Life of Plants* by declaring, 'A shoot kept in the dark will creep towards a single chink of light. The plant can see.'[15] Pardonable hyperbole. The pursuit of sunlight obtains even under extreme temperatures: certain polar animals regularly mark out their territory by locating the Sun against particular landmarks, while *Lecidea cancriformis*, an Antarctic lichen, can carry out photosynthesis at −4°F. The Arctic poppy will face east from midmorning, bending westwards from late afternoon (the motion is carried out by motor cells in a flexible segment just below the flower, the pulvinus). The high-altitude snow buttercup aligns itself in a similar way, sunlight helping the flower to maintain ideal levels of temperature and humidity, and enabling it to attract insects more effectively. Nature forever adapts.†

---

*Several flowers do not welcome sunlight. There are at least 3,000 species of non-photosynthetic plants, many of them parasitic on other plants, often feeding off fungi which in turn draw their energy from trees. Such plants are an odd tribe. Both peanuts and truffles ripen under ground (as succulent roots, such as potatoes and carrots, obviously do). In 1729 the French astronomer J. J. de Marian noticed that certain plants in darkened rooms opened up by day and closed at night independent of any sunlight. The ghost orchid, which entirely lacks chlorophyll, spends so much time under ground, flowers so irregularly and grows in so few places that in some countries it has been declared extinct, only to reappear. A Western Australian orchid, *Rhizanthella gardneri*, blooms under ground and never emerges above the soil at all. At the start of the autumn rains, it extrudes a tulip-shaped structure that grows upwards, lifting the soil above it so that cracks in the earth form, from which drifts a faint, insect-attracting perfume.

†In his fine memoir *Wildwood*, Roger Deakin talks about reflected sunlight and how insects are attracted to the brightness of books, dragonflies and damselflies in particular often settling on the page. 'What else could they be doing but resting and sunbathing, perhaps even catching up on some lost sleep?'[16]

In 1920 researchers working for the US Department of Agriculture discovered that the flowering of many plants was controlled by the amount of daily light they receive, and introduced the term 'photoperiodism' to describe this response. Plants were subsequently categorized as being short-day, which will not flower when daylight exceeds a certain critical length; long-day, which will not flower *unless* the days exceed a critical length; and intermediate, which bloom whatever the duration. It was next learned that, for plants to flower, the duration of daily *darkness* is also crucial. Thus, short-day plants bloom when nights are long, and long-day plants when nights are short or non-existent. One wonders if talking to any informed countryperson would not have led to the same conclusions, for the names of many plants have long reflected this sensitivity to light and dark. At least fifty varieties of flower observe regular hours of opening or closing, some of which are named accordingly: hence calendulas ('little weather glasses') became first 'gold-flowers' (as a quite different flower, chrysanthemums, are now known), then, through an association with the Virgin Mary, 'Mary's gold', and finally 'marigolds', celebrated for opening their flowers only during the brightest hours of sunlight. It is, as Perdita acknowledges in *The Winter's Tale*,

> *The marygold that goes to bed with the Sun*
> *And with it rises weeping.*

The 'shepherd's sundial' (also called 'the poor man's weather glass') opens in summer a little past 7 a.m. and closes just after 2 p.m.; when rain is imminent, it doesn't open at all. But many plants without such names also open and close on a schedule. The garden lettuce spreads itself at 7 a.m. and draws itself in at 10 a.m. – and so on. This orientation to the Sun of what are known as 'horary' ('hour-keeping') plants was put to practical use in 1751, when Carolus Linnaeus (1707–78) suggested that a clock be made of flowers, so the time could be reckoned by seeing (or smelling) which variety had just opened up. The rotation was complex, but at least one roster of openings went: 5–6 a.m. – morning glories and wild roses; 7–8 a.m. – dandelions; 8–9 a.m. – African daisies; 9–10 a.m. – gentians; 10–11 a.m. – California poppies; noon – morning glories

Flowers open and close at certain times of day. In 1745 the Swedish botanist Carolus
Linnaeus proposed a flower clock to complement sundials: it would indicate the time
within a half-hour of accuracy. This dial was painted in 1948

close, goat's-beard opens; 4 p.m. – four o'clocks open; 4–5 p.m. –
California poppies close; 6 p.m. – evening primroses and moonflowers
open; 8–9 p.m. – day lilies and dandelions close; 9–10 p.m. – flowering
tobacco opens; 10 p.m.–2 a.m. – night-blooming cereus opens. When
Linnaeus planted a version of the clock in the garden of his summer
home six degrees north of Uppsala, he took account of the difference in
latitude – for instance, he reckoned that goat's-beard would open there
at 3 a.m. to meet the first rays of the midnight sun – and shuffled his
flowers accordingly.[17] Such a clock is earlier described by Andrew
Marvell in 'The Garden' (1678):

> How well the skilful gardener drew
> Of flow're and herbs this dial new . . .
> How could such sweet and wholesome hours
> Be reckoned but with herbs and flow'rs!

In recent years, various forms of flower clock have been planted in
places as disparate as Tehran, Iran, and Christchurch, New Zealand, but

their accuracy is doubtful, as flowering time is so much at the mercy of weather.*

Research initiated in 1960 discovered that each variety of plant reflects light slightly differently, so a satellite in space should be able to identify the plant life of any given area on Earth. In 1972 the US, then under Nixon's presidency and preoccupied with Soviet plans for world dominance, created a task force to re-evaluate Soviet crop expectations. As Dan Morgan tells us: 'Information about Soviet crops was seen as vital economic intelligence with a bearing on the economic security of the United States.'[19] Several months later, LACIE (Large Area Crop Inventory Experiment) was set up, and by 1977 a US satellite accurately caught the Evil Empire's wheat crop six weeks prior to harvest. LACIE seems to have been curtailed shortly afterwards, but it is possible that some new form of agrospacial espionage has been initiated since. We do know that in 1995 the US navy researched whether blooms of bioluminescent algae could be used to track submarines. (They couldn't.) But between 1992 and 2001 a scientific group known as MEDEA (Measurements of Earth Data for Environmental Analysis) advised the federal government on environmental surveillance, and after energetic lobbying by Al Gore for its revival, it was reported in January 2009 that 'The nation's top scientists and spies are collaborating on an effort to use the federal government's intelligence assets – including spy satellites and other classified sensors – to assess the hidden complexities of environmental change'. So the use of satellites seems never to have stopped; only now it is espionage helping the environment, not the other way round. But who knows?[20]

☼

---

*During the 1920s a young scientist named John Nash Ott, experimenting with time-lapse photography, discovered that different wavelengths had differentiable effects on photosynthesis. He extrapolated that varied light frequencies might also affect human well-being.[18] Thereafter he worked on the links between light and cancer, and by the late 1960s the US Congress had passed a Radiation Control Act, its co-author crediting Ott with 'getting us all started on the road toward control of radiation from electronic products'.

Ott's work was sufficiently well known that several decades later he was asked by Paramount Pictures to make a time-lapse photo sequence of flowers for Barbra Streisand's film, *On a Clear Day You Can See Forever*, in which her singing is meant to cause the flowers in her house to bloom within seconds.

As with plants, so, often, with the animal kingdom. Sunfish, strikingly ugly, seemingly tailless creatures that can grow to six feet long – they are the heaviest bony fish in the world – live deep down in the oceans during stormy seasons (they are known as the 'couch potatoes of the sea'), then rise to sunbathe on the surface in bright weather. In the Sahara, foraging ants orient themselves by registering the polarization of light as well as the Earth's magnetic field and then employing their spatial memory; animals such as albatrosses and turtles that live almost all their lives in or over the sea use the Sun as a navigation beacon in much the same way. The minute beach crab *Talitrus*, with barely a millimetre of nerve substance, is able to calculate the time of day, even the course of an hour, from the angle formed by its body and the Sun's position. In response to changes in sunlight, scores of animals change colour with the seasons, altering pigmentation and camouflage along with their environment.

The Sun also plays its part in reproductive activity. As sunlight fades, herring shoals will swim ever closer together and move into shallow water, where they will develop their eggs, protected by their strength in numbers. As the Sun rises, so the shoals will move apart again.[21] Many of the more brilliant tropical birds live in the upper canopy of rainforest, where they are bathed in abundant light and can show themselves off to prospective mates to maximum effect. Other lushly plumaged creatures take advantage of the stippled sunshine that breaks through to lower parts of the forest to perform acrobatic mating displays, in which their colours glimmer in the fractured light, 'like dancers beneath a spinning disco ball'.[22]

Heliconius butterflies use polarized radiation to choose mates, their employment of visual cues in sexual selection being an example of light exploitation that may also have adaptive value in dense forest, where illumination varies greatly in colour and intensity. This raises the question of what Sun-inspired activities exist beyond our powers to observe. 'Some birds can simply see what humans can't,' says Dr Miyoko Chu of the Cornell Laboratory of Ornithology: for instance, blue tits differentiate among one another in ways inaccessible to humans. It has been known for some time that birds (along with some lizards, fish and insects) can see into the ultraviolet, but in 1998 researchers discovered that certain plumages reflect frequencies invisible to human eyes: while we have three cone types in our eyes, birds have four, admitting them to zones of the electromagnetic spectrum outside our ken, and greatly extending their range of colour combinations.

In the summer of 1944 Karl von Frisch (1886–1982), who in 1973 would share the Nobel Prize in Physiology with Konrad Lorenz, realized that when bees wagged their behinds in the hive they were signalling where to fly. Bumblebees do two dances, one circular, the other in a figure of eight, which Frisch interpreted to mean – to give one of his own examples, *nectar 1.5 km away and at 30 degrees from the Sun* – and identified the bees' modes of communication, showing their sensitivity to both ultraviolet and polarized light. They could take off on a bearing coordinated with the Sun's passage, even setting up flight paths to take in sunny resting places. This ability to keep at a constant angle to the Sun regardless of the passage of time Germans charmingly call *Winkeltreue*.[23]

When a bee arrives back in the hive after visiting a newly opened flower, she dances on the landing platform in front of the colony entrance (a colony will number 20,000 in winter, 60,000 in summer), first making a circle, then bisecting it, waggling her abdomen and buzzing energetically. Then she enters the hive and sets to again: sufficiently stimulated, she may carry on dancing for close to four hours. The further into the hive she goes before dancing, the further away the pollen source. As the combs of the nest are vertical, the waggle-steps cannot point directly to the flower and instead refer to the Sun. If the bee crosses the circle vertically, then the food source is in line with the Sun. If the target is, say, fifteen degrees to the right, then the dance will be fifteen degrees to the right of the vertical. The workers surrounding the dancer assimilate the message and fly away to find the flower. When they return with the pollen, they too will dance, so that most of the workforce will soon be actively gathering.[24] Flushed with his discovery, Frisch set himself to work out how bees could possibly convey the Sun's position. Hard as it was to believe, he concluded that they could even predict where the Sun would be all the way round the clock, so that if they kept dancing, changing their pattern regularly, they would reproduce its movements.*

*Among the insects that use the pattern of polarized light as an optical compass are ants and spiders, the latter being equipped with a specialized pair of secondary eyes just for this purpose. These do not 'see' in the normal sense but have a built-in filter that determines the direction of polarization. Spiders are mainly active after sunset, and use this aid to find their way back to their nest after foraging trips.[25]

Of the thousands of species of ant, several are similar to honeybees in the way they use the Sun to orient themselves, and have a similar spectral sensitivity: what the great biologist E. O. Wilson calls 'the almost fantastic ability' to 'memorize the path and angular velocity of the Sun'.[26] What goes on in the brains of both insects during foraging trips is extraordinary: an outward-bound worker ant typically winds and loops in 'tortuous searching patterns' until it finds food; during each of the twists and turns on its journey it registers the constant light of the Sun, and is aware of the angles it takes relative to it. On its return home, it simply reverses its mean angle by 180° – hardly a simple feat: a human traveller would require compass, stopwatch and integral vector calculus.

Other creatures orient themselves by the Sun when they migrate, a feature of life for many animals, from caribou, which travel some 2,000 miles, the longest such movement overland, to the baby loggerhead turtle, embarking on its 8,000-mile trek around the Atlantic, to the Zambian mole rat, whales, salmon (famously), eels, homing pigeons, the common toad, and birds that fly from Alaska to New Zealand, non-stop.[27] Every summer throughout North America, 650 different species feast and nest; come the autumn, 520 of those species migrate south, to return the following spring. The direction in which they fly depends largely on the length of daylight but also on the bird's temperature sensitivity.[28] They journey partly in pursuit of food; but as much an impetus is the decreasing (or increasing) daylight.

Gliding consumes only a twentieth of the energy required for sustained wing-beating, so soaring species like broad-winged hawks ride warm-air columns – 'thermals' – generated by the solar heating of the Earth, sometimes a mile above ground, covering as much distance as possible in long shallow sweeps.[29] They can gauge where thermals are powerful and reliable, and wait for the best conditions before setting off. Because thermals are fuelled by sunlight, they are most common on long summer days, although a blazing sun can seriously overheat flyers – a danger that geese avoid by travelling at night.

For centuries, scientists (like most people) thought birds unintelligent – hence the dismissive 'bird-brained' – but within the last five years we have come to understand the processes by which these creatures orient themselves and learned a new respect for them.[30] Day fliers, for example, which use the Sun to orient themselves while airborne, must possess some form of internal clock to keep track of time, because the position of a point on the Earth relative to the Sun changes by fifteen degrees every hour – so to orient itself consistently a bird must determine the Sun's apparent trajectory relative to its own direction at various times throughout the day; in other words, its solar compass has to be time-compensated.[31] To give a bird's-eye view: if the Sun at its present location is higher than it would be at the bird's goal at this time of day, the bird would have to fly away from the Sun; if lower, fly towards it.

☼

Just as some animals seek out the Sun, others avoid it. The tiny Selevin's mouse, or desert dormouse, a Russian rodent discovered only in 1939, is unable to withstand more than about eight minutes' continuous exposure without falling ill. Not possessing sweat glands, reptiles are especially sensitive, and will pack themselves tightly into any space where there is shade.[32] They are called 'cold-blooded', but this is a misnomer: they actually regulate their temperature quite precisely by moving in and out of the Sun. An albino king snake will often not survive in the wild because it can't get warm enough. Australian 'magnetic' termites build large, thin mounds with the flat sides facing south and north so that they can go to the north side to make use of the Sun's warmth, and then, when the heat gets to be too much, cool off by moving south.[33]

Certain animals seem to do without the Sun completely. One such, the olm or *Proteus anguinus*, is found in the subterranean waters of southern Europe, notably those of the Soča river basin near Trieste. It has lungs, four feet, tiny teeth forming a sieve to keep larger particles inside its mouth (and so is presumably a predator), an eel-like head and snake-like lower body, and no fins or eyes; it eats, sleeps and breeds underwater. A transparent fleshy white, if exposed to light it changes to an olive tint.

Several species have lost their useless light-adapted organs. Others (such as bears, bats and owls) much prefer darkness – one recalls Professor Higgins complaining of Eliza, 'She's an owl sickened by a few days of my sunshine' – but if completely deprived of sunlight, these animals would soon die.

The Texas blind salamander lives in underground streams in caves, and will never see
the Sun: fewer than a hundred are known to exist

In the early 1960s a British scientist, Dr James Lovelock (b. 1919), who
had been working for NASA on detecting life on Mars, advanced the
'Gaia Theory', named after the Greek goddess of Earth. 'The biosphere
is a self-regulating entity with the capacity to keep our planet healthy by
controlling the chemical and physical environment,' he argued in a best-
selling book, *Gaia: A New Look at Life*.[34] The living parts of the Earth, runs
the theory, regulate the inanimate parts (atmosphere, oceans) to their
own ends, stabilizing the environment and keeping it benign, a complex
system that can be thought of as a single organism, and thus our world
regulates itself to make itself hospitable to the amazing number of interacting species that constitute its 'life'. The globe, in other words, works
to keep everything in ideal balance.

Among a battery of criticisms from other scientists was that such a
theory assumed that evolution had foresight and planning, whereas the
evidence suggests that life is mainly a matter of chance. Lovelock refined
his ideas in a second book (he has since written four more), developing
a mathematical model he calls 'Daisyworld': on a hypothetically otherwise unpopulated planet, two different types of plant, black and white
daisies, are so organized as to maintain a balance of temperature and
atmosphere ideal for overall daisy growth: the black absorbs sunlight and

warms the Earth, the white reflects light and cools it, with each species growing or declining as necessary. In other words, living systems stabilize their global environment. It is not yet clear how well the Daisyworld formula can catch the full biospheric and climatic complexity of the Earth, and scientists, particularly evolutionary biologists, still regard Lovelock with suspicion. However, some of his arguments have been accepted. As Oliver Morton says, 'The idea that life is content passively adapting to environments over which it has no sway – which really was the dominant paradigm just forty years ago – has gone for good.'[35] We are now more ready to accept the idea of the Earth as the Sun's collaborator.

18

THE DARK BIOSPHERE

The ocean looks like a fairly homogeneous place; it is all
water, after all . . . Yet that seawater, like the seafloor, only *seems*
to be homogeneous . . .

Robert Kunzig, *Mapping the Deep*[1]

Why did Doctor Ballard, on his journeys to photograph the
Titanic, habitually play classical music during the descent and
rock music on the way back up? It is unthinkable that it could
have been the other way around.

James Hamilton-Paterson[2]

According to legend, Alexander the Great had himself lowered into the
Mediterranean in a glass cage whose door was fastened with chains, and
during his vigil observed a fish so large that it took three days and three
nights to swim past. He is said to have recorded: 'None of the men who
have been here before me, and none of those who shall come after me
shall see the mountains and the seas, and the darkness, and the light that
I have seen.'[3]

That darkness covers many unknown worlds. Even now, ninety-five
per cent of our oceans remain unexplored,[4] and oceanographers are
fond of pointing out that we know more about the Moon's surface
than the sea floor. We do know that throughout the great seabeds
wind a series of volcanic gashes – about 60,000 kilometres of carbon-
dioxide-spewing mountain ranges, pierced by immense scalding

springs. The oceanographer Bruce Heezen has named it 'the wound that never heals'.[5] Through these fissures surge great quantities of water at temperatures up to 750°F, almost four times boiling point at surface pressure, but here prevented from passing into steam by the down-weighing miles of ocean above. Laden with inorganic chemicals of all kinds – sulphates, nitrates and phosphates, along with compounds of hydrogen oxide, carbon dioxide and methane gas – the water smothers the seabed with untold amounts of insoluble salts that can never rise to the surface. The one exception to this comes through vents known as 'black smokers' – furious chemical reactors so hot that they leech copper, iron and zinc out of the Earth's crust and erupt above the surface to form columns of rock belching what looks like

A digital recreation of a hydrothermal vent. While nearly all life on Earth depends on solar energy, the many thousands of creatures that live round these vents survive on organic material produced through chemosynthesis. However, even at these depths some still depend upon oxygen liberated by photosynthetic organisms black smoke. The largest yet found, 'Godzilla', off the coast of Washington State, is sixteen storeys tall.[6]*

Some bacteria manage to exist in these immensely hot places, using iron as a means of respiration – just as we employ oxygen – metabolizing it into the black magnetic mineral magnetite. More than a hundred different species dependent on these bacteria have been discovered around the vents, not only microbes but holothurians – 'sea cucumbers' – eight-foot-long tubeworms that grow in thick clumps, with blood-red heads like rosebuds; shrimp and clams, each of which can be as much as a foot across; reefs of six-inch mussels; and scrabbling, porcelain-white crabs – all using the energy derived not just from the oxidization of ferrous iron salts but also from sulphur, hydrogen sulphide and molecular hydrogen. The iron-devouring bacteria feed the clams and shrimps, which in turn are eaten by the crabs and by a sinister-looking five-foot-wide whitish-grey hooded octopus. Nearly all these beasts were new to science, and cannot live anywhere else. This part of our world has been named the 'dark biosphere'.

What has all this got to do with the Sun? These creatures and habitats are unique, constituting the only ecosystem on Earth that derives its energy from chemosynthesis, and exists not on trickle-down sunlight but by blast-up geochemistry.[7] Yet even here photosynthesis plays a role, as the vast bulk of the oxidants used by these creatures comes from the light-driven ecosystems at the ocean surface. Even the sulphur-eating bacteria can live only in places where they also have access to oxygen, because they need it to oxidize hydrogen sulphide, stealing its electrons and latent energy. If the Sun were to disappear, most of these deep-sea communities would not long outlast those higher up (although a small number, bizarrely, would continue to exist, sustained by the light of the hot lava of the vents themselves).[8] So the Sun, indirectly, provides life even here.

The creatures photosynthesizing at the surface, which can extract

---

*One piece of subterranean research is worth mentioning. On 16 January 2006 the *Chikyu* ('Japan's Apollo mission') set off to drill 4.3 miles below the seabed, more than three times deeper than ever before, 125 miles off the coast of Nagoya, south-west of Tokyo. What it dredges up may show whether the energy that originated the first semblance of life on Earth was geothermal rather than solar. It may also help us understand why the Earth's magnetic poles repeatedly switch.

hydrogen from water, are blue-green algae – cyanophytes, or 'blue-growers' – found wherever there is constant moisture. 'The arrival of the blue-greens marked a point of no return in the history of life,' writes David Attenborough. 'The oxygen they produced accumulated over the millennia to form the kind of oxygen-rich atmosphere that we know today.'[9] These photosynthesizing organisms, made up of single cells called phytoplankton (Greek for 'wandering plants'), which measure at most a few thousandths of an inch across, are confined to a thin layer near the ocean surface; for light is quickly absorbed, and below about 700 feet normal photosynthesis is impossible. (The greatest depth at which conventional plant life has been found is 710 feet off the Bahamas, where a clump of maroon algae prospers in exceptionally transparent water.) But that thin layer is rich in photosynthesizing agents, and all animal life in the sea, from jellyfish to whales and even the denizens of sea-floor hot springs, depends on these cells. With the end of winter, chlorophyll, the molecule that plants use to absorb sunlight, turns the entire North Atlantic green. No one can number the many species of phytoplankton, but these seaborne plants (along with land-dwelling forests) draw out of the atmosphere half the carbon dioxide that we put in – and make life possible not just on land but at the greatest depths.

The oceans can be divided into two major realms, the shallower seas that fringe the continents and the deeper oceanic waters. The former, the continental shelf areas, are like submerged shoulders of the land-masses: it is here that the vast majority of marine fish live. At depths of 300–600 feet, the shelf suddenly gives way to great slopes that descend to what is called the abyssal floor, cut by winding V-shaped formations that, if exposed to terrestrial eyes, would rival the Grand Canyon. Huge trenches have been formed, in some cases plunging nearly seven miles down, and throwing up the Mid-Oceanic Ridge, foremost of all the mountain ranges on Earth, which stretches in an unbroken chain from the Arctic through the Atlantic to the Antarctic, Indian and Pacific Oceans, a distance of some 40,000 miles. The greatest depth to be sounded so far was measured in 1962 in the Mindanao Trench off the Philippines, where the sea floor lies 37,780 feet down, a gash that could drown Mount Everest with almost a mile and a half to spare. About eighty-six per cent of the world's oceanic water lies below 3,000 feet.[10]

One recalls that the Latin word *altus* translates as both high and deep, a meaning that lingers in the English phrase 'the high seas'.

In 1951 the naturalist Rachel Carson (eleven years later to publish her celebrated study of environmental dangers, *Silent Spring*) wrote *The Sea Around Us*. 'Down beyond the reach of the Sun's rays,' she begins poetically if not with total accuracy, 'there is no alternation of light and darkness. There is rather an endless night, as old as the sea itself;'

> If we subtract the shallow areas of the continental shelves and the scattered banks and shoals, where at least the pale ghost of sunlight moves over the underlying bottom, there still remains about half the Earth that is covered by miles-deep, lightless water that has been dark since the world began.[11]

Light is filtered out layer by layer, wavelength by wavelength, as it penetrates downward beginning with the ultraviolet and infrared, which are absorbed in the uppermost three feet.[12] In the night sky the lights of an aircraft are visible many miles away; even when overhead, the same lights under water fail to reach beyond 600 feet.[13] Below the first 300, the bottom of the 'euphotic' or 'well-lit' zone, red wavelengths are completely absorbed, and with them all the orange and yellow warmth of the Sun. By 500, only around one per cent of sunlight remains. Then the green goes too, although in the clearest waters blue-green illumination is still detectable at nearly 3,000 feet. By 1,000 feet down there is only a very dark blue. Hamilton-Paterson, in a diving bell that eventually descended to 4,978 feet, recorded at 700: 'Surprised there's any light outside, but there is . . . The sea's intense violet color is strangely piercing . . . It's a quality of light I've never quite seen before. It doesn't exist on the Earth's surface and perhaps can't even be produced artificially.'[14]

As sunlight falls away, myriad new creatures fill the darkness – inhabiting zones known respectively as twilight, sunless and bottom-living. In 1818 Sir John Ross, exploring the Arctic seas, brought up from 6,000 feet mud embedded with worms, 'thus proving there was animal life in the bed of the ocean'. In 1860 the survey ship *Bulldog* found evidence of life far beyond the Sun's assumed reach: shrimp, lanternfish, squids and arrow worms. Carson recounts the remarkable voyage of HMS *Challenger*, the first vessel equipped for oceanographic exploration, which set out from England in 1872 – although she is not specific about details. In 2006 the *New Yorker* writer David Grann, on a hunt for giant squid, retraced the *Challenger*'s voyage. The ship had roamed the globe for three and a half years, dredging the ocean floor in a zigzag stroll covering 68,930 miles at the rate of two knots – the equivalent of a slow walk. The work was repetitive and brutal – two men went insane, another killed himself – but by the journey's end the explorers had landed 13,000 different kinds of animals and plants. It took the next nineteen years to process this booty, which included 4,700 new animals, 2,000 of them living below 820 feet – about a tenth of the (even now) known species of fish. It was evident that between the waves and the floors of the deep ocean basins there thronged the largest and perhaps most remarkable biological communities anywhere, a diversity rivalling that of the tropical rainforests.*

The oceans total 320 million cubic miles of water and 140 million square miles of sea floor stretching over seven-tenths of the planet. Their depths have been inhabited for a comparatively short time, for they present awesome challenges to life: the extreme cold, as well as the crushing weight, forbid easy adaptation.[16] For the unprotected human body, 600 feet is the absolute limit; one-atmosphere-pressurized diving suits allow man to go as deep as 2,500 feet. But for the often-transparent creatures of the deep, the pressure inside their tissues is the same as that without, and so is not a problem. However, how sperm whales endure the pressure – close to 200,000 pounds per square foot in dives of more than 9,800 feet – remains a mystery. The concentration of oxygen in the lower depths is only one-thirtieth of that close to the surface, but within the zone of lowest oxygen a unique group of animals have adapted to meet such challenges.

Darkness might seem yet another obstacle, but the creatures of the deep have answers to that too. While at 1,500 feet the unaided human eye can see only a coarse pattern of silhouettes and shadows, fish can discern the subtlest differences. Such sensitivity may regulate their depth,

---

*In November 2009 scientists reported that the number of species known to be living beneath the ocean waves was now 17,650. Some 5,722 species live deeper than the 'black abyss' of 3,280 feet, including the whalebone eater, the see-through cucumber, the 'wildcat' tubeworm, the jumbo dumbo (a specimen of octopod) and the yeti crab, which lives on hypothermal vents roughly 7,200 feet down.[15]

with changes in light levels controlling the morning and evening migrations of various fish and krill. Even eyeless creatures may register the Sun's presence: water molecules polarize the Sun's light, which helps many animals in their hunting, as the tissues of their prey absorb or rotate the light passing through them. Squid, employing special cells embedded within their skin, also use polarized light (which would appear indistinguishably black to human eyes) to regulate their vertical migrations and to signal to one another.[17]

Where there is not enough light, many create their own. Between 600 and 700 species have this facility, but they employ it conservatively – as every time they create light, they expose themselves. However, some creatures can vary the wavelengths they emit, thereby using light itself as camouflage. Several have modified cells that act like torches, to be turned on or off at will (presumably to find or pursue prey), while others have rows of lights over their bodies or at the end of their legs or tentacles, another effective misdirection. Some fish have developed light-producing organs on their undersides, whose weak downward glow erases the shadow cast when they are viewed from below against the relatively better-lit waters above. Others use bioluminescence as a booby trap, coating an advancing predator with sticky, glowing tissue that makes the attacker vulnerable to other visually cued hunters, 'like bank robbers marked by exploding dye packets hidden in stolen currency', as one marine biologist colourfully puts it.[18]

Some eyeless creatures light up brightly when touched. Others concentrate red body pigment, which absorbs whatever blue-green light falls on them and reflects nothing, an effective 'visual stealth' strategy. The predatory *Malacosteus niger*, known as the Lightless Loosejaw, generates long-wavelength infrared (in addition to blue bioluminescence), invisible to other deep-sea animals, giving it vision in the same light zone as a military sniperscope. The fish's optic retinas contain derivatives of chlorophyll (harvested from its diet of tiny crustaceans), which in modified form allows it to generate infrared. During the Cold War, the US military considered using these creatures as sentries to signal the passage of Soviet submarines. In the end, however, Operation Lightless Loosejaw never materialized.[19]

Transparent animals, the better to conceal themselves from predators, have developed the smallest guts possible, as the intestines are the only

The deep-sea angler fish makes its own sunlight: searching for food at between a
thousand and fifteen hundred metres down, it will lure its prey with a bioluminescent
glow at the end of a long rod. The impressive teeth will then impale the curious

visible part of their anatomy.[20] In some, the stomach is needle-shaped
and always points down, whichever way the animal is oriented, mini-
mizing its presence against the surface light. Others cloak the stomach
in reflective tissue, which in the open ocean reflects light that is indistinguishable from that behind it.

All these highly varied creatures, be they fish, crustaceans, algae or
other organisms, inhabit the deep oceans well away from the Sun's glare.
Yet even the minutest micro-organisms living on or near the sea floor
owe something of their existence to the Sun. Swimming through H_2O
along the darkest frontier of the solar realm, they rely indirectly on the
oxygen released by the photosynthesis activated at the surface.*

*The Sun, however, primarily energizes organisms living in the shallows, not the depths.
Among the functions that certain coralline algae perform is to manufacture a chemical
that acts like a high-factor sunscreen shielding both themselves and their polyp hosts from
ultraviolet rays. Alginates from brown seaweed offer considerable protection from radiation, since they secrete iodine compounds that help treat cancer in humans. Japan has one
of the world's lowest rates of breast, endometrial and ovarian cancers, attributed to the kelp
konbu, which contains U-fucoidan, which interferes with tumour cell colonization, causing these cells to self-destruct. In protecting itself from the Sun, seaweed has provided us
with a way to safeguard ourselves.

☼

The ocean, writes Robert Kunzig in his magical book *Mapping the Deep*, 'is Earth's primary distributor of heat. Sunlight causes water molecules to strain at their hydrogen bonds, pushing and pulling their neighbors, and the heat that is stored in those vibrations can be transported great distances – by ocean currents that are driven by the heat, by variations in the amount of salt dissolved in the water (which affects its density), and by winds.'[21] The ocean's liquid topography is constantly shifting – which leads us on to those mighty agents of the Sun, currents and tides.

Currents, which operate at all sea levels, are of two types, surface and deep-water. Sometimes they flow simultaneously in different directions at different depths. There are nine major such ocean streams:

1. The Gulf Stream, which originates south of Florida and flows north along the East Coast of the United States, crossing the North Atlantic to the Norwegian Sea. It has a huge impact on the long-term weather of any landmass that it touches – the overall climate of Norway and the British Isles, for example, is about 18°F warmer in winter than continental interiors at the same latitude.

2. The Labrador Current, which cuts south from the Arctic Circle and travels along the Canadian coast, cooling the Canadian Atlantic provinces and, by journey's end, coastal New England. Some of it runs off into the Gulf of St Lawrence, but most of its chilling mass continues south-west.

3, 4, 5 and 6. The North Equatorial Current and South Equatorial Counter Current (3 and 4), which travel east–west near the equator. Both are about 600 miles wide and reach roughly four to ten degrees either side of the equator but never touch it, because deflected by their counterparts, the South Equatorial Counter Current and the Equatorial Counter Current (5 and 6).

7. The Kuroshio Current, which separates from the North Equatorial Current and sweeps the eastern coasts of Taiwan and Japan. It then splits to form an eastern branch flowing nearly to the Hawaiian Islands and a northern branch that skirts the coast of Asia to merge with the cold Oyashio Current to form the North Pacific Current.

8. The California Current, which runs from the Gulf of Alaska along

the West Coast and is partially responsible for the relative coldness of the water there.

9. The Doldrums, which are centred slightly north of the equator where the intensity of solar radiation makes the air especially humid; lowers air pressure (by the heated atmosphere expanding outwards); and creates clouds, light, variable winds, and thunderstorms, squalls and other severe weather patterns. The Doldrums are also characterized by long periods in which the winds disappear, becalming sailing vessels for days or weeks – thus giving the language one of the few nautical phrases that has a non-nautical meaning.

Two main factors determine how currents form and re-form: the Sun's great heat and the Earth's rotation (a third and lesser factor being the gravitational pull of the Sun and the Moon). By warming the air above the oceans, the Sun creates winds that drive the water along through friction (surface currents take up the upper quarter-mile of any ocean, about a tenth of our total hydrosphere). If water cools or is made saltier through evaporation, it becomes denser and sinks, setting off fresh currents, which in turn transport heat from one location to another, both upwards and outwards, further altering temperature patterns.[22] Aeschylus wrote of the 'multitudinous laughter of the waves', but the Sun's power to drive winds across open water can build ocean surges up to a hundred feet, trough to crest, and at any given moment there are ten such monsters churning through the oceans: not quite the light-hearted image conjured up by Aeschylus. Lying in front of such waves can be a deep depression, known as a 'hole in the sea', which can fling any ship that it traps on a fatal rollercoaster ride. These rifts form regularly in regions swept by powerful currents: the Agulhas off South Africa, the Kuroshio off Japan and the Gulf Stream off the eastern United States – which last also flows through the Bermuda Triangle, that mythical devourer of ships and aircraft.[23]

Some oceans' layers can be no more than a few inches thick, but each is distinguishable from its neighbours by sharp changes in temperature and saltiness, which together govern its movements up and down. Water nearer the equator – and thus more intensely heated by the Sun – can be 40°F warmer on the surface than further down. The warm currents at the equator, in part because they are lighter than currents to the north and south, sweep towards both Poles. The Earth's rotation acts on ocean currents through what is called the Coriolis Force – named after the nineteenth-century French engineer and mathematician Gustave-Gaspard Coriolis (1792–1843) – which deflects currents north-eastwards in the northern hemisphere and south-eastwards in the southern. Such movements are affected by friction with the Earth only at the sea floor. The Earth's eastward rotational velocity decreases from its maximum of 1,000 mph at the equator to zero at the Poles. Equatorial water is deflected at the speed of the planet's fastest rotation; as it migrates across progressively slower latitudes, it is still travelling more quickly than the waters through which it passes, and given its greater momentum, has to move diagonally north- or south-eastwards, in the direction of the planet's turning.

Beyond currents, there are monster waves, and these too have a solar cause, if indirectly. In 1997–8 people around the world watched satellite images showing winds change direction and a huge pool of unusually warm water roll across the tropical Pacific to mass into an area about one and a half times the size of the continental United States. This was the famed El Niño, now blamed for rearranging the weather patterns of countries as far apart as Chile and Australia and bringing storms, flooding, drought and brush fires. 'El Niño' means 'the little boy' in Spanish and refers to the baby Jesus (the effect usually reaches its peak around Christmas). It has a twin sister – a rebound of unusually cool water, called 'La Niña', which shows up on satellite images in brilliant blues and purples stretching across the mid-Pacific. Either sibling can bring extreme hot weather or extreme cold – thus the ferocious winter chill of 1941–2 that inflicted such a setback on the German invasion of the Soviet Union (when temperatures dropped to –40°F, machinery froze and a quarter of a million troops died of cold and disease) was set off by La Niña.[24]

☼

From these colossi it may seem literally a comedown to talk about tides – although those of the Minas Basin in Nova Scotia rise and fall by more than fifty feet twice a day. And the total power of waves breaking around the world's coastlines exceeds two trillion watts, enough to supply the daily electricity needs for 200–300 billion homes – more than a hundred times as many as there are in the world.

Many people see tides as the Moon's province alone, but, as is so often the case, the Sun cannot be ignored. The word 'tide' (from the Germanic root for time or season, not for anything watery) has two meanings: first, the variation of the sea level at any point along a coast which depends on shoreline topography and on currents near the shore; second, the deformation of the Earth's land and water caused by the pull of the Moon *and* Sun. Three hundred and ninety-one times further from the Earth but of thirty million times greater mass than the Moon, the Sun exercises by far the greater pull – by a factor of 178. However, because tides relate not to the strength of a gravitational field but to its gradient, it is the Moon that is responsible for fifty-six per cent of the Earth's tidal energy, the Sun for forty-four per cent – still very significant.

When the Sun and the Moon get as close to the Earth as each can (putting them 'at perigee'), they exert their maximum pull, causing the highest possible tides; when the Earth is at its greatest distance (apogee), tidal ranges are smaller; and at a new Moon, when Sun, Moon and Earth are all in alignment, the Sun and Moon pull on the Earth in the same direction (a condition bearing that Scrabble-player's dream word: syzygy), the solar tide augmenting the lunar, heightening high tides and lowering low ones.

The tidal cycle is twenty-four hours fifty minutes long, during which time most shores on Earth will experience two high tides and two low. Depending on local conditions, however, some places may have no observable tides, or a high tide that does not peak when the Moon is overhead, but lags behind by several hours. One can appreciate the dilemma of the Breton lady in *The Knight's Tale*, when, as Chaucer has it, having promised to give herself to her ardent suitor only if the rocks in the sea disappear, she looks over the cliff to find they have done so. Fortunately for her (since she is happily married), her suitor proves chivalrous and does not hold her to her promise.

Tides have been a source of speculation for millennia (although the oldest surviving tide table, for the river-spanning village of London Bridge, dates only from the twelfth century). The Romans, generally having known only the equable flowing Mediterranean, had little experience of them, and Caesar describes how his expeditionary forces twice paid a heavy price for their ignorance when invading England along the

Dover coast. The Greeks wrote about 'motions of the sea', but not until Kepler, in *Astronomia Nova* (1609), and Galileo, in *Dialogo* (1632), was there any concerted attempt to explain them. The first well-documented version was given in 1687 by Newton, although Laplace's *Mécanique céleste* (1799) is the major text before the nineteenth century; thereafter the clearest descriptions of tides are by two remarkable British scientists, Sir William Thomson (1824–1907) and Sir George Darwin (1845–1912), fifth child of Charles.*

Thomson was one of the most brilliant of all the great nineteenth-century men of science (he lies next to Newton in Westminster Abbey), and in 1882 addressed the British Association on the working of tides. 'If I were asked to tell you what I mean by the tides,' he began, 'I should feel it exceedingly difficult to answer the question . . . The truth is, the word "tide" as used by sailors at sea means horizontal motion of the water; but when used by landsmen or sailors in port, it means vertical motion.' He judged the mariners' notion correct, 'because before there can be a rise and fall of the water anywhere it must come from some other place, and the water cannot pass from place to place without moving horizontally, or nearly horizontally.'[26]

Often what we see are 'more properly wind waves than true tides', but there is another rise and fall, caused directly by the Sun's heat, which affects atmospheric pressure: when the barometric reading is high, the water is pushed down by the greater pressure of the air; it swells upwards when this drops. This sun-effected pressure affects land too; but because continents are much more rigid than oceans the effect is much smaller. Even so, Thomson explained, whole tracts of a continent may rise and fall as much as sixteen inches at high or low tide (we do not realize this, as everything around us rises or falls together). Local lakes, swimming pools, baths, cups of coffee or indeed the human stomach will not show

---

*A third great writer on tides was William Whewell (1794–1866), whose 'tidology', as he called it, charted the movements of the world's oceans, with the object of showing all those points where, throughout the globe, waters reached their high point at the same time. Whewell, a great polymath who went on to become a celebrated Master of Trinity College, Cambridge, introduced the term 'scientist' into English in 1834, in an article in the *Quarterly Review*. Spurning the alternative 'savant' as too unassuming and too French, he offered the word to denote the common enterprise of all those who studied the natural world, even as their disciplines were becoming more specialized and less unified.[25]

any signs of rising or falling, but still tides do funny things: for example, tidal friction is lengthening Earth days by 1.6 milliseconds per century, while tides themselves stretch each one of our bodies, making us taller.[27]*

Sir George Darwin's early education had been at his father's side, and although he was called to the Bar it was always likely he would turn to science. In 1898 he published *The Tides and Kindred Phenomena in the Solar System*, a scientific best-seller, and later contributed the long entry on tides for the eleventh edition of the *Encyclopaedia Britannica*. Thomson (or Lord Kelvin of Largs, as he became – the first scientist elevated to the peerage) had treated the Earth as an incompressible sphere, one that would not be drawn out of shape by solar and lunar pullings, whereas Darwin saw our world as having a liquid core responsive to both the Sun's and the Moon's powers. (A pound coin in Edinburgh, he liked to point out, is slightly heavier than the same coin in London, because slightly closer to the Earth's centre.) He calculated that in the remote past the Earth must have been rotating much faster and the Moon much closer.[29]

Darwin set out to apply Newtonian principles to calculating the Sun's effect on other planets. 'The efficiency of solar tidal friction is very much greater in its action on the nearer planets than on the further ones,' he concluded. Speaking of his own planet, he showed that the Earth's high tides occur at local noon and midnight, the low tides at sunrise and sunset, although in calculating tidal behaviour 'the irregular distribution of land and water and the various depths of the ocean in various places produce irregularities in the oscillations of the sea of such complexity that the rigorous solution of the problem is altogether beyond the powers of analysis' – a most un-nineteenth-century admission.[30]

Beyond the powers of analysis it may be, but sailors have not waited for the findings of science and have been navigating the seas for thousands of years as best they can; and for thousands of years that has meant learning how best to harness the Sun.

---

*One should not get too excited by this enhancement, though. High tides change our height by a factor of 10^{-16}, a distance a thousand times smaller than the diameter of an atom. In comparison, while in space, astronauts grow taller by up to three inches: as the skeleton is unrestrained by the pressure of gravity, the vertebrae relax, elongating the spinal column – which soon returns to normal on Earth. Equally, earthquakes make us lose weight: when the world is dilated by a giant shock, gravity at the Earth's surface decreases by about 0.0000015 per cent, meaning that a 150-pound person loses about 4×10^{-6} of an ounce.[28]

PART FOUR

HARNESSING THE SUN

When Elizabeth I of England received the Papal Bull of 1582, astronomers urged her to make the change to the new 'Gregorian' calendar, while Protestant bishops opposed it. This c.1641 engraving *Father Time Carrying the Pope Back to Rome* catches the popular mood. As the poem beneath the image puts it, 'This load of vanity, this pedler's packe/ This trunk of trash'

THE HEAVENLY GUIDE

Ahab soon calculated what his latitude must be at that precise instant. Then falling into a moment's revery, he again looked up towards the Sun and murmured to himself: 'Thou sea-mark! Thou high and mighty Pilot! Thou tellest me truly where I *am* – but canst thou cast the least hint where I *shall* be?'

Herman Melville, 1851[1]

The sailor cannot see the North,
but knows the needle can.

Emily Dickinson, 1862[2]

Captain Ahab can be forgiven his moment of frustration. 'The emptiness and homogeneity of the sea,' writes Daniel Boorstin, 'the vast sameness of the oceans on the surface, naturally drove sailors to seek their bearings in the heavens,'[3] and the Sun's role in guiding such travellers has a long history. The oldest proven sea travel anywhere is the migration of anatomically modern Homo sapiens to Australia, beginning about 60,000 years ago; but the most impressive ocean voyagers of old were the Polynesians, who by around 2000 BC had already made themselves formidable astronomers. The Sun was their principal marker, but usually in relation to other stars, the positions of which they memorized in the hundreds. The Chinese were not far behind, calling their blue-water junks 'starry rafts'. The ancient Greeks too became seafarers (although Homer engagingly introduces 'a talkative, bald-headed seaman who set the crew laughing and forgot his course', and the Greeks had no native word for 'sea' when their invasions broke into the Mediterranean). But sailors who left sight of land were constantly groping their way forward. Even on short voyages men were often 'at sea', as we still say. Seafarers gloomily swapped stories of those who had fallen foul of Mediterranean weather – from the Spartan king Menelaus, whose home-bound fleet was blown from Troy to Egypt, to St Paul's shipwreck on Malta. The lack of a means to determine longitude (east–west positioning) presented an often insurmountable challenge.

The directions by which we position ourselves – north, south, east and west – are set from two coordinates, horizon and zenith. The horizon (Greek, defining or separating) is the line where Earth and sky appear to meet. The zenith (Arabic, 'over one's head') is the point in the sky immediately above the observer. These two – lateral plane and vertical point – provide our initial frame of reference. Man learned to assign the values east and west by the risings and settings of objects in the sky, calling the directions of the line at right angles to their course 'north' and 'south'. He eventually invented 'meridians' – imaginary lines that curve through any given point on the Earth's surface to mark the shortest distance between the Poles.

One's own meridian is nothing more than the north–south line that runs through the particular spot where one happens to be. At any given moment, one of these meridians is exactly underneath the Sun, making it noon for everyone along it. For those to its east it is afternoon or later, for those to the west morning or before, and halfway round the world in either direction it is midnight. Points on the same meridian have the same longitude, longitude being measured by its distance east or west from a standard meridian (which, since 1884, has been accepted as passing through Greenwich, in southeast London).

Early navigators could roughly determine east and west by the Sun's rising and setting. Later, seafarers developed other aids. Knowing the latitude (i.e. north–south alignment) of a ship's position was easy, for the Sun's altitude was an almost perfect guide. (One degree of latitude is about sixty-five nautical miles, one nautical mile equalling 1.15 'statute miles', or 1.85 kilometres.) At the equinoxes, the noonday sun is directly above the equator, or at an altitude to it of 90°, while at the Poles it is invisible in a hemisphere's winter, but clearly seen in its summer. In between equator and Pole, the Sun's noon altitude can be established and compared with astronomical tables to determine one's position. The only device needed is a sighting instrument to measure altitude in degrees. The Greeks did not even use that, but simply measured the height of the circumpolar stars above the horizon at night.

The art of navigation evolved over the centuries, with certain notable breakthroughs, not all of which related to the orientation of the Sun. In AD 40 the Greek merchant Hippalus sailed from Berenice, on the Red Sea coast of Egypt, to Madras and back in a single year, a journey that had previously taken two, thus helping to transform oceanic commerce: he had discovered that monsoon winds (from *mawsim*, Arabic for 'season') reversed twice a year, and that by running directly before them, vessels travelling the China Sea and the Indian Ocean could cover vast distances at much greater speeds. Thereafter, the Greeks became so accustomed to using the names of the winds to indicate the directions from which they came that 'wind' became a synonym for direction, with the puffed cheeks and exhaled strong breath of the cherubs on early maps not mere decorations but the main direction-markers (hence 'Australia' – 'land of the south wind'). Columbus's Spanish crews thought of direction not in terms of degrees on a compass but as *los vientos*, the winds, while the Portuguese called their compass card a *rosa dos ventos*, a wind rose – duly marked off with the most significant aspects of the sky.*

As the Middle Ages opened (say by Charlemagne's time, c. 800), Arab navigators gauged latitude either by a *kamal*, a wooden rectangle on a string, or a crossbow on which two rods were hinged at one end to measure the angle of declination when the observer levelled the bottom rod on the horizon and the upper on the Sun (or a star). Then in 1086 the magnetic compass was pioneered by a Chinese waterworks director,

---

*Critics have pointed out several errors in Dan Brown's *The Da Vinci Code*, but his note about the Rose Line is to the point. 'For centuries,' he writes, 'the symbol of the Rose has been associated with maps and guiding souls in the proper direction. The Compass Rose – drawn on almost every map – indicated North, East, South, and West. Originally known as the Wind Rose, it denoted the directions of the thirty-two winds, and sixteen quarter-winds. When diagrammed inside a circle, these thirty-two points of the compass perfectly resembled a traditional thirty-two-petal rose bloom. To this day, the fundamental navigational tool was still known as a Compass Rose, its northernmost direction still marked by an arrowhead . . . or, more commonly, the symbol of the fleur-de-lis . . .'[4] Winds, of course, are primarily solar.

Shen Kua; at least, he offers the first written source. For the first time, Chinese sailors were able to find an absolute direction anywhere on the globe without complicated calculations, and were provided with – in Boorstin's words – 'a worldwide absolute for space comparable to that which the mechanical clock and the uniform hour provided for time'.[5]

Shen Kua further advised that if one rubbed a thin piece of steel on a lodestone ('the stone that leads') and hung it by a thread, the magnetized needle would point almost south. Though a lodestone is simply a piece of ore possessing magnetic properties, its remarkable power was for a long time associated with the dark arts (in China, such stones were used in fortune-telling), and common seamen were wary of it. St Augustine recounts his amazement on seeing that magnetite could not only attract iron but give it the power to attract yet more, creating a chain held together by an unseen force: hence its association with magic.

The origins of this wondrous new instrument are still hotly debated, historians disagreeing on whether it was invented more or less simultaneously in several places or whether the Europeans and Arabs learned of it from the Chinese. But wondrous though it was, its authority could be undone by extreme weather conditions, a truth that rang through the centuries.

From at least the eleventh century, most ships' pilots had manuals to help them, the best 'pilot book' to survive from such times being the *Konungs Skuggsja* (*The King's Mirror*), written as a dialogue between a Viking father and son. The youth is told he must 'observe the course of the heavenly bodies, recognize the quarters of the horizon, mark the movements of the ocean and understand the rise and fall of the tides'.[6] However, the ability to establish the Sun's position is not in itself sufficient for good navigation, but must be supplemented. When crossing the open sea, the Vikings would gauge latitude by means of a 'sun-shadow board' – a wooden disc with a pin or gnomon at the centre that could be adjusted upwards or downwards according to the season. The disc was floated in a bowl of water and the Sun's noon shadow noted. 'If the ship was on course,' explains Robert Ferguson in his panoramic history of the Vikings, 'then the shadow would reach a circle marked on the board. If it passed beyond, the ship was north of this latitude. Should it fail to reach the circle she was south of it and the skipper could make the necessary adjustments.'[7] Nevertheless, without a sure reckoning of latitude the navigation of the Vikings' great voyages principally depended on deep acquaintance with the seas they were crossing. For longer voyages, a pilot would put himself on the latitude of his destination, then try to hold course by keeping the observed angles of land (where available) and celestial bodies uniform. Not surprisingly, he often missed his destination, which is how Vikings ended up on the Long and Wonderful Beaches of Greenland and Labrador.[8]

Another skill the Vikings may have employed was using a 'sunstone' (a clump of natural crystal) to polarize light and thereby locate the Sun even when it was hidden from view. The Vikings' sunstone is believed by many to have been a form of cordierite, a mineral found almost exclusively on the pebbly beaches of Norway, which can split a light beam into two shafts of colours; when held up against a clear patch of sky and rotated, it changes from yellow to blue at the moment it is pointed towards the hidden Sun. A short story, 'Raudulfs thattr', preserved in a manuscript from the early fourteenth century, tells of a visit by Olav Haraldson, King of Norway, to a rich farmer. Olav asks his host's son, Sigurd, if he has any special talent, and the boy replies that he can tell the time even when there is no celestial body visible. Intrigued, the king next morning challenges Sigurd to demonstrate his skill against the overcast sky. Once the youth has done so, Olav orders him to hold up a sunstone in the direction that he believes the Sun to be. Sunlight streams through the prism, verifying Sigurd's gift.*

More conventional aids were improving all the time. A compass with a graduated scale that worked over 360° is mentioned by Peter of Maricourt in 1269. Judging from British ship account books, most vessels had at least two compasses; Magellan carried thirty-five replacement needles aboard his flagship. The fact that the north arrow of a compass

---

*There are several ways to determine direction using the Sun (or the stars). One is the shadow-tip method. Push a straight stick about three feet long into the ground so it stands upright. Mark the tip of the shadow so cast, then wait until the shadow moves one and a half to two inches – between ten and fifteen minutes – then mark the tip of the second shadow. Draw a line from the first mark and about a foot beyond the second, then stand with your left foot on the first and your right on the end of the line drawn: in the northern temperate zone, you will be facing in a northerly direction; if in a southerly, the opposite.

does not, in most places, point to the true Pole but either east or west
thereof by a spread that sailors term 'the variation' was early detected;*
but such inexactitude was not properly quantified until about 1490, and
precautionary bearings were still taken of the Sun at its rise and setting
well into the seventeenth century.

From c. 1480 on, a mariner's astrolabe was used to calculate the Sun's
noon altitude, so determining the observer's distance north or south of
the equator. Such a version – with a ring marked in degrees for measuring celestial altitudes – was followed by the cross-staff (a nearly
three-foot-long piece of wood marked with graduated measurements,
with a perpendicular vane which slid back and forth on it) and the back-
staff or back-quadrant, and together were the main navigational
instrument until the invention of the octant in 1731. Timothy Ferris
brings such sightings to life:

> At local noon on any clear day aboard a ship of the line, three officers
> could be seen helping to shoot the Sun – one holding the astrolabe
> steady, another sighting it, and the third reading the elevation – while
> deckhands stood by to catch the navigator when he fell or to retrieve the
> astrolabe if it were dropped and went scuttling across the rolling deck.[10]

In *Don Quixote* (1605), Cervantes makes his feelings plain about
people's general ability to use all this newfangled gadgetry. Quixote
(which surname, in Catalan, translates as 'horse's arse') is busy expounding to his long-suffering groom how Ptolemy remains the great authority
on navigation, and how well he, the Don, might pinpoint their position

---

*The Arctic has *five* North Poles: 1. The North Geographical Pole, at ninety degrees, the fixed
axis-head of the Earth; 2. The North Magnetic Pole: compasses point to it, but it wanders
about; 3. The North Celestial Pole, in the sky directly above (1): 4. The North Geomagnetic
Pole, which centres the Earth's magnetic field, and which also shifts; and 5. The Northern
Pole of Inaccessibility – the point in the Arctic Ocean furthest from land, approximately 680
miles north of Alaska, at 84° 05'N, 174° 85'W. Why the magnetic poles should be located
where they are remains a mystery, while the Earth's magnetic field has reversed its polarity tens of thousands of times over the billions of years. The field inverts about once every
300,000 years, the North Pole becoming the South and vice versa, although the last switch
occurred 780,000 years ago. Over the last 2,500 years the field has declined forty per cent
in strength, so another reversal soon is a real possibility. The field is generated by currents
of molten iron flowing within the Earth's outer core, in response to the gravitational pull
of the Sun. Traces of these ancient journeys are to be found in rocks worldwide.[9]

had he the right equipment to hand. 'If only I had an astrolabe here,' he proclaims, 'with which I could take the height of the Pole, I could tell you how far we have gone . . .' Sancho Panza gloomily retorts: 'My God, but your worship has got a pretty fellow for a witness of what you say, this same Tolmy or whatever you call him, with his amputation . . .'[11]

For once contemporary readers may well have laughed in sympathy with Quixote, not at him. The new instruments – possibly because, like compasses, they were associated with the black arts – were slow in coming into general use: it has been said that navigators were the only surviving upholders of the Ptolemaic theory, not because they believed the Sun went around the Earth but because it was simpler to consider that it did. And Jonathan Spence makes the point that by the beginning of the seventeenth century, 'no benefits from the exploration of the heavens opened up by Copernicus had yet been applied to the art of navigation'.[12]

Astrolabes passed out of fashion because they were never that precise, to be replaced by orreries, clockwork models of the planetary system. But be it compass, astrolabe, quadrant or sextant, telescope or orrery, none was perfect, and sailors still needed to know how to use their own first-hand observation of the Sun to calculate their position. Even the advent of the telescope and its sister instruments did not put an immediate end to naked-eye astronomy since it was not easy to make accurate positional observations until cross hairs were incorporated into telescopes in the 1660s.

Thus, a huge weight of responsibility lay on the lonely pilots. Spence goes on: 'Armed with whatever experience they had of winds and currents, fish movements and birds' flight, carrying simple maps and the narratives of previous voyagers when these existed, with compass, astrolabe and quadrant, the pilots took responsibility for vessels of a thousand tons or more, with more than a thousand passengers and crew jammed aboard.'

As Fernand Braudel describes sixteenth-century voyages, 'navigation in those days was a matter of following the shoreline, just as in the earliest days of water transport, moving crabwise from rock to rock, from promontories to islands and from islands to promontories . . . As a Portuguese chronicler puts it, "of travelling from one seaside inn to another, dining in one and supping in the next".[13] In exceptional circumstances a ship might lose sight of the coast, were she blown off course or set on one of the three or four direct routes long known and used; but rarely did ships take voluntarily to the open sea.* The threats to avoid under even greater penalty were storms and piracy (regarded for centuries as a quite honourable profession).[14] The Mediterranean, adds Braudel, 'has never been inhabited by the profusion of sea-going peoples found in the northern seas and the Atlantic'.[15]

But Braudel's view has come under challenge, historians pointing out that, as early as the days of Alexander the Great (356–323 BC), a new era in navigation was beginning with the construction of lighthouses, while during the same period rigging and steering gear were also improved, so that sea captains began to sail directly to their destinations instead of hugging the coast.†

But whatever the route, sea travel remained unpredictable: a 1551 record has a 'certain ship which, without ever striking sail, arrived at Naples from Drepana, in Sicily, in thirty-seven hours' (a distance of two hundred sea miles), the writer ascribing 'such swift motion' to the power of 'violent floods and outrageous winds'.[17] Without such unusual conditions, Sicily to Rome (eight hundred sea miles) still took twenty to twenty-seven days, and outside the Mediterranean traffic was even slower: the 2,760 miles from Berenice to peninsular India, for instance, consumed up to six months – much less than a mile an hour.[18] During such voyages, the Sun remained one's reliable helpmate, and the geographer

---

*Outside the Mediterranean, there existed one further difficulty that the sailor on the Inland Sea could ignore – tides. As George Darwin recognized, the construction of trustworthy tide tables is crucial to navigation. On the other hand, the use of a shipman's 'lodemenage' had by the sixteenth century developed into a fine art, and there would have been few spots on the main routes where the depth and nature of the ocean bed were unknown.

†Braudel's classic first appeared in 1949. In 1955 a commodore at the Admiralty, in a foreword to a history of sea navigation (not mentioning Braudel by name) delivered this broadside against the 'rock-scratching' theory: 'In these pages you will find disproved once and for all that persistent myth that the first sailors navigated by "hugging the shore". Those words could never have been written by a sailor. Nothing is more fraught with peril, and therefore the more assiduously avoided on a little-known coast, than hugging the shore. The myth is based on the assumption that the mariner has neither the means nor the ability to find his way out of sight of land. Now this assumption is shown to be groundless. Also it takes no account of the extra sense and undoubted ability of the small-ship sailor to know where he is with remarkable accuracy without instruments or observations. It is a fact that many deep-sea fishermen have this ability in some measure today, and it may well have been more commonplace in days gone by. Nothing is more sure, by whatever means they achieved it, than that the sailors of all ages have navigated in deep waters.' Nonetheless, the grand French historian made no alteration in later editions.[16]

Richard Hakluyt could still write in 1598 that 'no kind of men in any profession on the commonwealth pass their years in so great and continual hazard of life . . . Of so many, so few grow to grey hairs.'[19]

☼

Instruments that made use of the Sun had another ally in safe navigation: reliable – or semi-reliable – maps. Such aids had been employed in one form or another for centuries, yet the first mention of a chart being used aboard ship comes only in 1270, when the sainted Louis IX, King of France, attempting to cross directly from the south of France to Tunis on a crusade, found himself driven by storm to take refuge in Cagliari Bay, on the Sardinian coast. To allay his fears, his crew produced a map, showing him exactly where they were.

As the world shrank and trade grew, accurate versions of the Earth's surface were as prized as they were rare, and most map-makers mixed fiction, fact and supposition. Some 600 charts survive from the Middle Ages – Ecumenical Maps, as they are known, for they aimed to show the 'Ecumene', the whole inhabited world, all circular, with Jerusalem at the centre. Ptolemy's twenty-seven charts, drawn up for his *Geography* in AD 150, had been lost, but were rediscovered in the fifteenth century and widely copied; yet they treated the Earth as if it were a flat surface; these 'plane charts' (from which we erroneously get 'plain sailing') regularly led sailors astray, a failing for which not all the beautiful embellishments of artists could compensate. Gerard de Kremer, who became Gerardus Mercator Reupelmundanus (1512–94) when he matriculated at the Flemish university of Leuven, resolved the difficulty of how most nearly to reproduce a spherical world on a flat piece of paper – his famous projection extending the curves of latitude and longitude on to plane surfaces. There were other projections before Mercator's, but his allowed lines of fixed bearings to be rendered as straight lines, which made course-plotting easier. His maps, like those of his contemporaries, were often highly classified, considered by governments to be instruments of imperial power; to copy or share them with foreigners was treated as a capital offence.

By the year 1600 the revival of science and the development of printing had led to the widespread publication of astronomical maps, globes and books. In 1665 the Jesuit scholar Athanasius Kircher made the first world map – 'Mundus subterraneus' – to show currents, volcanic phenomena and valleys: thematic cartography was born. But even as late as 1740, fewer than 120 places in the world had had their coordinates accurately established; cartographers simply marked off whole areas with the single word 'Unexplored'.[20]* A French chart of 1753 – fifteen years before James Cook set sail in *Endeavour* – shows dotted shorelines accompanied by the words '*Je suppose*'.

Perhaps in part for that reason, throughout Western Europe astronomy was viewed not so much as a science that describes the universe but as an aid to navigation (a word which, taken from the Latin, *navis*, ship, and *agere*, to drive, for centuries meant the demanding art of conducting a ship across the sea). Astronomy was the first state science, and the story is told that King Charles II learned from the pillow talk of a mistress from seagoing Brittany that a Frenchman had devised a method of using the positions of the Moon to determine longitude at sea. No matter that there was scant evidence for this: Charles decided that, to seize the maritime initiative, he should invest in the latest astronomical technology, and in 1675 established the Royal Observatory at Greenwich. Navigation had already become so much part of the English consciousness that the poet laureate, John Dryden, could look back on its progress in 'Annus Mirabilis: The Year of Wonders 1666':

> *Rude as their ships was navigation then;*
> *No useful compass or meridian known;*
> *Coasting, they kept the land within their ken,*
> *And knew no North but when the Polestar shone.*
> *Of all who since have used the open sea,*
> *Than the bold English none more fame have won:*
> *Beyond the year, and out of heaven's high way,*
> *They make discoveries where they see no sun.*

Nonetheless, during the third Anglo-Dutch War (1672–4), the English navy's run of poor performances was ascribed to the low level of

---

*The power of print can be seen in the fact that the naming of 'America' (after the Florentine banker Amerigo Vespucci, rather than Columbus) seems to have originated in a woodcut globe of 1507.[21]

astronomical data available to enable sea captains to navigate and manoeuvre efficiently.[22] In 1731 John Hadley, Fellow of the Royal Society, and Thomas Godfrey, a Philadelphia glazier, independently invented the reflecting quadrant, or octant, in which a 45° arc is divided into ninety parts, the ends of the arc being joined by two arms. In 1757 John Campbell made the instrument less cumbersome by enlarging it to one-sixth (the *sextant*) of a circle – but in an instrument that was itself circular; and more accurate by adding filters and a telescope.[23]

Well into the nineteenth century, despite the profusion of instruments, observatories and maps, there continued to be times when the Sun was all that sailors had to go by. At one point in *Moby Dick* (first published in the UK in 1851 as *The Whale*), as Captain Ahab gazes through the ship's quadrant, waiting for the Sun to reach the meridian, he starts to rant at the strange equipment. His outburst reflects a ship's company's mixed awe and anger at the Sun's power, their doubts about how much science could ever help them and their dread belief that they might lose their way for ever:

Foolish toy! Babies' plaything of haughty admirals, and commodores, and captains; the world brags of thee, of thy cunning and might; but what after all canst thou do, but tell the poor, pitiful point, where thou thyself happenest to be on this wide planet, and the hand that holds thee: no! Not one jot more! Thou canst not tell where one drop of water or one grain of sand will be to-morrow noon; and yet with thy impotence thou insultest the Sun! Science! Curse thee, thou vain toy . . .[24]

Ahab well understood how fallible such 'toys' could be.* Later in the narrative, our obsessed captain asks his helmsman how the ship is heading and is told east-south-east. 'Thou liest!' shouts Ahab, smiting his man with his fist; but when he consults his two compasses they both point east, even though the *Pequod* is going west:

The old man with a rigid laugh exclaimed, 'I have it! It has happened before. Mr Starbuck, last night's thunder turned our compasses – that's all.'[26]

*During the Second World War, the Sahara's colossal deposits of iron ore seriously misdirected compasses, and soldiers had to go back to working out their positions by the Sun.[25]

Jonas Moore (1627–79), a leading mathematician of Charles II's time, helped
persuade Parliament to offer a large cash prize to anyone who solved the longitude
problem. An illustration from Moore's *A New System of Mathematiks* (1681) shows
navigators and astronomers at their instruments

Ahab is up to the task; he seizes a steel rod from the ship's rigging and
hands it to the mate with word to hold it upright, without its touching
the deck; then he re-magnetizes the needle to the Earth's field with a few
well-directed hammer blows before taking out two needles from the bin-
nacle and suspending one over his compass card. When finally it settles,
he points at it and exclaims, 'Look ye, for yourselves, if Ahab be not lord
of the level loadstone! The Sun is east, and that compass swears it!'

Because even the best compasses could be temperamental, the most
modern instruments faulty, science was always looking for better ways to
measure. In the eighteenth century the search was focused on the need
for more accurate determinations of longitude. While measurement of latitude (north–south positioning) was relatively simple, longitude (east–west

positioning, which was most easily measured as a function of time) was a far greater challenge. A typical ship's clock in the early seventeenth century was accurate to no better than a few minutes a day – which after several days at sea could mean a miscalculation of several miles. In her book *Longitude* Dava Sobel reminds us of the blood price of poor navigation. Sir Clowdesley Shovell, returning to England in 1707 from having captured Gibraltar three years before, piled four of his ships directly on to the Isles of Scilly at the far south-west tip of England, when he thought he was safely rounding the coast of Brittany about 120 miles to the west. Sixteen hundred of his men perished with him. The shock of so great a disaster so close to home aroused Parliament to establish in 1714 a special Board of Longitude, and also a £20,000 prize for the accurate determination of longitude.

As early as 1610 Galileo had proposed optimistically that absolute time could be measured at any point on the Earth by bearings from the moons of Jupiter, and he had even devised a helmet with a telescope attached that the observer could don while seated in a chair set in what were called 'gimbals' – similar to those used to keep a ship's compass horizontal. This proved effective for surveying distances on land, but never worked at sea. Around 1710 the Yorkshireman Jeremy Thacker introduced the word 'chronometer' into the English language. In *Gulliver's Travels*, published in 1726, Gulliver imagines himself living so long that 'I should then see the discovery of the longitude, the perpetual motion, the universal medicine and many other great inventions brought to the utmost perfection'. To Swift, an accurate measure of longitude obviously seemed about as impossible as the other marvels on the list.

Not to those who entered the competition, however. In theory, the problem sounded simple: one had to compare the local time with the time at a given place, such as Paris or Greenwich, relying on the regular, 'clockwork' nature of the motions of heavenly bodies, then make an easy mathematical calculation. But to do that one had to produce a clock that could maintain accurate time on a long, rough voyage with widely varying conditions of temperature, pressure and humidity.[27] In 1736, suitably motivated by the handsome prize – £20,000 then was worth nearly £3 million today – John Harrison (1693–1776), a lowly clockmaker, unveiled his 'marine timekeeper' or 'sea clock', accurate to a tenth of a second

In 1740–44, during one of England's wars with Spain, six British ships under
Commodore George Anson set out to circumnavigate the globe. Anson's map of his
journey movingly shows what befell his force without knowledge of longitude (their
east–west position). The straggly thin line to the west of the continent tracks the
squadron's progress, and shows how often Anson and his crew lost their way.
Only one ship – 500 men out of an original 1900 – completed the voyage.
The charts they carried typically placed Juan Fernandez
217 km (135 miles) west of Valparaiso on the South American coast. In fact it is
580 km (360 miles) west. Unsure of his maps, Anson set off in the wrong direction;
by the time he reversed his course, it took nine days to reach his starting point,
during which period 70 men died

per day. The board, set up specifically to assess claims for the prize, granted Harrison £500 for further development, and in 1759 he came up with an even better version – a succession of hardy portable clocks he called 'watches', a term taken from the practice of dividing a ship's day up into six watches of four hours each. Unfortunately for him, the board was dominated by well-established professional astronomers, who refused to believe that a man from so modest a background could have accomplished such a feat, handing over but half the prize, and that much only in 1765 – more than three years after tests had shown his invention lost just 5.1 seconds over 80 days at sea. When in 1773 Cook took HMS *Resolution* in one of the first crossings of the Antarctic Circle, among the voyage's signal accomplishments was the vindication of a chronometer based on one of Harrison's models. By the time the magnitude of his achievement was finally recognized, Harrison had but three years to live.

In 1884, after almost a century of nearly every great nation's running the meridian from which time was calculated through its own capital, a conference in Washington awarded the prime meridian to England, the consideration that tipped the scales being that Greenwich had developed the necessary instrumental facilities and compiled the observational data over 200 years. By then, as the jingoistic poet William Watson (1858–1935) put it in his ode on the coronation of King Edward VII:

> *Time and the ocean and some fostering star*
> *In high cabal have made us what we are.*[28]

But the development of the prime meridian is more than the story of sailors' use of the Sun. It is part of the whole history of timekeeping, and how the Sun was put to measure its own passing.

OF CALENDARS AND DIALS

Time must never be thought of as pre-existing in any sense; it is a *manufactured* quantity.

Hermann Bondi[1]

A calendar is a tool which cannot be justified by either logic or astronomy.

E. J. Bickerman[2]

'Even the most sophisticated people lose their heads over the mystery of counting time,' wrote Umberto Eco in the run-up to the current millennium. He was stating a general truth, although he was specifically talking about the confusion over whether the millennium would arrive on 31 December 1999 or the same day in the year 2000 – which went to the question of whether year one began at zero or a year from zero – the zero referred to, of course, being the moment of Christ's birth, that itself being debatable.[3] This system of counting time from the birth of Christ was initiated in the sixth century AD by Dionysius Exiguus; before him, dates were calculated from the reign of Diocletian (AD 284–305) onwards, or from the beginning of the world (measured with an unreal precision), but Dionysius got the year of the Nativity wrong: so calculations were off from the word go.

In 1956 many school examinations in Britain, as well as plays and other celebrations marking the 2,000th anniversary of Caesar's assassination, caused substantial embarrassment when people realized that without the year zero the deed was only 1,999 years past. By the same reckoning, Jesus died at thirty-two, not thirty-three. There are of course other systems for creating calendars and measuring years. On 12 September 2007 Ethiopians ushered in their own millennium, based on a calendar over seven years behind the Gregorian. Then there is the town of Hamelin, which long dated its municipal records not by the year of grace but from 26 July 1284, when, say its archives, 130 children were taken from the city – apparently by the Pied Piper – never to be seen again.[4]

There are yet still more possible sources of confusion. Calendars still divide about equally on when the seasons begin: is the vernal equinox 20 or 21 March? It is still a matter of contention among historians when Britain entered the First World War, since her ultimatum said hostilities would begin at midnight but neglected to say whether by London or Berlin time. More recently, the raid on Pearl Harbor in 1941, which is commemorated as having taken place on 7 December in Hawaii, occurred on 8 December in Japan, which lies on the other side of the International Date Line.

We can put to one side the business of correctly reckoning time zones, eras and millennia; but what of years and months? All calendric calculations involve some form of astronomy, whether observations of the Sun or the Moon or both, and every major civilization has developed at least one form of calendar acknowledging its own commemorations and festivals. Not to mark time is to be lost out of time: in the early 1940s an anthropologist studying the Sirionó tribe in Bolivia would tellingly note, 'No records of time are kept, and no type of calendar exists.' The tribe was, he concluded, 'man in the raw state of nature'.[5]

Almost all early civilizations started with lunar calendars – the Babylonians, Greeks, Jews and Egyptians in the Middle East; the Aztecs and Inca in the Americas; and the Chinese and Hindus in East Asia – then transformed them into lunar-solar hybrids. The Moon, after all, before the advent of sophisticated mapping systems, was more reliable than any star (Moon meaning 'the measurer'; thus month, the time measured), but it often differed from the natural calendar of farmer and shepherd. As a result, any particular society might employ three calendars, one for government use, one for religious observances and yet a third for day-to-day purposes.

Of all the myriad forms known to history, only four calendars have been based on the Sun alone: the Egyptian (eventually), the Achaemenian Later Avestan (as used in Persia from 559 to 331 BC), that developed by the Mayas and adopted by the Aztecs, and our own Julian/Gregorian. Even in these, the lunar element was never suppressed completely.[6] Whatever the community, its main religious dates shifted yearly against the seasons, be it Easter, Eid, Deepavali, the Chinese New Year or Yom Kippur.

The difficulties confronting the early calendar-makers, whether using Sun or Moon, were stupendous. The only way to tell the time was by the Sun during daylight hours and by the stars at night, and in the course of history each people, sometimes each generation, attempted its own solution. An added pressure came shortly before 3200 BC with the invention of writing: as literacy spread, people sought to assign dates to records, letters and inventories that would be commonly recognized and understood.

Whenever a calendar had to be created, such was its importance that it acquired a dimension of real power. In ancient and medieval China, since its emperors were considered the embodiment of the heavens' will, each change of reign – even more important, each change of dynasty – required that a fresh calendar be created, fixing different festival days, and yet new dates to plant or harvest, in order to show that a new disposition of celestial influences had asserted itself. (One might wonder just how many such changes could be made.) This tradition was well established by Han times (206 BC–AD 220) and the era between the early Han Dynasty and the Ming Dynasty of AD 1368 gave rise to some forty new calendars.* 'For an agricultural economy,' observes Joseph Needham of all this, 'astronomical knowledge as a regulator of the calendar was of prime importance. He who could give a calendar to the people would become their leader . . . The promulgation of the calendar by the emperor was a right corresponding to the issue of minted

*Chinese rulers had it both ways. Foreign calendars were regularly introduced, one version arriving from India during the Tang Dynasty (AD 618–906), a Muslim one under the Yuan (AD 1279–1368) and the modern Christian (Gregorian) calendar in the seventeenth century. But the Chinese systems were not officially put aside until 1912, and even today most Chinese calendars employ both the Gregorian reckoning and the old, sixty-year, cycle.

coins, with image and superscription, by Western rulers. The use of it signified recognition of imperial authority.'[7]

Time was a resource and, being a form of value, a ruler's exclusive
property – and not just in China. When around 305 BC Alexander's
former veteran marshal Seleucus became sovereign in Babylon, he fixed
an earlier and more legitimizing date for his accession, closer to the
great king's death, which over the years came to be described as his 'era',
a word that thereafter came into general use. Not until 1962 did the
British Parliament give up dating all its Acts from the day of the accession of whichever sovereign was then reigning; and one Japanese
calendar still resets the year number to the coming of each emperor to
the throne.

From about 2000 BC the Babylonians organized their calendar from
direct astronomical observation. The official day began at sunset, except
for the first of each new month, as reckoned from the sighting of its
young crescent moon. If this couldn't be seen – owing to poor visibility
or because it was too close to the Sun – the beginning of the month was
simply deferred, although common sense required that no month last
more than thirty days. Strictly, a lunar month is the time it takes the
Moon to pass through each of its phases – new Moon, half, full Moon –
and return to its original position: twenty-nine days, twelve hours, forty-
four minutes and three seconds. At some point in the fourth century BC
yet another system was devised, the Babylonians conceiving the notion
of a 'mean' sun – a fictitious sun moving at a uniform rate – which, without essential change, is what guides our reckonings today.

The Chinese, like the Babylonians, relied wholly on observation, and
seem to have drawn few deductions from the errors that resulted: their
almanacs were continually falling off course. Over two thousand years
they made more than fifty revisions, not all of which marked changes of
reign. Their first calendar went by the Moon, with alternating months of
twenty-nine and thirty days, short-changing the year by eleven days. By
about the sixth century their astronomers, again echoing the
Babylonians, recognized a nineteen-year cycle after which the phases of
the Moon recurred on the same day of the solar year, so they took to
adding extra days as necessary, to 'save the phenomena' – i.e. to keep calculation in step with nature. By the first century BC they had adopted a

system of twenty-four fortnightly periods, each such period corresponding to a 15° motion of the Sun along the ecliptic (its apparent path in the sky). Their year began on 5 February and the days of their months had names that, in imagery at least, put other cultures to shame – for instance, 'the Awakening of Insects' – 7 March; 'Grain in Ear' – 7 June; and 'Descent of Hoarfrost' – 24 October. By the sixth Christian century, the irregular apparent motion of the Sun had been taken into account, and eventually the motions of the Sun and Moon were incorporated.

The Islamic calendar, at first lunar, then changed to a lunar-solar mix (the leftward-facing crescent, the sign of a new Moon, appears on the flags of Muslim nations), also achieving a degree of accuracy by adding days where necessary. However, in AD 632 Muhammad, angered that certain communities were altering the sacred months in which fighting was prohibited, forbade all such intercalation, saying that the addition of days to make the calendar conform to the solar year violated the commands of God. He then introduced a purely lunar calendar but this makes the Islamic year only 354 or 355 days long, so that over the course of 34 or so of our years Muslim holidays will slip through all the seasons.

Strict adherence to the lunar calendar also means that the declaration of a new month depends on the actual *sighting* of its new Moon, so the calendar cannot be determined ahead of time and the months begin unpredictably. New Moons, says the Qur'an, 'are fixed times for the people and for the pilgrimage [to Mecca]'. Each month the official declaration of the Moon's sighting is eagerly awaited – although, as Islam has dispersed geographically, it has become harder to adhere to an exclusively lunar calendar. In 1971 the Shah of Iran switched his country from the Islamic (lunar) calendar to the Persian (solar), the better to celebrate the 2,500th anniversary of the pre-Islamic Peacock Throne, prompting an intensive programme worldwide to create an international Islamic calendar, a mixture of calculation and sighting; but no version has yet been universally accepted.

In Jewish observance, the day divides into six watches – as reflected in the Psalms – from midnight to midnight. To keep pace with the solar year, Jewish leap years add an extra month in the third, sixth, eighth, eleventh, fourteenth, seventeenth and nineteenth years of each nineteen-year period. Every twenty-eighth year (known as a Nisan, when the position of the Sun is calculated to have been the same as it was on the day of Creation), Orthodox Jews engage in a blessing of the Sun (the last one was in 2009). As with the Muslim calendar, the months of the Jewish register are reckoned by the Moon, beginning when its crescent first appears in the western sky.

The Egyptians began with a lunar model, but it, like others of its time, proved so inaccurate that – just as later – a year of twelve lunar months was a full eleven days apart from a solar year, and festival dates again drifted badly. They elected to divide their calendar into three seasons – the flooding of the Nile, its subsidence and the harvest – each lasting four lunar months. To make a full year, they added another month whenever Sirius rose late in the twelfth month. Eventually, once they recognized that the length of the year was close to 365 days, they simply tacked five extra days on to the final month.

After moving to a solar year, they devised a more sophisticated calendar using a roster of thirty-six stars located around Sirius, the fresh sighting of each star marking a new day. Each of the thirty-six 'decans' (now so called because they came into view ten days apart) would be invisible for seventy days before rising. At any one time, eighteen decans covered the period from sunset to sunrise; three each were assigned to the in-between periods of dusk and twilight, and the remaining twelve marked total darkness (a word rendered in Egyptian by the powerful phrase 'what grips the bowels'). From this emerged the twenty-four-hour day, with hours that varied in length through the year – daylight hours being longer in summer, for example – until a new dynasty, beginning the Egyptian New Kingdom (1539 BC on), introduced the sixty-minute hour. They added an extra day every four years.

The Mayan method of measuring time was impressive but mystifying. They had a calendar that, by AD 800, used both agricultural and solar cycles, one of which was made up of eighteen months, each of twenty days (followed by an intercalary 'month' of five days, called the *uayeb*), with names such as Pop, Zip, Zec, Mol, Yax and Zac that have a fine ring; but these were later subsumed into a calendar of thirteen twenty-eight-day months, based on the Moon. Their initial, solar version was accurate to within about three seconds per year – closer to the true year than the

Gregorian.[8] Its 'long count' ends on 21 December 2012, which coincides with various unusual alignments in our solar system sufficiently well to have inspired myriad theories (under the general rubric of 'the Mayan Prophecy') about a coming apocalypse.

The Aztec calendar consisted of a 365-day cycle called *xiuhpohuali* (year count) and a 260-day ritual cycle called *tonalpohuali* (day count). The former made up the agricultural calendar, since it had to take account of the Sun; the latter was the sacred calendar. The Inca likewise had two versions. When in 2004 I visited the mountain city of Cuzco, I was told that its massive sun pillars (sadly destroyed by the Spaniards) had been used to fix the planting of crops. More recently, a range of thirteen stone towers within a 2,300-year-old ruin 250 miles north of Lima has been revealed as a solar index (something like Stonehenge), employed by the Inca to help run their empire; but there also coexisted a star-determined agricultural calendar consulted by the lower classes.

In astronomy, the Inca, like the Aztecs, were not as developed as the Mayans, yet they had an elaborate calendar of twelve lunar months, which they corrected from time to time according to sightings from Cuzco. The four months bracketing the solstices honoured the Sun; those around the equinoxes were governed by the cults of water and the Moon goddess; and the four left over were dedicated to agriculture, death, the thunder god (deity of war and of all climatic events) and the goddess of the planet Venus.[9] Otherwise, the Inca ways of measuring time were not that different from those of many of their counterparts – imaginative, imprecise, regularly tinkered with and driven by a mix of solar and lunar observations.

The innumerable Greek states had a whole mix of indices. One, used in Athens in the fifth century BC, organized itself from the summer solstice, beginning the New Year with the following new Moon. In Athens – as in most Greek jurisdictions – the day began at sunset, a natural arrangement with a system that reckoned time by the Moon; but magistrates were free to repeat dates – several times if they so wished, such as having a succession of 25 Decembers. There were limits to this power, for the year had to end on the right date; were one day repeated, another had to be omitted, and by the last month there was little freedom left for manipulation. The calendar was finally reformed around

432 BC by Meton of Athens, the lunar month being aligned with the solar year by intercalating seven months in every nineteen-year cycle, and by varying months between twenty-nine and thirty days, thus making the lunar month overlong by only two minutes.*

The first significant change in European calendars following the Athenian innovations of 432 BC came about as a mixture of practical necessity and political factors (and perhaps a little out of lust). Under the Roman Republic, the endlessly recurring need for a correcting month was assessed annually, often in the light of blatantly political considerations. Julius Caesar was both Pontifex Maximus (a senior priest of the state religion) and a proconsul (provincial governor), and in his former role he oversaw the calendar. However, he was at grave odds – quarrels that ultimately passed into civil war – with the faction led by his son-in-law and former ally, Pompey; and during this time the calendar year had contracted anarchically from a 365-day span, with January falling in the autumn.

Once he had triumphed over his enemies, Caesar commissioned a Greek astronomer, Sosigenes of Alexandria, to devise a new calendar, and he suggested increasing the lengths of some months by one day each and February by one day every three years only (a 'leap year'), creating a new calendar, the Julian, of c. 365.25 days per annum, with the year starting on 1 January. One difficulty with this was that too many leap days were added with respect to the astronomical seasons. On average, solstices and equinoxes advance by eleven minutes per year against the Julian calendar, causing the Julian to drift backwards one day about every 128 years. Caesar happily dubbed 46 BC 'the last year of confusion', but with 45 BC lasting a full 445 days the citizens of Rome wittily reversed this to 'the year of confusion'. They were nevertheless jubilant because they believed that Caesar had extended their lives by an extra three months.

---

*If one takes the year 432 BC as reckoned in the Gregorian calendar, it is daunting to realize that this same year appears in other calendars as follows: *Ab urbe condita* (from a massive history of Rome, c. 753 BC) – 322; Bahai – 2275/4; Berber – 519; Buddhist – 113; Burmese – 1069; Byzantine – 5077–8; Chinese (sexagenary cycle) – 2205–66; Coptic – 715–4; Ethiopian – 439-8; Hebrew – 3329–30; Hindu – Vikram Samvat -376–5; Hindu – Kali Yuga 2670–1; Holocene – 9569; Iranian – 1053–2 BP; Islamic – 1085–4 BH; Korean – 1902; and Thai – 112. This list omits the Armenian, Japanese, Mayan, Aztec or Inca calendars, which do not cover such early dates.

Several of our months reflect the Julian counting method – the seventh month being September (*septem*), the eighth October (*octo*) and so on. The new version mirrored the Greek-Egyptian calendar of 238 BC, which Caesar was happy to introduce, as at the time he was passionately involved with Cleopatra, and also because before the civil war the Senate had refused to intercalate the extra months that the old calendar needed, as that would have prolonged Caesar's term of office.

The new calendar was still seriously out of kilter, for it mis-reckoned every third rather than fourth year a leap year, so that by 11 BC, a mere thirty-three years after Caesar's death, the year was starting three days late. Caesar's great-nephew, adopted son and more tactful successor as dictator, Augustus, corrected this by skipping three leap years, not having another one until AD 8. During his reign, the fifth and sixth months, 'Quintilis' and 'Sextilis', were renamed 'July' in Caesar's honour, for regularising the calendar, and 'August', for Augustus.*

The advent of Christianity made new demands on the reckoning of time. The Julian calendar was used to 'fix' certain events such as Christmas, the Epiphany and the Annunciation, and also to determine the movable-feast sequence of Easter, Pentecost and Lent. Easter is assigned as the Sunday after the first full Moon to fall on or after 21 March; at least a dozen other festivals follow from that dating. As the Gospels state unequivocally that Jesus was crucified at Passover, Easter depended on the complicated lunar calculations by which Jewish authority set that festival. Many early Christians held that Jesus had died on a

---

*Over the centuries other attempts to rename the months have been made, but all have failed. In 1793, for example, the French Republic declared a new Revolutionary calendar, with twelve thirty-month days beginning with Year One of the Revolution, 22 September 1792. The new months were: Vendémiaire, Brumaire, Frimaire, Nivôse, Pluviôse, Ventôse, Germinal, Floréal, Prairial, Messidor, Thermidor and Fructidor. Every tenth day was a holiday, and there were five *sans-culottides* (intercalary) days (named after the Revolutionary lower classes, who did not wear fashionable culottes). It lasted just over a decade before it was abolished, since a ten-day week gave workers less rest; also, every New Year started on a different date (because fixed to the equinox), a major source of confusion for almost everybody, not least because it was incompatible with the secular rhythms of trade fairs and agricultural markets. The Revolution also attempted to introduce a decimal clock, with each *heure* twice as long as the traditional, pre-Revolutionary one. On Fructidor 22, Year XI (9 September 1803), this innovation too was repealed. (More successfully, the English, from 1880 on, were to yoke together the last day of the Christian week – Saturday – and the first – Sunday – into 'the weekend', which after the Great War more and more countries adopted both as a term and as a practice.

Friday and risen two days later; but if one followed the Jewish calendar there was no assurance that Easter would fall on a Sunday. This led to a major schism between the Eastern Orthodox Church and Rome, the former observing Easter on the fourteenth day of the lunar month, regardless of the day of the week. Different faiths still celebrate Easter on different Sundays.

How best to determine an ideal calendar was exhaustively examined in *On the Theory of Time-Reckoning*, written in 725 by the Venerable Bede, who calculated that the 365.25-day Julian year was longer than the solar year by eleven minutes, four seconds, and that a more accurate measurement was closer to 365.24 days. Still nothing was done, and over time Caesar's calendar got more and more out of step with the seasons. When at last reform came, it took a pope to do it. The story that has been passed down is that when it became apparent that the Easter of 1576 would be particularly ill-timed, Gregory XIII went to the Vatican's Tower of the Four Winds, where a papal astronomer showed him the solar image on the meridian line across the floor of the Calendar Room, and the Pontiff could see that the Sun was ten days away from where it had to be to reach equinox on 20–21 March. At that moment, it is said, he decided that the calendar must be aligned with the unyielding heavens. Maybe the story is true, but the need for reform had long been recognized: aside from Bede, as early as the thirteenth century Roger Bacon had sent Clement IV a treatise on the calendar's shortcomings.

In a matter of weeks a new plan, devised by a well-known Calabrian physician and amateur astronomer named Aloysius Lilius (c. 1510–76), was presented to the Pope by the astronomer's brother (Lilius himself having recently died). Gregory asked the Jesuit mathematician Christopher Clavius, a Bavarian living in Rome, where he was acclaimed as 'the Euclid of the Sixteenth Century', to review it. Clavius endorsed the innovations and added some of his own. Over the next few years all Catholic countries were instructed to omit the extra ten days. The Pope agreed to 1582 as the changeover year, and October was chosen as the changeover month, as containing the fewest Church feasts, so imposing the minimum disruption.

There was also a good political reason for the choice, for Easter 1583 would thus fall on the same day in both the Julian calendar (31 March) and the Gregorian (10 April), a happy conjunction that would not repeat itself for many years. One might ask why Gregory didn't delete fifteen days rather than ten, shifting the spring equinox back to its traditional date of 25 March. Had he done so, however, the winter solstice would have moved to 25 December, by this time a major Christian feast day. As Duncan Steel notes: 'By allowing Christmas and the solstice to coincide once more, the Church would have walked into trouble. Christianity had successfully pinched the solstice festival of the pagan religions over twelve hundred years before . . . and was not about to hand it back.'[10]

In Spain, Portugal and parts of Italy, the new calendar was taken up immediately; France and the Low Countries followed by the end of that year. In what is now Belgium, the calendar went from 21 December 1582 straight to 1 January 1583, depriving everyone of Christmas. Catholics in Germany followed suit in 1584, Denmark-Norway by 1586 – although Sweden held out until 1753. Most other non-Catholic Christians scorned the new calendar; not until 1700 did the German Protestant states adopt it. In Grisons, most easterly of the Swiss cantons, French- or Italian-speaking Catholics and German-speaking Protestants, even those living on the same street, kept different calendars, beginning their years on 1 January and 25 March respectively, a situation that continued down to 1798, when the French invaded and imposed the Gregorian on everyone – even though they themselves kept to the new Revolutionary version.

Protestant England was likewise suspicious of any pronouncement from Rome, and although Elizabeth I was not unsympathetic to the proposed change, the Spanish Armada of 1588 effectively scuppered any chance of its being accepted. Voltaire would scoff: 'The English mob preferred their calendar to disagree with the Sun than to agree with the Pope.'[11] More likely, well aware of the Gregorian calendar's flaws, English scientists thought that if they developed a superior calendar it would help effect a rapprochement with those European nations fence-sitting in the quarrel with Rome. One interpretation of *The Taming of the Shrew* is that Katharina represents the Protestant religion, Petruchio pre-Reformation Catholicism; thus his order that she should call the Sun the

Moon no longer appears senseless when one considers that at the time the play was written (1592) England was already ten days out of sync with the Roman calendar.*

Britain and her colonies delayed two centuries before making the adjustment, by which time their calendar had to be moved forward a further day, Wednesday 2 September 1752 being followed by Thursday 14th. When the British Parliament debated the change, rumours spread throughout the country that salaried employees were losing eleven days' pay, and that everyone was surrendering eleven days of their life. The protests were not unreasonable: people who had paid rent for a full month found they received no rebate, while bankers refused to pay their taxes on 25 March and put off payment by eleven days – hence the British tax year still begins on 5 April. As recently as 1995 an Ulster politician attacked the Vatican for meddling with the calendar. Yet ultimately it is not the objections that are remarkable but the fact that there was so much *agreement.*

Japan waited until 1872 to adopt the Western calendar, when the reform led to peasant riots. Turkey was last, accepting the inevitable in 1927. Russia may boast the most confusing history of all. Up until the end of the fifteenth century, new years began every 1 March, then from 1 September until 1700, when Peter the Great changed the date to 1 January. In 1709 the Julian calendar, as favoured by the Orthodox Church, came in, more than 127 years after the Gregorian had been introduced into Western Europe. For most of the nineteenth century the Department of Foreign Affairs used the Gregorian, as did the expansionist Russian navy; finally, in 1918 Lenin decreed the adoption of the Gregorian throughout the country – but this held only until 1923, when both Gregorian and Julian were discarded in an attempt to eliminate the Christian reckoning. In their place came the 'Eternal Calendar' (with special cards distributed to workers, to increase production). By 1929 a

---

*According to the Gregorian version, Shakespeare died on Friday 3 May 1616, his fellow writer Cervantes on Tuesday 23 April (said to be Shakespeare's birthday). But by the older Julian calendar that England used, Shakespeare had died on its 23 April, St George's Day. Although the Spaniard's death was thus ten days earlier than Shakespeare's, the two men are often said to have perished together. In honour of this conjunction, UNESCO established 23 April as International Day of the Book. Anyway, Cervantes probably died on 22 April (Gregorian) but was buried on the 23rd.

The 1751 bill bringing in the Gregorian calendar was so unpopular that William Hogarth introduced a stolen Tory placard 'Give Us Our Eleven Days!' – it lies on the floor, under the foot of the man with the knob kerry – into *The Election Dinner* (1755), his painting of a tavern meeting organized by Whig candidates, with Tories protesting outside

week of only five days and therefore months of six weeks were in place, the former remaining so until 1934, when the Gregorian calendar was reimposed, although the seven-day week did not make a return until 1940. The potential for confusion was amply realized.*

The new calendar had both virtues and drawbacks. It did not perfectly track the equinoxes and solstices, but it did more accurately indicate the time of year in respect of the seasons, being based on the tropical/solar year (i.e. the time it takes the Earth to orbit the Sun, measured between

*The great variety of calendars over history was to provide Tolkien with rich material. *The Lord of the Rings* devotes a seven-page appendix to the 'Shire Year', with its twelve months: Afteryule, Astron, Afterlithe, Winterfilth (winter full moon), Solmath, Thrimidge, Wedmath, Blotmath, Rethe, Forelithe, Halimath and Foreyule. Each year begins on a Saturday, and Mid-Year's Day has no weekday name. 'In Middle-Earth the Eldar also observed a short period or solar year, called a *coranar* or "sun-round" when considered more or less astronomically' and so on; J. R. R. enjoying himself.[12]

two spring equinoxes) of 365 days, 5 hours, 49 minutes (approximately). As each year is reckoned as 365 days exactly, it means that the following New Year begins five hours and forty-nine minutes ahead of the Earth's completion of its circuit, so that after any four years the calendar is four times five hours forty-nine minutes ahead of the Sun. To get back in sync with the Sun, every fourth year, the so-called leap year, an additional twenty-four hours (the intercalary day) is added, yielding a year of 366 days. (Islam makes no such adjustments, and so Ramadan, for example, floats through spring, summer, autumn and winter: particularly confusing, since it is from the Arabic word for August – *rams*, meaning 'parched thirst' – that Ramadan takes its name.)

In fact, the calendar has really been held back eleven minutes too much each year – forty-four minutes every four years. In the course of a hundred years, this adds up to nearly a full day; if allowed to accumulate, four hundred years would embrace 146,100 days instead of the actual 146,097. Consequently, the three centurial years out of every four whose first two digits are not exactly divisible by four – so far, 1700, 1800, 1900 – are not leap years; by this arrangement the calendar again inches back towards the path of the Sun, and all is well with the world.*

Gregory's calendar was meant to get Easter back to its old sequence – an adjustment that was made for devotional rather than scientific purposes. Yet working out exactly when that festival (and others) will fall continues to engage our attention. In October 2003 I travelled to Heidelberg to see Dr Reinhold Bien at the Astronomisch Rechen-Institut. Comfortably got up in brown-checked shirt, pleated trousers and scuffed shoes, he was small and rounded, a little like a friendly hedgehog. At the institute, Dr Bien measures the heavens, charts the motions of the stars, and each day posts the times for the morrow's sunrise and sunset. But he also advises the German government as to when

*One should not forget that impressive English legal fiction, 'time immemorial', thought up by common lawyers in 1276 to solve problems of land ownership. This phrase, meaning 'a time before legal history, and beyond legal memory', was fixed by statute as 3 September 1189, the day Richard I came to the English throne. In 1832 the dating of legal memory from a particular time was modified; instead, it was held that rights exercised for twenty years (or as against the Crown thirty years) should not be open to question merely because they had not been enjoyed before. 'Prehistory', another word with a grand ring, refers simply to the time before written texts.

certain key celebrations should take place each year. 'One can't frame a calendar for all time,' he patiently explained. 'The Earth's orbit is slowing down, so what is accurate today will not be in the future.' He was soon telling me with gusto stories about Christopher Clavius, and the railleries of John Donne, a great poet but also an Anglican convert, who dismissed Clavius as a glutton and a drunkard. As Dr Bien defended Clavius, I found my gaze wandering to a small notice pinned just above his cluttered desk. It read, in English: THIS YEAR CHRISTMAS WILL BE ON 25 DECEMBER. A joke; but only just.

☼

The calculation of the year's passage is one kind of timekeeping. Measurement of the hours, and of lesser units of time, probably began much later, though nobody knows quite when. So long as people lived by raising crops and herding animals there was little need for measuring smaller units of time. From at least the ninth century on, most cultures measured in weeks, marking out the year with dates drawn from folklore, their own needs and observations, and (in the more sophisticated cities) the liturgical calendar. 'The understanding of a savage,' wrote William Hazlitt in 1827, 'is a kind of natural almanac, and more true in its prognostication of the future.'[13] Leaving aside what is implied by that loaded word 'savage', we know that the Konso people of Central Africa still reckon their day by function rather than by calculation: 5 p.m. to 6 p.m., for instance, is *kakalseema* ('when the cattle return home'), with periods of the day named after the activity performed then.*

The simple concept of an 'hour' as having a uniform duration took over two of the five millennia of known history to develop. For the Egyptians, an hour in January and an hour in August, or an hour in northerly

---

*Many sense small passages of time naturalistically. I have corresponded with several blind people about their experiences of the Sun, and certain responses seem widespread: the aviator Miles Hilton-Barber, for instance, uses the feeling of the Sun on his face to help when flying (but not at the controls), to get an idea of time. The writer Ved Mehta told me that when he left his native India in 1949, aged fifteen, to attend a college for the blind in Arkansas, many at the school didn't wear a Braille watch, as they preferred to reckon the time from the atmosphere, 'and the Sun of course is part of the atmosphere'.

Alexandria and in southerly Memphis, had markedly different lengths. The most natural division of time is into two parts – 'day' and 'night'. The Romans, up until the end of the fourth century BC, divided the former into before midday (*ante meridiem*) and after midday (*post meridiem*). Since all court work took place before noon, they posted a civil officer to detect the moment of the Sun's crossing the zenith, then proclaim it in the Forum.* They also distinguished between *dies naturalis*, the natural day that runs from sunrise to sunset, and *dies civilis*, the civil day that runs from the completion of one Earthly rotation to another – for them, from midnight to midnight. The word 'day' was ever ambiguous.

Finer subdivisions were made: the night into four 'watches', each named after its last 'hour' and proclaimed by constables. Were greater accuracy needed, a range of time-based words came into play: *occasus solis* (sunset), *crepusculum* (twilight, hence 'crepuscular'), *vesperum* (appearance of the evening star), *conticinium* (the falling of silence), *concubium* (bedtime), *nox intempestum* (timeless night, when nothing is done) and *gallicinium* (cockcrow), among others.

Before the Industrial Revolution, and the advent of better lamps and lanterns, work hours for most European societies were set by sunrise and sunset. From about the twelfth century on, church bells were sounded, indicating when to begin and end toil, announcing curfew (from the French, *couvre-feu* – cover fire, i.e. lights out) and so on. For centuries in Europe, the period between midnight and cockcrow was seen as 'dead time', thus inspiring the phrase 'the dead of night'.

Regardless of the era, we have remained transfixed by the instant of noon. Noonday demons were the merciless torturers of the hermit Desert Fathers of the early Christian Church. Both France and Italy have whole regions – the Midi and the Mezzogiorno – named after the noonday sun. While ancient Rome had timekeepers to shout out its arrival, one Parisian inventor fitted a lens into his sundial to act as a

---

*In court proceedings *klepsydrae* or clepsydras (meaning 'water thieves', in which a pierced, measured vessel sank in a tub of water) were used like eggtimers, and their span measured the time that advocates could speak: the phrase *aquam dare*, 'to grant water', meant to allot time to a lawyer, while *aquam perdere*, 'to lose water', meant to waste time. If a speaker in the Senate spoke out of turn or for too long, his colleagues would shout that his water should be removed.

Ethiopians celebrating the new millennium – our 12 September 2007

burning glass that precisely at noon set off a small cannon, and a 'noon gun' is still fired daily in places as far apart as Cape Town and Santiago (Chile). For years the lighthouse-keeper at Brockton Point in Vancouver marked midday by detonating a stick of dynamite. During the nineteenth century, to give navigators a precise visual signal by which to reset their chronometers, certain major harbours would drop a huge time ball at 1 p.m. (not at noon, as that is when observatories took their readings).*

To tell the time – more or less – throughout daylight hours, not just at noon, or sunrise or sunset, was a challenge, but once again the Sun provided the necessary means. Gnomons (from the Greek, 'indicators') were first used to measure height, but later employed as proto-sundials, with the length of the shadow cast by the gnomon determining the hour of the day. Gnomons could be anything vertical, including the human figure. As Chaucer writes:

*The duel fought out in the American West was traditionally held at 'high' noon so that the Sun's glare would not be in either man's squinting eyes: part of the myth of fairness. Fred Zinnemann's classic 1952 film *High Noon* eschews images of dwindling shadows while Sun and tension mount, and instead we see inexorable clock hands, as time drains away – a sign of how these machines had usurped the Sun's ancient role.

It was four o'clock, according to my guess,
Since eleven feet, a little more or less,
My shadow at the time did fall,
Considering that I myself am six feet tall.[14]

To measure the hours after dark, man turned to water clocks, used at night from at least 1450 BC on in Egypt, a millennium before they came to Rome. However, they were not particularly accurate until the third century AD, when the Alexandrian Tesibius (285–222 BC) invented a device to ensure a uniform flow. China had water clocks from at least 30 BC, and ended by adopting a whole series of small vessels on a rotating wheel. 'And so,' writes Joseph Needham, 'the great breakthrough in accurate time measurement came about.'[15]

Besides clepsydras, notched-candle clocks (invented by Alfred the Great, says legend), sandglasses (the necks of which became worn from repeated use, passing the grains too quickly, thus shortening their hours), fire clocks, the Egyptian *merkhet* (made from plumb line and palm leaf), incense clocks banded in different aromas (by which one could 'smell' the hour) and a score of other methods all served to reckon the time. But the most popular was the sundial. Made of stone or wood, it fixed a piece of metal parallel to what we now know to be the Earth's axis and at right angles to a circle divided by as many lines as were deemed necessary, so that when the Sun shone the shadow of the metal blade moved around the circle (the 'heliotrope' – sun-turning) according to its motion through the sky.

The earliest known example is Egyptian, dated about 1500 BC. By the sixth century BC sundials were being used in Greece, and it was Anaximander who invented the discipline of 'gnomonics' as a science. For at least the next ten centuries the sundial would be the world's most accurate timekeeping device,[16] but it took time for it to spread, partly because it was not always used properly. No one seems to have understood, for example, why a dial looted in the sack of Syracuse (latitude about 37° north) in 212 BC failed to show the right time when shipped to Rome (about 42° north).[17] Dials, it began to be realized, had to be specially made for different latitudes, because the Sun's altitude decreases as one moves Polewards, producing longer shadows. 'In order

Borneo tribesmen measure the length of the Sun's shadow at summer solstice
with a gnomon. Joseph Needham included this photo in his
Science and Civilization in China (1953)

to make the fallen shadow even approximate the correct time,' explains
Dava Sobel, 'the dial must be laid out with regard to latitude north or
south of the equator where it is to be used, respecting the changing high
point of the Sun in the sky from day to day over the course of the year
and the variable speed of the Earth's movement around its orbit. There
is nothing obvious about the construction of a proper sundial.'[18]* Of
course, for centuries it was not realized that these were the transactions
involved.

---

*Even so, Boy Scouts learn to tell direction by their watches, holding them face upwards and
pointing the hour hand towards the Sun. In the northern hemisphere, the south alignment
will then lie halfway between the hour hand and twelve o'clock: a rough but ready reckoner.

Not until about AD 1371 would there be a polar-oriented sundial – at the Great Mosque in Damascus – tilting its gnomon at an angle corresponding to its latitude and with this, as we now know, the curvature of the Earth. Thus, time was measured not by the gnomon shadow's length but by its *angle*. This was a major breakthrough, but one not actually needed until mechanical clocks were invented. Until then, almost everyone used unequal hours.

The approximations used to design ancient portable dials could introduce errors of up to a quarter of an hour. The inconsistency of measurement between two or more sundials may be guessed from the words Seneca (c. 4 BC–AD 65) puts into the mouth of one of his characters, talking of the death of the Emperor Claudius: 'I cannot tell you the hour exactly: it is easier to get agreement among philosophers than among clocks.'[19] Inaccurate as they were, sundials struck some people as only too modern, changing their relationship to time in ways that made them nostalgic for the 'happy life' that existed before time could be measured. A play attributed to Plautus (c. 254–c. 184 BC) has one character exclaim:

> *The Gods confound the man who first found out*
> *How to distinguish hours! Confound him too*
> *Who in this place set up a sundial,*
> *To cut and hack my days so wretchedly*
> *Into small pieces! When I was a boy,*
> *My belly was my sundial – one surer,*
> *Truer and more exact than any of them.*

Evidently, sundials were still something of a novelty. But it was not long before people learned to love them, and appreciate what they brought.

Around 1400 the chiming clock was invented, and by the 1600s it had become so pervasive that theatregoers might not even have registered the anachronism in *Julius Caesar*, when, in response to Brutus's 'I cannot, by the progress of the stars / Give guess how near to day', and asking the time, Shakespeare has his fellow conspirator Cassius reply: 'The clock hath stricken three.' A similar mistake appears in *Cymbeline*, when again a clock strikes three; but in *Richard II* it is a dial, not a clock, that Shakespeare has his king invoke, to make the passing of time a visible movement:[20]

> *I wasted time, and now doth time waste me;*
> *For now hath time made me his numb'ring clock:*
> *My thoughts are minutes; and, with sighs, they jar*
> *Their watches on unto mine eyes, the outward watch,*
> *Whereto my finger, like a dial's point,*
> *Is pointing still, in cleansing them from tears.*

The increase in the number of weight-driven clocks intensified a pre-occupation with time, paradoxically sparking a boom in sundials, whose construction flourished particularly during the Renaissance and was so profitable a business that design methods were closely guarded. The 'art of dialling' formed an important branch of mathematics and was the subject of many textbooks.[21] The making of sundials remained the province of astronomers rather than clockmakers, as one had to take into account the Earth's rotation and elliptical motion as well as the angle of its axis.

Even the arrival of precision timekeeping in the form of pendulum clocks and the balance spring did not dent the sundial's popularity: that took the large-scale production of watches. As Dava Sobel notes, 'A clock or watch may keep time, but only a sundial can *find* time [by interrogating the external world] – a distinctly different function.'[22] Charles I (1600-49) carried a silver sundial, which he entrusted to an attendant on the eve of his execution as a last gift to his son, the Duke of York (after whom New York is named). Thomas Jefferson in old age would distract himself from his chronic rheumatism by calculating the hour lines for a dial. Instead of a watch, George Washington used a silver pocket dial given him by Lafayette.

Over the years sundials have been T-shaped, portable, perpendicular, sunken, cubical and flat (the common or garden variety). Vitruvius, architectural theorist of ancient Rome and a contemporary of Julius Caesar, counted at least thirteen styles already in use in Greece by 30 BC and says he could not invent any new types, since the field was exhausted. But that was by no means the case. Universal sundials, adjustable for use in any latitude, came in during the eighteenth century. The design of sundials became ever more demanding as standards of clockmaking rose; many became prized works of art.

The Samrat Yantra, the giant sundial at the Jaipur Observatory, one of a family of
massive instruments built under Maharajah Jai Singh II (1686–1743). These devices
had no telescopes and relied upon naked eyesight and super-precise construction

Where once there had been nostalgia for the era before sundials, the
persistence of the sundial may have had something to do with the pastoral sense of rustic peace that seemed to attach to them in the era of the
clock. Shakespeare's stricken Henry VI exclaims: 'O God! Methinks it
were a happy life /To carve out dials quaintly, point by point, /Thereby
to see the minutes how they run.'[23] 'Of the several modes of counting
time,' wrote Hazlitt, 'that by the sundial is perhaps the most apposite and
striking, if not the most convenient or comprehensive. It does not
obtrude its observations, though it "morals on the time", and, by its stationary character, forms a contrast to the most fleeting of all essences.'[24]

'Morals on the time' is a reference to the custom of sundials being
inscribed with mottoes. There are myriad such sayings. An eighteenth-
century English verse runs: 'Read the riddle that I've found, /Come,
answer it to me, /What is it travels o'er new ground, /And old continually? Answer: a shadow.' Two others much quoted are: 'I tell only sunny

hours' and 'The clock the time may wrongly tell, /I never if the Sun shines well', although the latter, while celebrating its dial's accuracy, underlines its one drawback: it works only when the Sun is out. Nonetheless, when NASA landed a spacecraft on Mars in January 2004, it deposited a suitable timepiece . . . two aluminium sundials, no larger than a human palm, built into two Mars Exploration Rovers and carrying the motto: 'Two Worlds; One Sun.'[25]

Back in the 1930s, the flamboyant film mogul Sam Goldwyn visited his New York bankers and spied a sundial. Swinging round on his companions, he exclaimed: 'What *will* they think of next?' *Not* a joke; but only just.

21

HOW TIME GOES BY

Somewhere in the east: early morning: set off at dawn. Travel round in front of the Sun, steal a day's march on him. Keep it up for ever, never grow a day older technically.

Leopold Bloom in *Ulysses*[1]

PETRUCHIO: It shall be what o'clock I say it is.
HORTENSIO: Why, so this gallant will command the Sun.

Shakespeare, *The Taming of the Shrew*[2]

In the early 1960s my father retired from his family firm to run a pub in Cornwall, and during the school holidays I would work alongside him. Twice each day, at two-thirty in the afternoon and again at eleven in the evening – in keeping with the licensing laws whereby pubs had to close first for three hours, then again for nearly twelve hours until reopening at ten-thirty the following morning – he would intone in his basso profundo: 'Time, gentlemen, please!' It was a peculiar social ritual, a polite command to stop drinking that had the feel of a teleological pronouncement. Patrons would know they had a minute or two to drink up. In contrast to that sixty seconds of silence imposed by my English teacher at the end of his classes, which we'd so longed to be over, the hardened locals stretched out every last moment of tippling as long as they could, for the announcement was never welcome. As Oliver St John Gogarty

(1878–1957), the Irish poet who inspired the character of Buck Mulligan in *Ulysses*, wrote:

> *No wonder stars are winking*
> *No wonder heaven mocks*
> *At men who cease from drinking*
> *Good booze because of clocks!* . . .
>
> *It makes me wonder whether*
> *In this grim pantomime*
> *Did fiend or man first blether*
> *'Time, Gentlemen, Time!'*[3]*

'Time' has continued to vex mankind since the beginning of . . . well, the word is difficult to avoid. Its complexities feed on mankind's subjective experiences of time versus our objective measurement, and the impossibility of ever reconciling the two. Anthony Burgess hints at this in his essay 'Thoughts on Time':

In the first great flush of the establishment of universal public time [Greenwich Mean reckoning], there were certain works of the imagination which . . . said something about the double reality of time. Oscar Wilde wrote *The Picture of Dorian Gray*, in which the hero transfers public time, as well as public morality, to his portrait while himself residing in a private time which is motionless . . . The experience of time during both the First and Second World Wars was of a new kind to the average participant . . . The start of battles had to rely on public time, but soldiers lived on inner time – the eternities of apprehension that were really only a minute long, the deserts of boredom, the terror that was outside time.[4]

Time as experienced subjectively may indeed be mysterious: in the *Iliad*, its quality has one value for the victors, another for the vanquished.

---

*The modern history of pub restrictions in Britain begins with DORA, the Defence of the Realm Act, passed during the First World War to reduce hangovers among workers in munitions factories.

St Augustine said sourly that he knew what time was, until he was asked to explain it. But no matter how it is parsed, it is the Sun that determines time, and our employment of that guiding star to monitor time's passing has been the most pervasive of all the ways in which civilizations harness the Sun.

It has ever been critical for two groups – astronomers and navigators – to measure time with precision, but throughout the Christian world it has been the Church that has provided the main impetus towards timekeeping. Much the same is true for Muslims and Jews – Islam requires its faithful to pray five times a day, Judaism three. Christians would pray according to the movements of the heavens, and it was St Benedict who, in his *Rule* of AD 530, laid down exact times for devotion: Matins, Lauds ('There you have placed a tent for the Sun /It comes forth . . . like a champion to run its course'), Prime (first), Tierce or Terce (third), Sext (sixth), None (the ninth hour after sunrise, which survives, though three hours backtimed, as noon), Vespers and Compline. Lauds and Vespers, the services of sunrise and sunset, are specifically Sun-related, while the others are linked to set hours. This timetable was widely adopted, so much so that Pope Sabinianus (605–6) proclaimed that church bells should be rung to mark the passing of the hours. In the years that followed, a large number of forms of civil life came to be regulated by time. 'Punctuality,' writes Kevin Jackson, 'became a new obsession, and constant research into timekeeping mechanisms eventually resulted in the making of the clock.'[5]

In the later Middle Ages, from roughly 1270 to 1520, by far the best-selling volume throughout Europe was not the Bible but *The Book of Hours*, a guide to Benedict's *Rule*. During these years, the practice was maintained of defining one hour as the twelfth part of the day or the night, so that in the summer daytime hours were longer than those of darkness, the opposite being true in the winter, a custom ending only when the newly invented mechanical clock, with its uniform motion duplicating that of the heavens, gradually made people familiar with the 'mean solar' way of measuring used by astronomers. Mechanical clocks driven by weights and gears would seem to have been the invention of an eleventh-century Arab engineer and were introduced into England around 1270 in experimental form; the first clocks in regular use in Europe and for which we have sure evidence were those made by Roger

An illuminated manuscript page from *Les Très Riches Heures du Duc de Berry* (1412–16) by the three brothers Limbourg, showing June, a curious time for haymaking, with the Hotel de Nesle, the Duke's Parisian residence, in the background. The book is a collection of texts for each liturgical hour of the day

Stoke for Norwich Cathedral (1321–5), and by Giovanni de' Dondi of Padua, whose one-metre-high construction of 1364, with astrolabe and calendar dials and indicators for the Sun, Moon and planets, provided a continuous display of the major elements of the (Earth-centred) solar system and of the legal, religious and civil calendars. These clocks were made not to *show* the time but rather to *sound* it. The very word 'clock' comes from the Latin word for 'bell', *clocca*, and this hour-checking machine was for years also known as a horologue ('hour-teller') –

although medieval striking clocks were thoughtfully designed to remain silent at night. This device, Daniel Boorstin notes, was

> a new kind of public utility, offering a service each citizen could not afford to provide himself. People unwittingly recognized the new era when, noting the time of day or night, they said it was nine 'o'clock' – a time 'of the clock'. When Shakespeare's characters mentioned the time, 'of the clock', they recalled the hour they had heard last struck.[6]

In 1504, after a brawl in which a man was killed, the Nuremberg locksmith Peter Henlein (1479–1542) sought sanctuary in a monastery, staying there for several years, during which time he invented a portable clock – the first watch – devised, as the Nuremberg Chronicles of 1511 record, 'with very many wheels, and these horologia, in any position and without any weight, both indicate and strike for forty hours, even when carried on the breast and in the purse'.[7]

Yet until well into that century, people had to set their timepieces according to the shifting sunrise each day and stop to adjust them every time they were caught running fast or slow. The expected accuracy was no better than to the nearest fifteen minutes – Tycho Brahe's clock, typically, had just an hour hand. Cardinal Richelieu (1585–1642) was showing off his collection of clocks when a visitor knocked two on to the floor. Quite undismayed, the cardinal observed, 'That's the first time they have both gone off together.'

Towards the end of the sixteenth century the Swiss clockmaker Jost Bürgi constructed a piece that could mark seconds as well as minutes, 'but that,' observes Lisa Jardine, 'was a one-off, difficult to reproduce, and reliable seconds' measurement had to wait another hundred years.'[8] That is probably an exaggeration: by 1670 minute hands were common, and the average error of the best timepieces had been reduced to about ten seconds a day. (The noun 'minute', derived from *pars minuta prima*, 'the first tiny part', enters the English language in the 1600s; 'second' comes from *pars minuta secunda*.) By 1680 it was common for both minute and second hands to appear.

Punctuality and timekeeping soon became the fashion, even fetish: Louis XIV had not one clockmaker but four, who, along with their armoury of devices, accompanied him on his 'progresses'. Courtiers at Versailles were expected to order their days by the fixed times of the Sun King's rising, and of his prayers, council meetings, meals, walks, hunts, concerts. One of the six classes of French nobility was the *noblesse de cloche* – 'nobility of the bell' – largely drawn from the mayors of large towns, the bell being the embodiment of municipal authority. Timepieces were exact enough that philosophers from Descartes to Paley took to using clocks as a metaphor for the perfection of the divine Creation. Lilliput's commissioners report on Lemuel Gulliver's watch: 'We think it is his God, because he consults it all the time.' With both Frederick the Great (1712–86) and that hero of Trafalgar and Navarino, Admiral Codrington (1770–1851), having their pocket watches smashed by enemy fire, it seemed the mark of eminent commanders 'up at the sharp end' to have their accessories thus shattered.

In order to meet the new demand for individual timepieces (not only watches, but clocks small enough to fit into a modest house or a craftsman's workshop), clockmakers were forced to become pioneers in the creation of scientific instruments: their products, for instance, required precision screws, which in turn necessitated the improvement of the metal lathe. The mechanical revolution of the nineteenth century was the result in large part of ordinary people's wish to tell the time for themselves. And yet there continued to be holdouts, even into the twentieth: Virginia Woolf, in *Mrs Dalloway*, lets out almost a cry of pain at the ubiquity of the clock:

> Shredding and slicing, dividing and subdividing, the clocks of Harley Street nibbled at the June day, counselled submission, upheld authority and pointed out in chorus the supreme advantages of a sense of proportion, until the mounds of time were so far diminished that a commercial clock, suspended above a shop in Oxford Street, announced, genially and fraternally, as if it were a pleasure to Messrs Rigby and Lowndes to give the information gratis, that it was half-past one.[9]*

---

*When in 1834 a clock was commissioned as part of the rebuilding of the fire-gutted Palace of Westminster, the government called for 'a noble clock, indeed a king of clocks, the biggest the world has ever seen, within sight and sound of the throbbing heart of London'. The Astronomer Royal further insisted that it should be accurate to within a second. The outcome was Big Ben (strictly speaking, a bell), finally completed in 1859.[10]

The calendar used by an Inuit hunter in the 1920s, following the introduction of
Christianity to the Canadian Eastern Arctic: days of the week are represented by
straight strokes, Sundays by Xs. The calendar was also used as a hunting tally,
including counts of caribou, fish, seals, walrus and polar bear

But once people could tell the 'exact' time, what time were they
telling? There were so many to choose from. In 1848 the United
Kingdom became the first country in the world to standardize time
across its entire territory – to the signal of the Greenwich Observatory
(with Dublin Mean Time set just over twenty-five minutes behind). That
year saw the publication of *Dombey and Son*, in which a distraught Mr
Dombey complains, 'There was even a railway time observed in clocks, as
if the Sun itself had given in.' In 1890 Dr Watson records travelling by
train with Sherlock Holmes to a case in the West Country and being star-
tled by the Master gauging minute fluctuations in the train's velocity by
timing the telegraph poles, set at standard sixty-yard intervals, as they
flashed past their window, with the train acting as the Sun, the poles as
longitudes – a powerful image of the way in which the concept of the

standardization of time had insinuated itself into people's consciousness. Not everyone's, of course. Oscar Wilde (1854–1900) once arrived exceptionally late for a dinner, his hostess pointing angily at the clock on the wall and exclaiming, 'Mr Wilde, are you aware what the time is?' 'My dear lady,' Wilde replied, 'pray tell me, how can that nasty little machine possibly know what the great golden Sun is up to?' But it did.[11]

Einstein, of course, tells us there is no such thing as absolute time: salutary to recall that the office where he worked for so long was a clearing house for patents on the synchronization of clocks. And he himself recalled: 'At the time when I was establishing how clocks functioned, I was finding the utmost difficulty in keeping a clock in my room.'*

While the Sun appeared to have been replaced, in fact it continued to hold sway throughout the civilized world. Even by the mid-nineteenth century the towns and cities of most countries still had individual, Sun-based systems of timekeeping. Every French city, for instance, held to its own local time, taken from reading the Sun at its equally local zenith. Time bowed down to space; there was nothing God-given in the starting point of a second, or a minute, or an hour. The result was that the prospect of a globe united by the telegraph, fast trains and steamships had so far been stalled, because the spinning, angled Earth and the apparently moving Sun made nonsense of the idea of universal time. The clock on the wall told the hour for the family, the town hall clock that for the citizenry; but on the other side of the hill the clocks were strangers, and the imposition of universal time standards might even be dangerous: Field Marshal Helmuth von Moltke (1800–91), head of the Prussian, then of the German, army for nearly thirty years, lobbied for a unitary time system for Germany as an aid to making the trains run on time, so that its troops might be mobilized more effectively, but this was opposed by those who feared that an integrated rail system would provoke Russia to invade. Nevertheless, it was becoming obvious that

---

*Similarly, in *The Sound and the Fury* Faulkner has the Harvard student Quentin, desperate to escape civil time, shatter his pocket watch, 'because Father said clocks slay time. He said time is dead as long as it is clicked off by little wheels; only when the clock stops does time come to life.'[12] Others have found the very ticking of a clock a thing of solace: T. E. Lawrence, tied down and viciously lashed by his Turkish captors during the Great War, yet notes, 'Somewhere in the place a cheap clock ticked loudly, and it distressed me that their beating was not in its time'.[13]

uniformity could not be put off indefinitely. 'Societies were moving faster than their ability to measure,' as the historian Clark Blaise puts it.[14] Writing of the First World War, when wristwatches came into common use (they were much favoured by sentries), Burgess says:

> It was a war of railway timetables. For the first engagements in August 1914, about 2 million Frenchmen were deployed in 4,278 trains, and only 19 ran late. The wristwatch, which before the war had been considered effeminate, became the badge of masculine leadership. 'Synchronize your watches.' And then over the top.[15]*

In the United States, the same problem that had bedevilled the rest of the world was troubling individual states. Following the Civil War, railroad expansion had been prodigious. In the forty years after 1860 (by which year the US already had the world's largest railroad system) the amount of train track that had been laid down multiplied sixfold. By the turn of the century nearly every town of any size had its own railway station – if not several. However, as in most of Europe, the reckoning of time was a local matter, set by local noon, which, at the latitude of New York, is about one minute later for each eleven miles one travels westwards. Noon in New York was 11.55 a.m. in Philadelphia, 11.47 in Washington, 11.35 in Pittsburgh. Illinois used twenty-seven different time regions, Wisconsin thirty-eight. There were 144 official times in North America, and a traveller in the 1870s going from the District of Columbia to San Francisco, had he set his watch in every town through which he passed, would have had to adjust it over 200 times. If a passenger wondered when he might arrive at his final destination, he had to know the time standard of the railroad that was taking him there, and make the proper conversion to the local times at his boarding and at his eventual descent. Two cities, set a hundred miles apart, maintained a ten-minute temporal separation, even though a train could cover that distance in less than two hours – so

---

*Early in the war, Thomas Dinesen, the brother of Karen Blixen of *Out of Africa* fame, turned up at a recruiting office to enlist, only to be asked if he were homosexual, 'the evidence being that he was wearing a wristwatch'. He said he was not, and there were no further questions.[16]

which town's time was 'official'? The train itself might have set out from a city five hundred miles away, so who 'owned' the time – the towns along the route, the passengers or the railroad company? No wonder that Wilde noted that the chief occupation of a typical American was 'catching trains'. He visited the States in 1882: one wonders how often he missed his connection.

When people had travelled no faster than the speed of a horse, none of these considerations came into play, but in a railway-framed economy scheduling was a nightmare. As Blaise writes, 'It was the slow increase in speed and power – the fusion of rails and steam – that undermined the standards of horse and sail power and, eventually, the Sun itself in measuring time.' That quotation comes from Blaise's biography of the enterprising nineteenth-century Canadian Sandford Fleming, who in June 1876 happened to miss his train in Bandoran (set on the main Irish rail line between Londonderry and Sligo), because the timetable he consulted misprinted '5.35 p.m.' for '5.35 a.m.', condemning him to wait for sixteen hours.[17] Fleming was, among other things, the chief engineer of the Canadian Pacific Railway, and his 'monumental vexation' at this delay inflamed a desire to number the hours from one to twenty-four. 'Why should modern societies adhere to ante meridiem and post meridiem, why double-count the hours, one to twelve, twice a day? . . . [Hours] ought not to be considered hours in the ordinary sense, but simply twenty-fourth parts of the mean time occupied in the diurnal revolution of the Earth.' He would make it his mission to bring about a twenty-four-hour clock – by which 5.35 p.m. would become 17.35 h. On further reflection, he set himself the much greater task of relating all time zones, worldwide, to their relationship to longitude, and of introducing 'terrestrial, non-local' time.*

Standard time was the best gauge in the world: it would convert celestial motion to civic time. By 1880 England had been on standard time for more than thirty years, reform having started with the railways; why not

---

*Written before the twenty-four-hour clock had entered popular consciousness, the opening line of Orwell's *1984* – 'It was a bright cold day in April, and the clocks were striking thirteen' – was intended to make a British reader, unused to continental measurements, feel uneasy. The standard Italian translation (clocks that strike all twenty-four hours having been commonplace in Italy from as early as the fourteenth century) runs: '*Era una bella e fredda mattina d'Aprile e gli orologi batterono l'una*' – in English, '. . . were striking one'.

America? Because the US Congress, fearing a municipal uproar if it initiated change, had procrastinated, and the railroad industry, though well aware of the negative impact of a lack of standardization on its profits, dithered too. From 1869 on, this question was debated, until at last popular discontent forced the railroad barons' hand: as of Sunday 18 November 1883, they bypassed Congress and adopted the Greenwich system, dividing the country into four zones, eastern, central, mountain and Pacific, with noon falling an hour later in each. That day came to be known as 'the Sunday of Two Noons', since towns along the eastern edge of each new time belt had to turn their clocks back half an hour, creating a second midday, in order to mesh with those closer to the western edge of the same belt. Within a few years, this system became standard, but not without considerable bitterness, and some towns, such as Bangor, Maine, and Savannah, Georgia, still would not cooperate, either from religious principle or stubbornness. Detroit, perched on the crack between the eastern and central time zones, could not make up its mind, so that for many years its inhabitants had to ascertain, in making appointments, 'Is that solar, train or city time?' Congress itself did not ratify standard time until forced to do so by war in 1918.

As the historian Mark Smith records, 'The telegraph, not the Sun, now communicated time to a temporarily unified nation and, in the process, helped pave the way for the globalization of abstracted, decontextualized world time.'[18] Similarly, people were coming to accept that if a calendar were to serve a practical, global use, all dates had to be interpreted through a solar dateline.* The question was: where would that date line be? In 1884 twenty-five countries sent representatives to a conference in Washington, DC. Eleven national meridians (through St Petersburg, Rio, Lisbon, Tokyo, Copenhagen, Stockholm, Cadiz, Greenwich, Rome, Berlin and Paris), as well as additional contenders fixed on Jerusalem, the pyramids of Giza, Pisa (to honour Galileo), the Naval Observatory in Washington, and the

---

*When the United States bought Alaska in 1867, before standard time was implemented, the Russian Orthodox inhabitants found themselves suddenly having to observe the Sabbath on the American Sunday, which was Monday by Moscow reckoning. They were forced to petition for guidance as to when to celebrate Mass – on the Russian Monday or the American Sunday.

Azores – the original point of definition for the age of exploration – competed for primacy. The French, represented by their chief of mission and by the great astronomer Jules-César Janssen, were intransigent, insisting (without much evidence) that their *ligne sacrée* was a better scientific choice.

Fleming argued for a notional clock set 180° from Greenwich, in the middle of the Pacific, thus bypassing national sensibilities but still making use of Greenwich, though *without* involving England; but every longitudinal meridian touches land at some point in its arc, and, in addition, were Fleming's solution to be adopted, England would be split every noon between two days. His proposal soon foundered, and the debate ended only when Sir George Airy, Britain's irascible Astronomer Royal, wrote that the Prime Meridian 'must be that of Greenwich, for the navigation of almost the whole world [even then, ninety per cent] depends on calculations founded on that of Greenwich'. France abstained from voting, and true to its delegates' defiance 'Greenwich' has never appeared on any of its charts. (Surely by pure coincidence, the anarchist who tried to blow up the Observatory in 1894 was French.) As for the chosen site, once the vote came through, the meridian was moved from the famous obelisk on Pole Hill to a point nineteen feet further east. Sundial time was thereby banished, and a sophisticated abstraction rose to take its place.

And so in 1884 the Earth was sliced into twenty-four time zones, each an hour apart, with eastbound travellers adding successive hours, westbound ones losing them. Of course, there has to come a moment – which happens to confront us in the West Pacific – when the logic of the system subtracts a day from eastbound travellers and gives one to those heading west (Verne's Phileas Fogg learns this just in time to win his bet in *Around the World in Eighty Days*). One by one, countries adopted the Greenwich Mean Time (GMT) – even France, in 1911. However, in 1972 the French, still unhappy at what they perceived as Britain's unjustified supremacy, carried a resolution at the United Nations, to put GMT alongside Coordinated Universal Time or UTC – to be regulated by a signal from Paris (naturally). While GMT is reckoned by the Earth's rotation and celestial measurements, UTC is set by caesium-beam atomic clocks, which are less accurate but easier to consult.* In point of fact, the two systems are rarely more than a second apart, as UTC is augmented with leap seconds to compensate for the Earth's decelerating rotation.

☼

In *Waiting for Godot*, Pozzo asks in his usual desperation, 'What's the time?' His tramp companions answer, not quite in unison:

VLADIMIR: Seven o'clock. Eight o'clock.
ESTRAGON: That all depends on the time of year.

Those who sought to meddle with the Sun for their own purposes and pleasures were not finished: saving daylight was next. This was a project that dated back to Benjamin Franklin (1706–90), who on 26 April 1784, when serving as US minister to France, proposed (at the age of seventy-eight and in a moment of whimsy) that Parisians conserve energy – in the form of candle wax and tallow – by rising with the Sun rather than sleeping in with their shutters closed against the day.

The idea never caught on, and more than a century would pass before it would find a receptive audience. Then in July 1907 the successful London builder William Willett (1857–1915), a keen golfer and horse-man, self-published *The Waste of Daylight*, in which he argued that more people should enjoy the early-morning sunshine as he did, and complained about how annoying it was to have to abandon a game of golf because of fading light. Might clocks be shifted forwards or backwards by twenty minutes on four successive weekends, to make the change easier? It was not just a question for sports enthusiasts:

*Caesium clocks measure time by counting the tick-tock cycles of atoms of the metallic element caesium, each cycle being the exceptionally fast vibrations of the atoms when exposed to microwaves in a vacuum. The fifty-five electrons of caesium-133 are ideally distributed for this purpose, only the outermost being confined to orbits in stable shells. The reactions to microwaves by the outermost electron (which is hardly disturbed by the others) can be accurately determined. As the microwaves hit, the electron jumps from a lower orbit to a higher one and back again, absorbing and releasing measurable packets of light energy. This corresponds to a time measurement inaccuracy of two nanoseconds per day or one second in 1.4 million years. The experts make the necessary corrections.[19]

Physicists Jack Perry and Louis Essen adjusting a caesium-beam atomic clock, which they developed in 1955. This clock led to the replacement of the astronomical second with the atomic second as the standard unit of time

Everyone appreciates the long light evenings. Everyone laments their shrinkage as the days grow shorter, and nearly everyone has given utterance to a regret that the clear light of early mornings, during spring and summer months, is so seldom seen or used.[20]

Scientists and astronomers were divided on the question, although the press chirruped, 'Will the chickens know what time to go to bed?' and the editors of *Nature* ridiculed the idea by equating the time change with the artificial elevation of thermometer readings:

It would be more reasonable to change the readings of a thermometer at a particular season than to alter the time shown on the clock . . . to increase the readings of thermometers by 10° during the winter months, so that 32°F shall be 42°F. One temperature can be called another just as easily as 2 A.M. can be expressed as 3 A.M.; but the change of name in neither case causes a change of condition.[21]

But Willett was not easily put off, and within two years a Daylight Saving Bill had been drafted, finally achieving passage as a wartime economy measure in 1916. Earlier that year the Germans had already passed such a law, hoping it would help them conserve fuel and allow factory workers on the evening shifts to work without artificial light. Willett himself had died the previous year; but his neighbours erected a handsome memorial to him – a sundial, positioned to read an hour in advance of the time on any ordinary face.

The 1916 law was made permanent throughout Britain in 1925. America had passed a similar measure in 1916 too, but it proved so unpopular that Congress had to repeal it three years later. (Farmers, whom the measure was designed to help, hated DST, since they had to wake with the Sun no matter what time their clock said, and were thus inconvenienced by having to change their schedule to sell their crops to people who observed the new system.) Then in 1922 President Harding issued an executive order mandating that all federal employees start work at 8 a.m. rather than at 9. Private employers could do as they pleased. The result was chaos, as some trains, buses, theatres and retailers shifted their hours and others did not. Washingtonians rebelled, deriding Harding's policy as 'ragtime'. After one summer of anarchy, he backed down. It wasn't until the Second World War that DST was adopted again, with only the governor of Oklahoma holding out. However, it was repealed with victory, and in the decades following remained in use in the US only by local option, with predictably deranged results: one year, Iowa alone had twenty-three different DST systems. Jump to 1965, and seventy-one of the largest American cities had adopted DST; fifty-nine had not. Confusion reigned, particularly over transport routes, radio programmes and business hours. The US Naval Observatory dubbed its own country 'the world's worst timekeeper'.

The problem was finally resolved by the Uniform Time Act of 1966, although Indiana, (most of) Arizona and Hawaii still don't observe DST. In 1996 the European Union, faced with a puzzle-map of time zones, standardized DST, while in the US, as part of the Energy Policy

Act of 2005, DST was extended to begin on the second Sunday of March instead of in April, and to end on the first Sunday of November. As of today, DST has been adopted by more than 1 billion people in about seventy countries – slightly less than a sixth of the world's population.*

Like Robinson Crusoe notching a stick, or inhabitants of the Gulag scratching a line for every day of their imprisonment, we remain enmeshed in time, unable to leave it alone. As recently as August 2007, Venezuela's president Hugo Chávez announced that, to improve the 'metabolism' of his citizens, he had ordered the country's clocks to be moved forward half an hour 'where the human brain is conditioned by sunlight' – reversing an opposite decision made in 1965, but one that aligns Venezuela with Afghanistan, India, Iran and Myanmar, all of which offset time in half-hour increments from Greenwich reckoning. Gail Collins, writing in *The New York Times* about Chávez's *Diktat*, likened it to the scene in Woody Allen's film *Bananas*, in which a revolutionary hero becomes president of a South American country and announces that from that day on 'underwear will be worn on the outside'.[23] But Newfoundland is also on the half-hour system, defying the rest of Canada, while Nepal is fifteen minutes ahead of India, five hours forty-five minutes ahead of Greenwich. Saudi Arabia supposedly has its clocks put forward to midnight every day at sunset. As one commentator has quipped, 'Keeping one's watch properly attuned aboard the Riyadh–Rangoon express must be an exhausting experience.'[24]

Nor has fine-tuning stopped there. A second used to be defined as 1/31556925.9747 of the solar year. But it is nearly sixty years ago now that the National Physical Laboratory at Teddington invented the atomic clock, happily to discover that it was more accurate to base timekeeping on vibrating atoms than on the orbiting of the Earth. 'It was slightly embarrassing,' recalls David Rooney, curator of timekeeping at the Royal Observatory. 'When clocks diverge, it isn't good. By the 1970s we needed another fudge factor. So the leap second was introduced, to push together Earth rotation time and atomic vibration time.' Such seconds

---

*When clocks move forwards or backwards an hour, the body's clock – its circadian rhythm, which is governed by daylight – takes time to adjust. In a study of 55,000 people, scientists found that on days off from work, subjects tended to sleep on standard, not daylight time.[22]

are not inserted every year: the decision to add or subtract one (so far, it has always been added) is made by the International Earth Rotation Service in Paris – the last insertion was on 1 January 2006. It added an extra pip to the BBC time signal.

Now that measuring precise time is the responsibility of agencies such as the US Naval Observatory in Washington, DC, the International Earth Rotation Service at the Paris Observatory and the Bureau International des Poids et Mesures in Sèvres, France, all of which define a second as 9,192,631,770 vibrations of the radiation (at a specified wavelength) emitted by a caesium-133 atom, the Sun has been officially deprived of its long-term role as our timekeeper. This definition of a second – the first to be based not on the Earth's motion around the Sun but on the behaviour of atoms – was formally endorsed in 1967. But those 'leap seconds' that have periodically to be added to keep our clocks in sync with the planet's turnings, because the Earth goes its own way in space, heedless of atomic time, ensure that we can never entirely turn our backs on the Sun's guardianship.

One can go on fiddling almost indefinitely (the old joke runs that even a stopped clock is right twice a day). Back in 1907 Einstein came up with the equivalence principle, which states that gravity is locally indistinguishable from acceleration, and diminishes as distance increases from the centre of mass, so that time goes faster in, say, Santa Fe, high up in New Mexico, than in Poughkeepsie, down low in New York – by about a millisecond per century. A recent experiment on a westward round-the-world jet flight showed that clocks gained 273 nanoseconds, of which about two-thirds were gravitational.[25] Meanwhile, a clock eight feet in diameter has been constructed atop Mount Washington in Nevada, made to last 10,000 years (the period of time in which caesium-ion atomic clocks are said to 'lose' a second); and a French clock, made by the engineer and astronomer Passement, exhibits a perpetual calendar, showing the date until 9999.[26] An advertisement lauds 'the Ultimate Time-keeper', a watch based on 'complex astronomical algorithms' that provides 'local times for sunrise and sunset, moonrise and moonset, lunar phase as well as digital, analog and military time for wherever you are', in a.m./p.m. or 2400 format. It is programmed for 583 cities worldwide and automatically adjusts for DST. Constructed in titanium or steel with a sapphire crystal, it offers 'the broadest interpretation of time that money can buy', and can be purchased for $895.[27]

The Swiss watchmakers Swatch have proposed a planet-wide Internet Time that enables users throughout the world to bypass individual zones and rendezvous in the same 'real' time. Meanwhile, the scientists who run the atomic clock at Teddington, as well as competitors in the US and Japan, are at work on an even more accurate machine, known as the 'ion-trap' and due to be realized by 2020. Experts say that if it were activated now and were still running at the time predicted for the end of the universe, it might be wrong by half a second – if that – twenty times the accuracy of the current most advanced model.[28] In 2006 the US suggested that world time should be switched entirely to the atomic clock, which would involve the dropping of leap seconds – a notion that had the British Royal Astronomical Society up in arms. Had the proposal been accepted, says David Rooney, it would have been the first time in history that time was not dependent on the rising and setting of the Sun.[29]

Then there is that other, famous exchange from *Godot*:

VLADIMIR: That passed the time.
ESTRAGON: It would have passed anyway.

THE SUN IN OUR POCKET

The first man I saw . . . had been eight years upon a project for extracting sunbeams out of cucumbers, which were to be put into vials hermetically sealed, and let to warm the air in raw inclement summers. He told me, he did not doubt in eight years more, that he should be able to supply the Governor's gardens with sunshine at a reasonable rate.

Jonathan Swift, *Gulliver's Travels*[1]

He will literally have the Sun in his pocket.

The villain in Ian Fleming's *The Man with the Golden Gun*, talking about buyers of his solar energy converter

The Man with the Golden Gun was Ian Fleming's last novel, published posthumously and incomplete. Nevertheless, it became the ninth James Bond film, released in December 1974, at the height of the 1970s energy crisis, as worldwide enthusiasm for alternative forms of energy was coming to a boil. Bond has to recover a solar agitator, vital to a special converter, 'ninety-five per cent efficient, a device that will harness the Sun's radiation and give awesome power to whoever possesses it'. His principal adversary is a professional assassin played by Christopher Lee (Fleming's cousin and the author's original choice to play Bond); the story climaxes in the destruction of the solar installation on an island off the China coast.

In the decades since the film was released, solar power has become an ever more popular subject. The Sun's role as supreme timekeeper may have been overtaken by the atom, but it is still an open question whether the atom will also pre-empt the Sun in mankind's search for a usable source of energy. The Sun is the great self-renewing resource, the creator of coal, peat, oil, hydroelectricity and natural gas (methane). It raises moisture into the atmosphere, to return as the downpours that drive turbines; it powers the winds and the waves, and all their effects; it shows no signs of dying out (except to those thinking in numbers of astronomical magnitude); and it lavishes itself over the entire planet, delivering to the Earth's surface more energy in just forty-four minutes than we use in a year. About thirty-five per cent of what actually reaches our world is reflected back into space by clouds, and nineteen per cent is absorbed by the atmosphere; but this still leaves 12,000 times as much energy as that used in all man-made devices. Only two forms of renewable energy are not the result of solar radiation: geothermal and tidal (the Sun raises tides by its great mass, not its radiance). Yet only in the last thirty years has this plentiful source been taken seriously by those in power. Ian Fleming was ahead of his time.*

The notion of harnessing the Sun occurred to human beings as soon as they began to experiment with their surroundings. By the third century BC both Greeks and Romans employed 'burning mirrors' – hand-held concave reflectors – to focus sunlight on enemy warships. Archimedes (287–212 BC) reportedly devised just such a battery in 212 BC to defend Syracuse from a blockading Roman fleet by burning the enemy's sails 'at the distance of a bowshot' (about fifty yards). While the story is likely a myth,† it does show that the Greeks knew early on that sunlight was of its nature pure energy – and dangerous.

---

*As late as the 1960s, *Punch* cartoonist Rowland Emmett (1906–90) published a drawing of a solar locomotive, arousing gentle amusement – as did the eccentric 'inventions' of Caractacus Potts, a character he created for the 1968 Ian Fleming movie, *Chitty Chitty Bang Bang*. Even in 1987 the TV host David Letterman could win an easy laugh when he included in his 'Top Ten New York City Science Projects' 'No. 10. Summer Heat and Rotting Garbage: Fuel of the Future'.

†In 1992 researchers at Leicester University noted that Roman galleys traditionally furled their sails in battle, so setting them alight was not an option. The combined efforts of 440 men and shields could ignite a damp patch of wood at fifty yards, the project concluded, but not do serious damage, and would have been a poor use of manpower. A 'mirror platoon' of fifty men could have inflicted deep burns on targeted steersmen, or even the Roman high command (whose purple cloaks would have made them stand out); but such a tactic, if successful, would have been used again, which no historian has reported.[2]

According to legend, around 212 BC the Greek astronomer and mathematician
Archimedes used mirrors to focus the Sun's rays on to the Roman fleet

Around AD 100 Pliny the Younger (61–113) built a home using glass for
the first time to retain heat, and over the next several hundred years Roman
public baths were designed with large south-facing windows. The Romans
were also the first people to build greenhouses. In the sixth century the
Emperor Justinian even enacted a law protecting public and domestic sunrooms from the erection of buildings that obstructed sunlight.

The great tenth-century Persian scientist al-Hatham (b. 965) produced an important study, *On Spherical Mirrors*, which retold the legend
of Archimedes and the reflectors at Syracuse as fact; by 1270 it had been
translated into Latin and so came to the notice of Roger Bacon, who
warned his Pope that curved reflectors might be used by the Saracens
against Crusaders in the Holy Land. 'The idea of transforming the beneficent rays of the Sun into a fierce military weapon to incinerate human
beings,' writes Frank Kryza of the Vatican response, in his history of solar
power,[3] 'was seen as a deviant and evil idea, the work of witchcraft and
the devil'.

Early in the sixteenth century Leonardo da Vinci proposed a monstrous mirror four miles wide, to be employed commercially as a source of heat, not as a weapon. For whatever reason – lack of funds, of capacity (he was proposing to use more glass than then existed!), or of time (as ever, he was engaged on many other projects) – nothing came of this idea, but his concept marked a shift in focus from destructive to industrial uses, and also boosted solar research; interest in mirrors and lenses flourished.

During the reign of Louis XIV – perhaps inspired by the Sun King himself – numerous solar experiments were made. In 1747, under Louis XIV's successor, Georges Buffon (1707–88) used 140 flat mirrors to ignite wood placed 200 feet away – so demonstrating that Archimedes's feat was at least a possibility. Then came the Industrial Revolution and the new thinking it engendered. 'Mastery of the Sun,' Kryza observes, 'seemed tantalizingly close in the age of steam . . . The engineers of the nineteenth century worked with forces large enough to give them the sense, for the first time in history, that they were the masters of nature, in possession of the instruments they needed to change the conditions of life profoundly. Why not tame the Sun's energy?'[4] Solar pumps, heat engines and stills were just some of the secondary products.

By the 1830s, during his stay in South Africa, Sir John Herschel invented the 'actinometer', essentially a bulb of water that, when exposed to sunlight, could be used to calculate the energy received from the Sun. Further, as Stuart Clark relates:

He also performed some rather more eccentric experiments such as the day he placed a fresh egg in a tin cup and then laid a pane of glass across the top. Returning with his wife and six children some time later, he retrieved the cooked egg, burning his fingers in the process. Ceremoniously, he cut the egg into pieces and doled it out, so that all could say they had eaten an egg boiled hard by the South African sun. Suitably impressed by his newly found culinary skills, a week later he cooked a mutton chop and potatoes the same way. 'It was thoroughly done, and very good,' he recorded in his diary.[5]

Herschel went on to build a solar cooker out of blocks of black-painted mahogany, yielding a maximum temperature of about 240°F, eleven per cent above sea-level boiling point.

One long-running ambition had been to produce a viable solar-powered engine. Since the early seventeenth century, when Salomon de Caux, employing lenses, a frame and a metal vessel containing water and air, had constructed an early prototype, there had been attempts to make such a machine, but people considered them more games than practical projects. In 1861, however, a French mathematics teacher, Augustin Mouchot, poured water into an iron bucket, which he surrounded with reflectors. As the water evaporated, it produced enough steam to propel a small motor. Within four years he had succeeded in creating a conventional steam engine. A few months later he displayed this device to Napoleon III, who, impressed, offered financial assistance. Mouchot enlarged his invention's capacity, refining the reflector into a truncated cone, like a dish with inward-slanting sides. He further constructed a device that enabled the entire machine to face the Sun continuously. After six years' research he amazed spectators with his brainchild, one reporter describing it as an inverted 'mammoth lampshade . . . coated on the inside with very thin silver leaf' with the boiler sitting in the middle as an 'enormous thimble' made of blackened copper and covered with a glass bell. At the 1878 Universal Exposition in Paris, he displayed a solar-powered printing press using a parabolic mirror, steam engine and piston; it was to take another 122 years for solar energy to make its comeback at a world's fair – at Expo 2000, presented by the city of Freiburg.

Eager to exploit these inventions, the French government decided that the most suitable venue would be Algeria, a colony bathed in almost constant sunshine yet entirely dependent on coal, which was prohibitively expensive there: the willing Mouchot was immediately dispatched. 'Eventually industry will no longer find in Europe the resources to satisfy its prodigious expansion; coal will undoubtedly be used up. What then?' He soon invented a portable solar oven for French troops and a solar engine that could operate a printing press; but the high cost of these contraptions, coupled with the increasing cheapness of English coal, condemned them to idleness, and the Industrial

Revolution roared on unimpressed; in those days no one anticipated global warming.*

In 1891 the Baltimore inventor Clarence Kemp, 'the real father of solar energy in the US', patented the Climax, the first commercial Sun-powered water heater, which combined the old practice of exposing metal tanks to sunlight with the scientific principle of the hot box, so increasing their capacity to absorb heat.[6] By 1897, his biographers proudly state, 'thirty per cent of heaters used in Pasadena, California, were of Kemp's design', but this merely underlines the truth that such inventions were never successful outside the state; and when in 1902, based on a reflector made up of 1,788 individual mirrors, he created enough energy for a fifteen-horsepower† solar pump to irrigate a Pasadena ostrich farm, it was reported as no more than an offbeat experiment.

For centuries already, the rich of Europe had taken to setting trees against 'fruit walls' that built up solar heat, then slowly released it when the Sun went down, and England and Holland pioneered the development of greenhouses with south-sloping faces. The first commercial use of solar energy was probably evaporating salt from brine, and the first large-scale application was the distillation of drinkable water, by evaporation from brackish wells or isolated tracts of seawater. A still erected in Chile in 1872 produced 6,000 gallons of water per day from 51,000 square feet of collecting surface for over 40 years.

The story of harnessing the Sun continued to advance by fits and starts. In the late 1870s William Grylls Adams, deputy registrar for the Crown in Bombay, wrote an award-winning book, *Solar Heat: A Substitute for Fuel in Tropical Countries*, and tried to introduce the technology into

*In the 1850s the French political economist Frédéric Bastiat famously reduced protectionism to an absurdity in *A Petition of the Candle-makers*, in which that calling petitions to put out the Sun, as its radiance was bad for (their) business.

†The word 'horsepower' dates back to the Scottish inventor James Watt (1736–1819), who created the modern steam engine. Finding that potential buyers had difficulty understanding what the devices could achieve, he restated their relative strengths in terms that could be more easily understood. His customers were mineowners who used horses to haul coal and pump groundwater from the mines. By testing a number of horses at mine sites, Watt calculated that the average English horse could haul coal at the rate of 22,000 foot-pounds per minute for about ten hours a day (a foot-pound being that amount of work needed to lift one pound a distance of a foot). Watt arbitrarily increased the figure by fifty per cent, and so was born the 'horsepower' unit – a rate of work equal to 33,000 foot-pounds of labour. It is still used to compare the relative power of anything from lawnmowers to spacecraft.

Augustine Mouchot's solar-powered printing press at the
Paris World Exhibition of 1878

British India – without success. The torch then passed to France and
the engineer Charles Tellier, the 'father of refrigeration', who in 1885
installed a collector on his roof similar to the solar panels used today.
Instead of using water to produce steam, he chose liquid ammonia,
which passes into vapour at a lower temperature. Exposed to the Sun,
the containers emitted enough gaseous ammonia to drive a water
pump that could lift 300 gallons per daylight hour. But Tellier decided
to pursue his refrigeration interests instead (there was more money in
preserving food), and France bade adieu to the last major development of solar mechanical power on her soil until well into the
twentieth century.[7]

A few years later, in 1900, an enterprising Bostonian, Aubrey Eneas,
formed the first solar power company, and began building sun-powered
machines to irrigate the Arizona desert. In 1903 he moved to Los
Angeles, closer to his most promising customer base, and the following

year sold his first complete system for $2,160. After less than a week, a windstorm tumbled the boiler rigging into the reflector. Accustomed to setbacks, Eneas constructed another pump. In the autumn of 1904 a rancher in Wilcox, Arizona, bought the improved model, but it too was destroyed, this time by a hailstorm. It became clear that the massive parabolic reflector was too vulnerable, and the company folded. Other entrepreneurs were to follow (notably Henry E. Willsie, operating out of St Louis and Needles, California, who created a system that allowed his machines to function at night, using the heat gathered during the day), but their companies also failed to turn a profit.

Despite this dismal history, proponents continued to believe that if they could find the right combination of technologies, they could produce unlimited energy. One who shared that dream was a Brooklyn-born engineer, Frank Shuman (1862–1918), whose first solar motor, built in 1897, performed poorly, because even at respectable pressures the steam exerted insufficient force. Rather than trying to generate more heat, he replaced the boiler pipes with a flat metal container similar to Tellier's original design and devised a low-cost reflector: two rows of mirrors strung together to double the amount of sunshine intercepted. He then constructed the largest conversion system ever built, capable of delivering 55 horsepower and driving a water pump that gushed 3,000 gallons per minute – at a cost of $150 per horsepower, compared with the $80 of a conventionally operated coal system: a respectable outlay, he reckoned, considering that the investment would be quickly recouped because the fuel was free. Another reason he was not concerned about the fact that the energy delivered by his machine cost more than that from coal- or oil-fired engines was that, like the early French entrepreneurs, he planned to ship his invention to vast sunburned North Africa.

In 1912 he started work on the world's first solar power station – fittingly in Egypt, once the centre of sun worship. The site was Meadi, then 15 miles south of Cairo, and boasted 7 curved reflecting troughs, each 205 feet in length, with the thousand-horsepower steam engine producing over 55 horsepower. But the beginning was also the end. Two months after the final trials, the Archduke Franz Ferdinand was assassinated, igniting the Great War. The engineers operating Shuman's plant returned to

A revolving solarium at Aix-Les-Bains in September 1930

their respective countries to perform war-related tasks, and before the armistice was signed Shuman was dead. After the war, with the fall in the price of oil, interest in solar experiments once again evaporated.

By this time, oil and coal companies had established massive infrastructures, stable markets and ample supplies of carbon fuels. Solar pioneers, by contrast, were still trying to perfect their technology, and they had the additional difficult task of convincing sceptics to see energy from the Sun as something more than a curiosity. In North America, the discovery of huge quantities of natural gas in the Los Angeles Basin during the 1920s and 1930s killed off the local solar water-heater industry there. George Gamow, writing in 1940, could comment dismissively: 'The direct utilization of solar heat . . . is employed only in a few tricky devices – to run the refrigerators of cold-drink stands in the Arizona desert, or to heat water for the public baths of the oriental city of Tashkent.'[8] Solar water heaters did find some renewed enthusiasm in Florida, and by 1941 some 60,000 were in use in the Sunshine State. *Your Solar House* (1947), a best-selling book by forty-nine solar architects, reflected a real demand. However, after the Second World War, Florida Power and

Light campaigned aggressively to increase electricity consumption by offering water heaters at bargain prices, and again solar power retreated.

Elsewhere, the situation was similar. In Japan, where rice farmers urgently needed cheap hot water, a company began marketing a simple heater made up of a basin covered by glass, and by the 1960s more than 100,000 were in use; but the industry collapsed under a glut of cheap oil, just as it had in California and Florida. Even in Australia, with its pervasive light, solar heating devices numbered just a few thousand. In Israel, in the early days after its founding, electricity was rationed, so people sought to meet their needs in other ways, and by the mid-1960s one household in twenty had solar heating; but then cheap oil from the fields captured in the Six Days' War once again put solar power on the back burner.

In the US, the first solar-heated office building had been erected in the early 1950s, the first solar-heated and radiation-cooled house had started up (at a retrofitting cost of nearly $4,000 – about $30,000 in today's money), and some companies had returned to producing solar cells and water heaters. Then in 1953–4, researchers at Bell Telephone (now subsumed into AT&T) made a startling discovery based on an earlier technology. In 1839 the French physicist Alexandre-Edmond Becquerel had established that if two electrodes were immersed in acid and one of them was activated, a current flowed between the two. In 1873 the British engineer Willoughby Smith observed that the element selenium changed electrical resistance when subject to sunlight, but achieved only moderate efficiency. Bell Labs, experimenting with various materials, discovered that silicon possesses five times that efficiency, so the most effective way to convert the Sun's rays into electricity would be through silicon photovoltaic cells.

Bell was soon producing thin wafers of ultra-pure silicon, to which were added small amounts of arsenic and boron, to improve conductivity. When struck by the Sun's rays, electrons within the silicon were knocked loose by the penetrating heat and moved towards the wafer's upper surface, creating an imbalance between the front and back of the cell. When the top and bottom surfaces were joined by a connector – usually a wire – a current passed between them. 'A solar cell is altogether a much simpler structure than a green leaf,' wrote the solar chemist Mary Archer (wife of popular novelist Jeffrey), 'but it is like a green leaf in that one side is designed for looking at the Sun.'[9]

The New York Times lauded the Bell Labs discovery as 'the beginning of a new era, leading eventually to ... harnessing the almost limitless energy of the Sun for the uses of civilization'. And indeed it marked a significant improvement: in bright sunlight a cell had an energy conversion rate as high as twenty-two per cent. Even so, photovoltaic cells were still not economical, costing $300 ($2,200 in 2010 dollars) per kilowatt. But this was the time of the space race, and the US government's budget for research on solar cells skyrocketed once it was realized that satellites could generate electricity from solar panelling, which needed no restocking. In 1958 Vanguard 1, the first satellite to be equipped with cells, lofted into orbit. In the decades since, costs have dropped dramatically – on average, four per cent per annum over the past fifteen years.

Photovoltaic cells protect pipelines from freezing; they power lights, radios, roadside emergency telephones, refrigerators, air conditioners, water pumps and village electrification. They are to be found in even the smallest devices, such as pocket calculators and watches, iPod chargers, cameras and car mirrors. In 2003 around fifty per cent of all such cells were manufactured in Japan;* while the US accounted for about twelve

---

*Japan's history is illuminating. During the 264 years after 1603, known as the 'Edo Period', a policy of near isolation from the rest of the world was enforced. Edo at the time was the largest city on Earth, with between a million and a million and a quarter people (compared with London's 860,000 in 1801). Virtually nothing could be imported (with the notable exception of goods passed through the Dutch trading post in Nagasaki harbour), so the country had to be self-sufficient. Japan has few fossil fuels – coal, for example, being mainly used for salt extraction. Because of this and other scarcities, the entire population was forced to re-employ and recycle all it could. Therefore, everything was treated as a potential resource, even ash, candle drippings and human waste, ever a valuable fertilizer.

Tinkers repaired pots and pans, kettles and buckets, ceramic specialists glued broken china and glass, cobblers stitched and soldered: hardly anything was thrown away, but carefully put together again or given a new use. There were used-paper buyers, used-clothes dealers (all clothes at the time were hand-woven, so specially precious: there were about 4,000 old-clothes dealers in Edo alone), used-umbrella-rib buyers, used-barrel brokers, and tinkers who worked the city singing 'Let's exchange, let's exchange,' offering toys and sweets for old nails and other scraps of metal. Rubbish dumps didn't exist. Above all, Edo had one source of energy – the Sun. Almost everything was made directly or indirectly from it, with the exception of stone, metal, ceramics and other mineral-based materials.[10]

This was even true of lighting. Commercial power started up in November 1887, twenty years after the fall of the shogunate, with the first fossil-fuel-driven generator, but until that moment all artificial light came from paper lanterns and wax candles. Oil was extracted mainly from sesame seeds, but also from those of camellia, rape and cotton. In coastal areas, whale and sardine oil were used. Firewood was widely employed, but people consumed less than the annual increment of growing trees, so they were never eating into their energy supply. Was Edo an irrelevant, impoverished Shangri-La or a model for today?

per cent. In 1985 annual installation demand worldwide stood at 21 megawatts (i.e. 21 million watts); in 2005 it was 1,501 MW, an increase of more than 7,000 per cent.

☼

After collecting about a dozen books and more than 140 articles in just five years, I realized that solar initiatives are being made everywhere, in ever greater numbers, from China to Tanzania, from South Africa, where solar-powered traffic lights free traffic from the whims of an enfeebled power grid, to Abu Dhabi, capital of the United Arab Emirates, which, despite its reputation as an oil metropolis and the highest per-capita emitter of CO_2, is planning a research facility and a 500-megawatt solar power plant. So how far has the revolution gone?

In 2004 and 2006 I made two research trips to find out. My first stop was Freiburg, a city of roughly 215,000 people in Baden-Württemberg, between the Black Forest and the Rhine Valley. It was ravaged during the Second World War, in 1940 German planes mistakenly dropping sixty bombs near the railway station, then in November 1944 an Allied air raid destroying eighty per cent of the inner city. But this meant that after 1945 an extensive programme of reconstruction was put in place, and in recent years Freiburg has attracted a wealth of solar industries and research; no other German city has as many environment-related entities. Given that, and the fact that it is the country's sunniest city, it is not surprising that Freiburg proudly calls itself 'the environmental capital of Germany'.

At SolarRegionFreiburg,* the centre for solar energy in the area, I talked to three of the town's experts: Franziska Breyer, a trim blonde woman in her late thirties, a forester by training; Tom Dresel, a sociologist-turned-publicist who oversees solar projects; and Otto Wöhrbach, the director of the local planetarium.

---

*During my visit I encountered a research group from Madison, Wisconsin, Freiburg's US sister city, there to learn from Freiburg's initiatives so as to implement them back home. Another of Freiburg's nine twins is Besançon, a town of 130,000 inhabitants, one of the few in France to manage its own energy projects. In 1991 the city installed photovoltaic panels on cars in its highways department fleet. Beforehand, the warning signs indicating road-works were connected to car batteries that required regular recharging – usually the motor was kept running during the day to prevent the battery from going flat. Solar panels have entirely solved this problem and save the money previously spent on petrol, battery recharging, repairs and maintenance.

'It really all began by accident,' Frau Breyer explained. 'There's a village about fifteen kilometres from here called Wyhl, and in the early 1970s it was planned to build a nuclear power plant there. Students, farmers and vintners (we're a wine-growing area) staged a sit-in, and the moat of the building site became a discussion centre. Eventually, the plans were withdrawn, but not before somebody said to the protesters, "You don't want nuclear energy, fine; but what are you going to put in its place? Being destructive is easy; try to be constructive." That set people thinking.'

The first hesitant steps towards solar power began in 1976; then in 1981 the Fraunhofer Institute of Solar Energy Systems was founded in Freiburg. Ridiculed at the time by the science community, who saw it as little more than an offshoot of hippie enthusiasms, the centre is now the largest of its kind in Europe, with a staff of more than 350. Its success was fuelled in part by changing political realities. In 1983, for the first time in thirty years, a new political party won the five per cent of the vote necessary under West Germany's proportional representation system to obtain seats in the federal Parliament, and the Greens, 'Die Grünen', entered the Bundestag. Pictures of bearded, long-haired deputies without ties sitting next to the very proper Chancellor Kohl in Parliament swept around the world. The Greens' success galvanized Freiburg's citizenry, and an entire solar economy developed. In 1992 the city council resolved that only the construction of low-energy buildings would be allowed on municipal land. In addition to solar panels and collectors on roofs, many passive features, such as superinsulation, south-facing windows with low-emissivity glass, and foam insulation panels, became popular. 'The future starts every day,' said Tom Dresel. 'You see the movement of the city step by step.'

Back in 1945 the town had been occupied by the French. In the late 1990s, on the site of a former French army base, construction began on a new neighbourhood of 6,000 people, Vauban (so-named by the French occupiers after an eighteenth-century marshal), with the aim of making it a 'sustainable model district'. Solar energy was used to heat the water of many households, and the neighbourhood was developed according to a principle of sustainability. 'The road to sustainability is paved with innovation,' said Frau Breyer, smiling. Plainly, Vauban works.

The four of us talked for nearly two hours, then I set off to walk around the town. It seemed at first glance to be like any other university city in the area – prosperous, clean, filled with students. Then I began noticing all the innovations. On the outskirts, atop a long-disused silver mine, there now stands a solar observatory. Five wind turbines operating within the city boundaries have doubled the percentage of renewable electricity. By the end of my walk I had counted thirty installations: a technology park, several solar power stations, a 'zero-emissions' hotel, the solar-powered railway station and a number of houses with solar panels on their roofs. Thirty-year-old buildings are undergoing solar-friendly redesign with backing from the savings and loan societies, and private companies and public facilities are making their roofs available for solar modules. Locals buy shares in the panels and are reimbursed when the power is sold to the city electricity grid. There are 70,000 square metres of cells on Freiburg's roofs, while large solar scanners are programmed to realign with the Sun every twelve to fifteen minutes to maximize intake. The city has also constructed the world's first autonomous solar building, not connected to any main grid. The schools have a solar training centre and 'sun power stations', twenty-five of which (a third) maintain their own solar projects: this in a country that gets an average of only 1,528 hours of sunshine a year.

'Freiburg has been a pacemaker for other German cities,' Franziska Breyer told me. Gelsenkirchen, for example, on the northern Rhine, was the most important coal-mining and steel town in Europe early in the twentieth century – the 'city of a thousand furnaces'. It is now reinventing itself as the 'city of a thousand suns', borrowing many of its innovations from Freiburg. 'Our changes are ones that China and the two Koreas could adopt too.' On that note, Beijing and Freiburg already have one feature in common: twice as many bicycles as cars. Only the reasons differ. In Beijing, the bicycle is the primary means of transport. In Freiburg, the use of bicycles is a symbol of an environmentally conscious community. It's only a matter of time, I was told, before students are roaring off to their classes on solar-powered motorbikes.

☼

My next such trip, to Almería in southern Spain, took place nearly two years later, in July 2006. Less than an hour's drive from the city lies

Tabernas, the only surviving sand desert in Europe. Not only does the terrain experience 355 sunny days a year; it is ideally suited to action movies: among those shot here have been *Patton*, *The Magnificent Seven*, *The Wind and the Lion*, *Indiana Jones and the Last Crusade* and, memorably, *Lawrence of Arabia*. Fellini's masterpiece *8½* was partly filmed here, as was Sergio Leone's trilogy of Westerns, *A Fistful of Dollars*, *For A Few Dollars More* and *The Good, the Bad and the Ugly*.

Little of this success benefited Almería. By the early 1970s the area was one of the poorest in Spain. Then underground water tables were suddenly discovered, and an agricultural revolution began. Hundreds of greenhouses were put up to take advantage of the combined possibilities of the water and the ever-present sunlight ('Almería is where the Sun spends the winter' went the saying: the average temperature is 62.6°F); soon locals were claiming that it was not the Great Wall of China that could be seen from outer space but their greenhouses, one vast sea of plastic. By the end of the century the city had been transformed into the richest in southern Spain, and settlers were flooding in.*

The period directly overlapped the oil crises of the 1970s, and by the early 1980s the International Energy Agency, with the participation of nine countries, set up a small power station there to test two separate solar energy installations: one, a field of ninety computer-controlled mirrors (heliostats) which followed the Sun's rays and concentrated them on a single central tower, where they were then converted into thermal energy; the other, three fields of curved troughs, which tracked the Sun and reflected its energy on to metal tubes filled with oil. These tubes would be slowly heated to 290°C, and the oil fed into a steam generator. A third project was added, entirely a Spanish affair, in which a central tower was supplied by 300 heliostats whose mirrors reflected the concentrated heat on to black absorption panels atop the tower, with a water/steam receiver and a thermal storage system of molten salt tanks. The hope was that at least one project could be taken

---

*Since my visit in 2006, Spain's property bubble has burst and its construction industry has crashed. Almería now has an unemployment rate of almost twenty-five per cent, one of the country's worst. High-rise apartment buildings lie unfinished on the city's outskirts in an area known as 'Pueblo de Luz' – Town of Light. Yet half the global solar capacity in 2008 was in Spain.

up commercially, but by the late 1980s Spain's partners had become frustrated with the lack of progress and left; only Germany stayed the course. Since 1999 just one project kept going, under separate management, but that has thrived: the Plataforma Solar de Almería (PSA) has since gone on to become the largest R&D centre for solar energy in Europe (worldwide, its only rivals are the Weizmann Institute in Israel, the Sandia Laboratories in Albuquerque, New Mexico, and the nearly 2,000 giant mirrors in the Mojave Desert outside Barstow in California).

One of the first lessons I learned on my trip to Almería was that concentrated solar power and photovoltaic panels represent very different technologies. PV panels use solar photons to excite electrons into generating a current, while solar thermal energy, the kind in which Almería has been a pioneer, uses photons to heat fluid molecules. Such conversion requires long metallic mirrors that focus sunlight on to a pipe, whose water runs through a heat exchanger, generating enough steam to turn a turbine. Since the whole process requires a lot of land and copious sunlight, its ideal location is in sun-baked deserts.

I was taken around the site by José Martinez Soler, a cheerful member of the staff in his early thirties who was completing a doctorate on the marketing of solar energy. He proudly announced that PSA would soon be celebrating both its twenty-fifth anniversary and the fact that two Spanish companies had at last made commercial ventures out of the station's research. A town like Freiburg can process only relatively low heat from its sunlight, but the state-of-the-art silicon panels and silvered mirrors used in Almería can intensify sunlight to temperatures up to a thousand times its original level. (As José was explaining all this, we passed an outdoor sign reading, in English, DANGER CONCENTRATED SUNLIGHT.)

I left, impressed by all I had seen, but unconvinced that solar energy can be more than a minor contributor to the world's needs. The next day I met Alfonso Sevilla Portillo, a local expert on energy matters who had been the first director of PSA, having earlier worked for several years in California. Fashionably turned out and in his early sixties, he disagreed profoundly with his old company's strategy. PSA was developing a technology it could sell, when 'what we have to do is *shape consumption*. If we continue to live as we do now, we will never be able to supply the energy for all our needs.' Solar energy R&D should be integral to an overall philosophy of how to live, he said. Dr Sevilla had left PSA to work on a project in Kronsberg, on the outskirts of Hanover, where some six thousand units would house fifteen thousand people, making up five compact neighbourhoods. The town will be using forty per cent less energy for the same quality of life.

Is this the way of the future? Whereas in the past inventors were more interested in the scientific and philosophical ramifications of 'capturing the Sun', much of the present enthusiasm rises from fears about global warming and the exhaustion of natural fuels – exhaustion, or the fact that most oil is found in politically unstable regions such as the Persian Gulf, Nigeria and Venezuela. The sources of fossil energy may have increased – petrol, kerosene, propane – but techniques of converting other sources have multiplied, from solar power to nuclear fission, fusion, wind, waves and biomass processing (translating material from dead plants and animals into ethanol, biogas and biodiesel fuel). Nearly every day newspapers carry a fresh story about such alternatives, not all of which are necessarily feasible. As one sceptic noted about wind, 'Not since Don Quixote have so many windmills presented such an orgy of illusion.'[11]

But even the most fantastic-seeming ideas have a way of becoming reality. In the 1980s the great science-fiction writer (and master skindiver) Arthur C. Clarke argued that we could obtain unlimited electricity from the sea 'without huge masses of spinning hardware', making a case in 'The Shining Ones' for heat engines 'to tap the thermal gradient between the warm surface layers and the near-freezing waters of the abyss'.[12] Sure enough, the 'Pelamis', a snake-like machine – at 500 feet almost as long as a passenger train – now generates energy by absorbing wave motion, while horizontal turbines are being mounted on the seabed, similar to underwater windmills. Britain alone could generate up to twenty per cent of the electricity it needs from waves and tides.[13]

In 1981 a solar-powered plane flew the English Channel, and now Omega, the watch company, is developing a sun-powered aircraft (which accumulates energy in lithium batteries on its wings) to circle the globe. In March 2007 the Swiss boat Solar21 crossed the Atlantic in sixty-three days; and sailing of a more ambitious kind is now part of space exploration, which uses sunlight beating on a giant reflective area to power the craft. Meanwhile, scientists at the celebrated furnace of

Odeillo Font-Romeu in the Pyrenees, the largest in the world, have found that reflectors taken from anti-aircraft searchlights can concentrate temperatures to 3,500°C. In Udaipur, Rajasthan, the local maharaja has introduced sun-powered rickshaws, while for the last two decades engineers have been working on a solar-powered car, and although such vehicles are not yet commercially viable, a biennial 3,021-kilometre race for these cars takes place across central Australia from Darwin to Adelaide.[14] Hybrids, such as Volkswagen's Eos (named after the Greek goddess of dawn) and the French car-maker Venturi's Eclectic, have retractable hardtops with energy-translating sunroofs. However, sunlight can provide only so much – enough power for about fourteen miles a day; even then, the vehicles have to be extremely light and aerodynamic, and normally carry just one person.

Ingenuity is certainly not wanting wherever one looks. In May 2009 *Visionaire*, the fashion magazine, published a 'solar' issue, with a black-and-white cover whose photochromic inks broke into full colour when exposed to the Sun. 'Solio', a portable charger about the size of a mobile phone, fans out a photovoltaic trefoil to capture sunpower, which it can then transfer to mobile phones, hand-held computers, game consoles or music players. Such personal devices can also be wired into sports tags or into a 'solar jacket' (made from a fabric called Microtene), which has two-by-three-inch panels embedded in its removable Nehru-style collar.[15] Rubbish bins being introduced on American beaches have solar-powered sensors that send an email to the public works department when a container is three-quarters full. While chemists are currently working to produce a paint that can transform sunlight directly into electricity, in 2008 scientists at the Idaho National Laboratories came up with a plastic that did just that: 'Solar Skin' uses a thin film of copper-indium-gallium selenide that can be applied direct to glass or metal. And a self-healing paint on cars and furniture is being developed that would restore a damaged coating in a matter of minutes by exposure to the Sun.

On a larger scale, engineers in New Jersey have patented a device for switching the system from conventional power to back up solar power within seconds of a failure. Solar ovens are now able to cook 600 meals twice a day;[16] methane that builds up in refuse landfills is being sold as a power source; and biological systems are being developed to use sunlit algae to convert exhaled carbon dioxide and water into oxygen and protein-rich carbohydrates – and into fuel.

It was recently argued that the Moon's slow axial rotation, unclouded skies and abundant local materials would enable installations to be built there to harvest the Sun's energy. Solar power, collected through hundreds of lunar panels, would provide a clean and reliable energy for space-based applications and ultimately for Earthlings as well.[17] Another initiative, space-based solar power, or SSP, would entail launching satellites equipped with massive photovoltaic surface areas, that would unfold (or inflate; the technology is still experimental) once in orbit. Sunlight is roughly eight times as intense in our neighbourhood of space than it is on Earth; but full-scale tests of SSP have yet to be conducted, and an attempt in 2005 to launch the first solar sailing craft, *Cosmos 1*, failed.[18]

All these initiatives make it difficult to know how to gauge the progress of solar energy – after all, by the early years of the Reagan administration, government enthusiasm for solar power had largely disappeared, along with the panels Jimmy Carter had installed on the White House roof. Reagan also slashed the budget of the Solar Energy Research Institute and allowed tax incentives for renewables to collapse. It was not just the Gipper: between 1980 and 2005 the fraction of all US research and development spending devoted to energy declined from ten to two per cent, and the 2007 budget committed just $159 million to solar R&D, barely half that on nuclear energy ($303 million), and a third that on coal ($427 million). At the end of 2009 legislation was introduced into Congress to scuttle thirteen solar plants and wind farms planned for the Mojave Desert in California, 'arguably the best solar land in the world', according to one environmentalist. In a world expected to add 2.5 billion people by mid-century, government and industry investment in energy technologies has been falling, not rising.[19]*

---

*In 2009 half the electricity in the US was generated from coal, which yields energy more cheaply than do oil or natural gas and is competitive with the uranium used in nuclear power plants. America, often called the 'Saudi Arabia of coal', has enough, at present rates of consumption, to last 250 years: every day, on average, each person in the US consumes twenty pounds of coal to keep electricity flowing. Yet China consumes more 'black gold' (coal, not oil!) than the US, the European Union and Japan combined. India too is stepping up its construction of coal-fired power plants. Ironically, the consequent sulphur pollution is so pervasive as to exercise a temporarily beneficial side effect, its tiny airborne particles deflecting the Sun's radiation back into space and so slowing global warming.[20]

There is also the question of how long subsidies will continue: solar projects of all kinds have been hugely helped by tax credits, grants and programmes in which power companies must compensate subscribers who channel energy back into the national grids from solar appliances. But in Denmark, where seventeen per cent of electricity comes from wind turbines, new projects have virtually ground to a halt since subsidies fell victim to fluctuating political priorities. In Spain, whose government funding is more generous than anywhere in Europe, a programme of cutbacks was set in place in 2008. Other nations may follow – indeed, in October 2009 companies importing solar panels to the US were hit by up to $70 million in unexpected tariffs.

Germany too is politically divided by squabbling on this matter (conservatives arguing that subsidized solar power is growing so fast that it will raise electricity bills). Despite the row over subsidies, however, the country continues to be a leader in solar power: fifteen of the world's largest twenty plants are based in Germany,[21] generating 750 megawatts of sunpower, more than five times the 2006 US figure. The Japanese are not far behind – a million and a half buildings in Tokyo have solar water heaters, more than the entire US. In January 2010 it was reported that China, barely a player in the industry five years ago, had leapfrogged Japan and the West to emerge as the largest manufacturer of solar panels, on track to produce more than half the world's panels by end 2011, as well as the largest maker (just under fifty per cent) of wind turbines.[22]

In 2005 a Spanish law decreed that all new dwellings should provide for solar appliances. Israel uses solar water-heating systems in about thirty per cent of its buildings, and these are mandatory in all new homes. China imposes similar provisions, and Sweden intends to give up fossil fuels entirely.[23] Solar power plants are under construction in Mexico, South Africa, Egypt, Algeria and Morocco. Several US states have legislated to protect the 'right to light' for town gardens – a late bow to Justinian* –

---

*Justinian's law protecting sunrooms from constructions that block their light is echoed in the 'ancient lights' provision of English law, which gives house owners the right to a certain level of natural illumination. In effect, the owner of a building with windows that have received daylight for twenty years or more is entitled to forbid any obstruction that would deprive him of that illumination. (In the centre of London, near Chinatown and Covent Garden, particularly in back alleys, signs saying 'ancient lights' mark individual windows.)

while California, home of the 'Million Solar Roofs' initiative, alone produces fifty-four per cent of the world's wind electricity. A decade ago, only 500 Californian rooftops bore solar panels; today, there are nearly 50,000 such installations, providing the equivalent energy of a major power plant.

Worldwide, removable roof panel insulation is a relatively new phenomenon, having become part of the original architecture of houses only in the 1980s, on the initiative of the remarkable American engineer Harold R. Hay, who got the idea on a mission for the US government to India in the 1950s, when he noticed that many people were living in rusty sheet-metal shacks that were hot by day and cold at night. He designed roof panels that could be taken off during the day and replaced at night in the winter months, and the reverse could be done during the summer. This one idea made a huge difference to hundreds of thousands of Indian homes. Back in 1976 Hay offered some memorable advice:

> We're a Mediterranean – not an omni-climatic – animal. We belong in the Earth's temperate zones ... but our technology has made it possible for us to heat the planet's Arctic regions and cool its tropics enough to make ourselves comfortable there. That, in one sense, is what the energy crisis is all about. We've learned to use enough energy to make ourselves comfortable in areas that we're not really physically adapted to live in ... We can both use and misuse even solar energy.[24]

Whatever the philosophical implications, solar energy evangelists point out that at least costs are coming down. 'Thirty years ago solar

---

The question 'Who owns sunlight?' has become pertinent worldwide. In 1959 a Florida appellate court stated that the 'ancient lights' doctrine has been unanimously repudiated in the US, but in 1975 a committee of the California legislature decided that if persons had rights to sunlight and were able to sell them, institutions would be more willing to finance solar projects. Although several witnesses warned that the creation of such rights would create problems, such as disputes over how much these rights were worth, in 1978 California enacted a law protecting homeowners' investments in rooftop panels. Trees that impeded the panels' access to the Sun brought on their owners fines of up to $1,000 a day. California now has a history of litigation on the issue. Meanwhile, in 2005 three heliostats were imported into New York City, at a cost of $355,000 each, to redirect sunbeams into a vacant lot in Lower Manhattan. Teardrop Park lies in the shadow of three skyscrapers, but the heliostats, installed on the roof of a twenty-three-storey building nearby, follow the movement of the Sun, intercept its light and redirect it into the park.

power was cost-effective on satellites,' says Daniel Shugar, president of SunPower Systems, a California company. 'Today it can be cost-effective for powering houses and businesses.'[25] What in 2007 was an $11 billion market is growing at more than twenty-five per cent a year. But solar panels are still comparatively expensive.

Early in 2008 *The New York Times* asked four companies to provide quotes on installing such panels on a Manhattan apartment-house roof.[26] The lowest price, after negotiation, was $370,000 for a 50-kilowatt system. State grants and tax credits brought in $265,000, and the balance was financed with a low-interest, ten-year loan: that left $105,000 still to be found for the 266 panels, which would however run for a good 25 years. Then there is what is known as 'net metering': homeowners with solar electric systems can see their meters spin backwards when the Sun is shining, as they earn credit that erases part – sometimes all – of their power costs. So solar power, while initially expensive, may arguably make sense in the long run.

It may be a long run indeed. In 2009 solar energy generated 0.5 per cent of Germany's electricity, renewable resources as a whole 14.2 per cent; in the US in 2001, renewables accounted for 6 per cent of energy consumed, a figure that is expected to remain about the same for the next quarter-century. The US currently has about 128 million homes, only about 80,000 of which have installed solar power. Solar input worldwide meets just nine per cent of humanity's primary energy demands, and is projected to decline to eight per cent (these figures vary according to source: *National Geographic* puts the figure at less than one per cent).

Some long-hoped-for options are still at the planning stage. Fusion – producing energy by combining hydrogen atoms into helium, the process that fires the Sun – has been talked about for decades as a potentially limitless energy source, but has still to be harnessed.[27] While the harvests from renewable sources of energy may be growing rapidly, they are starting from such a small base that their overall place in the energy supply chain remains minimal.

'The big problem is big numbers,' says Neil de Grasse Tyson, of the Hayden Planetarium. 'The world uses some 320 billion kilowatt-hours of energy a day.'[28] Within the next century we will be using three times that much. Americans are particularly heavy consumers; while constituting less than five per cent of the people in the world they consume more than a fifth of its energy resources. A true switch to sunpower in the US would require $420 billion in subsidies – a huge figure, but only about eleven days' US gross domestic product. In his Inaugural Address of January 2009 President Obama declared, 'We will harness the Sun and the winds and the soil to fuel our cars and run our factories,' and the following month signed a $787 billion stimulus package, intended to double the amount of renewable energy produced over the next three years. By October 2010, the first of nine multi-billion-dollar solar farms in California, Nevada, and Arizona were at last given the green light, while twenty-seven US states now require their utility companies to generate an increasing percentage of electric power through the use of renewable energy sources. Enthusiasts speak of at least 250,000 square miles of land in the south-western US alone that would be suitable for solar power plants, or point out that if only 0.35 per cent of the Earth's land surface (an area the size of France would do) were covered with solar cells, this would be enough to cater for all our energy needs. If only . . .

Dr Tyson puts the whole question into startling perspective with one simple story. In 1964 the Russian astronomer Nikolai Kardashev proposed three different levels of civilization in respect of energy use. A type I civilization harnesses the energy on or within its home planet. It controls all the sunlight that falls on the surface, and can, if it so desires, also reach into a volcano or a hurricane to tap their respective energies. A more advanced type II civilization can utilize the entire energy production of its host star – 10 billion times more power than a type I; and type III uses all the energy of all the stars in its host galaxy, increasing its power 10 billion times over type II. 'So where do Earthlings fit in?' concludes Tyson. 'Sorry to break the news, but any civilization so fragile that its members must stockpile fossil fuels, run away from erupting volcanoes, evacuate cities in advance of hurricanes and rush to high ground during tsunamis is not in charge of its own planet, and can be none other than type zero.'

Perhaps there is some hope. As the biologist Oliver Morton suggests:

The challenge that faces us is to find new technologies that sit in the space between the photovoltaic cell and the leaf – new hybrids of industry and nature. To make leaf-like things that generate alternative fuels, or even, conceivably, electricity . . . We need to work on a whole gamut of solar conversion technologies.[29]

Finding such solutions is not impossible. In 1931, not long before he died, Thomas Edison, the godfather of electricity, told Henry Ford: 'I'd put my money on the Sun and solar energy. What a source of power! I hope we don't have to wait until oil and coal run out before we tackle that.'[30]

PART FIVE

INSPIRED BY A STAR

Swastikas date back to the Neolithic (c. 9500 BC), and were used all over the world as a sun symbol long before the Nazis appropriated it. This 'black sun' or 'sun wheel' is a symbol of Nazi mysticism

THE VITAL SYMBOL

I see the Sun, and if I don't see the Sun I know it's there. And there's a whole life in that – in knowing the Sun is there.

Mitya, in *The Brothers Karamazov*[1]

'What,' it will be Questioned, 'When the Sun rises, do you not see a round disc of fire something like a Guinea?'
   'Oh, no, no, I see an Innumerable company of the Heavenly host crying, "Holy, Holy, Holy is the Lord God Almighty."'

William Blake, *A Vision of the Last Judgement*[2]

Anne Frank would regularly confide to her diary how inspired she felt when she looked at the old white horsechestnut outside the small house where her family were in hiding from the Nazis, especially when the tree was lit up by the morning Sun: 'As long as this exists, and I may live to see it, this sunshine, the cloudless skies, while this lasts I cannot be unhappy.'[3] The Sun, either in itself or as an embodiment of what is good in nature, inspires in all kinds of ways. It is the most ubiquitous of symbols, appearing frequently in heraldry (when represented as giving light, it is a 'sun radiant'; with a human face, a 'sun in splendour'), and as one of the most popular institutional emblems, for everything from oil conglomerates to the Smithsonian. In the UK alone, 497 trademarks employ it, against 366 that use the crown and 239 the Union Jack.

One wonders which world culture has the most elaborate solar imagery. Twenty nations display the Sun on their flags: Antigua,

Argentina, Bangladesh, Ivory Coast, Japan, Macedonia, Malawi, Namibia, Nepal, Nicaragua, Niger, the Philippines, Rwanda, Taiwan, Tunisia and Uruguay (a full-faced sun in splendour), as well as little-known countries such as Bergonia, Kazakhstan, Kiribati and Kyrgyzstan. The newly designed 'Ausflag', the national ensign of Australia, features a black kangaroo leaping across the Sun, while several countries have solar stripes. India's has a golden wheel representing the twenty-four hours of the day, while the Oval Office boasts a huge yellow sunburst on its carpet.

Japan takes its flag from a rising-sun banner first displayed at court in AD 701. In 1870 the scarlet-on-white design, known as *Ninomaru* – the sun flag – was adopted. The imperial Japanese navy (the army too, until 1945) was given its own emblem, showing the Sun and sixteen rays: *Kyokujitsu-ki*. Yet among the several hundred family crests in Japan, while fifty-one depict the stars and seventeen the Moon, only seven bear the Sun – as if even its greatest noblemen felt they could never be on such intimate terms with so imperial an entity.

Royalty feels no such qualms. Early on 3 February 1461, during the fifteenth-century struggle between the rival houses of Lancaster and York known as the Wars of the Roses, Edward IV (as he was about to become; 1442–83) and his army reached Mortimer's Cross, a hamlet in south Shropshire, ready to do battle against the Lancastrian foe. Edward, only twenty-one years old, had just heard that his father and youngest brother had been captured, mocked and slaughtered. His troops were battle-weary and in low heart. As dawn broke, three suns appeared in the sky and suddenly joined together in one.

Hilary Mantel, in her novel *Wolf Hall*, brings what happened to dramatic life: '[Edward could see] three blurred discs of silver, sparkling and hazy through the particles of frost. Their garland of light spread over the sorry fields, over the sodden forests of the Welsh borderlands, over his demoralized and unpaid troops.'[4] Taking place was most likely a parhelion, the dispersion of light through ice crystals, although it could also have been the morning sun multiply reflected off the wetlands to the east. Whatever the cause, his men cowered back from so awesome a sight, but Edward was inspired to cry, 'This is a good sign, for those three suns betoken the Father, the Son and the Holy Ghost!' At this, his whole army, reportedly between 5,000- and 10,000-strong, sank to their knees

Louis XIV, aged fourteen, as Le Roi Soleil in *Ballet de la Nuit*, performed in Paris on 23 February 1653 – the first time Louis was called the Sun King

This Pre-Columbian mask was probably worn as a head ornament or diadem. The Sun was the principal god of the La Tolita culture

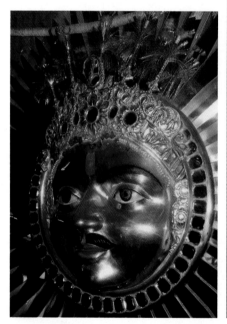

Detail of ornamental sun in the Surya Chopar room in the City Palace Museum, Udaipur, India. The moustache is a symbol of virility

Three masks from the Museum of the Sun in Riga, Latvia's capital. The museum, the only one of its kind, was opened in 2008, after its owner had collected 319 solar masks over ten years

The Sun (2002) by Yue Minjun (1962–) is unusual in taking a humorous approach – all this award-winning Chinese artist's paintings show laughing characters

Girl with a sunflower (1962), from the Croatian Naïve School by Ivan Rabuzin

Inti worshipping the Sun, while a band of conquistadors look on from a distance

New Year celebrations at the Futamigaura seashore, Japan. In the background can be seen the same rocks as appear in the first colour insert. As part of the ceremony, bronze frogs are specially stationed to combine Sun with water to symbolize fertility

Sunset at the Crimean Shores (1856), by Ivan Aivazovsky (1817–1900). He was the most prolific Armenian painter of his time, with more than 6,000 works – and the most forged ever as well

Turner's *Heidelberg Castle* (1840–45). The castle ruins are among the most important Renaissance structures north of the Alps. Turner completed this painting in the decade before his death, when he saw the Sun not so much as a resplendent yellow but an inspiring white

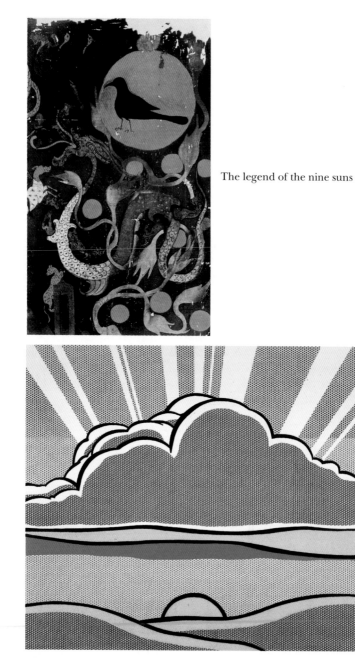

The legend of the nine suns

Roy Lichtenstein's *Sinking Sun*, his rendering of a Californian sunset, was acquired by the actor/director Dennis Hopper and his then-wife Brooke Hayward in 1964; in 2006 it was sold at Sotheby's for $14 million

Van Gogh made a number of variations on the theme of a man sowing before a setting sun. With *The Sower* (1888), he finally captured 'the most vital yellow' of sunlight

Rooms by the Sea, Edward Hopper (1951), uses sunlight to emphasize loneliness. Hopper produced many works where a figure is near or gazing out of a window, all of them preoccupied with sunlight, while, according to one of his patrons, 'He always said that his favourite thing was painting sunlight on the side of a house'

David Hockney also has favourite subjects, one being Californian swimming pools. *A Bigger Splash* was painted in the summer of 1967. One critic noted its 'salient eruption of white chaos', but set next to the paintings of Turner's last years it suggests the two artists had remarkably similar visions in prayer – then rose, to storm to victory. 'His whole life took wing and soared. In that wash of brilliant light he saw his future.' Edward went on to have the sun in splendour woven into his standards.[5]*

The royal personage most closely identified with the Sun is certainly Louis XIV, *Le Roi Soleil* (1638–1715). In 1653, at just fifteen years of age, he appeared before his court wearing a golden wig, an embroidered tunic and a pink-and-white-plumed headdress, crusted with rubies and ablaze with solar rays, while little suns blazed forth from his garters and the buckles of his high-heeled shoes.[6] Nine years later he staged an equestrian ballet between the Louvre and the Tuileries, before five thousand guests. The opening extravaganza featured a troupe of noblemen turned out in costumes representing the great civilizations, Louis himself dressed as a Roman emperor, with a golden cloak and a shield carrying a solar emblem. After him came the Persians, led by his brother, whose escutcheon bore the emblem of the Moon and the motto *Uno soli minor* – 'Lesser to the Sun alone'. Eight years on, Molière depicted Louis as Apollo in *Les amants magnifiques*, in which the king, once more caparisoned as the Sun, appeared on stage in an opera that climaxed in volcanoes and fireworks.

On another occasion, in the hours before dawn of 18 August 1674, Louis led his court down through the palace gardens, where they suddenly came upon a rocky island some seventy feet across, crowned by an eighty-foot obelisk topped by a sphere of fiery light. A bas-relief depicted the king, crossing a river at the head of his army. Below, to one side, sprawled a lion crushed by defeat, to the other an eagle in submission.[7] The whole display burned silently, until at a signal 1,500 charges of powder were detonated, as a firework fleur-de-lys blazed along the canal. The triumph of both Sun and the Sun-of-Justice sovereign was complete.

Identifying himself with the Sun was a politically astute move, for it shed a quasi-divine magnificence over Louis – after all, the Sun was the

---

*Edward also construed the suns as foretelling the happy reunion of the three sons (suns) of York (himself and his brothers George and Richard) and incorporated them into his ensign. For centuries, people have punned on 'Sun' and 'son'; in *Richard III*, the opening lines 'Now is the winter of our discontent /Made glorious summer by the Sun of York' play with the fact that the badge of Edward IV, son of a Duke of York – whose title he had inherited – was a sun. Shakespeare certainly knew the story of Mortimer's Cross, as Edward exclaims in *Henry VI Pt. III*: 'Dazzle mine eyes, or do I see three suns?'

Fireworks on the Grand Canal at Versailles during the 1674 festival that Louis XIV
organized to celebrate one of his military victories

heavenly body associated with Apollo, god of peace and the arts, as well
as a symbol of military glory, which suited the warrior-monarch's ambitions. Louis's decision was received with joy by contemporary
chroniclers, the historian Father Claude-François Ménestrier, perhaps
the most famous Jesuit in France, rejoicing:

> What grander and more heroic effort could be wished for than the
> action of the Sun lighting up the whole world and working incessantly
> to maintain all that lies upon it? What could be chosen more worthy
> of a king who promised such great things as soon as he started gov-
> erning his states as his own?[8]

Then there was Versailles, the spectacular palace that Louis made his
supreme residence. The Sun was not the only symbol on display, but it
cast all others into shadow. Throughout, decoration combined images
and attributes of Apollo (laurel, lyre, tripod) with the king's portraits and
emblems. The Grand Apartment glorified this Most-Christian King as the

Five young gypsies perform their salute to the Sun in the mid-1950s

Sun, and within it the Apollo Salon was designated the main room. The regularity of the king's work habits and his regular risings and retirings were another point of comparison: one thinks of the journal entry three centuries on of Wallace Stevens (1879–1955): 'The day of the Sun is like the day of a king. It is a promenade in the morning, a sitting on the throne at noon, a pageant in the evening.'[9]

Yet another point of comparison between monarch and Sun is that both the star and the institution of royalty have been devalued in the intervening years. From at least the nineteenth century on, the diminishing influence of the diurnal cycle on the developed world's daily life, and the evolution of scientific knowledge, have made the Sun less and less a mythological object. No longer regarded as a god with divine powers (the easiest way to channel one's respect for it), it became associated with a new, more impersonal pantheon: it was the clock of our existence, maker of our weather (a late discovery), regulator of the seasons and a force with the power over life and death – but in an impersonal sense, not as an anthropomorphized deity that humans could appeal to or placate. And as we have come to understand it differently, its symbolic value has changed too.

The Sun has always played a role in religion. For Jews, the seven-branched menorah originally represented the Burning Bush seen by Moses on Mount Sinai, and thus the light of God, but in later readings its arms signified the five then-known planets, plus the Moon and, as the central candle, the master star. In Japan, the devout are advised that the best time for meditation is in the moments before sunrise or after sunset. To describe the Transfiguration, evangelists could find no better simile than the Sun's brilliance, the whiteness of its light an apt image of the transition of the Son of God from the world of the flesh to that of the spirit.[10]

The Sun has also appeared in visions. Between 13 May and 13 October 1917, 'a beautiful lady who said she came from heaven' was said to have appeared six times to three young shepherds near the town of Fatima, on the outskirts of Lisbon, the children reporting a 'dancing sun' cavorting across the sky, a sure instrument of God. As Fatima grew into one of the staging posts of Christian pilgrimage, on the following 13 October crowds of more than 25,000 are reported to have witnessed how the atmosphere became 'orange' as the Sun dimmed, then the clouds suddenly parted and a bright light, like a halo, emanated from the gloom.[11] The Sun began to spin and change colour, then hurtle Earthward.[12]*

From the 1960s on, sun-inspired visions of another kind arose from that part of the West's counterculture involved in taking psychedelic drugs. The 'father' of LSD, Albert Hofmann (1906–2008), for instance, became fascinated by the mechanisms through which plants turn sunlight into substances that transform mental states. 'Everything comes from the Sun,' he said, proclaiming it the gateway into new forms of consciousness.[14] Hofmann took to studying hallucinogenic substances found in Mexican mushrooms. The songwriter/singer Donovan wrote one of the first psychedelic songs, 'Sunshine Superman', while the hippie musical *Hair* ended with 'Let the Sunshine In'. And people did.

Of course, drug-enhanced celebrations of the Sun are not unique to

*Other less famous apparitions or mirages include the testimony of a nineteenth-century Anglican clergyman, who was leaning idly against his father's carriage when he saw a group of fairies playing a few yards away. 'Get inside,' ordered an unsympathetic parent. 'It's just the Sun.'[13]

any one group or country. Mircea Eliade describes the mythology of the Desanas, a small tribe living in the equatorial forests of the Vaupes River in Colombian Amazonia that persists in its hunting-based culture, and whose religious mythology revolves around the creative power of the Sun-father. 'To the Desanas,' says Eliade, 'the soul . . . possesses its own luminosity, bestowed by the Sun at the birth of every human being.' This association is reinforced by the psychedelic drug *yagé*, the drinking of which brings the initiate to a subterranean world inhabited by luminous beings. 'Taking *yagé* is expressed by a verb meaning "drink and see", and it is interpreted as a regressus . . . to the primeval moment when [the] Sun-father began the creation.' Light, says Eliade, is also associated with sexuality – another aspect of the creation myth involves souls originating from semen that falls with the Sun's rays:

> If everything which exists, lives, and procreates is an emanation of the
> Sun, and if 'spirituality' (intelligence, wisdom, clairvoyance, etc.) par-
> takes of the nature of solar light, it follows that every religious act has,
> at the same time, a 'seminal' and a 'visionary' meaning. The sexual
> connotations of light-experiences and hallucinatory visions appear to
> be the logical consequence of a coherent solar theology.[15]

Perhaps. Solar myths confirm, reassure, inspire. It is up to individual taste whether they do more than that. There have ever been plenty of innocent and practical ways of taking the Sun as inspiration. The wheel, first seen on a pictogram in Uruk, in Mesopotamia, around 3500 BC, is one example. Yoga practitioners perform the Sun salutation. Ted Hughes demanded that his publisher bring out his poems only on days when the Earth's magnetic field was propitious. Mata Hari (1876–1917), the Dutch erotic artist shot by the French as a German spy, took her stage name from the Malay for Sun, literally 'eye of day'. The game of polo descended from a competition among Mongol tribesmen, who would whack a human head up and down a field, a pastime said to have a solar origin. A similar version of the game appeared in Persia in the fifth century BC – horsemen driving not a skull but a 'lighted fireball' across a 'sky-field' in 'the Sun game'.

I realize that the list above will appear arbitrary, but that is how it is with the Sun. It finds its way into odd and uncategorizable areas. The rest of this chapter details four of them: how the Sun has affected language, and its common symbols: blond hair, gold and mirrors.

☼

Words based on the Sun, or influenced by it, reflect cultural attitudes. The names of both people and places frequently have something of the solar. In Persia, such first names as Afrouz, Afshid, Dalileh, Dori, Farimehr, Jala, Jahantab, Khorshed, Kurshid, Mehrasa, Mehrshid, Shams, Shidoush and Talayeh, all meaning 'sun', are all still used, and there are as many again meaning 'light' or 'dazzling'. In Sanskrit, 'Asia' means 'sunrise', 'Jayaditya' translates as 'victorious sun' and 'Khorvash' as 'sun-lovely'. Both 'Ravi' and 'Ravindra' mean sun, the syllable 'Ra' in many languages being used both for people's names and to cover important sun-related concepts. For the Japanese, 'Nippon' translates as 'the origin of sun'.

In the northern hemisphere, the further north or west one goes in the Old World, the less likely one is to find such equivalences. Sun-related positional vocabulary is still used in celestial navigation, but geographical terms – Orient/Occident/Levant – have died away, though so much of our culture is Mediterranean-connected that there is still a repository of such metaphor, in English especially, not because of the climate but because its writers enjoyed thinking up new solar phrases. Shakespeare coined several, 'to burn daylight', an expression for wasting time, occurring in both *The Merry Wives of Windsor* and *Romeo and Juliet*.

Street slang makes full use of the Sun. The chilling phrase 'a Harlem sunset' describes a wound given in a razor fight – introduced by Raymond Chandler in a 1940 novel and used since in a poem by Seamus Heaney. A sundowner is both a tramp in the Australian bush – someone who comes to a sheep station at sunset for food and shelter – and a strict naval captain, originally one who compelled midshipmen to return from shore leave by sundown. It is said to be time for drinking when 'the Sun is over the foreyard' or 'over the yardarm': since in home waters and northern latitudes the Sun would be over the yardarm – one of the horizontal spars mounted on the masts of old sailing ships – towards noon, the drinking hour evidently started earlier.[16]

Many proverbs revolve around the Sun. Some of the more imaginative include 'The same sun that will melt butter will harden clay' and 'The Sun doesn't shine on both sides of the hedge at once'. 'Flying into the Sun' means being so passionately committed as to destroy oneself, while a 'sunflower' is a pretty girl, i.e. someone unfolding for the sons/Suns (c. 1959); and 'sunbeam' means tableware that hasn't been used and so doesn't have to be washed (c. 1950). 'Sunfisher' is slang for a bucking horse, but to 'get the Sun into a horse's coat' is to allow a horse a rest from racing.

Most derivations are easy enough to tease out – 'solar plexus' is so named because the ganglia radiate from the stomach area throughout the body; and when Archie Goodwin in the Nero Wolfe books describes things as 'sunfast', he means that they are certain, because sunfast dyes do not bleach out – but others do mystify: to be intensely nervous is said to be 'walking on the Sun'. The legal world has its own way with words: a 'sunshine law' is one that forbids government to pass edicts without there being a certain amount of time for them to be considered by the public; 'daylight' is a legal term stipulating that the period of time after sunrise and before sunset is to be considered part of the day and not the night, often invoked when defining burglary. One of the oddest names for the Sun is 'Spanish faggot', which appears in *Grose's Dictionary* of 1811: this may have come about because from the fifteenth century and continuing through the reign of Carlos I (1500–58), coins from Spain or its dominions were stamped with the image of faggots – sheaths of arrows bound loosely together – which were seen as similar to the Sun's rays.

If there is one colour that symbolizes the Sun it is yellow. 'Just look at that yellow Sun!' exclaims Nellie Forbush, the heroine of *South Pacific*. The Sun can be orange, of course, or red, or white, but yellow is its primary colour. Given that fact, however, the various shades that the Sun can take on have challenged human ingenuity. The varieties of solar yellows stocked by paint suppliers range from the time and weather tints (high noon, low sunrise, sun shower, sun porch and so on) and the golds (from fool's through to golden pond) to such possibly solar colours as Yellow Brick Road, mellow yellow, and ho-hum yellow (I list, I do not explain). It seems hardly surprising, in the light of such an inventory, that the British artist Sir Terry Frost (1915–2003), captured by the Germans when they overran Crete in 1941, spent his captivity experimenting with yellow and came up with 381 shades. All are part of the Sun's wardrobe.

Human beings come nearest to the Sun in the colour of their hair. The women of ancient Rome would bleach themselves to blondeness with quicklime and wood ash, jealous of the German slavewomen brought back by soldier-husbands. A blonde head typically has many more strands than dark or red, and blondes have long been thought more fertile. In her cultural history of the species, Joanna Pitman points out that the majority of heroines in fairy stories are blonde, from Beauty and the Beast to Cinderella, Goldilocks and Rapunzel. 'The men of [ancient] Greece were bewitched by blonde hair,' she adds, for it represented fantasy and wealth:

> Above all, it was sexy. Luminous and bright, it shimmered and glowed, producing, in a country of few natural blondes, a rare and exciting contrast to the mass of Mediterranean dark hair all around . . . 'Golden' was the master epithet for Aphrodite in all of Homer's work.[17]

Blondeness, says Pitman, became a prejudice in the Dark Ages, an obsession in the Renaissance, a mystique in Elizabethan England (Elizabeth herself was auburn-haired), a subject of myths in the nineteenth century, an ideology in the 1930s – both for Hitler's Germany (where in fact at most only eight per cent of women were natural blondes) and Stalin's Russia (where blondes were almost as scarce) – and a sexual invitation in the 1950s, with Hollywood's glorification of glamorous but unintelligent blondes. As Raymond Chandler points out in *Farewell, My Lovely,* 'There are blondes and blondes and it is almost a joke word nowadays' (and he wrote that in 1940).[18] In the early 1960s California set the trend of tanned blonde beauties, and the Beach Boys rhapsodized over such golden girls. We seem to be back in a never-ending realm of colours, but the message is the same: yellow, of whatever hue, is the Sun's colour, evoking awe, even worship.

The cultural historian Hugh Aldersey-Williams, writing of the shining yellow metal gold, adds a further gloss: 'The word gleam stems from an Indo-European root, *ghlei-, ghlo-* or *ghel-,* meaning "to shine, glitter or glow," which is also the origin of the word yellow.'[19] Just as yellow is the colour, gold is the prime metal of the Sun. Pitman makes this connection: 'The colour gold had long been established in the classical canons of beauty and power. Almost two thousand years before Homer, in the time of the Proto-Indo-Europeans, the colour was connected to the worship of Sun and fire, and to the adoration of a yellow dawn goddess.'[20] The association has endured through the centuries. Thus James Joyce fuses solar rays, blondeness and gold into one perfect tableau when Leo Bloom, suddenly bathed in morning sunshine, reflects: 'Quick warm sunlight came running from Berkeley Road, swiftly, in slim sandals, along the brightening footpath. Runs, she runs to meet me, a girl with gold hair on the wind.'[21] He was writing truer than he knew. Gold can indeed be found in human hair (almost the only part of the body to contain it). On average, it is concentrated at the rate of around $8/1,000,000,000$ of a gram: not enough to make a barber smile – and more in adults than in children, and more in men than women.[22]

The Sun's association with gold is easy to understand, and not just as a matter of colour. Long before the days of alchemy, gold was regarded as the most perfect thing on Earth, and thus indissolubly linked with the great star.[23] The alchemist in Victor Hugo's *The Hunchback of Notre-Dame* exclaims: 'Gold is the Sun; to make gold is to become God!'[24] The only metal that never tarnishes, gold is a suitable symbol of purity; it can be beaten out so thinly that it has hardly any solidity left, approaching a state of pure evanescence.[25] It has often been associated not just with divinity but, because of its rarity and worth, with royalty.

Every culture seems to have valued gold. In 624 BC Periander, tyrant of Corinth, invited the city-state's nobility to a dinner and had his soldiers strip the women of their gold ornaments and gowns adorned with gold thread, which booty financed his rule for decades. By 560 BC the first gold coins were minted in Lydia, in what is now western Turkey, and in 1284 Venice introduced the gold ducat, which became the most popular coin in the world. Columbus, sending back gold to his queen, exulted: 'O most excellent gold! Who has gold has a treasure with which he gets what he wants, imposes his will in the world and even helps souls to paradise.' The Inca considered gold 'tears wept by the Sun' (or more coarsely, its sweat), employing it only for ceremonial purposes, although Atahualpa sued for his freedom from the Spanish by having his cell filled with gold higher than a man could reach, and Montezuma II sent Cortés a sun-like gold disc the size of a cartwheel. The fabled city of El Dorado was named for its king, 'the gilded man', whom the sixteenth-century chronicler Gonzalo Fernández de Oviedo (1478–1557) describes as going about 'continually covered in gold dust fine as ground salt'.[26]

Durable, flexible and beautiful, gold is also capable of a myriad uses: it resists the fiercest fire, in minute quantities makes windows less conductive of sunlight; and gold visors carried on Apollo 11 helped to protect the astronauts' eyes from the intense radiation beyond the atmosphere. It has conjured cities out of the ground almost overnight, from San Francisco to Johannesburg. Above all else, it is extremely scarce: a troy ounce of the stuff is today (May 2010) worth more than nine hundred dollars. During mankind's time on Earth, only an estimated 300,000 tonnes have been won from the Earth, nearly all since 1848, and no more; only enough, if piled up, to fill the Eiffel Tower. Yet it is estimated that 8 trillion tonnes roll dissolved in the oceans alone, still more in the Earth's crust. As for the Sun, about six ten-billionths of its mass consists of gold, which totals 1,200,000,000,000,000 tonnes – enough to cover Scotland half a mile deep.

In an essay he wrote in 1930 on the return to the gold standard in international markets, John Maynard Keynes considered what associations the metal evoked. In the days before the evolution of representative money, he argued, it was natural to choose a metal as the most suitable commodity for representing value. Quoting Freud, he says there are reasons deep in our subconscious why gold should 'satisfy strong instincts' and serve as a symbol. 'The magical properties, with which Egyptian priestcraft anciently imbued the yellow metal, it has never altogether lost.'[27]* It does duty, so to speak, for the Sun on Earth. The Latin word for gold, *aurum* (thus the chemical symbol Au) is cognate with the Greek, Aurora – the golden goddess of dawn.

It is no surprise that the coin considered the most beautiful ever struck in the US (known as a double eagle, as it was twice the value of the $10 coin

---

*Money has a long association with the Sun. The first coin issued by the US government, on 6 July 1787, was called the 'Fugio' ('I flee'), with the Sun at its zenith. Over the centuries Muslim sovereigns have held solar ceremonies in which they were weighed against a quantity of gold (representing the Sun), and the poor received money accordingly.

known as the eagle), contained an ounce of twenty-two-carat gold. This coin was in circulation until 1933, when, in response to the Depression, Roosevelt banned private ownership of gold. It was revived in 2009, slightly smaller than the original but containing exactly an ounce of twenty-four-carat gold, with Liberty striding towards the viewer on one side, the Sun gilding the Capitol dome behind her, and, on the other, an eagle in flight over a blazing sun. Flip the coin, and the Sun still lands heads up.[28]*

☼

Mirrors give us back the Sun, and are another of its most potent symbols. The Egyptians invariably styled their mirrors after the solar disc, a slightly flattened circle representing the Sun as it often appears on the horizon. This divine connection made the mirror a religious symbol that was used during festivals and ceremonials and even placed in tombs, often in front of the face or on the breast of the corpse, to assure the presence of Ra. The Chinese hung mirrors from temple ceilings, reflecting the wisdom of heaven and drawing fire from the Sun. Other cultures linked mirrors to the Sun through language – the Aztecs, for example, named one of their sun gods Tezcatlipoca, meaning 'Smoking Mirror'.

The Japanese were unique in the prominence and ubiquity of their sun-linked mirror imagery. According to legend, when their sun goddess Amaterasu hid herself in a dark cave, a mirror hung on a tree outside

---

*A tradition stretching back 300 years dictates that British coins (and later stamps) show an incoming monarch's profile facing in the opposite direction to his or her predecessor's. On the death of George V in 1936 the appropriate mandarins set to work. George V had faced left, so his successor Edward VIII would be looking the other way – but the new king refused, believing that his left side showed him to better advantage. As a compromise, it was suggested that his left profile be deployed but photographically reversed, so that the royal head would still be gazing right. Again the new king exercised his veto: the manner in which he parted his hair on the left side was one of his better points. All this created turmoil at the Royal Mint and at the Post Office, but the first issue of stamps emerged with his left profile facing left, as decreed. Unfortunately, the background prepared for the stamps had already been engraved to show the monarch looking right; when the authorities learned of Edward's wishes they changed the direction of his gaze, but not the background. Hence the king is shown peering into the shade rather than the Sun – the opposite of an inspiring omen. After Edward abdicated that December, *The Times* commented: 'Even the strong minds can yield to this weakness, the superstitious anxiety of those who shook apprehensive heads at the new stamps, because the head of King Edward VIII was turned away from the light, and looked forward into the gloom – apt symbol of a reign that began with everything in its favour and moved onwards into calamity.'[29]

For thousands of years Indian women longing to get pregnant have stood
naked in water facing the Sun

gave her a glimpse of her own light, encouraging her to come out.
Thereafter she ordered her grandson to Earth to create the 'Cradle of
the Sun' – Nippon – giving him a mirror to hand down to his descendants (and sending with him a whole company of mirror-makers). In this
myth the mirror's identification with the reflection of the disembodied
spirit of divine light made it the terrestrial representation of the Sun, and
thus the most important treasure in the imperial regalia.

Burnished gold shines back well enough, and some of the earliest Roman
mirrors were lined with it, but generally more reflective surfaces have been
used: bronze, tin (mixed with mercury) and eventually silver. By the sixteenth century the main Venetian mirror-makers would back a plate of
glass with a thin sheet of reflecting metal, which remained the dominant
method for 300 years. The modern process – backing a glass surface with
metallic silver – had to wait until the 1830s, when the discovery of immense
deposits of silver in Australia, Central America and Europe (which drove

down its value against gold by a factor of five) and the inventive powers of a great German chemist, Justus von Liebig (1803–73), combined to make it practical. Until then mirrors, particularly large ones, had been extremely expensive, one reason why, at a cost of 654,000 *livres*, Louis XIV created the most famous of all halls of mirrors, the Galerie des Glaces at Versailles. With seventeen windows overlooking the gardens matched by seventeen vast mirrors along the wall, all crafted in a Paris workshop in an attempt to outdo anything Venice could produce, it further affirmed the greatness of France – and its king.

When in 1682 the hall was displayed to the awed court, it was a commanding success, 'a dazzling mass of riches and lights, duplicated a thousand times over in just as many mirrors, creating views more brilliant than fire and where a thousand things even more sparkling came into play'.[30] There were 306 panels in all, blending together so that each seemed part of a much larger sheet. The Gallery was not only the symbolic focus of the kingdom during the Ancien Régime but also continued to play a role after the Revolution – in 1919 the Treaty of Versailles was signed there. In her history of the mirror, the historian Sabine Melchior-Bonnet links the hall to the Sun. 'Everything at Versailles,' she writes, was

> specular magic, not only the castle reflected in the water of the canal, not the symmetry of an unfolding architecture, nor the repetition of gesture in the mirror, but first and foremost all the rules of etiquette by which the audience of the courtiers bowed in unison . . . The court thought itself a spectacle. Each person wanted to see, to see himself and be seen, to be narcissistically bedazzled as all gazes converged upon the eye of the Sun King, from whom all light radiated.[31]

To a contemporary observer, standing inside the Gallery on a bright summer day, it could seem that the room had been merged into the Sun, light bouncing endlessly from mirror to mirror.

DRAWING ON THE SUN

Light is not only glorious and sacred, it is voracious, carnivorous, unsparing. It devours the whole world impartially, without distinction.

> J. M. W. Turner: note appended to
> *The Angel Standing in the Sun,* 1846

The essential thing about Matisse's painting is not to judge it except by eye. You have to look at it as you would look at sunshine through the window.

> Jules Flandrin (1871–1947)

In the art of the Western world, it was a long time before painters depicted the Sun with any real interest. For centuries they illuminated their canvases with indirect light, all objects near and far equally detailed and vivid, so that without internal clues it is hard to determine the time of day. Sunshine, in any real sense, is out of the picture. Painters explored candle-glow long before they turned to daylight.

In medieval times artists rendered landscapes as idealized backgrounds that represented the panoply of God's abundance, a filling-in behind the main subject. They generally ignored the world of clouds, storms, rain and sun except for freaks of nature – an earthquake, a volcanic eruption, or a bolt of lightning, all of which they could use to evoke divine displeasure or to illustrate some event from the Bible. The Sun often appears in the sky but it is left hanging there as talisman or accessory. In the Sistine Chapel, Michelangelo's Sun is a blotchy yellow ball being willed skyward by the pointing finger of omnipotence – it certainly has no majesty of its own.

Between 1300 and 1650, however, as Renaissance man began to analyse the world around him, the Sun gradually lost its power as a religious and astrological symbol. What light consisted of, how it operated, what its effects were, all came under investigation. A similar change occurred in the world of art, as painters of the era began to take an interest in nature for its own sake. By the mid-1600s landscape (the very word being a coinage of that era) had become a fit subject for art. While Rembrandt (1606–69) often used the Sun's rays to intensify his portraits, by the 1650s artists such as Jacob van Ruysdael (1628/9–82) and Claude Lorraine (1600–82) had become interested in depicting sunlight itself. Ruysdael's attention to meteorology was extraordinary – his *Winter Landscape with Two Windmills*, for instance, shows a shaft of light radiating from a Sun pillar created in convincing detail by ice crystals in the freezing air.[1]

Claude (one of the last artists to be known by his first name alone) was a friend of Nicolas Poussin, and they would travel the Roman Campagna together, sketching; but whereas Poussin subordinates the landscape to biblical and allegorical figures, Claude does the reverse: his subjects are the land, the sea, the air. Light, the key feature of his seaport pictures, derives from a Sun just above the horizon, a feature he introduced into *Harbour Scene* (1634) – the first time in the history of art that the Sun clearly illuminates an entire picture.[2] He had so little interest in anything besides the natural world that he engaged others to paint in the small figures that appear on his canvases, and he once remarked to a potential buyer that what he was selling was the landscape; the figures came as an extra.

Elsewhere, artists were beginning to depict natural phenomena with an almost scientific concern for accuracy. In the early eighteenth century, for instance, the German portraitist Aegid Quirin Asam, having been commissioned by the Benedictine abbey of Weltenburg to paint their order's founder St Benedict (said to have been inspired to initiate his *Rule* during a solar eclipse), pursued eclipses all around Europe. His portrait shows the saint on a high tower, a thin ray emerging from an otherwise occluded Sun. No other painter has gone to such lengths to get eclipse details right.

An altar painting in the Benedictine abbey of Weltenburg, in southern Germany,
showing their founder St Benedict at the moment he was inspired to initiate his *Rule*,
during a solar eclipse

☼

Claude and Ruysdael had in common that they were much admired by
an artist who, perhaps more than anyone else, was to make the Sun his
main subject: Joseph Mallord William Turner (1775–1851). John Ruskin
recorded that, weeks before dying, the old artist announced, 'The Sun
is God,' a remark that sums up his painting so appositely that, although
perhaps apocryphal, it has been little questioned.

Our star was Turner's primary inspiration from a very early age. At the
evening classes on the nature of perspective that he attended from 1783
to 1789, he insisted on putting reflections in the windows he drew on to
an architect's perspective sketch of a mansion. When the architect told
him to paint the panes a plain dark grey and the bars white, as this was the

established practice, Turner remonstrated, 'But it will spoil my drawing.' The *Annual Register* would later note that the young artist was 'always on the alert for any remarkable phenomenon of nature. He could not walk London streets without seeing effects of light and shade and composition, whether in the smoke issuing from a chimneypot, or in the shadows upon a brick wall, and storing them in his memory for future use.' By his early twenties, rather than producing imitations of his elders, as other young artists did, he was turning his canvases into visions of air and light.

But he was still not the equal of the master. In 1800, invited by the picture's owner to view Claude's *Seaport with the Embarkation of the Queen of Sheba*, he was left on his own to appreciate it. When the owner returned, breaking the spell, Turner burst into tears: 'I shall never be able to paint anything like that.' By then he had completed 300 watercolours and was already an extraordinarily tactile artist: the fingertips were for him as capable of creating expression as any brush. When years later a young man claiming to be a painter called on him, Turner commanded, 'Show me your hands.' They were clean. 'Turn the fellow out,' he snapped to his manservant. 'He's no artist.'

In 1802 he made his first trip to the Continent. In the brief peace between Napoleon's wars, Paris was packed, 'fluttering all the gay colours of the rainbow, and harmonized and softened by the glowing tints of the setting Sun'. He came home, to produce a series of wonderful watercolours and oils. At the Royal Academy exhibition of 1807, he showed *Sun Rising Through Vapour: Fishermen Cleaning and Selling Fish.* As one art historian has remarked, the first part of the title 'could easily be the name of half his work'.[3]

How to put on to canvas the true nature of light and shadow? A sketchbook of 1806–8 illustrates Turner's musings: 'Reflection in water[:] tho' the real shadow is nearly the same from the plane of the Horizon in near Objects, yet when the whole of the light lays behind it frequently streaks a shade 3 times its height.' In 1808, seeing an object floating on the River Dee, he observes: 'Yet the reflection of the white body had not any light or white reflection but on the contrary had its reflection dark.' There are many times, he realizes sadly, 'When panting art toils after truth in vain'.[4]

In 1810 he bought a house in Twickenham, south-west of London, the

Thames a short stroll away. Impelled by his desire to depict such elemental forces as water and air, he devoted himself to analysing sunlight
and its portrayal on canvas, reading Goethe on the subject, who argued
that colour was a product of both light and darkness, and that yellow and
red carried active and affirmative symbolic associations. He himself
would write several essays on the nature of light, drawing on recent
research to show that 'light is colour'. Around 1817 he adopted a fresh
palette, helped by a range of new pigments, particularly shades of yellow.

He stayed put in Britain until 1819, then began a series of forays to the
Continent, particularly to Italy. The quality of light he encountered there
changed his art. He began to paint 'golden visions, glorious and beautiful', as Constable put it.[5] Around this time his friend, the amateur
painter James Skene, contributed an empathetic article to Brewster's
Edinburgh Encyclopedia, in which he wrote:

> Painting can but approximate to all the niceties, combinations and
> intricacies, of direct and reflected light, . . . modified by the almost
> imperceptible gradation of intensity as it recedes from the eye. . . .
> [Turner's] scrutinizing genius seems to tremble on the verge of some
> new discovery in colour.[6]

Nevertheless, not all was praise. In 1826 the *British Press* cavilled at
Forum Romanum and *Cologne; the Arrival of a Packet Boat*: 'All is yellow,
yellow, nothing but yellow, violently contrasted with blue.' Another
reviewer said he was 'desperately afflicted with what we may call a "yellow
fever"'. Turner cheerfully dismissed such criticisms, tripping off to a
friend's wedding in yellow stockings and quoting Malvolio (depicted in
Twelfth Night as 'cross-gartered' and 'yellow-stockinged') – perhaps that
unfortunate's speech, 'I'll be revenged upon the whole pack of you.' By
now yellow was famously his favourite colour, and he rejoiced (or so he
told friends) in the nickname 'the Yellow Dwarf'.

By autumn 1828 he was back in Rome, the inspiration for his great
painting *Regulus*. As in many of his later works, the Sun here consumes
nearly all other elements, including humans. The painting refers to the
grim story of the Carthaginians destroying the intransigent Roman
Regulus by keeping him for days in a darkened room, then cutting off

his eyelids and turning his face to the Sun. As one critic commented, the star has become 'a lump of white standing out like the boss of a shield', while another noted that the painting could not be looked at from up close; one had to step well back, and even then one was greeted by 'a burst of sunlight'. Viewers found themselves compelled to use words such as 'blazing', 'fiery' and 'resplendent'.

Canaletto's cityscapes had captured the special light of Venice; Turner, in the paintings of his final years, caught another quality of that luminous city – the way in which it casts its light over water. Some who saw these works were bothered and bewildered by these explosions of colour: 'Here is a picture that represents nothing in nature beyond eggs and spinach.' Mark Twain was to describe one canvas as 'like a ginger cat having a fit in a bowl of tomatoes'. But Turner's most discerning admirer, Ruskin, argued that his works were distinguished by the special intensity

of the light which he sheds through every hue, and which, far more than their brilliant colour, is the real source of their overpowering effect upon the eye . . . as if the Sun which they represent were a quiet,

The Fighting Téméraire – the painting Turner called 'My Darling'

and subdued, and gentle, and manageable luminary and never dazzled anybody, under any circumstances whatsoever.[7]

A particularly glorious representation of sunlight, even by Turner's standards, is his 1839 painting *The Fighting Téméraire, tugged to her last berth to be broken up, 1838* – one of the best-loved works of all time. Behind the old warship a low Sun fires the sky and, in reflection, the river. As Anthony Bailey has observed, to paint that sun Turner packed his palette with 'his most ferocious pigments: lemon yellow, chrome yellow, orange, scarlet, vermilion and red lead, hot paints that he laid over an already warm ground of earth colours'.[8] The *Téméraire*, which had served heroically at Trafalgar, offered not only an occasion for elegy but also a political point. As Turner's other main biographer James Hamilton describes the work:

> It brings and balances facts together, sail and steam, air and water, past and present, setting Sun and new Moon; it balances qualities: [the] old age and the new, dignity and presumption, silence and noise, steadiness and urgency, the temporal and the eternal; and it balances geometric forms: the horizontal, the vertical and the diagonal. Where these lines rush towards the setting Sun, the black tug and its ghostly white charge move inexorably out into our space.[9]*

Turner referred to the *Téméraire* as 'My Darling' and refused to sell it. As he grew older, he became ever more obsessed with the Sun, getting up early to watch it rise downriver from his rooftop balcony: the darker the

---

*A letter to *The Times* of 24 August 2005 pointed out that the *Téméraire* was being towed upstream, so Turner, in sketching the scene, must have been facing east as the ship was tugged west. Thus the painting may not portray the nostalgic sunset of the age of sail at all, but symbolize instead the dawn of the exciting age of steam. The first *Téméraire* was a French seventy-four-gun ship of the line captured in 1759. It was considered deadly bad luck to rename a taken vessel (hence the presence of a *Swiftsure* in the French fleet, from which it was recaptured in 1782 at the Battle of the Saints). Names were then passed on, and a second *Téméraire*, launched in 1798, was twice painted by Turner, the first time shown closing with the enemy at the height of the Battle of Trafalgar. It is interesting to compare this powerful narrative with Dickens's painterly vision of the morning Sun rising over the London docks in *Our Mutual Friend*: 'The white face of the winter day came sluggishly on, veiled in a frosty mist; and the shadowy ships in the river slowly changed to black substances; and the Sun, blood-red on the eastern marshes behind dark masts and yards, seemed filled with the ruins of a forest it had set on fire.'[10]

day, the more he craved its brightness. A friend who happened to be with him on one of those mornings saw him gazing unflinchingly at the fully risen star. When he expressed wonder, Turner replied, 'It hurts my eyes no more than it would hurt yours to look at a candle.' Now his yellows were becoming white, as in white-hot: he wanted to capture pure light and seemed to understand what astronauts and space cameras have only recently confirmed – that although even the midday Sun seems yellow to us, it is actually white. Once he tried, without success, to sketch an eclipse.

Twelve years on from painting the *Téméraire* on her way to the breakers, Turner too was reaching his final berth. In December 1851 he is recorded as saying, 'I should like to see the Sun again', and it was this request that Ruskin (not present at the time) turned into the declaratory 'the Sun is God'.*

☼

Turner was far from being the only important painter to take an interest in the Sun. Artists like Jean-Baptiste-Camille Corot (1796–1875) were particularly fond of sunrises and sunsets, moments rich with colour harmonies and symbolic potential. Sunrise, for example, tended to speak of hope, while the intensities of sunset could imply such emotions as anguish or desire. Sunset also suggested the fleetingness of this world, or perhaps the life eternal beyond death, and portraitists sometimes used it to suggest something elevated about their subject: thus Gainsborough's *Mrs Sheridan* shows her at sunset to emphasize her contemplative side, and Henry Raeburn's *Sir John and Lady Clerk* (1792) sets them against the evening light to impart an other-worldly glow.

This culminated in the romantic nature paintings of Europe and America in the nineteenth century. In contrast to Turner's vertiginous sunlight, the German Romantics preferred silent vistas in which luminosity pervades vast space. Influenced by the pantheistic doctrines then popular in Germany, Caspar David Friedrich (1774–1840) and his peers sought to express a transcendental vision through the effects of sunlight as well as moonlight. The landscapes of the American Luminists of the

---

*In 1993 I asked the art historian James Hamilton if he would write Turner's life. When four years later his book came out, one of the points he made was that, given the painter's religious preoccupations, he could have said, 'The *Son* is God': we have no way of knowing.

Impression, Sunrise – Claude Monet's 1872 painting hung in the first exhibition of the
Impressionists in 1874. From its title a French critic coined the disparaging term
'Impressionism' – which its followers immediately
took up with enthusiasm

mid-nineteenth century are also filled with atmospheric effects capturing the wonders of the natural world, though in a very different mood.
Although they did not articulate any particular philosophy, their paintings of vast mountains suffused with light expressed the optimism of
their new country, and the feeling that America was the 'New Eden', a
mighty nation specially blessed by God.[11]

Similar schools were to be found in Britain. Samuel Palmer (1805–81)
was the period's most successful realizer of solar light, an artist who,
while not a successor to Turner in style or early subjects, was still inspired
by him. Living in Shoreham, on the West Sussex coast, in his twenties he
was one of 'the Ancients' – a group that included Blake and Edward
Calvert (1799–1883) – whose night-time wanderings led locals to call
them 'extollagers', a rustic version of 'astrologers'. Palmer achieved his

most famous paintings before he was thirty, most of them showing their subjects under moonlight, but he was fascinated by sunlight too, especially that of the fading day: *Late Twilight* (1825), in which the Sun has already set, but its gloaming suffuses the sky, *Yellow Twilight* (c. 1830), *The Golden Valley* (c. 1833) and *Landscape Twilight* (c. 1824).

Departing from such works, the Impressionists of the 1860s–90s had something quite different in mind: the depiction of light as a natural phenomenon. They had learned from scientific studies to consider white sunlight as the composite of the colours of the spectrum and, realizing that sunlight and its refractions reached even into shadows, they banished black from their palettes. Nor did the Sun itself appear often on their canvases: rather, the Impressionists devoted themselves to capturing the way in which light changed at different times of day and different seasons. They made a point of going outdoors – '*en plein air*' – to paint, until then a surprisingly rare practice: artists of earlier times might sketch out their landscapes in the open, but they then carried the canvases back to the studio to be completed. When in 1876 Pierre-Auguste Renoir (1841–1919) exhibited his masterpiece *Le Moulin de la Galette, Montmartre*, critics recognized that 'the effect of strong sunlight falling through foliage on the figures' was the artist's primary concern.

Of all the Impressionists, Claude Monet (1840–1926) was the most obsessively concerned with the varying qualities of sunlight: it was his *Impression, Sunrise*, exhibited in 1874, that earned him and his circle their nickname. Little is to be seen of the Sun's disc in Monet's picture of the sea reaching up to Le Havre, but a good deal of its light spills across the canvas. Monet discovered that he required a whole sequence of paintings to record the nuances of light, working on each for just half an hour a day, even then finding it difficult to capture its fleeting changes, and writing in despair: 'The Sun goes down so quickly that I can't follow it.' He came to devote himself to two series of paintings, and in both the *Haystack* and *Rouen Cathedral* series (the latter twenty-seven paintings, the former thirty, all made between autumn 1890 and summer 1891) scrutinized the nearly imperceptible shifts in the quality of light, hour by hour, from dawn to dusk, throughout the seasons. When in 1879 his first wife Camille was dying, he spent hours sitting by her bedside, until one afternoon he realized with horror that he was preoccupied not with her health but with the changing light on her bedspread – and fled in anguish from her room. But his self-reproach was unjust: he was simply following his lifetime's passion. As his obituary in *The Times* declared: 'A pioneer in the discovery of "colour in shadow", he painted colour for the sake of light rather than light for the sake of colour.'[12]

Light was the main subject for Paul Cézanne (1839–1906) too – not the Sun itself, but its effects. 'I have been satisfied by my work since the day I realized that the Sun cannot be portrayed,' he wrote, 'but must be represented by another medium, by colour.'[13] When in 1876 he moved to the Riviera town of L'Estaque, he urged his friend Camille Pissarro (1830–1903) to join him.* 'The Sun is so startling, it makes it look as if objects could be lifted off in their outlines, which are cast not just in black and white, but in blue, red, brown, purple.' Pissarro didn't go, but Cézanne's own love of the region – its red roofs and blue sea, the greens of pine and olive and the glare of the Mediterranean sun – prolonged his experiments. Exploring sunlight in its many iterations, he depicted it as filtered through leaves, on water, in shadows cast by the noonday sun and as the background to labourers returning home at sunset, or fishermen chatting in the shade.

In an essay on painters and their relation to the Sun, the art historian Valerie Fletcher argues that as the Impressionists exhausted the possibilities of sunlight, a general reaction set in. The Neo-Impressionists (a title coined in 1886 by the critic Félix Fénéon) were led by Georges Seurat (1859–91), whose fascination with the principles of optics and longing to depict sunlight more accurately led him to experiment with a complex mix of colours. Seurat had at first worked only in drawings, without using colour at all, deliberately leaving patches of paper unmarked. The values would alternate between dark and light, resulting in a glow or halo around a figure, an effect he called 'irradiation'. But his fullest explorations of light came through his radical use of colour. Believing that his predecessors, while looking at a coastal sunrise or a mountain sunset, had

---

*Pissarro painted with a dry, hard brush, developing a granulated, thickly encrusted surface that captured the refractory character of dappled sunlight as it filtered through dense foliage – an approach that couldn't have been more different from that of the Luminists, who believed in concealing their brushwork. The ways in which sunlight was rendered really did vary hugely during these years.

lost sight of the fact that colour is merely the brain's interpretation of variable wavelengths, he determined to make painting more objective by taking it back indoors. Thus, in his great work *Sunday Afternoon on the Island of La Grande Jatte*, which was painted entirely in his studio, he simulated the vivid mid-afternoon sunlight by meticulously scattering tiny dots of a high-keyed solar orange among the other colours to induce a shimmering sensation, with more orange in the sunlit areas and fewer in the shade. (Constable had done something similar with small points of red, but as subliminal foci, not to emphasize sunlight.) His purpose was not immediately understood, and when his *Models* was exhibited in 1868 one critic wrote of Seurat's figures: 'daubed in all the colours of the rainbow . . . suffering, it would seem, from some horrible skin disease.'

Dedicating himself to the translation of light into scientifically determined combinations of pure colour, Seurat (still only thirty-one when he died) would wander the poorer districts on the outskirts of Paris at dawn and at dusk. He called the paintings that this produced 'Chromo-Luminism', to emphasize his commitment to catching the Sun's colours. By treating his subject analytically, however, he removed himself from the immediacy of experience. A close contemporary of his, Vincent Van Gogh (1853–90), would take a very different approach, his paintings of the Sun – his central subject – exploding with the intensity of his inner life. Although early on he was, like Seurat, enthusiastic about grey, 'or rather the absence of colour', in February 1888 Van Gogh left his native Holland aged thirty-four for southern France in search of a 'stronger Sun'. Thereafter it is a constant, in both his pictures and his letters. 'The fact is,' he wrote, 'the Sun has never penetrated us people of the North . . . Nature and fine weather are the advantages of the South . . . The difference in the stronger light and in the blue sky teaches you to see, and especially, or even only, when you see it for a long time.'

While living in his little sunlit house at Arles (whose outside walls he painted sunflower yellow), he painted a series of golden sunflowers, as if through them he could gather the Sun's life-giving powers into his home. The flowers were a vehicle for the Sun's energy – as were cornfields – as comes through in his description of these paintings: 'Under the blue sky the orange, yellow, red splashes of the flowers take on an amazing brilliance, and in the limpid air there is something or other happier, more lovely than in the North. It vibrates . . . '[14] He adds: 'The peony is Jeannin's, the hollyhock belongs to Quost [two painter contemporaries], but the sunflower is mine.'

One of Van Gogh's tenets was that an artist should 'exaggerate the essential'. Accordingly, his drawings suggest the Sun's heat and light by concentric circles of dashed lines. Similar visual shorthand brings to life his painted Suns, in such works as *Wheatfield with Reaper* (1889), in which a disc of intense golden yellow dominates the horizon. Ascribing a strongly optimistic significance to the golden colour of the Sun, he experimented endlessly with colour combinations. Writing to Émile Bernard, he details his paint choices: 'Field of ripe wheat in a yellow ochre tone with a little crimson. The chrome yellow 1 sky almost as bright as the Sun itself, which is chrome yellow 1 & 2 mixed, very yellow, then.' In his quest for 'the most vital yellow' of sunlight, he began to take stimulating drugs – mainly digitalis, derived from foxglove leaves and at that time a treatment for his epilepsy (he also suffered from, besides much else, acute intermittent porphyria); camphor; and terpenes, a dangerous chemical used in paints. These he would augment with absinthe, the liquor that contains thujone, known even then to poison the nervous system. Some doctors believe these various concoctions brought on xanthopsia (yellow vision), whose sufferers see everything as if through a yellow filter. But it was passion, not sickness, that produced his great late paintings.

Like most of the Romantics, he had moved away from traditional Christianity – he had once been a Protestant preacher – and, though he still aspired to believe in an all-powerful deity, his canvases were not religious statements. He painted the Sun stripped of allegory or personification, to capture, he said, the *sensation* of its intensity. As Robert Mighall observes: 'Even Turner, self-pronounced sun worshipper, swathed his god in mists or trailing clouds of sublimity; approached it with a due reverence of poetic allusion, yet never quite rent the veil of its temple aside.'[15] But if the Sun was for Van Gogh the source of life, he was no mere idolator: he caught the threat of this great burning mass, the pitilessness of the universe, the evanescence of our existence.

In a letter to his brother Theo, Van Gogh explains what he is trying to accomplish (his easel anchored with iron pegs against the Mistral winds):

I am struggling with a canvas begun some days before my indisposition, a 'Reaper'; the study is all yellow, terribly thickly painted, but the subject was fine and simple. For I see in this reaper – a vague figure fighting like a devil in the midst of the heat to get to the end of his task – I see in him the image of death, in the sense that humanity might be the wheat he is reaping. So it is – if you like – the opposite of that sower I tried to do before. But there's nothing sad in this death, it goes its way in broad daylight with a Sun flooding everything with a light of pure gold.[16]

Then: 'There! The "Reaper" is finished, I think it will be one of those you keep at home – it is an image of death as the great book of nature speaks of it – but what I have sought is the "almost smiling" . . . I find it queer that I saw it like this from between the iron bars of a cell.'*

Because we know of his bouts of derangement and his confinements to a sanatorium (where the scientific treatment for his illness was to leave him soaking in a tub for hours at a time), it is easy to forget that in his letters (he was a compulsive letter writer, his *Collected Letters* taking up more than 1,500 pages) Van Gogh comes across as the best and most lucid of correspondents. Humorous, endlessly curious, sympathetic, wise about literature as well as painting ('There is no writer, in my opinion, who is *so* much a painter and a black-and-white artist as Dickens: his figures are resurrections'), indeed about all the arts, he was wonderfully sane on the page. His letters are almost unbearably moving, the more so because of the serenity of the writing. His last years saw him paint some of his greatest works, in such an outburst of creativity that he seems to know his time is short. Then one high summer day, 29 July 1890, the Sun's blaze on a cornfield near Auvers finally consumed his will to live. After he shot himself, a note was found stuffed inside his

---

*There is a surprisingly appropriate passage in Aristotle's *Poetics* linking sowing with the Sun: 'To cast forth seed corn is called "sowing"; but to cast forth its flame, as said of the Sun, has no special name. This nameless act, however, stands in just the same relation to its object, sunlight, as sowing to the seed corn. Hence the expression in the poet "sowing around a god-created flame".' Van Gogh was an omnivorous reader of the classics but seems not to have come across this observation.

shirt: 'The truth is, we can only make our pictures speak . . . Well, my own work, I am risking my life for it, and my reason has half-foundered because of it – that's all right.'

☼

Many works by twentieth-century artists centre on the Sun – Georges Rouault's *Crucifixion*, in which the Sun stands for the blood of Calvary, Paul Klee's *Leaf from the Book of Towns* and *Ad Marginem*, Max Ernst's *Dada Suns*, Paul Nash's enraged painting of the Western Front in 1918, *We Are Making a New World*,[17] several paintings by both Wassily Kandinsky and the Australian impressionist Lloyd Rees (1895–1988), Graham Sutherland's *Sunrise between the Hedges* and Joán Miró's *People and Dog in the Sunlight*, to cite just a few. But one artist actually advanced our understanding of what painting could achieve with the help of the Sun: Henri Matisse (1869–1954). His childhood was spent under the cool light of Normandy, and only when he moved to Nice in 1917 did colour truly erupt into his work; yet those first years shaped him, and the whole story of his artistic development resides in that overcast beginning, in how the luminous subtlety of his originally limited, brown-and-grey northern palette was transformed by his years in the sun into a radical new way of depicting light.

Matisse and Van Gogh were not the first painters from the north to experience an epiphany once translated to the vivid south: Delacroix had recharged his vision eighty years before beneath the intense heavens of Morocco. But Matisse was arguably the most transformed by the experience. Writing about his move to the south of France, he mused: 'What made me stay was the great coloured reflections of January, the luminosity of the days.' This is more than a matter of intensified sunlight: there is a real cultural split based on solar proximities. The history of agriculture, for instance, distinguishes between wine and beer. The former is further south, and integrates place and work; in the latter, a given beer does not stem from a unique source of flavour and tone but is a chemically uniform product. These facts mark out cultures: beyond a certain distance from the equator, one finds almost anything but the grape used for alcohol (Japan, for example, is too far north to produce a decent wine – sake is a beer), and a whole ecology of malt/junipers/potatoes (as used in vodka) radically contrasting with that of the lands of wine.[18] When Matisse moved south he was uniting his northern culture to one almost completely foreign to him, and his response brought colour to painting in a way never achieved before.

Even into middle age, he was constantly discovering new modes. A letter from his daughter Marguerite, written in 1925 (when he was already fifty-five), describes her response to his latest paintings: 'The subtlety in tone of the new canvases in harmonies of mauve and pink is extraordinary; it's the light sliding over and caressing the objects . . . that constantly astonishes you.'

At the beginning of 1930 he travelled to Tahiti and other Pacific islands, where he would 'find dawn and darkness unlike anything he had seen before'. Visiting the atoll of Fakarava, he went snorkelling and discovered 'that underwater light which is like a second sky'. In this new world seen through goggles, he experimented with focus, depth and angles, staring down into the green floor of the lagoon, looking up at a watery ceiling opaque and wavy as medieval glass, and plunging repeatedly between the two, schooling his retina to compare the different luminosities.

Back in France, his painting became increasingly abstract, his canvases multiplying and dividing light, making it glimmer and glow in compositions that reduced their minimal elements to strips, swathes and swatches, broad bands and tilted panes of pure colour. Pierre Bonnard, to whom he lent some of his paintings, staring in Matisse's studio at the expanses of flat uninflected colour, complained wonderingly: 'How can you just put them down like that, and make them stick?' But so he did.

Sometimes even Matisse himself needed time to absorb the implications of what he was doing. *Interior with Bars of Sunlight* (1942) expunges his model almost completely, leaving her as a young-woman-shaped space of bare canvas in an otherwise geometrical arrangement of coloured bars and rectangles. He kept this apparently unfinished work by him, staring at it from time to time, 'as if pondering a problem', finally signing off on it in 1945. 'The painting is a time capsule,' glosses Hilary Spurling, his greatest biographer, 'whose contents would make sense only long afterwards in the light of optical

Olafur Eliasson's faux-Sun installation *The Weather Project* at the Tate Modern. Visitors would lie beneath it as if sunbathing

experiments conducted by a generation of abstract artists who had not yet emerged.'[19] Matisse wrote a note to any future owners of the painting, urging them not to colour in the unpainted figure in the armchair at the bottom right of the canvas: 'The figure, just as it is, has its own colour, the colour I wanted, projected by the optical effect caused by the combination of all the other colours.' Like Turner before him, who had etched out rainbows in *Kilchurn Castle, Scotland, with a Rainbow* (1802) and *Crichton Castle, with a Rainbow* (1818) by

rubbing away pigment and letting bare canvas conjure up their searing luminosity, Matisse knew that white was often the Sun's most
vibrant shade.*

His oculist in Nice (who had also treated Monet) explained that the eye
could not synthesize pigment fast enough to keep up with the speed and
intensity of Matisse's response to colour. The artist's last pictures, thirty
views of his studio at Vence, just outside Nice, project light and colour with
a vitality so intense as to seem almost independent of its physical origins.

In 1991, when I was in book publishing, I asked Hilary Spurling to
write a biography of Matisse. The resulting two transformative volumes
would consume her efforts for more than fifteen years. In May 2003
Hilary invited me to the New York opening of the Matisse–Picasso exhibition. She was the ideal guide, pointing out here a Picasso woodland
scene, there the same theme by Matisse, the great blocks of black that
massed to form the trees somehow drawing one's eye even more surely
to the shafts of light that illuminated the forest floor. It was the same with
Matisse's interiors: blocks of furniture – table, bed, chair, a slash of black
representing a curtain rod – expressing how sunlight lit up the room.
Then she told me a story.

At the end of December 1917 Matisse, just a week shy of 48, had
moved to Nice, carrying little more than a suitcase, his paints and his
violin, leaving his wife and three children behind in Paris.[21] He knew
that Pierre Auguste Renoir lived a few miles up the coast at Cagnes, and
in his first week in town resolved to pay the master a visit, but as he
approached Renoir's house, a roll of canvases under his arm, he found
himself walking up and down the street, trying to summon up his
courage. In the end he tossed a coin, which came down in favour of
knocking on the great man's door.

*The Danish-Icelandic artist Olafur Eliasson (b. 1967) filled the Turbine Hall of Tate
Modern for six months in 2003–4 with his installation *The Weather Project*. An 18,000-watt
bank of streetlight bulbs behind a translucent semicircular screen produced an unearthly
light, while generators wafted plumes of smoke. A huge ceiling mirror doubled the
exhibit's apparent size, casting its reflection on the viewers below, but also suggesting a
huge circular sun. Eliasson has an ingenious take on the white walls of art galleries: 'Chalk
is white and chalk was used as a disinfectant and so early modernists decided on white walls
as symbols of purification. But if chalk had been yellow, maybe all our galleries would be
yellow today, and we would interpret yellow as a neutral colour.'[20]

Once inside, he found Renoir hardly pleased to be interrupted. His host was in his late seventies, widowed and so wasted that he had to be carried from his bed to his studio in a special chair. He lived to paint, but here he was – frail, gaunt and so crippled by arthritis that he could not hold a paintbrush, and every day a padded handle had to be wedged between his right thumb and forefinger. The dying artist was not pleased to be forced into conversation with a vigorous rival at the height of his powers. Nevertheless, the two talked, and after a while Matisse spread out his paintings for inspection. The old man stomped from canvas to canvas, muttering to himself, then looked up and said, 'I knew from the way you were carrying on that you were no good, had no talent.' Matisse stood aghast. Then Renoir motioned to a couple of canvases lying side by side, the very same forest scene and apartment interior that would be displayed so many years later at the New York exhibition. 'Until I saw these,' he added. 'You use black to capture the quality of light. I could never manage that. You're the real thing.'

☼

Exactly twenty years after Matisse's cardinal move south in 1917, a major British artist was born in the northern industrial city of Bradford and made a similar journey. Had he painted a century earlier, David Hockney might also have moved to southern France, but instead he migrated to Los Angeles in 1964, transforming his art and his reputation. 'I was brought up in Gothic gloom,' he explained in a 1993 interview conducted at his house in Malibu, under drenching California light. 'All the buildings are absolutely black, and it rains practically every other day. I'm a bit like Van Gogh . . . He thought there was more joy in the Sun, and I tend to think that as well.' *Sun*, part of Hockney's 1973 'Weather Series' and one of his most celebrated paintings, shows a potted plant on a window sill, bathed in colour-awakening light. His hair dyed its signature blond, and dressed in gold-cuffed corduroys, a yellow cardigan and shirt and an olive-coloured knitted tie with pink polka dots, he explained all this while walking through his bedroom, which he had painted green and pink after a Matisse drawing. 'Van Gogh,' he added, 'is one of the few artists who used yellow very well.'[22]

Before leaving for the West Coast, Hockney had been at the forefront of British pop art (the term having been first used by Lawrence Alloway in a 1958 issue of *Architectural Digest*). The movement sprang from a rebellion against Abstract Expressionism, which pop artists thought pretentious and over-intense. In California, Hockney became known for his depictions of waterhole-like swimming pools, tropically lush Los Angeles canyons and beautiful young men. He also designed the sets for several operas – from Glyndebourne to the Metropolitan in New York – poignant, as for the last thirty years he has struggled with hereditary deafness. But, like Scriabin and Sir Arthur Bliss, he is a synaesthete, in his case perceiving music visually then translating it into cascades of colour. His deafness, he says, has afforded him compensations, changing the way in which he perceives space and giving him a sharper sense of light and shade.

In the late 1990s, in order to look after his ailing mother, Hockney bought a house in the old seaside resort of Bridlington, close to where he spent his childhood, 'East Yorkshire's answer to Malibu', as a journalist who visited him for the *Observer* described the town.[23] Just before that interview, Hockney had been lying in bed trying to capture the Sun as it rose over the bay outside his window by using an Iphone application called 'Brushes' – and his thumb. When he finishes such drawings, he told the *Observer*, he emails them to twenty or so friends. On good days each recipient might receive half a dozen original works well before breakfast. Visitors to the house routinely find themselves woken before dawn to join him in looking at the way first light falls on a particular copse – the obsession of his recent work.

Hockney has been invited to fill an entire gallery at the Royal Academy during the Olympic year, 2012, and is planning a 'very big' sunrise for the show. 'I'm well aware that most pictures of sunrises are clichés, but I'm also aware that a sunrise is never a cliché in nature. So that's the challenge.' Now seventy-three, he is still energetically seeking out inspiration, and some months ago travelled to the far tip of Norway, near Tromsø, where it 'never got dark . . . There's a place where you can watch the Sun at midnight that is like the edge of the world.' He also went to see Edvard Munch's *Sunrise* at Oslo University, and was fascinated by the techniques Munch employed. 'He's got lines in it that cameras could never see, but we could – and, of course, in Oslo in June Munch could look at the Sun for a lot longer than Van Gogh could at Arles.'

In 1949, *Life* photographer Gjon Mili visited Pablo Picasso and showed him some photographs of ice skaters with tiny lights affixed to their skates jumping in the dark – and Picasso was soon using a small flashlight in a dark room

Like many other artists before him, Hockney has long been fascinated by the processes of vision, which has led to an interest in photography – so much so that for a while he stopped painting altogether. There is something pleasingly apt in seeing an artist whose works are so light-suffused turn to a medium that actually means 'writing with light'. Hockney's idol (and Matisse's long-time friend and great rival) Picasso did just that one day in 1949, drawing figures in the air with a torch to create remarkable but ephemeral works of art that could be captured only by slow-exposure film.[24] It is predominantly during Hockney's lifetime, of course, that the two art forms of photography and film have established themselves, each offering artists new ways to express the Sun's power. Those seeking to investigate the optical effects of the Sun now have another set of tools at their disposal.

NEGATIVE CAPABILITIES

> That star . . . that star. It's so far away, by the time the light
> from it reaches here it might not be up there any more.
>
> Eli Wallach, in John Huston's *The Misfits*, 1961

> Architecture is the skilful, exact and magnificent play of vol-
> umes assembled in light. Our eyes are designed to see forms
> in light.
>
> Le Corbusier, 1923[1]

Long before the first photograph, the great Persian scientist Ibn al-
Haytham (AD 965–1040) invented the camera obscura, a 'darkened
chamber' into which light enters through a tiny aperture. If a reflecting
surface is so placed that it intercepts the entering rays, the viewer will
see a full-colour, moving image – upside down – of the scene outside.
Kepler improved this basic version by adding a lens to enhance the
image's quality, as described by an English diplomat to Francis Bacon
in 1620:

> He hath a little black tent set up, exactly close and dark, save at one
> hole, about an inch and a half in the diameter, to which he applies a
> long perspective trunk, with a convex lens fitted to the said hole,
> through which the visible radiations of all the objects without are intro-
> mitted, falling upon a paper.[2]

The next hurdle on the way to moving pictures would have been to capture on a screen precise images of distant figures, in full colour and motion, but that was too great a leap, many centuries away. The invention of the magic lantern was another stepping stone, and although he didn't invent it, a Jesuit alchemist, Father Athanasius Kircher (1601–80), did publish a study of some of the principles involved in its construction. In 1646 his book on optics, *Ars magna lucis et umbrae* ('The Great Art of Light and Shade'), appeared. Among the novelties he describes are a projecting device with a lens that focused sunlight cast by a mirror, and the use of light from a candle or lamp to project images – the very technique that was later refined into the magic lantern, forerunner of film projectors. By century's end portable instruments of this kind were on sale in the major cities of Europe.

Given that the chemical, as well as the optical, principles of photography were known by the early eighteenth century, it is one of the enigmas of cultural history that photography took so long to be invented. Not until 1826 did Joseph Nicéphore Niépce (1765–1833) produce the first permanent photograph (*View from a Window at Le Gras*) with a process he called 'heliography' – 'Sun writing'. The exposure time was at least eight hours, so that sunlight illuminates the photo from both the right and the left sides.

For a few years thereafter only objects that remained immobile for several hours could be photographed, then in 1829 Niépce teamed up with Louis Daguerre (1789–1851); and although Niépce died four years into their collaboration Daguerre went on to introduce a process – involving a thin iodine layer applied to a silver substrate in order to create a light-sensitive surface – that required a much shorter exposure time. By 1839 these 'daguerreotypes' took only about ten minutes, and by 1842 improvements in the sensitizing step, and more effective lenses, had reduced it to fifteen seconds. Daguerre also created what he called a 'diorama' – a room surrounded by newly devised equipment that produced complex light illusions. One of these machines was the phantasmagoria (literally 'marketplace of ghosts') – one of whose first achievements was to 'reproduce' the effects of a sunrise.

The Englishman William Fox Talbot (1800–77) was Daguerre's main rival. To create *Botanical Specimen, c. 1835*, the result solely of the interaction of sunlight and chemicals, Talbot placed a leaf on a sheet of photosensitized paper and left it out in the Sun: the image of the leaf appeared as a faint, almost evanescent, impression against the Sun-darkened background. Produced without the intervention of a camera, these 'photograms', as he called them, seem more the work of the Sun itself than of the photographer.

Working along the same lines as Daguerre, Talbot had earlier discovered another means of fixing a silver-process image, but kept it secret. After reading about his competitor's invention, he refined his technique to make it fast enough to register people, and developed what he called the 'calotype process' (from the Greek *kalos*, beautiful), which creates negative images. The earliest surviving paper negative, dated August 1835, is of the oriel window in the South Gallery at Talbot's country house, Lacock Abbey, in Wiltshire. Talbot describes how he created it:

> Not having with me . . . a camera obscura of any considerable size, I constructed one out of a large box, the image being thrown upon one end of it by a good object-glass fixed at the opposite end. The apparatus, being armed with a sensitive paper, was taken out in a summer afternoon, and placed about one hundred yards from a building favourably illuminated by the Sun. An hour or so afterwards I opened the box and I found depicted upon the paper a very distinct representation of the building, with the exception of those parts of it which lay in the shade . . .[3]

The pace of invention now accelerated. The great astronomer John Herschel was another important contributor, making extensive experiments on the light sensitivity of various metal salts and vegetable dyes, developing techniques for colour photography and advancing terms now standard, such as 'negative', 'positive', 'snapshot' and 'photographer'. The first such images, still suggesting the notion that the Sun was their author, were called 'heliographs'.*

---

*In the 1850s William Barnes, a Dorset peasant who got a 'ten-year' Cambridge degree, tried to purge English of Latinisms, and advocated replacing the word 'photograph' with 'Sun print'. He failed, although for many years the *Chambers English Dictionary* included the definition: 'Sun-picture – a picture or print made by the agency of the Sun's rays; a photograph.' 'Heliograph' held its meaning until the early 1900s; thereafter it signified a military signalling device.

A rivalry soon sprang up between photography and painting. Talbot, for instance, was fascinated by the camera's ability to catch the qualities of light reflected on glass, a phenomenon that he believed challenged the best painters. Another early photographer, Roger Fenton, pointed his lens straight at the Sun – seemingly in contravention of all known photographic principles – to produce an early Turner effect. This desire to challenge painting, a new technology confronting the techniques of an earlier art, continued throughout the early days of photography, as its practitioners struggled to establish their medium's identity. Fenton's *Salisbury Cathedral, the Nave, from the South Transept, 1858* illustrates how the play of light and dark in an interior space could be captured by the photographer: by leaving the shutter open so long that the edges of the shadows blurred under the Sun's motion, he both made the cathedral more ethereal and registered the nuances of semi-darkness peculiar to Gothic space.[4] No wonder Roland Barthes suggested that cameras were originally 'clocks for seeing'.[5]

Exposure time was beginning to be as much opportunity as nuisance. When in 1867 the great portraitist Julia Margaret Cameron took her celebrated photo of John Herschel, such was the poor quality of the indoor light that he had to sit still for nearly nine minutes; yet this gave a greater depth to his craggy features. However, the next chapter in the story of photography and the Sun was to be set not in England but in Paris. Over the years, several cities have styled themselves 'the City of Light', from Varanasi in North-Central India to Lyons in southern France. By the 1890s Paris had asserted the prime claim. The term *lumière*, which embraces 'light', is much stronger than in its English equivalent, additionally implying happiness, enthusiasm and the world of art. Yet there is an irony in the title, for Paris has for most of its history been a city of darkness. In the thirteenth century it boasted a total of three public lamps; in the fifteenth and sixteenth the law required each household to place a candle in a street-facing window, but it was never enforced. Paris, the City of Light, was in reality one of shadow.*

---

*James Joyce (1842–1941) spent his final days in Paris, steadily losing his sight, sitting by his window wearing a dentist's jacket to reflect the Sun on to the open page. He would lay his head on his desk the better to see what he was writing – his last manuscript slants heavily to the right, poignant evidence of his attempts to catch every particle of light that he could.

However, in the run-up to the 1889 World Exhibition, whose literal highlight was the Eiffel Tower, Paris was transformed by 20,000 gas lamps, making night-time photography possible, while on the outskirts of Lille, in northern France, a new factory was mass-processing photographs. It was in the newly illuminated capital that the great Eugène Atget (1857–1927) now became one of the first to exploit photography's potential as an art form, doing much to capture the great buildings and gardens of France. Early on in his career, he would photograph at midday, so that his 'documents', as he called them, would catch the finer details. In his later studies, by contrast, he worked in the early morning, when a soft luminosity suffused each subject with a sense of peace. In Roman literature 'lux' meant not only 'light' but 'life'. 'Light' is a term not of physics but of physiology. Atget's work gave these words new meaning.

Of living photographers who have employed the Sun to greatest effect, one might cite the Swiss-American Robert Frank (b. 1924), whose classic book *The Americans* captures a cross-section of his countrymen in the mid-1950s. In his introduction, the Beat writer Jack Kerouac acclaims Frank's capacity to convey not only 'the strange secrecy of a shadow', but also 'that crazy feeling in America when the Sun is hot on the streets . . . and music comes out of the jukebox or from a nearby funeral'. Commenting on one particular shot (*Restaurant – US1 Leaving Columbia, S. Carolina, 1955*), Kerouac describes how it shows 'the Sun coming in the window and setting on the chair in a holy glow I never thought could be caught on film'.[6] Frank is the nearest that American photography has come to Atget.

☼

Heliography, heliograms, Sun pictures, photograms – all such terminology underscores the degree to which the Sun was a co-creator of that art form in its early days. Amplifying its role, in 1932 Edwin Land (1909–91) perfected the first usable material to polarize light artificially, which in time yielded not only inexpensive filters for that purpose (in 1953 Land's company was producing 6 million pairs of card-board-frame 3-D movie glasses *each week*) but also the Polaroid camera. Yet that is the last of the significant innovations in the technology of photography that concern the Sun,[7] because with the coming of digital photography in the 1970s a drastic change occurred in the nature of the medium. Digital

The great Hungarian artist Brassaï (1899–1984) photographed this man on the Riviera in 1936. All the light in the picture – and all the shade – does not seem to fall on the scene but to radiate from the umbrella[8]

cameras record and capture images as binary data, which facilitates the storage and editing of images on personal computers, and the deleting of unsuccessful takes. According to one historian, an 'existential doubt' has 'crept in to our understanding of these images', as the digital world 'would no longer permit a distinction . . . between the real impression . . . and a "representation" generated in the darkness of a computer'. David Hockney has even described Gustave Le Gray's combining of two negatives, in the 1850s, to make a single print as a 'willed deception' that amounted to a form of Stalinist photography.[9] Another commentator has pointed out that 'the largely unquestioned assumption that the intrinsic and essential nature of the photographic apparatus is to record factual reality has been completely overturned'. In traditional photography,

it is possible to trace a physical path from the object represented, to the light that reflects off it, to the photographic emulsion . . . that the light hits, to the resulting image. In digital imaging this path is not traceable, for an additional step is added: converting the image into data, and

The Great Wave, Sète (1857), the groundbreaking photograph by Gustave Le Gray (1820–82) that knits together *two* exposures, taken at different times of the day. It is just possible to see the join where sky meets sea

thereby breaking the link between image and physical referent. Any iteration of the image may be altered, and there is no 'generational' difference to alert us to the stage at which the change occurred.[10]

Such innovators as Atget and Fenton might have replied that photography has ever been something of a trick; but others might feel that, while the Sun and photography have yet to be formally divorced, they are surely going through a trial separation.

☼

Photography has had many great moments in the centuries since Daguerre, but the dominant medium of our time has proved to be the moving picture – the one truly new art form to emerge during the twentieth century. While photography was coming into its own, the technologies that would make possible the art of the moving image were also being developed. In 1824 the English polymath Peter Mark Roget had published a paper arguing that the human eye retains an image for a fraction of a second longer than that image is actually present: this inspired inventors to demonstrate such a principle in practice, and it was found that if sixteen pictures were made of a movement that occurred over the span of one second and were shown in comparable time, the eye would register such images as one continuous movement.[11]

In 1832 a Belgian, Joseph Plateau, invented the 'Fantascope', a device that simulated motion. A series of pictures depicting a sequence of movements (juggling, dancing) was arranged around the edges of a slotted disc. When this was put before a mirror and rotated, a spectator looking through the slots 'perceived' a narrative sequence. Exactly forty years later, the British photographer Eadweard Muybridge (1830–1904) employed twelve cameras to take twenty-four consecutive pictures of a trotting horse, Occident, to see whether at any time it had all its feet off the ground simultaneously. By capturing Occident's four hooves in mid-air, thus verifying the 'unsupported transit' theory, the experiment also proved a major breakthrough for photography.* In 1879, building on this and other earlier experiments, Muybridge invented the Zoopraxiscope, which projected a sequential series of photographs, taken by multiple cameras in order to capture very small increments of motion, that could be linked together to create moving pictures.

In 1882 the Parisian physiologist Étienne-Jules Marey constructed a 'photographic gun' to take multiple images per second (the term 'shooting a film' possibly derives from this), and by 1888 he was producing the first successful moving-picture cameras, capable of recording a flow of very brief shots of a subject's movement on the same plate, rather than the individual images Muybridge had produced. That same year, Thomas Edison (1847–1931) completed the first-ever film, just ahead of the French brothers Louis and Auguste Lumière, who in 1895 combined printer, camera and projector to cast moving pictures on to a screen for an audience.[12]

---

*Muybridge, who had become known for his landscape photographs of Yosemite, had published them under the pseudonym 'Helios' – perhaps an ironic way of acknowledging his co-creator.

The clouds part after the tornado in *The Wizard of Oz*, and the next moment the
screen bursts into vivid Technicolor. Not only was this a seminal moment in the
story, but sunshine after the storm represented people's hopes after the Depression.
Ironically, the film was premiered on 12 August 1939, three weeks before the
outbreak of the Second World War.

In its early years film-making depended almost entirely on the Sun –
on natural light falling through open-roofed studios. Edison formed
the first movie unit in the US in 1893 by building the Black Maria, a
tarpaper-covered shack at his New Jersey laboratories; its stages were
mounted on a platform that revolved to follow the Sun's progress
across the sky. In bad weather he resorted to greenhouse-like glass shel-
ters, which retained whatever light was to be had.

In 1897 Biograph, the first US company devoted entirely to the new
medium (its star director, D. W. Griffith, made 400 films for it), opened
a studio on the roof of a Manhattan office block, moving a few years later
to a converted brownstone at 11 East 14th Street, Biograph's first indoor
studio and the first in the world to rely exclusively on artificial light. Early
in 1910 Biograph sent Griffith with a troupe including Lillian Gish, Mary
Pickford and Lionel Barrymore to film *Ramona* in Los Angeles, then a
town fewer than 100,000 people – the move prompted in part by the

search for better light, but more by the need to escape the patent disputes that often disrupted production back east.* Filming completed,
Griffith and cast gravitated north to a small community famous, if at all,
for its flowers – Hollywood – and were much charmed by it.

By 1911 another fifteen film companies had arrived in Hollywood,
travelling 'westward, to the land of haunted romance where the Sun
shines so fraudulently bright', as the film historian David Thomson puts
it – and the area proved so invigorating that by the end of the First World
War it had become the industry's capital.[13]

The Sun dictated much of the early grammar of film because in those
pioneer days it was the only adequate source of light. Later, special 'sun
arcs' were developed to simulate sunshine on demand. One such, featured
in *Popular Science* in January 1923, measured a foot in diameter, making it
the largest light bulb in the world. These 30,000-watt monsters consumed
a third as much power as was required to run a trolley car, and each contained enough tungsten filament to operate 55,000 household lamps.

At first the goal was simply to have enough light to film by. However,
as film-makers became more concerned with the aesthetics of what they
were shooting, it wasn't just the quantity but the quality of the light that
mattered. Cinematographers fondly referred to 'the magic hour' – the
brief period between twilight and full darkness when the light has a
deep, warm tone, sometimes known as 'sweet light' because of its wondrous effects. As the cultural historian Michael Sims puts it: 'The most
overlooked ordinary scenes – a rusty basketball backboard, a sand-buried
hurricane fence, a bedraggled starling on a dirty street – suddenly gain
new significance, merely because they seem abruptly to be touched with
grandeur from the Sun's first or last light.'[14]

As had happened in painting, the Sun itself started to play a role in the
movies. Letting the camera linger on a rising or a setting sun became a
beloved motif for generations of directors. As Sims observes, 'The symbolic new beginning of dawn is as irresistible to film-makers as it is to

---

*At the time, Edison owned almost all the patents relevant to motion-picture production,
and often sued film-makers acting independently of his Motion Picture Patents Company
in the east. In California, film-makers could work beyond reach of his harassment; when
he did send agents there, word usually arrived ahead of them, and the competition could
escape to Mexico.

writers. It provides powerful visual metaphors (and blissful cinematography) in such films as F. W. Murnau's *Sunrise* . . . and even in the light of a new black-and-white day washing over Baghdad-on-the-Hudson in Woody Allen's *Manhattan*.'[15] Allen has shown himself particularly sensitive to the light that bathes his hometown: New York, as the critic Holland Cotter has noted, is 'an island city . . . with an island light, alternately obdurate and romantically moody. It can be too candid . . . but its toughness is democratic: it falls on everybody and everything.'[16]

Not only dawn and sunset draw the camera's lens. Many notable films have drawn directly on the Sun: the cinematographer Kazuo Miyagawa (1908–99) was the first to turn the camera directly into its glare, albeit through forest foliage, in Akira Kurosawa's *Rashômon* (1950); Conrad Hall was the first to do so in colour, in *Cool Hand Luke* (1967). Other films that feature the Sun for purposes of mood and meaning include Erich von Stroheim's *Greed* (1924), with its final scenes in California's Death Valley superbly conveying the full force of desert heat and the wretchedness and despair of death from thirst; Nicolas Roeg's visually dazzling *Walkabout* (1971), shot in the Australian outback; Michelangelo Antonioni's *The Passenger* (1975), which lands Jack Nicholson in the African desert; even Dennis Hopper's 1969 elegiac road movie *Easy Rider*, to choose but four. My most cherished example of a cinematic depiction of the Sun is the opening scene of 'Festival of Beauty', part two of Leni Riefenstahl's *Olympia*, her chronicle of the 1936 Games in Berlin. The shots of athletes preparing for the day against the backdrop of a moody, then resplendent, Sun movingly evoke the harmony of sportsmen and sportswomen with nature. The film is in black and white.

☼

All the arts connect with nature, but this is particularly true of architecture, for better and sometimes for worse. I have mentioned the ways in which certain buildings have been constructed to record the Sun's movements, reflect its rays, or catch them as an aid to good health. But architecture can also prove the enemy of the Sun. As early as 1902, for instance, Manhattan boasted 181 edifices ten to nineteen storeys high and three over twenty. John Tauranac describes the result in his history of the Empire State Building:

Streets that had once been bathed in midday sun were becoming dark
and narrow canyons, and the Financial District came to be described
as the 'canyons of Wall Street' . . . The canyons appeared, canyons
where sunlight seldom reached the sidewalks below, canyons that rose
straight up for two hundred, three hundred feet or more, canyons created not by erosion but by economics.[17]

The Roman town of Bram, in France, founded in AD 333. Its plan recalls the ancient
Zoroastrian townships in Central Asia, which were conceived either on a circular
design or in the shape of a crossed circle

If the problem was at its worst in America, it was repeated elsewhere,
and architects worldwide duly took notice. Le Corbusier (1887–1965),
one of the most influential architects of the last century, wrote: 'The
materials for urban design are: Sun/sky/trees/steel/cement – in this
order of importance . . . Space and light and order. Those are the things
that men need just as much as they need bread or a place to sleep.'[18]

In 1922 he proposed a 'Contemporary Town' for 3 million inhabitants,
the centrepiece of which was to be a group of sixty-storey skyscrapers. In
the years that followed, he reformulated his ideas of interior design, par-

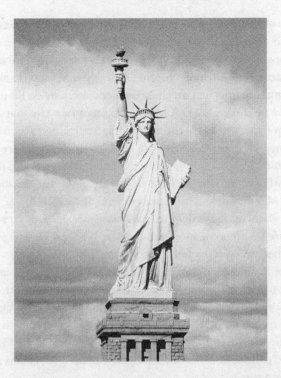

The 151-foot Statue of Liberty, bearing the crown of Mithras. Since the 1940s, it has
been claimed that the seven spikes on Liberty's diadem epitomize not Mithras but
the Seven Seas and seven continents; but equally it could represent the crown
worn by Roman emperors as part of the cult of Sol Invictus, or the
headpiece of the solar deity Apollo. Nobody knows for sure:
so why not accept it as a crown inspired by the Sun?

ticularly in the Villa Savoye, his 'Temple to the Sun' (in Poissy, on the outskirts of Paris), completed in 1929; and of urbanism, finally giving them
concrete shape in the Unité d'Habitation at Marseilles, built between 1946
and 1952: an eighteen-floor apartment block that shared many of the features of a heliopathy clinic. 'Doling out cosmic energy, the Sun's effects are
both physical and moral, and they have been too much neglected in
recent times,' he declared – revealing his concern about tuberculosis. 'The
results of that neglect can be seen in cemetery and sanatorium.'
The Marseilles Building was open to sunlight at all times of day, with its

balconies serving as sunscreens that were designed to allow sunshine into the flats in winter and shade in summer.*

He took those precepts a stage further in the Indian city of Chandigarh, a conurbation covering 114 square kilometres at the base of the Himalayas, much of which he designed, working on his plans from 1951 until he died in 1965. Chandigarh was a new city that had been built from scratch to house the many refugees fleeing Pakistan after the catastrophic partition of British India in 1947. Its designs were meant to embody the ideals of an independent nation, but it was also important to Le Corbusier to take account of the demanding Punjabi climate, where in May and June temperatures can reach 115°F.

Three elements in his designs, partly sculptural, partly architectural, underscore his preoccupation with the Sun: '24 Solar Hours' is inscribed on the 45-metre-wide inclined face of the city's 'Geometric Hill'; the 'Tower of Shadows' occupies a square of 15.5 metres, its facades designed to give maximum shade; and the 'Course of the Sun' consists of two parabolic steel arches standing in a pool of water. Other parts of the city are equally Sun-conscious – the High Court building is designed specifically for the tropically rainy climate: it even has Le Corbusier's signature 'parasol' roof – an overhanging roll-up lid of concrete to fend off rain and sun.†

Of all modern architects Le Corbusier was the most aware of the Sun's effects on people in their daily lives. Current practitioners may wish they were more so. A case in point is Frank Gehry's $274-million Walt Disney Concert Hall in Los Angeles, where the glare from the stainless-steel facade has imperilled drivers and raised the temperature

---

*The designs of buildings in hot-weather countries have often included sun baffles, or sun breakers, also known as *brise-soleils*, placed on the outside of windows or over the entire surface of a building's facade. Different countries have tried different means to reduce the glare: lattices (*shīsh*, or *mushrabīah*), pierced screens (*qamariyah*), as used at the Taj Mahal, or blinds of split bamboo as used in Japan (*sudare*). One of Le Corbusier's most lasting innovations was a more substantial *brise-soleil* in 1933, taken up around the world.

†The longing to live in houses facing the Sun has continued over the ages, sometimes with unexpected results. In the eighteenth century, when the city of Bath was almost completely reconstructed, one of its leading architects, John Wood the Younger, created the Royal Crescent, several of whose thirty otherwise uniform Palladian houses could not face south, and for years, such was the social importance of receiving the morning sun, that no one would live in them. (On the other hand, some had rather too much sunlight, the 'white glare' that Jane Austen noted in *Northanger Abbey*.)

Frank Gehry's concert hall in Los Angeles

on nearby pavements to as high as 138°F – hot enough to melt plastic traffic cones and set off spontaneous blazes in rubbish bins. Ten minutes' exposure to this state-of-the-art building, under a bright sun such as LA famously boasts, can inflict a bad sunburn. In 2005, just two years after the concert hall opened, its surfaces had to be sandblasted to dim the brightness. It was all so unnecessary: the Sydney Opera House, built between 1959 and 1973, has its roof made up of precisely 1,056,056 glazed white granite tiles, specially treated to minimize glare. Similar buildings, especially those with concave surfaces, now have to be tested at the planning stage to ensure against overheating. Architects have learned to grade the Sun's effects according to three levels of increasing intensity: veiling reflections, discomfort glare and disability glare. As of this writing, Gehry is constructing an extension to the Philadelphia Museum of Art – entirely underground.

26

TALK OF THE DAY

At the rising of the Sun, I shall triumph! I shall triumph!

The Prince in Puccini's *Turandot*, 1926

There is a house in New Orleans
They call the Rising Sun.
It's been the ruin of many a poor girl,
And me, O God, for one.

Attributed to Georgia Turner and
Bert Martin, c. 1934

In the autumn of 2003 I submitted a letter to *The New York Times Book Review* outlining my search for information about the Sun. A few days after this was published, the telephone rang and I found myself talking to Gabriela Roepke, Professor of Opera at the New School – the university founded in Greenwich Village in 1919 – who with Old World courtesy said she would be happy to advise me about the Sun in opera and classical music. I thanked her, noticing her strong foreign accent, and we agreed to meet. Could I come to where she lived, on New York's Upper West Side?

Her apartment turned out to be three small rooms, with books and memorabilia stacked everywhere. A bird-like Chilean in her mid-eighties, she had been a dramatist: one of her nine plays, *White Butterfly*, appears in the 1960 *Best Short Plays of the Year*, but such writing was behind her now, and opera her love. Not being in the best of health, she found it hard to get around, so over the next few months I would make the trip to the twenty-second floor of her apartment building, waiting while she slowly made tea before we settled down to talk about opera and watch her many video recordings, which she expertly fast-forwarded to the scenes she wanted to discuss.

She explained that opera was a rarefied art form, the preserve of composers from just a handful of countries – Italy, France, Germany, England and Russia, with the mid-to-late nineteenth century its golden age, dominated by Wagner and Verdi. Their works and those of their contemporaries often invoke the Sun, she said, generally in the context of young love. None is more tragic than Nikolay Rimsky-Korsakov's *The Snow Maiden* (1880–1), and no others that she could think of make the Sun an actual character in the action. The doomed heroine, Winter's daughter, falls catastrophically in love with the Sun. In Act IV, as dawn is breaking over the valley of the Sun god, the Snow Maiden calls on her mother, who warns her to stay out of his light. But before the Snow Maiden can enter the protecting forest, Mizgir, a local merchant,

Judith Jamison (1943–) as the Sun in the 1968 revival of Lucas Hoving's *Icarus*. Although very few ballets feature the star *per se*, Isadora Duncan (1877–1927) wrote that dance 'is the rhythm of all that dies in order to live again; it is the eternal rising of the Sun'

appears, and soon he is presenting her as his bride to the tsar; just as she declares her love a ray of sunlight falls over her, and she melts away.

In his memoirs, Rimsky-Korsakov mentions his interest in folksongs, and how he 'was captivated by the poetic side of the cult of sun worship, and sought its survivals and echoes in both the tunes and the words of the songs . . . These occupations subsequently had a great influence in the direction of my own activity as a composer. But of that later.' Sadly, that is all he writes of his solar enthusiasms until the last of his fifteen operas, first performed in 1909, *The Golden Cockerel* (from a folktale retold by Pushkin), when the Japanese queen sings a hymn to the star.[1]

To lead me further into the Sun's association with tragic love, Gabriela put on one of her recordings of *La Bohème* by Giacomo Puccini (1858–1924). Soon I was watching Act I, in which Mimì, the seamstress heroine coughing her life away, has just met Rodolfo, the impoverished poet from the room above. Mimì's candle has blown out; she asks Rodolfo to relight it, then goes back to her room, only to return a few seconds later to say she has lost her key. Both their candles are extinguished by draughts, and the young pair stumble about in the dark, with inevitable consequences. 'Yes, they call me Mimì,' she begins, 'but my name is Lucia [Light] . . . When the thaw comes, April's first kiss is mine. I am the first to see the Sun . . .' They vow to remain together when the world returns to life, come spring. In Act IV, in her dying aria, *Sono andati? Fingevo di dormire* ('Have they gone? I was pretending to sleep'), Mimì relives the moment of their meeting ('The first Sun is mine!'). The scene must have been thrilling performed in La Scala, with its five tiers done up in sunrise red and yellow. Nor is *La Bohème* Puccini's only work to use solar imagery. More than any other composer, he scatters his operas with references to the Sun – such arias appear in *Manon Lescaut* (1893), *Tosca* (1900), *Madama Butterfly* (1904) and *Turandot* (1926).

When I next visited Gabriela, she was recovering from a fall and finding it an effort even to make her way across the few feet from her chair to her video collection. This time I made the tea while she put on Puccini's single-act work *Sister Angelica* (1880). The title character is a girl who has borne a bastard child and been sent to a convent, to spend the rest of her life doing penance. The opera opens on a lovely evening in

May. Sister Angelica and the other nuns have gathered to celebrate one of the three days of the year when the setting Sun turns the fountain in their courtyard to gold – a representation of the 'fair smile of Our Lady'. By the end of the opera, Angelica, hearing that her child has died, takes poison – suicide, of course, being a mortal sin. But just as the opera's opening scene is lit by the Sun, so too is the climax, as Angelica implores the Virgin for forgiveness and a blaze of light appears as a sign of the mercy granted her.

The Sun as a sign of the infinite goodness of God is a theme taken up by the other great composer who made the star his leitmotif: Richard Wagner (1813–83). In *Das Rheingold* (1869), the first of the four parts of *Der Ring des Nibelungen* (*The Nibelungs' Ring*), the lustful dwarf Alberich sets out to woo the three maidens who guard the gold at the bottom of the Rhine: 'Look, sisters, the Sun shines into these depths . . . Nymphs, what gleams and glitters there?' The maidens know all too well what they are protecting, and, even as they repel the dwarf's advances, intone: 'Rhine gold! We delight in your glow!' His desires thwarted, Alberich steals the maidens' precious hoard, from which he casts an enchanted ring. Thus Sun and gold blend into a talisman of infinite power. Significantly, Wagner wrote the libretto of the cycle at the time of the first great gold rushes.

When the hero, Siegfried, takes centre stage in Part III, he comes upon the sleeping Brünnhilde and wakens her with a kiss, just as the Sun rises. Most of *Siegfried* is orchestrated in gloomy minor chords, but at this point the violin section enters on high notes and wind instruments take up the first two chords, leading on to strings in a yet higher register and the notes of a harp, as Brünnhilde, dazzled by the morning light, cries out: 'Sun, I greet you. Light, I greet you, Radiant Day.' The vivid combination of instruments Wagner employs to celebrate the Sun in all its glory is extraordinarily rare in his music.[2]

And yet in *Tristan und Isolde* (1857–9) Wagner – alone of all opera composers – has his hero inveigh against the Sun:

> *Accursed Day with your light!*
> *Will you for ever be witness to my anguish?*
> *Will it burn for ever, this Light,*

CHASING THE SUN

> *Which even at night kept me from her?*
> *Ah, Isolde, sweet beauty!*
> *When at last, when, oh when will you extinguish the spark,*
> *That I may know my fortune?*
> *The light – when will it be extinguished?*

The Sun may indeed be the symbol of civilization and right living, but Tristan rejects it all, giving himself to *Nachtzicht*, night vision. Here Wagner invokes a familiar conflict, going back at least to Egyptian myth, which opposes the forces of light (Horus) to those of sexuality (Seth).*

Tristan and Isolde's love is doomed by a world that cannot allow it to flourish. Only under cover of darkness can they meet as lovers – amid many exchanges about their hatred for daylight's false values, and their devotion to *das Wunderreich der Nacht* (the wonder-realm of night). As the conductor Charles Mackerras has pointed out, the music matches the libretto: 'The orchestration is bright and glaring for the vanity and deception of the day, then suddenly it becomes nocturnal and shadowy.'[3]

Gabriela explained that the lovers' refrain of cursing daylight so pervades the opera that it acquired its own name: *Tagesgespräch* – talk of the day. Wagner had been initiated into this light/darkness dialectic by the writings of Arthur Schopenhauer (1788–1860), in particular *The World as Will and Representation* (1819), which Wagner describes in his autobiography as having an impact on him that 'was extraordinary and decisive for the rest of my life'.[4] *Tristan*, which he composed just two years after he read Schopenhauer, equates light and darkness with two of the book's most important concepts, those of Phenomenon and Noumenon: the former being our (false) representation of the world, while the real world is 'Noumenon', where all things are indivisible and one. Most people live in 'the world of day'; but a small elite, of whom Tristan is one, see beyond that chimera into another world, at once more terrifying, more real and ultimately offering a deeper satisfaction,

*Others followed his lead. In Charles Gounod's *Roméo et Juliette* (1867), the chorus sings of Romeo, 'Love thrives in the shadows, let it guide his steps!' Then, to take instances from popular music: 'The day is my enemy, the night my friend. When the Sun goes down . . . I'm all alone . . . with you . . . All through the night'; Sky Masterson in *Guys and Dolls*: 'My time of day is the dark time'; and that creature of the night Édith Piaf, catching sight of the morning Sun, *'Voilà, le salop!'* ('There's the son of a bitch!').

The Sun and Moon dance in traditional Japan

for it is there that he can be for ever united with his love. Tristan is aware that the realm of night can be shared only in its fullest sense when he and Isolde die: thus his is a *Tod-geweihtes Herz* (a death-consecrated heart). He rapturously sings:

> *Now we are become the devotees of night.*
> *Spiteful day, with envy armed,*
> *Could still deceivingly keep us apart*
> *But never again delude us.*

So one of the greatest pieces of music in all Western culture is a direct attack on the Sun? Possibly, which may be why *Tristan* has been denounced as a 'work against civilization'. Gabriela suggested that at our next meeting we move backwards in time to Mozart (1756–91), who went so far as to change many elements of the fairy tale that inspired one of his greatest operas, in order to wrest it into being a celebration of the Sun: *The Magic Flute*.

☼

The set for the final scene of the second act of *The Magic Flute*, c. 1730

Mozart composed this opera the year he died. The libretto, by his old friend Johann Josef Schikaneder (1751–1812), is based on a story, set in Egypt, in which the Queen of the Night represents the power of good and the black-skinned and black-hearted Monostatos that of evil. Mozart and Schikaneder reversed the queen's character, making her a symbol of wrongdoing, and added the figure of Sarastro, the all-wise high priest of the Sun (Isis). Sarastro (his name derives from Zoroaster/Zarathustra) has taken the Princess Pamina to the Temple of the Sun, to release her from the evil influence of her mother the Queen of Night, who in turn induces the noble young Prince Tamino to search out and liberate her daughter with the promise that he will be given her hand in marriage. Tamino, after manifold trials in which his character is tested by Sarastro, not only learns the truth about the fell queen but gains both Pamino's freedom and her love; in the climactic scene the couple are then welcomed to the Temple of the Sun.

The South African director William Kentridge has declared that there are 'references throughout the opera of turning darkness into light, of the light of the Sun banishing the night', and certainly the Sun has more than just a 'shine on' role. Act II opens in a forest of palms with golden leaves, amid which rises a great pyramid, surrounded by thrones for the eighteen priests of the Sun; and for the finale, the stage transforms into a sun. The star even has its own key, C major being used throughout to embody solar radiance.

Mozart and Schikaneder were Freemasons, and Masonic elements appear throughout – none of them more central to the Masonic system than the Sun. The music historian William Mann argues that Mozart was interested 'in forging a link between the Freemasonry of his time and ancient Egypt [from] which Masonry claimed its original derivation'.[5] This opera, of all works of art, exalts the Sun specifically because of its significance to a long-established secret society. A Masonic encyclopedia, published in 1912, explains why:

> Hardly any of the symbols of Masonry are more important in their sig-
> nification or more extensive in their application than the Sun. As the
> source of material light, it reminds the Mason of that intellectual light
> of which he is in constant search . . . the Sun is then presented to us in
> Masonry first as a symbol of light, but then more emphatically as a
> symbol of sovereign authority.[6]

Precisely such thinking informs Mozart's opera – although one may be forgiven for escaping into the music.

☼

Although many other works find one way or another to pay homage to the Sun, Gabriela and I largely kept to classical opera (partly because that was her preference, partly because she was recovering slowly from her fall and I was anxious not to tire her): *The Sorrows of Young Werther* by Jules Massenet (1842–1912), from Goethe's novel, in which the main aria in Act II is Sophie's *Du gai soleil, plein de flame*; Pietro Mascagni's *Iris* (innocence and desire in turn-of-the-century Kyoto, written in 1897, several years before *Madama Butterfly*), with its aria *L'Aurora – Son lo la Vita*; Umberto Giordano's 1898 opera *André Chenier*, about a brilliant young poet who perishes under the Reign of Terror; and *The Cunning Little Vixen*, by Leoš Janáček, with its sunrise sequence. Memorably, in Beethoven's *Fidelio*, the prisoners emerge from their dungeons blinded by the Sun's light: it is seen as life-giving, so that in Act II Florestan sings of how its deprivation makes him feel he is in hell.

It is, of course, easier to track connections to the Sun in opera than in musical forms without words – the Sun is hard to represent in notes alone – but many compositions invoke it in their titles. Nature permeates the music of Frederick Delius (1862–1934), whether the ocean in *Sea Drift* or the mountains and meadows in the *Mass of Life*. He was a Sun-worshipping pagan, at home writing about the passing of nature (*To Daffodils*), severed relationships (*Songs of Sunset*), or simple mortality (*Songs of Farewell, A Late Lark*), and equally obsessed with spring and summer: a sense of the Sun seems present in even his untitled pieces.

Delius was composing at a time when it was considered vulgar to produce sensory effects for programme music; but that did not daunt his contemporary, Jean Sibelius (1865–1957). In the final movement of his Fifth Symphony, there appears to be a powerful representation of the Sun. Alex Ross, in an extended critical appreciation in *The New Yorker*, describes the climax of the 'swan hymn' that runs through the symphony:

> The swan hymn, now [in the final movement] carried by the trumpets, undergoes convulsive transformations, and is reborn as a fearsome new being. Its intervals split wide open, shatter, and re-form. The symphony ends with six far-flung chords, through which the main theme shoots like a pulse of energy. The swan becomes the Sun.[7]

The star inspired other works by Sibelius, including a song to sunrise. Solar elements appear in many other classical works, including music by Maurice Ravel (1875–1937), Claude Debussy (1862–1918) and Igor Stravinsky (1882–1971). One afternoon Gabrielle and I made up a list: it ranged from *Also Sprach Zarathustra* (the brief opening prologue, 'Sunrise') by Richard Strauss, 1896, to *Eternity's Sunrise* by John Tavener (1999), based on the poetry of William Blake, and my own favourites, the opening sunrise movement from Dvořák's Ninth Symphony (1893), Elgar's 'Chanson de Matin' (1899) and the 'Helios Overture' by Carl Nielsen (1903), all composed within ten years of one another.

The majority of the works are inspired by sunrises rather than sunsets, or by the star's internal works; and just because a work mentions the Sun in its title does not necessarily mean that it actually relates to it – Haydn (1732–1809) titled one of his string quartets, Op. 20 (1772), *Sun Quartets*, though it contains nothing particularly sun-like. Gustav Holst (1874–1934) excised the Sun from his best-known work, the orchestral suite *The Planets* (1914–16). The music was partly inspired by his horoscope chart, which dealt with the 'seven influences of destiny and constituents of our spirit', but he chose to replace Sun and Moon with Uranus and Neptune. In his earlier Choral Hymns from the Rig-Veda (1908–12), however, he did include the *Hymn to Vena* (Sun rising through the mist).

I left it for several weeks before I next went to see Gabriela, but had come across something during my researches that I knew would interest her: a curiosity from Sergei Prokofiev (1891–1953), although not a musical work. In 1916, shortly after composing his first two piano concertos, he acquired a sun-coloured notebook in which he asked people to write their answer to a single question: 'What do you think about the Sun?' – and kept up his enquiries until 1921, by which time *The Gambler* (1916) and *The Love for Three Oranges* had brought him fame. There are fortyeight pages all told (now in a Moscow museum), which include answers not just from his friends and acquaintances but from people he encountered in his travels, and from the likes of Artur Rubinstein, Feodor Chaliapin, Vladimir Mayakovsky and Igor Stravinsky. I enjoyed many of the answers, such as Mayakovsky's 'From all of you, who soaked in the Sun for plain fun, who spilled tears into centuries while you cried, I'll walk away and place the monocle of the Sun into my gaping, wide-open eye'; and also Stravinsky's direct 'It is very stupid that in German the Sun is of feminine and not masculine gender'.

Looking forward to showing Gabriela some of the material from the notebook, I checked in at the desk on the ground floor of her block and asked if I could go up to her apartment. The receptionist looked at me perplexed. 'Miss Roepke's not here,' he said. 'She had all her things packed three weeks ago and returned to Chile. She won't be coming back.'

I turned and headed for the door. I knew why she had gone home.

☼

Thanks to the power of Google, we know that 2,462 copyrighted songs have 'Sun' in their title – more by now, as this figure is from May 2009. The roster of composers ranges from Beethoven to Irving Berlin, and includes ninety-nine songs featuring the phrase 'rising Sun' alone. When in the early 1970s the Beach Boys recorded 'I'll Follow the Sun' they were reflecting at least a decade of a craze for doing exactly that, principally on the great surfing beaches of southern California.

To the Beach Boys, the Sun stood for freedom. From the mid-1960s to the mid-1980s it became the metaphor of choice for expressing the ups and downs of love. Katrina and the Waves' anthem 'Walking on Sunshine' (1985) seized upon the charge of sunlight to express the intensities of love; Bill Withers' 'Ain't No Sunshine When She's Gone' captured the agonies of love lost; while the Beatles' 'Good Day Sunshine' joined being in love with a feeling of being merged with the world.

One of the best-known songs of those decades uses a different kind of metaphor. 'Here Comes the Sun', composed by George Harrison in 1969 to become the opening track to side two of the Beatles' *Abbey Road* album, had its genesis in a collaboration with his friend Eric Clapton and celebrated the escape he longed for during a time when so many things were sapping his energy – primarily the gruelling shifts to finish the album before the Beatles broke up: the phrase reflected Harrison's relief when each session was over. As he put it, the song 'was written at the time when Apple was getting like school, where we had to go and be businessmen: "Sign this" and "sign that". Anyway, it seems as if winter in England goes on for ever; by the time spring comes you really deserve it. So one day I decided I was going to slag off Apple and I went over to Eric Clapton's house. The relief of not having to go see all those dopey accountants was wonderful, and I walked around the garden with one of Eric's acoustic guitars and wrote "Here Comes the Sun".'[8]*

Along with all the music that either invokes the Sun or has been

*A second great Sun classic of the pop era is 'The House of the Rising Sun'. Originally an English ballad, it took its title from a euphemism for a brothel, so to that extent is not about our Sun at all; but as a cultural reference point it is unique. In 2005 a poll ranked it Britain's fourth favourite song of all time, and worldwide its popularity has not been far off that. What *was* the 'Rising Sun'? Possibly a metaphor for the slave pens of the American South, or the plantation house, or even the plantation itself. As for its being a particular place in New Orleans, nothing is certain.[9]

inspired by it, a slew of modern-day compositions attempts to capture how it actually sounds. The Stanford Solar Center in California lists a dozen artists who have attempted representations: Stephen Taylor, for instance, has composed *Shattering Suns*, a symphony 'inspired by images of celestial catastrophe' with music 'built on the breathing – the sounds of the Sun, as recorded by Stanford Solar researchers and played by a synthesizer'. Musicians have moved from the inner realms of their imagination – Mozart denoting C major to represent the Sun, for example, presumably because he felt that that key resonated with a kind of solar energy and optimism – to representations drawn from science. This doesn't necessarily mean such compositions have any claim to accuracy or that anyone even now, with all our sophisticated technology, actually knows what the Sun sounds like.

In 1992 the prolific Danish composer Poul Ruders, inspired by Carl Nielsen's 'Helios Overture', created the symphonic drama *GONG*, with a score meant to evoke what scientists have told us. He explained that recent research has shown that the Sun's surface reverberates like a gong, in four different, simultaneous tempi. Formally, his composition follows the Sun throughout its lifespan, from explosive birth through the hyperactive release of energy we know today to final collapse, concluding with a chord taken from the middle of the piece and sustained over several bars, from virtual nothingness to full force. I confess to finding the symphony difficult to listen to; and it turns out that it is not an accurate rendering in any case.

From 2006 on, the new discipline of helioseismology has been measuring the super-low-frequency soundwaves that echo inside the Sun, caused by the waves bouncing around inside the giant ball of gas, huge convective bubbles of different notes and vibrations, nearly all thousands of times lower in pitch than earthly bells. These oscillations trigger small motions – small for the Sun, that is – on its surface, creating tremendous up- and downdraughts that generate a broad range of low-pitched noises, just as air blown across a bottle's mouth sets off a few notes – only on the Sun they range into the millions, ringing like bells or vibrating like organ pipes. Researchers have been tracking the waves as they resonate from one side of the star to the other – a journey taking at its fastest about two hours, at four hundred times the speed of sound.[10]

When in 2003 I visited the famous old observatory on the outskirts of St Petersburg, its director asked, 'Would you like to hear the Sun?' A moment later I was in the main observation tower listening to a record that sounded like a nest of alarmed and angry snakes – nothing gong-like at all, or indeed like breathing. Months later I read in Kenneth Tynan's *Diaries*, 'The Sun has echoes that are not entirely gentle. It hisses.' So maybe in years to come we will be given a musical work for timpani and cobra.

BUSY OLD FOOL . . .

Down sank the great red Sun, and in golden, glimmering vapors
Veiled the light of his face, like the Prophet descending from Sinai.

Longfellow, *Evangeline*[1]

Apollo still raged away in the brazen sky – as one might say if
one had a taste for that sort of thing.

Dick Francis, *Smokescreen*[2]

One twentieth-century novelist, poet and essayist has made extraordinary use of the Sun. He considered himself a scientist – an expert on
butterflies: Vladimir Nabokov. I hadn't known his work well, but remembered that in his final novel, *Ada* (1969), the heroine and her brother
Van Veen wander through the gardens of their ancestral home and begin
to play a game. Ada explains the rules:

> The shadows of leaves on the sand were variously interrupted by
> roundlets of live light. The player chose his roundlet – the best, the
> brightest he could find – and firmly outlined it with a stick; whereupon
> the yellow round light would appear to grow convex like the brimming
> surface of some golden dye. Then the player delicately scooped out
> the earth with his stick or fingers within the roundlet. The level of that
> gleaming *infusion de tilleul* would magically sink in its goblet of earth
> and finally dwindle to one precious drop. That player won who made
> the most goblets in, say, twenty minutes.[3]

This interested me because Kurt Vonnegut's 1963 novel *Cat's Cradle* takes its title from a game played by Eskimos, wherein children try to snare the Sun with string: same idea, different method. Then, as I read more of Nabokov, I realized the Sun recurs with particular urgency throughout his work. The deceived lover in *Laughter in the Dark* cries out for sunlight at the very moment he discovers his mistress's infidelity; in *Bend Sinister* the protagonist, Krug, perceives the word 'loyalty' as being like a golden fork lying out in the Sun; and *Pale Fire* glows with solar reflections.[4] There are many other instances.

Nabokov was unusually acute about the quality of sunlight, writing in his memoirs: 'All colours make me happy, even the grey blood-orange Sun.'[5] His consciousness of the Sun was heightened by its ability to lessen the effects of a case of psoriasis on occasion so severe that its 'indescribable torments' almost drove him to suicide; sunbathing and radiation treatments helped in later life.[6] He was also synaesthetic; his biographer Brian Boyd mentions that as a child Nabokov had a 'love of color and light' – both he and his mother saw the letters of the alphabet in vivid tints – and that he linked his first attempts at poetry 'with his mother's jewels and with stained glass, prisms, spectra, rainbows':

> It [the theme of materials affected by sunlight] shines in all its glory in the scene of his first poem. After taking shelter from a thunderstorm in a pavilion at Vyra, Nabokov sees the Sun return and cast luminous colored rhomboids on the floor from the colored-glass lozenges of the pavilion windows, while outside a rainbow slips into view, and at that moment his first poem begins.[7]

Boyd points out how Nabokov would pay attention 'to details of an order not expected: the haze over a frying pan, the colors and shapes of shadows . . . He had a painter's sense of light.' Even so, I was quite unprepared when I took up his masterpiece, *Lolita*: a novel quite brimful with solar references, almost as if its author were having one of his private jokes. But this profligacy is not intended facetiously at all: drawing on the Sun to an extent unequalled anywhere else in all literature, he uses it to remarkable effect.

Vladimir Nabokov (1899–1977), Sun enthusiast, ardent lepidopterist and wordsmith:
even the fictional narrator of *Pale Fire* (1962) is called John Shade

This large claim needs furnishing. The first third of the novel is saturated with images of sunlight, dictating the narrative's mood and
directing us to the easy optimism of Humbert Humbert's early hopes.
In the opening sentence, Humbert presents Lolita as the 'light of my
life' and goes on to talk of the 'sun of my infancy' having set.[8] When
he first spies Lolita (through 'sun-speared eyes') it is a 'sun-shot
moment'. She is 'a photographic image upon a screen . . . already
riding into the low morning sun'. While her mother engages in 'some
heliotropic fussing', Humbert focuses on a sunbathing Lolita 'through
prismatic layers of light', but he can also take in the 'sun-shot' pavements and 'the reflection of the afternoon sun, a dazzling white
diamond with innumerable iridescent spikes [that] quivered in the
round back of a parked car'. When, a few pages on, Lolita creeps up on
him, putting her hand over his eyes as he reclines in a low chair, 'her
fingers were a luminous crimson as they tried to blot out the sun'.

Playfully, she tosses an apple 'into the sun-dusted air', and Humbert is soon lost 'in the pungent but healthy heat which like summer haze hung about little Haze' – Haze being Lolita's surname. 'The implied sun pulsated in the supplied poplars'; 'the sun was on her lips' – all this in the first sixty pages.

Such references are very different from the metaphor bank of the Romantics, in which the Sun is primarily a standard-bearer of classical mythology. Nabokov is observant to a fault, alive to anything that the Sun lights upon. He may use the Sun for its symbolic power, as when he writes that Humbert dreams of 'the red sun of desire and decision (the two things that create a live world)', but – naturalist that he is – he is at his best as a subtle observer of the physical world around him, of the 'beauty and animation of the sun and shadows of leaves rippling on the white refrigerator', or 'those southern boulevards at midday that have solid shade on one side and smooth sunshine on the other'. 'I knew the sun shone,' says Humbert, 'because my ignition key was reflected in the windshield.' Nabokov's interest in light melded with his lifelong fascination with butterflies in at least one scene: we note that 'a big black glossy Packard' – the one that later kills Lolita's mother – mounts the lawn and stands 'shining in the sun, its doors open like wings'.

The web of images is effective both in and of itself and for the way it serves as part of a larger purpose: to employ sunlight in all its many forms, and the world of shadow and darkness, to mirror what is happening to the central characters. Astonishingly, references to the Sun appear about every three pages in the first third of the book. 'Movie-ladies' have 'sun-kissed shoulders'; a moment of sexual epiphany is 'the ultimate sunburst'; while Lolita at her summer camp appears 'a sun-coloured little orphan'. But such sunny days are numbered, and as Part I ends the language reflects the darkening mood. 'But somewhere behind the raging bliss, bewildered shadows conferred' – this at the very time when Humbert first exults 'She was mine, she was mine'. His 'rainbow blood' may be in full flow, but the images of sunshine disappear, and those shadows close in. Significantly, there comes 'a tremendous sunset which the tired child ignored', and before the short, brutal near-rape scene Humbert wildly pursues 'the shadow of her infidelity'. It is a sign that he cannot see his love object clearly: when a sunburst strikes the highway along which he is driving, he has to pull in at a filling station for new sunglasses. But Clare Quilty, her eventual 'saviour', is now in hot pursuit – 'our shadow'. (Quilty's co-playwright is Vivian Darkbloom – an anagram of 'Vladimir Nabokov' – while a member of one of Quilty's drama groups makes his debut in a play entitled *Sunburst*.)

Though it is nothing as simple as Lolita representing the Sun, or the Sun her, once she seems to tire of Humbert he confesses: 'I hear myself crying from a doorway into the sun.' And before long he is admitting 'The sun had gone out of the game'. After the love of his life has absconded, Humbert has a moment when he thinks he has glimpsed her – 'a trick of harlequin light', maybe – and he writes a poem to her to remind her of an old perfume called (inevitably) Soleil Vert. But she has left him for ever, and he is forced to accept 'that I smugly did not know a thing about my darling's mind and that quite possibly . . . there was in her a garden and a twilight'. Sunsets now are 'blotched', forcing Humbert to drive on and on 'through the drizzle of the dying day'. When, more than two years later, he finds Lolita again, she is worn out, vast with the pregnancy that will kill her in her teens, her smile 'a frozen little shadow of itself'. Failing to get her to go back with him, he takes off into the night, intent on tracking down and killing Quilty, and when in ghastly fashion he does so, 'the sun was visible again, burning like a man', leaving behind the corpse of her latest abductor while he walks away 'through the spotted gaze of the sun'.

All this may suggest that the second half of the novel succumbs to an obsessive symbolism, but it doesn't come across as such – there are at least as many other references to the Sun and to shadow as I have quoted that have no obvious symbolic content (I must be the only person to have read *Lolita* for its Sun images; curiously, in all the 565 pages of Nabokov's published letters, there is not one reference to the Sun). We can simply delight in Nabokov's fascination with sunlight and his awareness of what it conveys.

☼

If Nabokov is a special case, we should recall the judgement of Max Müller that from the time that man first started to tell stories he has used the Sun as an image. By the days of the epic poets, the whole heavens were being drawn on as metaphors and symbols. In the *Iliad*, for instance, Homer identifies some 650 characters with stars, and finds them in 45 constellations or star patterns, each identified with a particular warrior, while such items as armour, chariots and spears are linked to objects in the sky. He even harnesses the precession of the equinoxes: the epic battles and the duels between various heroes have been shown to be allegories of the passage of the equinoctial rising. The *Iliad* may be considered the world's oldest substantial astronomic text, using the zodiac to allegorize the movement of the heavens. Such, at least, is the remarkable argument of *Homer's Secret Iliad*, written in rough draft over a period of some thirty years by Edna Johnston, a librarian turned amateur literary critic from south-east Kansas, and published posthumously in 1999.*

Johnston's research convincingly suggests that the Homeric epics might be among the first works of literature to incorporate a knowledgeable study of the Sun's apparent motions, as if Homer were intent on setting down what was known about the heavens for future generations. And it is not only the *Iliad* that displays an extraordinary grasp of astronomical knowledge, she says. In the *Odyssey*, the symbolism of Odysseus's return home after nineteen years may show Homer's grasp of the nineteen-year Metonic cycle, over the course of which the sequence of the Moon's phases returns to the same days of the year. In one of the *Odyssey*'s most memorable metaphors, Homer uses the Sun's position to denote the time of Odysseus's departure from home: 'and when the child of morning, rosy-fingered Dawn, appeared they again set sail.'[10]†

The year after *Homer's Secret Iliad* appeared, the work of another amateur critic on another classic would show that it too dealt at least in part

---

*Johnston was born in 1916 and studied at Kansas State Teachers' College, taking mythology as one of her special subjects and learning ancient Greek. Over the next three decades she read and reread the *Iliad* and the *Odyssey* until she could recite long passages from memory. At some point she made her discovery – she had no wish, she said, to diminish the poems as narratives, dense with well-drawn characters, convincing plots, 'pathos, horror, excitement, calm; philosophy, history and so on',[9] but she saw an overarching purpose in the two great works: to preserve astronomical knowledge. Fearing she might be out on a limb, she never made her findings public. At her death her daughter Florence, and her son-in-law, spent seven years editing and expanding the various notes until they could finally publish the results.

†Over 400 years later, at least one major Greek dramatist was to make keen use of the Sun. Euripides has Orestes flee after killing his mother, Clytemnestra, and search out a place on which the Sun did not shine when he committed the deed, as a way of escaping the Furies; then, in *The Bacchae*, the young King Pentheus of Thebes is driven mad by Bacchus and sees two Thebes and two Suns.[11]

with astronomy. Dolores Cullen first encountered Chaucer when she was a middle-aged college student in California, and became caught up by his work. In 1998 she published a book about the religious allegories in *The Canterbury Tales*. But still she was not satisfied. 'As I read, I was distracted by questions about the pilgrims. What made this precise group necessary? Chaucer's reputation was too well known, his skills too well recognized for me to think that the group was a haphazard collection. Why was there one pair of brothers – not from a religious order, but two men related by birth? Why not three brothers or no brothers? Why was there a wife – but no husband and wife? Why no children? Why so few women?'[12] In *Chaucer's Pilgrims: The Allegory*, she turns to the poet's interest in astronomy and argues that each pilgrim corresponds to a body or bodies in the night sky, which reflect, in name and appearance, the mythical characters after whom the constellations are called. The study was well received.

It is generally known that Chaucer was highly versed in astronomy, and in 1391 even penned a treatise on the astrolabe for his younger son Lewis; Cullen shows that this interest suffuses *The Canterbury Tales*, influencing not only the identities of the pilgrims but also their stories. From 'The Franklin's Tale', with its theories of tides, and of the effects of perihelion (the time when the Earth and Sun are closest to each other), through to 'The Parson's Tale' (about the uses of trigonometry), 'The Canon's Yeoman's Tale' (about a charlatan alchemist), 'The Squire's Tale' (which takes in the precise relationship between the Sun and Mars) and onwards, Chaucer puts astronomy and its related disciplines to intense use. But unlike Homer he does not seem to be writing for posterity so much as responding to a subject that interests him, confident that his audience, at a time when familiarity with the movement of the heavens was taken for granted, would share that interest.

Today's analyses of astronomical references in classical works have nothing over the ingenuity of those early biblical writers who brought such energy to explaining the motions of celestial bodies. For instance, the two 'lights' mentioned in Genesis – Sun and Moon – were interpreted by the Catholic Church as representing the papacy (the greater light) and the empire, i.e. the old Roman Empire or the current Holy Roman Empire (the lesser). Metaphorically, this meant that just as the Moon receives its light from the Sun, so too did the empire receive its authority from the Church, and thus was subject to being overridden. Dante Alighieri (1265–1321), in *De Monarchia*, bravely rebuts this interpretation, and in his case gets away with it.* But in his greatest work he shows himself to be steeped in the Christian symbolism of his time – and the contemporary understanding of astronomy that provided much of its imagery. All three books of *The Divine Comedy*, *Inferno*, *Purgatorio* and *Paradiso*, rest on a structure of medieval astronomical doctrine, so that when in the *Paradiso* the souls are placed in rank, their order reflects the doctrine of the planetary spheres as outlined by Aristotle and Ptolemy; while the organization of the *Inferno* almost exactly mirrors the assumed ninefold layout of the heavens. The Sun itself receives due attention – early on in the *Inferno*, for instance, it indicates times of day, through an oblique reference to the dawn light on the hillside ('The hour was morning's prime, and on his way aloft the Sun ascended . . .').

At another moment Dante realizes that his journey through the heavens has completed a quarter of the daily revolution of the stellar sphere around the Earth. He has therefore spent six hours in Gemini, roughly the same amount of time that Adam enjoyed the pleasures of Eden before being cast out. Now directly beneath Cadiz, close to the westernmost shores of the Mediterranean, he sees an expanse of the Atlantic and, to the east, nearly as far as the Phoenician coast. Only this portion of the Earth's surface is visible to him, because the Sun lies two zodiacal signs behind (further west), beneath Aries.

The *Purgatorio* refers more than once to the Sun's position, since Dante uses it as the central device to establish the timing of events in the poem. In the *Inferno*, by contrast, he takes only the positions of the Moon and stars for this purpose, since the Sun, as an expression of God's power, will not tell the time in hell, the infinity of which is part of the horror. In the *Paradiso* he attains the heaven of the Sun, whose special virtue is wisdom. Two moments bear mention: Canto 33 of the *Inferno* has the first reference in literature to the Sun's being fundamentally a

*'Lucifer' (Light-bearer, with all his positive connotations) caused a special problem for early Christians. The name appears in the fourteenth chapter of the Old Testament Book of Isaiah, verse 12 – and nowhere else. It was popularized by its appearance in the King James translation, although it appears first in a fourth-century version. The explanation is that in the original Hebrew text it is not an angel who falls, but a Babylonian king, who during his lifetime had persecuted the children of Israel: an early Christian scribe made the change.

star ('I said nothing /In answer all that day nor the next night /Until another star rose on the world'); and in the *Purgatorio* (pictured as a huge island-mountain in the uninhabited southern hemisphere), the various souls Dante meets are shocked to discover that he is still alive by noticing that, unlike them, he casts a shadow.

☼

Over the centuries, two main problems had to be faced in writing about the Sun. First, it is such a major theme that the temptation to grandiosity is hard to resist. Poets employed the Sun at their peril; but some did make use of it with greater finesse than others. Thomas Malory (1405–71) introduces Gawain (a translation of 'bright hair') into *Le Morte d'Arthur* (published posthumously c. 1485) as a Sun spirit, the knight gaining and losing strength with the Moon's waxing and waning, so losing the Sun's light. Edmund Spenser (1552–99) trod the 'big theme' tightrope with great skill in *Epithalamion* ('Of a Wedding', 1595), a poem that can, like Homer's and Chaucer's works, be read as an astronomical text. By contrast, Milton's 'On the Morning of Christ's Nativity' comes up with an image that is simply absurd:

> *So when the Sun in bed,*
> *Curtain'd with cloudy red,*
> *Pillows his chin upon an Orient wave,*
> *The flocking shadows pale*
> *Troop to th'infernal jail . . .*[13]

The same simile does service in Marvell's 'Upon Appleton House',[14] but he made up for this with 'To His Coy Mistress': 'Though we cannot make our Sun /Stand still, yet we will make him run'). Robert Herrick (1591–1674) was again decidedly pedestrian in his 'The glorious lamp of heaven, the Sun /The higher he's a-getting, /The sooner will his race be run /And nearer he's to setting'.[15]

The second part of the challenge in earlier times was that references to astronomy had to conform to Church doctrine on matters pertaining to the heavens. In his account of the Creation in *Paradise Lost*, Milton has to resolve the biblical conundrum of where light came from if the Sun were created only on the fourth day, and gets round it by having light shine from its nook under God's throne for those first few days until it can find its fit place in the Sun. He probably decided to use Ptolemaic astronomy because it was more in accord with scripture, as well as being more worthy of epic; but he is obviously torn, and the possibility of a Sun-centred universe is left open.[16] He can be a master of prevarication: in Book X, man's Fall thrusts the Sun 'from the Equinoctial Road' – which can be read as either deflecting the Sun from its course around the celestial equator (as the Ptolemaic system requires) or tilting the Earth on its axis (as in the steadily gaining Copernican system).

During the 1,500 years that Ptolemy's doctrines held sway, writers were generally content to employ the Sun for occasional grandstanding (John Suckling's 'The Sun upon a holiday is not so fine a sight' – about a beautiful girl dancing*); as an excuse for retelling famous myths (Sir Philip Sidney's sonnet sequence *Astrophel and Stella*); or as a reference point when examining the heavens.

One exception is the Italian Giulio Camillo (1480–1544), whose influential *L'idea del Teatro* ('The Idea of the Theatre') is nothing less than a history of the universe, in which the Sun holds pride of place. Frances Yates, in her study of that writer, explains how his work 'shows within the mind and memory of a man of the Renaissance the Sun looming with a new importance, mystical, emotional, magical, the Sun becoming of central significance. It shows an inner movement of the imagination towards the Sun which must be taken into account as one of the factors in the heliocentric revolution.'[17] References to Camillo's work have figured in the texts of writers from Rousseau to Ted Hughes.

The Sun was soon seized on throughout Europe as a royal emblem, kings being portrayed in allegory as the embodiment of the Sun's glorious attributes – notably in *City of the Sun* by Camillo's countryman, the Dominican Tommaso Campanella (1568–1639), a philosophical dialogue fashioned after Plato's *Republic* and written in 1602, just after Campanella had been condemned to life imprisonment for sedition and heresy. The city of the title, set on the island of Taprobane (now Sri Lanka), is benignly presided over by a priest-prince called Sun.[18]

*I have a soft spot for Suckling, for during his short life (1609–42) he invented the game of cribbage.

Both Shakespeare (1564–1616) and John Donne (1572–1631) were Campanella's contemporaries. The former's use of the Sun characteristically defies generalization: he is ever a keen observer of the heavens, referring to the Sun more than forty times in the late plays alone. Many of these references occur in a literary form that had its origins in classical poetry but which was brought to perfection by the troubadours of medieval Provence and the Minnesängers of Germany: the aubade, a short lyric of love fulfilled, supposedly uttered at dawn (as opposed to the serenade, sung by a hopeful suitor at evening), and generally lamenting that the lovers must bid farewell to the joys of the night. It has been suggested that the form was an ingenious development from the cry of nightwatchmen announcing the coming day. Shakespeare made glorious use of aubades, typically planting them as hinge points in his plays, classically in *Romeo and Juliet*.[19]

Over the centuries the aubade was adapted for other purposes, the Victorians, for instance, using it for a graver kind of lament, as Tennyson did in *In Memoriam*. Perhaps the best known of all English aubades is Donne's 'The Sunne Rising', which opens seductively:

> *Busy old fool, unruly Sun*
> *Why dost thou thus,*
> *Through windows, and through curtains, call on us?*
> *Must to thy motions lovers' seasons run?*

If this is the first instance of a poet mocking the Sun (I have found none earlier), in Donne's case there was reason. In 1601 he had secretly married the seventeen-year-old Anne, daughter of Sir George More, the diminutive and irascible Lieutenant of the Tower of London (a 'short man with a short fuse', Donne's latest biographer calls him). He wrote to the livid father: 'Sir, I acknowledge my fault to be so great as I dare scarce offer any other prayer to you in mine own behalf than this, to believe that I neither had dishonest end nor means ... I humbly beg of you that she may not, to her danger, feel the terror of your sudden anger.' Sir George replied by having him thrown into the Fleet Prison for some weeks (along with the priest who performed the marriage), attempting (unsuccessfully) to annul the union and getting

Donne's employer to dismiss him. 'John Donne, Ann Donne, Undone' went the popular epigram, attributed to Donne himself. For the next decade the poet and his family struggled in near poverty; not until 1609 were he and his father-in-law reconciled. But from the perspective of immortality, Donne definitely had the last word. Sir George had published his own vision of sunrise, borrowing from the Psalms, a very different one from Donne's:

> Who seeth not the glorious arising of the Sun, his coming forth as a bride-groom out of his chamber, and his rejoicing like a mighty man to run his race? . . . The course of the Sun goeth round the Earth, and his light will have entrance, wheresoever the body of man can have passage.[20]

With 'The Sunne Rising' Donne achieved the ultimate poetic payback.

He was to write a number of other poems deploying the Sun as a metaphor, but he also wrote a satire, *Ignatius His Conclave*, which in the course of attacking the Jesuits (for Donne was a fervent Anglican convert from Rome) ridiculed many of the new scientific theories, and in particular the argument that the Earth circled the Sun. He was to express his views even more forcefully in verse:

> *And new Philosophy calls all in doubt,*
> *The Element of Fire is quite put out;*
> *The Sun is lost, and th' Earth, and no man's wit*
> *Can direct him well where to look for it.*[21]

This appeared in 1611, the same year in which Galileo reported the existence of spots on the Sun and valleys on the Moon.

By the 1690s Newton had explained the operation of gravity, the make-up of the Sun's rays and the origin of rainbows. The scientific revolution was in full flood, whether the world of poesy liked it or not, and certainly there were those in that world who did not. William Blake (1757–1827), for instance, had no use for science, and viewed its advances as one of the evils of the day. One poem mentions the Sun, but without any post-Copernican trappings:

Ah, sunflower, weary of time,
Who countest the steps of the Sun;
Seeking after that sweet golden clime
Where the traveller's journey is done . . .

This has been variously interpreted, the image usually taken to represent man's yearning to escape the transience of this world for the golden eternity of the next. Blake may also have been alluding to the Greek legend in which the sunflower was created when a woman 'pined away with desire' after the sun god, and so turned into his flower. But whatever meaning we finally assign, here was a major poet using the Sun for symbolic purposes that were quite separate from any scientific findings.

Samuel Coleridge (1772–1834) was a more typical man of his time. 'I shall attack Chemistry, like a Shark,' he wrote enthusiastically in 1800, on the occasion of his going into London to attend lectures on science, primarily to feast on a new stock of metaphors.[22] He, like many of his fellow Romantics, sought to break into science's magic circle, be it fashioned by mathematicians or physicists, chemists or biologists. Far from ignoring the new truths of cosmology, they wanted to understand them. Of those nineteenth-century poets writing in English, for example, Lord Byron (1788–1824) was vitally interested in scientific matters, especially in astronomy and geology; so too was Robert Browning (1812–89), who called Shelley 'the Sun-treader' and commended him for his 'Hymn to Apollo', with its radiant sun god. Percy Bysshe Shelley (1792–1822) did work as a chemist, introducing scientific material into such poems as 'Queen Mab' and 'Prometheus Unbound', while in his 'Hymn to Apollo' the god calls himself 'the eye with which the universe /Beholds itself and knows itself divine'. John Keats (1795–1821), who studied medicine, regularly employs astronomical images, his long poems *Hyperion* and *The Fall of Hyperion* both being named after the classical sun god.

Besides them, Emerson, Holmes, Poe, Tennyson, Arnold, Hardy ('The sun rested his chin upon the meadows'), Coventry Patmore and Charles Dickens (often in the context of the weather in general, but with particular grimness in *Bleak House* and the opening chapter of *Little Dorrit*) wrote on specifically solar themes. Emily Dickinson even composed a poem that poked fun at the post-Copernican understand-ing of the cosmos: 'The Earth upon an axis /Was once supposed to turn,' runs her 'Sic Transit Gloria Mundi', 'By way of a gymnastic /In honor of the sun!'

But the truth is that, for all their professed interest in science, from 1750 on, for almost 150 more years, when poets sat down to write they continued to prefer the safer pastures of the old myths. The Sun was still decked out as Apollo, Helios, Hyperion or Phoebus; its identity as a roil-ing ball of gases that was but one of many stars in the galaxy was not – yet – the stuff of poetry. But the new findings of science could not be ignored, and overall exerted what was thought to be a negative influ-ence. Even Macaulay, who was to become the very apostle of progress, accepted the decline of poetry as the unavoidable result of scientific rev-elations, while Hegel argued that as societies advance in rational achievement so they tend to lose those great skills that rest on imagina-tion. By the turn of the century *the* great scholar of the classical world's knowledge of the heavens was A. E. Housman (1859–1936). In old age, he wrote to *The Times Literary Supplement*: 'Quintilian says you will never understand the poets unless you learn astronomy.'[23] That may have been true of the ancients, Housman's field of study, but not of his fellow poets.

At least one of the old myths, however, has persisted through the cen-turies. The legend of the phoenix is a metaphor for solar rebirth, the Sun dying its daily death in the west and being reborn in the east, cross-ing the sky like a heaven-arching bird. Hesiod first assigned the phoenix its great longevity, and he also introduced the idea of cyclical time: as the bird was reborn, so history repeated itself. Until the seventeenth century, many believed that the phoenix really existed. Shakespeare (notably in 'The Phoenix and the Turtle'), Apollinaire (by his name making himself a son of the Sun), Byron, Nietzsche (who signed himself 'Phoenix' and wrote of himself as a creature reborn), Mallarmé and Yeats all drew on the myth. E. Nesbit, in her children's novel *The Phoenix and the Carpet*, makes her bird a creature of great vanity and considerable eccentricity. In more recent times Harry Potter encounters the phoenix Fawkes, owned by his headmaster Albus Dumbledore, while the X-Men mutant comic-book superheroine Jean Grey acquires the powerful Phoenix Force. Most notably, Cyrano de Bergerac's unfinished work *Histoire des estats et empires de la lune et du soleil* (c. 1661) recounts a journey whose narrator reaches the Empire of the Sun, where he meets a phoenix. The tale underlines Cyrano's belief that there is a burning soul within human beings that is in touch with the Sun, the great soul of the world.

Such a theory would have appealed strongly to D. H. Lawrence (1885–1930), whose poem 'Phoenix' explores just that idea. Lawrence brings the myth into his novels, and at one point he even adopted the bird as his personal symbol. He could be ridiculous and even repulsive (Bertrand Russell thought he heralded the Nazis, in his anti-Semitism and phallus-centred philosophy), but he did celebrate the Sun to memorable effect. His preoccupation began early (Nottingham, his birthplace, was a grimy coal-mining city) and endured throughout his life. The final page of his greatest novel, *Sons and Lovers* (1913), describes the night when Paul Morel comes to terms with having renounced his mistress and buried his mother:

> Everywhere the vastness and terror of the immense night which is roused and stirred for a brief while by the day, but which returns, and will remain at last eternal . . . Night, in which everything was lost, went reaching out, beyond stars and Sun. Stars and Sun, a few bright grains, went spinning round for terror and holding each other in embrace, there in a darkness that outpassed them all and left them tiny and daunted. So much, and himself, infinitesimal, at the core of nothingness, and yet not nothing . . . But no, he would not give in. Turning sharply, he walked towards the city's gold phosphorescence. He would not take that direction, to follow her. He walked towards the faintly humming, glowing town, quickly.[24]

Here, as always for Lawrence, the Sun is regenerative, its gold echoed in the phosphorescence of the city calling Morel back to life. Nowhere do we see this more clearly than in the short story 'Sun', a paean to its life-giving powers, which Harry Crosby's Black Sun Press published in 1928. It tells of a refined New York lady dispatched to the Sicilian coast to recuperate from post-partum depression. 'You know, Juliet,' her mother advises, 'the doctor told you to lie in the sun, without your clothes. Why don't you?' So she does, and soon the Sun is 'gradually

Aldous Huxley and D.H. Lawrence (1894–1963) – in their different ways
both devotees of the Sun

penetrating her to know her, in the cosmic carnal sense of the word',[25]
and is not to be denied. A local, about her age, suddenly appears at the
ravine close to her rented garden. Ah, the Lawrentian peasant . . .!

> Now the strange challenge of his eyes had held her, blue and over-
> whelming like the blue Sun's heart. And she had seen the fierce stirring
> of the phallus under his thin trousers: for her. And with his red face, and
> with his broad body, he was like the Sun to her, the Sun in its broad heat.

Juliet is loath to make the first approach, and her admirer waits endlessly on her initiative with 'the dogged passivity of the earth'. Into her
sunny retreat there, her husband suddenly arrives, the 'utterly sunless'
Maurice, a man who is 'the soul of gentle timidity in his human relations', so that Juliet 'being so sunned . . . could not see him, his
sunlessness was like nonentity'. But she realizes that it is her grey spouse
that she will have to settle for: 'she had not enough courage, she was not
free enough . . . ' Even though her lust remains unconsummated, this
story must rate as the purest expression of the erotic power of Sun
exposure in all literature.

But Lawrence was not finished. During Easter 1929 he saw in a Tuscan village shop window a toy white rooster bursting out of an egg, this inspiring *The Escaped Cock*, retitled against Lawrence's wishes *The Man Who Died* (1929), the story of a resurrected and sexually potent Son of God. The image of the lotus stands in for the female sex organs: 'No other flower . . . offers her soft, gold depths . . . to the penetration of the flooding, violet-dark Sun that has died and risen.' There is something wonderfully daft about the punning here, in which blasphemy is the least of the provocations:

He crouched to her, and he felt the blaze of his manhood and his power rise up in his loins, magnificent.
'I am risen!' Magnificent, blazing, indomitable in the depths of his loins, his own Sun dawned . . .

Shortly after this, Lawrence wrote 'The Middle Classes', a poem in which he describes the Maurices of this world as 'sunless. /They have only two measures: /mankind and money, /They have utterly no reference to the Sun'. But there is hardly a work by Lawrence that does not touch on these themes. In his last book, *Apocalypse* – written in the winter of 1929–30, when he was dying amid general condemnation of his works – he wrote:

Don't let us imagine we see the Sun as the old civilizations saw it. All we see is a scientific little luminary, dwindled to a ball of blazing gas. In the realities before Ezekiel and John, the Sun was still a magnificent reality, men drew forth from him strength and splendour, and gave him back homage and thanks.[26]

In this final work, Lawrence urged mankind to re-establish its primal connection to the cosmos – without the mediation of science, or what he saw as its reductiveness. 'Start with the Sun and the rest will slowly, steadily happen,' he ends the book.

An even more fervent apostle of the Sun (and not coincidentally a fan of Lawrence's and a fellow Sun worshipper) was a millionaire American, a hedonist poet of the 1920s who died in a suicide pact at thirty-one – 'a fugitive from a bad Scott Fitzgerald novel,' as one obituarist described him.[27] This is a theatrical way to re-introduce Harry Crosby, but everything about Crosby springs from the pages of melodrama. Both Malcolm Cowley, in *Exile's Return*, and Geoffrey Wolff, in *Black Sun*, wrote at length about him, while to Paul Fussell he was 'the ultimately mad American' whose life showed 'the power of the Sun to take over entirely a malleable mind'.[28] Crosby's own writings – avant-garde, experimental, surreal, deeply off-putting – were supported by great wealth, wholly inherited and effectively multiplied by his living in Europe when the franc was very weak. The money also allowed him to refurbish a medieval mill (inevitably, 'Le Moulin du Soleil') outside Paris into a luxurious country house, travel to exotic destinations whenever he wished, experiment with photography, push his Bugatti to its limits and even learn to fly solo at a time when the airplane was 'a gadget so new . . . that no one had agreed on its spelling'.

Descended from old New England money (J. P. Morgan was an uncle), he enlisted in the Field Service Ambulance Corps early in the First World War, serving at Verdun and on the Somme, receiving the Croix de Guerre and surviving a wound that should have killed him. After an 'accelerated' degree at Harvard and a scandalous marriage to a woman six years his senior whom he induced to leave her husband, he settled in France, buying racehorses, sampling opium, drinking to excess and travelling. And he started to write. In 1927 he launched the Black Sun Press, had a solar face tattooed on his back as an expression of devotion and elicited his first short story from Lawrence, whom he paid in 'sunny' twenty-dollar gold pieces.

Crosby was developing an obsessive interest in imagery centred on the Sun, which he introduced into his own writing with a vengeance. The 'black sun' symbolized his attempt to unify the forces of life and death, but as a visual design it also carried sexual significance. 'Every doodle of a "black sun" that Crosby added to his signature,' writes one biographer, 'also includes an arrow, jutting upward from the "y" in Crosby's last name and aiming toward the center of the sun's circle: a phallic thrust received by a welcoming erogenous zone.'[29]

Crosby comes across as almost totally unpleasant, yet he is not so far off in his views from Lawrence's cri de coeur in *Apocalypse*:

What man most passionately wants is his living wholeness and his living unison . . . The magnificent here and now of life in the flesh is ours, and ours alone, and ours only for a time. We ought to dance with rapture that we should be alive and in the flesh, and part of the living, incarnate cosmos.

By the mid-1920s Crosby had published an elaborate diary of the years 1922–6, 'Shadows of the Sun', as well as two volumes of poetry, the first entitled *Chariot of the Sun*. Other Black Sun projects included work by Hart Crane, James Joyce, T. S. Eliot and Man Ray. As a spotter of talent, Crosby was not without gifts.*

By now he was interested in the philosophical applications of suicide, and indeed had been openly obsessing about death for at least five years ('Time is a tyranny to be abolished.'). One last outré venture was called for. His second book of poetry, *Transit to Venus*, had been inspired by Josephine Rotch (newly Mrs Albert Smith Bigelow), the woman with whom he would spend the last week of his life. On 9 December 1929, the day before they died, Josephine wrote Crosby a passionate verse letter that ends by insisting 'that the Sun is *our* God / and that death is *our* marriage'. He was found with a .25-calibre gunshot in his right temple; she had been shot in the left and died first.

☼

During and after the years in which Crosby was gallivanting around his adopted country, the literary world was at the height of its love affair with the Sun. A host of writers now reclaimed it, but without the heavily philosophical trappings. The new message was: the Sun gets into *everything*. W. H. Auden's 'Take Icarus, for instance' from '1 September, 1939', about the boy who flew too close to its fire, must be one of the

---

*The Black Sun has other connotations. Russia's great symbolist poet Osip Mandelstam (1891–1938) records: 'I have recalled the image of Pushkin's burial [Pushkin fell in a duel in 1837] to arouse in your memory the image of a night Sun.'[30] This language recurs throughout Mandelstam's *Tristia* (1922), with such lines as ' . . . for the mother in love /The Black Sun will rise'. The Sun's role seems to vary, finally linking with the Christian apocalyptic tradition. The 'Black Sun' image also connects with Mandelstam's ambivalence about the Bolshevik Revolution, which he wanted to endorse but which seemed to be bringing in darkness, not light; it not only represents the death of specific people but becomes a symbol of Russia itself – the Sun as a force of despair.

most famous lines about the Sun in literature. Yeats too, in 'The Song of Wandering Aengus' – 'The silver apples of the Moon /The golden apples of the Sun' – 'Lines Written in Dejection' and, unforgettably, 'The Second Coming' – 'A gaze blank and pitiless as the Sun' – invoked the star; while in the realm of light verse pride of place goes to John Betjeman's 'A Subaltern's Love Song' (1941), with its unforgettable lines 'Miss J. Hunter Dunn, Miss J. Hunter Dunn /Furnish'd and burnish'd by Aldershot sun'. Of all Betjeman's poems, none so successfully converted suburban trivia into hilarious, helter-skelter eroticism.[31] Maybe, as Josef Brodsky has said, 'all the best poets are solar-powered'.*

The literature of almost every culture contains its share of solar allusions. Arthur Rimbaud (1854–91) and Paul Verlaine (1844–96), for

---

*Among novels offering solar passages of note are *Moby Dick* (dying sperm whales turn their heads towards the Sun)[32] and *The Wind in the Willows*, in which Rat and Mole are granted an epiphany at sunrise. Playwrights too have recorded the Sun as an overwhelming presence, from Tennessee Williams to Peter Shaffer, whose *Royal Hunt of the Sun* (1964) attempted to recapture the world of the Inca and the Spanish Conquest – and whose *Equus* would have delighted Lawrence. *Hyperion* (1797–9), the first novel by the German lyric poet Friedrich Hölderlin, is in large part an analysis of the Sun's role in Western philosophy. Colette (1873–1954) reveals in *Sido* that during her childhood her mother would reward good behaviour by letting her see the dawn.

The division into poets and novelists is an artificial one, but the list of fiction writers would continue: John O'Hara ('Against the Game'), William Faulkner (*Light in August* and 'That Evening Sun'), Patrick White and Doris Lessing (both particularly good on tropical relentlessness), Elizabeth Bowen (*A World of Love*), William Golding (*The Scorpion God*), Frank O'Hara, Garrison Keillor (both *Wobegon Boy* and his parody of 'Casey at the Bat'), John Updike (stories such as 'Deaths of Distant Friends' and 'Leaves'), Norman Rush (*Mating* is about a Utopian community in the Kalahari that survives on solar power) and J. G. Ballard (especially *Empire of the Sun*). Among poets one adds: Dylan Thomas, in *Under Milk Wood*, Matthew Arnold ('Is it so small a thing /To have enjoy'd the Sun'), Walt Whitman ('Give Me the Splendid Silent Sun'), Ezra Pound, Louis MacNeice ('The Sunlight on the Garden'), Wilfred Owen ('Exposure', 'Futility'), Wallace Stevens ('Sunday Morning'), Philip Larkin ('Solar' and 'High Windows'), Thom Gunn ('Sunlight'), Richard Eberhart ('This Fevers Me'), Simon Armitage (whose collection *Tyrannosaurus Rex Versus the Corduroy Kid* offers light as a metaphor for poetry), Updike again ('Seven Stanzas at Easter', 'Cosmic Gall'), Lisa Jarnot (*Ring of Fire*) and Amy Clampitt (particularly 'The Sun Underfoot Among the Sundews', 'A Baroque Sunburst', 'What the Light Was Like' and 'Winchester: the Autumn Equinox'). Deliberately omitted here, mainly for reasons of space, is the tradition of the 'pastoral', in which the writer looks back to an idealized past, when either it is forever summer, or the characters are bathed in sympathetic light. Novelists such as Evelyn Waugh (in *Brideshead Revisited*), L. P. Hartley (*The Go-Between*), George Orwell (*Coming Up for Air*) and, supremely, Marcel Proust have all contributed to this form of solar reverence, but it is an indirect form of homage. There also exists a host of bad writing about the Sun; one has some sympathy with the bartender in the American TV series *Cheers* who picks up his girlfriend's book and remarks, '*The Sun Also Rises* – well, that's real profound.'

instance, shared an ambivalence about the Sun – as the Verlaine scholar Martin Sorrell comments: 'Where Rimbaud's quest for a kind of cosmic revelation made him worship the Sun's power to absorb and take him over, Verlaine's poems . . . reveal almost an equal fascination with the Sun, but as a star whose force is shrouded, guarded, veiled and intangible. In a fascinating way, the two poets' concern with the Sun is the two sides of one coin.'[33]

For Rimbaud, who produced nearly all his best work before he was twenty, the Sun was either the unmasker of social evils or the bringer of misery. 'The hearth of affection and life, it pours burning love on the delighted earth,' he wrote, but his last years were spent in the desert heat of East Africa, which seared him into a decidedly less receptive frame of mind. 'I saw the low Sun, stained with mystic horrors,' he confided wretchedly to a friend. 'We're in the steam-baths of springtime. Skins are dripping, stomachs turning sour, brains becoming muddled.'[34] But in the end the Sun 'which shines like a scoured cauldron'[35] won through, and on his deathbed he wept to his then companion, Isobel, that never again would he feel its light. 'I shall go under the earth, and you shall walk in the Sun.'

When Norman Mailer was courting Barbara, his sixth wife, he wrote to her: 'And you, stand up lady, are golden as the sun.' She later replied in verse: 'You were there and/ I was there/ in a pocket of sunshine . . .' *The New York Times* commented: 'These two lived large, sun-drenched lives in almost every regard'

Thinly veiled autobiography continued with the novels of André Gide (1869–1951), whose first book, *The Immoralist* (1902) is narrated by Michel, a scholar who spends his early years in northern France. He marries and honeymoons in North Africa – and finds his world changed. 'I came to think it a very astonishing thing to be alive, that every day shone for me.' As it did for Lawrence, the Sun becomes a liberating power. Leaving his bride behind, Michel embarks on a Mediterranean adventure. Immediately drawn to the 'beautiful, brown, sunburned skins' of Italian boys, he is 'shamed to tears' at his own whiteness and, opening his body to the Sun, sets about his transformation: 'I exposed my whole body to its flame. I sat down, lay down, turned myself about . . . Soon a delicious burning enveloped me: my whole being surged up into my skin.' Freed from restraint, he becomes the brown immoralist of the novel's title. Oscar Wilde had once mentioned to Gide that the Sun hates or deters thought, and sure enough Michel is soon concluding that 'existing is occupation enough'.[36]

In *The Outsider* (1944), Albert Camus has his protagonist Meursault kill a North African on a blazingly hot beach, apparently for no reason – that being the point, and 'Unreason' the name, that Camus attached to his sun-blasted Algerian experiences. Declaiming this philosophy in *The Rebel* (1953), he wrote: 'Historic absolutism, despite its triumphs, has never ceased to come into collision with an irrepressible demand of human nature of which the Mediterranean, where intelligence is intimately related to the blinding light of the Sun, guards the secret.' When Sartre was asked if Camus, a close friend, was also an existentialist, he replied: 'No, that's a grave misconception . . . I would call his pessimism "solar", if you remember how much black there is in the Sun.' Camus' own comment was: 'To correct a natural indifference I was placed halfway between misery and the Sun. Misery kept me from believing that all was well under the Sun, and the Sun taught me that history wasn't everything.'

What becomes clear is that by the mid-twentieth century the Sun could serve almost any literary purpose: symbol, metaphor, inspiration, dramatic force, intimate companion, intransigent adversary, comic butt, tragic endgame, source of redemption, or source of philosophical belief. *Lolita* is so remarkable because Nabokov makes use of the Sun in nearly all these ways.

And still the literary exploitation of the Sun goes on. In 2008 the American novelist Elizabeth Strout published *Olive Kitteridge*, thirteen interconnected stories about a retired teacher in the (fictional) coastal town of Crosby, Maine: it won a Pulitzer Prize. An extraordinarily perceptive narrative about love and acceptance, it does not use the Sun metaphorically in the way that Nabokov does but as a presence – or absence – that affects everyone in Crosby, particularly the prickly, truculent Mrs Kitteridge. Strout can be wonderfully observant about, say, 'the abandonment of a sunbather', or light emerging through the fog, or making a red glass on a bureau glow.[37] A strip of wooden floor shines 'like honey' under 'a block of winter sun', while later the Sun lingers 'across a snow-covered field and made it violet in color'. But it is more than that: in a short book, there are seventy solar references, with the Sun acting as a guide to character and to what characters should do. By the novel's end, Olive, newly widowed, suddenly finds a local widower who seems to love her. Will she seize her chance? She is about to turn away when the Sun gives her 'a sudden surging greediness for life . . . She remembered what hope was, and this was it . . . She pictured the sunny room, the sun-washed wall, the bayberry outside. It baffled her, the world. She did not want to leave it yet.'

28

THE RISING STAR OF POLITICS

If we allow ourselves . . . a glimpse into that shadowy place we call our soul, isn't that why we're here now . . . ? The two of us? Looking for a way back? Into the Sun?

Richard Nixon, in Peter Morgan's play *Frost/Nixon*[1]

The light was here before. Why should it be dark again? We had the morning. We must act now as though the light has come again.

Alexander Dubček, welcoming the Velvet Revolution in an address to the crowds in Wenceslas Square in 1989

During his time at Yale, George W. Bush received one D – in astronomy.[2] On his being elected President, the following tale went round: one day, keen to improve his standing in the polls, he told his advisers his new idea. 'We'll land on the Sun.' It is gently pointed out that the Sun is awfully hot. The President considers this information. 'No problem,' he says. 'We'll land at night.' Only this joke did not start with Bush; a decade earlier, it was told about Mikhail Gorbachev, and before that Nikita Khrushchev.*

---

*Jokes combining the Sun and politicians are, unsurprisingly, thin on the ground, but the award-winning German film *The Lives of Others* (2006) successfully recycles one that made the rounds in the old East Germany. In the film, a junior official ill-advisedly tells his superior officer the following tale (for which he is later demoted: there is no such thing as a harmless joke): Erich Honecker (who led the GDR from 1971 to 1989) wakes up, throws open his doors and welcomes the rising Sun with a hearty 'Good morning!' 'Good morning, dear Erich,' the Sun replies. After lunch he greets the Sun again. 'Good day, dear

No. 10 Downing Street, complete with a symbolic rising sun

Throughout history, the Sun has been appropriated for political or religio-political effect, and hardly a generation has failed to provide examples. In modern times, Sri Lanka has witnessed the rise of a leader of the 'Liberation Tigers' who styles himself a solar deity. Sendero Luminoso (the Shining Path) a terrorist offshoot of the Peruvian Communist Party, was set up in the early 1990s by a scrofulous professor of philosophy who had to keep out of the Sun for health reasons. While Louis XIV was the most extravagant in conscripting the Sun, the star

Erich,' it replies. At bedtime he bids the Sun goodnight. 'Screw you, Erich' it answers. 'I'm in the West now.' Political cartoonists have also picked up on the Sun to make their points. Ralph Steadman, for instance, records his efforts during Ted Heath's election campaign against Harold Wilson in 1970: '. . . I followed with a desert landscape showing a sun burning down on the figure of Edward Heath dying of thirst. In the sun was written "A Better Tomorrow" and foolishly I thought that would nail him. It didn't, of course, because he won.' Steadman followed that with another landscape called 'Early Morning Scene' showing Heath in his pyjamas watching the Sun come up with Harold Wilson's face on it.[3]

lived on after the French Revolution as a potent symbol of a new social order – as seen in the work of a radical aristocrat, Constantin-François de Chasseboeuf, Comte de Volney (1757–1820), a member of the new National Constituent Assembly. In *Les Ruines, ou méditations sur les révolutions des empires*, part novel and part tract, de Volney urged his readers to replace the remnants of a decayed civilization with a world order of reason and equality whose dominant symbol would be not Christ but the Sun. The work was banned in France, guaranteeing it cult status and numerous foreign editions. When in 1792 an anonymous translation was published in Britain thirty-five Anglican prelates condemned it. No less a figure than Thomas Jefferson, who had befriended Volney while in France, translated the first twenty chapters for an American edition, only to lose his nerve and ask that the manuscript be burned.

The book (in several volumes) charges that Christianity had purloined solar beliefs, and urges a return to the Sun as the prime symbol of self-determination – ideas that would find their way into a wide range of works. In the nineteenth century many artists and thinkers, inspired by the Romantic movement and the rise of nationalism, proclaimed that man would achieve his ultimate fulfilment by breaking free from the confines of religion. In Britain, the 'Sun-as-God' theory became a staple of radical pantheism, and the solar nature of all deities was preached up and down the country. 'The history of the Sun is the history of Jesus Christ,' declared one typical firebrand, Godfrey Higgins (1772–1833)[4] – by which he did not mean that the Sun 'equalled' Christ, but rather that Christianity had appropriated solar symbolism. By rejecting that secondary faith, human beings could rediscover what was most of value in themselves.

In the last quarter of that century, no one was more influential (Darwin excepted) than the great philologist-turned-philosopher Friedrich Nietzsche (1844–1900). In the winter of 1876–7 Nietzsche, then a sickly scholar of thirty-two, travelled to Sorrento, south of Naples, where the sensual culture inspired him to put the human body at the centre of his thinking. 'I shook off nine years of moss,' he enthused after visiting a cave that housed a relief portraying Mithras, the Persian sun god.*

*Mithraism (from the Persian for 'light of day') is a recent term: in antiquity its Roman adherents referred to their religion as 'the secret doctrines of the Persians'. The cult reached Rome around the first century AD, with its apogee in the fourth, finding particular

During and after his time in Italy, he would write poems praising the Sun as his lover. In *Ecce Homo*, his last book before insanity overtook him, he writes: 'On this perfect day when everything is ripening and not only the grape turns brown, the eye of the Sun just fell upon life: I looked back, I looked forward, and never saw so many good things at once.'[6]

By the time he wrote *The Gay Science* (1882), Nietzsche was arguing that the Sun's rise and culmination at its zenith be employed as symbols for what men must become.[7] (This was a little disingenuous, since his migraines often forced him to hide away in the shadows.)[8] In *Thus Spake Zarathustra* (1885), he introduced the *Übermensch*, the superman who has attained self-mastery and self-direction, all of which he linked to the Sun.

These books were part of a general movement of artists, musicians and writers who lauded the Sun, Nietzsche's contemporary, the painter Edvard Munch (1863–1944), making a vital contribution. A member of the radical Naturalist movement in his native Kristiania (as Oslo was then known), he was invited in 1892 to mount an exhibition in Berlin and remained there for three years. His friends included Henrik Ibsen, whose portrait he painted on several occasions, the first time in 1897 as a poster for a production of *John Gabriel Borkman*. The playwright's powerful head dominates the picture, while on the right a lighthouse casts its beams into space. The symbolism is forthright: Ibsen is the light-bearer, irradiating the hidden realms of human experience. Munch also completed a series of sketches for *Ghosts*, whose protagonist, the painter Osvald Alving, yearns for life yet lies under the fatal curse of inherited syphilis. Munch, a fellow sufferer, saw his own feelings reflected in Osvald's yearning for light and redemption, and heard in Osvald's cry 'Give me the Sun!' the voice of his own torn spirit. He described his painting *Spring* as 'the craving of a mortally ill person for light and warmth, for life. The Sun . . . in *Spring* was the sunshine in the window.

---

support among soldiers. Ceremonies centred on a *mithraeum*, an adapted natural cave or an imitation, arranged as an 'image of the universe'. The faithful were divided into seven ranks, the sixth of which was that of 'heliodromus', or sun courier, and the cult connected strongly with the Sun, whose passage from solstice to solstice was seen as paralleling the soul's journey from pre-existence, briefly into the body and beyond into an afterlife. However, Mithraism quickly declined, as Christianity began to assimilate the pagan deities. Apollo the sun god was seen as a precursor of Christ, while Apollo the bringer of sudden illumination was transformed into the Holy Spirit.[5]

It was Osvald's Sun.'[9] By this time he had come to believe that his art should aspire to 'the heaven of the Sun's realm'.

While in Berlin he also met the leading German interpreters of Nietzsche, and in 1905–6 painted the great man's demonic sister Elizabeth. He collaborated with another friend and Nietzsche intoxicate, August Strindberg, on a series of paintings and poems that took over an entire issue of the magazine *Quickborn*, that was devoted to glorifying the Sun. Strindberg even bragged about having exchanged books with the philosopher: 'Nietzsche has shot such a monstrous load of semen into my spiritual life that I feel I have the belly of a whore. Nietzsche is my spouse!'[10]

In 1909 Munch was commissioned to paint a cycle of eleven murals for the University of Oslo's Festival Hall. His first idea, *The Mountain of Mankind*, an image taken directly from *Thus Spake Zarathustra*, was rejected, so he replaced it with the monumental *The Sun*, also inspired by Nietzsche, in which the main painting is flanked by panels showing how the Sun burns at the core of all forms of creativity. In an undated entry in his notebook, he elaborated: on one side of the hall, the rays become primitive invisible forces, on the other, intellectual impulses –

> The first pair [of panels] are overflowing with light – It travels into the bodies – and in and out of the crystal . . . There is light that travels like X-rays – the other side painting – Chemistry – represents the hidden energies – the workplace of fire and warmth – On the other side the Sun sends its rays to even greater distances . . .[11]

It took seven years to finish, the completed panels showing a tousle-headed creator god sitting before the Sun to which he has given life. Munch's most recent biographer describes how 'the enormous burning globe is so Sun-like that your eyes instinctively flinch away from the central white circle, just as you avoid directly looking at the original'.[12] On seeing the painting, Richard Strauss, author of his own homage to Nietzsche, the tone poem *Also sprach Zarathustra*, exclaimed that it corresponded exactly with what he was trying to achieve in music.

In 1893 Munch completed his masterpiece, *The Scream*, later recording how inspiration had struck him after a sunset walk, high up a hill to the east of Oslo. The city's main slaughterhouse stood there, as did its madhouse, and the shrieks of the animals being butchered combined with the howling of the insane into a cacophony terrible to hear. Munch walked on, across the bridge leading into the city. At that moment:

> The Sun was setting. I felt a breath of melancholy. Suddenly the sky turned blood-red. I stopped and leaned against the railing . . . stood there, trembling with fear. And I sensed a great, infinite scream pass through nature.[13]

It is impossible to know how much of an influence Munch's paintings or Nietzsche's books had on the general bubbling up of solar-based philosophies of the time, but it has to have been considerable. Music also had its part to play, not just Wagner's operas or Strauss's tone poem but also the erotic compositions of Alexander Scriabin, whose fourth and most famous symphony, composed in 1905, was entitled 'Orgiastic (or Orgasmic) Poem' – yet from the outset he saw it as a work about lust not just for the flesh but for the Sun, telling a friend, 'When you listen to it, look straight into the Sun's eye.'

Like Munch and Strindberg, Scriabin was drawn to Nietzsche, and took the hero of his third sonata directly from the Superman. His fifth and last symphony, 'Prometheus: The Poem of Fire', written between 1909 and 1910, also portrays a superman hero. Scriabin instructed that the finale should open in a completely dark auditorium, gradually to be pervaded by light as Prometheus's quest ends with the Sun.

Nietzsche, Munch, Strindberg and Scriabin were part of a movement of artists and thinkers all proclaiming a similar message: the immanence of the Superman, the emblematic importance of the Sun, and man's will to power. Throw in the mysticism of George Ivanovich Gurdjieff (c. 1872–1949), a distorted interpretation of the already dubious doctrine of the 'survival of the fittest', Wagner's notions about the outer reaches of art and the philosophical tracts of Schopenhauer, and where was it all leading? – inevitably to the Nazis. But one must add some other ingredients.

The second half of the nineteenth century saw the rise of spiritualism and occultism. Astrologers wrote of the Age of Aquarius, the notion that a new epoch would be inaugurated when the Sun rose in Aquarius on 21 March (the year was left vague), to herald Christ's second coming. Then there was *The Secret Doctrine* (1888) by the larger-than-life Madame Blavatsky, founder of the Theosophical Society, which made a deep impression in Germany and elsewhere in Europe. Her admirers included Mohandas Gandhi, Alfred Kinsey, Rudolf Steiner, Aleister Crowley, James Joyce and William Butler Yeats (who described her as 'the most human person alive').*

'In the shoreless oceans of space,' she wrote, 'radiates the central, spiritual and *invisible* sun.' Our inner beings emanate from the 'eternal central sun', to be reabsorbed by it at the end of time.[15] The Sun was the lens through which the spiritual light of the sun unseeable (i.e. God) reaches our senses. Her theories were elaborated by the British writer Gerald Massey (1828–1907), who attacked Christianity for turning the Sun from a symbol of life into one of death. Another devotee, Alan Leo (1860–1917), moved the Sun to the centre of astrological interpretation, almost single-handedly reviving that most dubious activity after its sharp decline in the seventeenth century.

The final ingredient in the cauldron is the *Völkisch* movement. Even before the advent of Nietzsche, Wagner, Munch et al., the northern and central German states, which were finally unified in 1871, were aflame with pride in the ancient German archetype, as described by the Roman historian Tacitus in his great pioneering work of ethnography, *Germania*:

> Personally I associate myself with the opinions of those who hold that in the peoples of Germany there has been given to the world a race unmixed by intermarriage with other races, a peculiar people and

---

*The Ukrainian aristocrat, adventuress, supposed spy and undoubted occultist Hélèna Petrovna Hahn, as she was born (1831–91), was married off at seventeen to a provincial vice-governor of forty. Stealing a horse, she rode over mountain ranges back to her grandfather in Tbilisi. Within weeks she had married the skipper of an English tramp ship taking cargo in Odessa, and embarked on a decade of travel. She smoked up to 200 cigarettes a day, ate like a horse and swore like a trooper. While in Cairo in 1871, she formed a society for the occult. Her next port of call was New York, where she was reputed to have demonstrated physical and mental powers that included levitation, clairvoyance, clairaudience, out-of-body projection, telepathy, and materialization (producing physical objects out of nothing). In 1875 she founded the Theosophical Society, a specimen of New Age thinking that bundled all religions into one, plagiarizing from each. (She had also read Volney.) She moved the society first to India, then London, where she presided over a range of disciples. As a biographer wrote, the world debated furiously whether she was 'a genius, a consummate fraud, or simply a lunatic . . . An excellent case could have been made for any of the three.'[14]

pure, like no one but themselves, whence it comes that their physique, so far as can be said with their vast numbers, is identical: fierce blue eyes, red hair, tall frames, powerful.[16]

The self-styled *Volk* gloried in Tacitus's descriptions of their ancient woodland forebears who dwelt among babbling brooks and picturesque mountains, and, from the 1870s on, the notion of the *Volk* gained currency. It was a wide church. As Simon Schama notes, it embraced the *Wandervögel* youth movement as well as the Ramblers, 'who communed, Siegfried-style, around bonfires on forested hills', and also many who would later ardently support the Third Reich.[17] But at the outset this was all still fairly innocent.

Völkisch enthusiasms included nudism (as we have seen, a form of sun worship), as extolled in the aptly named Heinrich Pudor's *Naked Mankind: A Leap Into the Future* (1893), which also happened to be brazenly anti-Semitic. Nietzsche was particularly close to the *Volk* tradition, which he treated with surprising respect. It was reciprocated. 'Taking their inspirational cue from Nietzsche's proclamation that "God is dead",' writes Richard Noll, '[the *Volk*] created their own forms of personal religion.'[18] Occultism, the ideas of ancient theocracies, secret societies and the mystical apparatus of Rosicrucianism, Cabbalism and Freemasonry were all woven in.[19] Whatever worked.

The Sun was about to be moved centre stage. By the 1890s many *Volk* believed it was 'the sole God of the true Germans',[20] and began resurrecting pagan holidays to replace Christian holy days. Noll notes:

Perhaps the most central neo-pagan element in German *völkisch* movements was Sun worship [which] was extolled as true ancient Teutonic religion, and while it was primarily a literary device and a powerful rhetorical metaphor for the experience of God, actual solar-worship rituals did take place . . .[21]

After Germany's defeat in the First World War, a number of *völkisch* movements were reorganized along propagandizing lines, and members became ever more viciously chauvinistic; *völkisch* thinkers began to contrast Germany, as they saw it a nation of heroes, with Britain, 'an island of tradesmen', which epitomized the degeneracy of modern materialism. Hitler, who allied himself with the movement – 'the basic ideas of the Nationalist-Socialist movement are populist (*völkisch*),' he wrote in *Mein Kampf* – also embraced solar imagery in the form of the swastika,* which, on 7 August 1920, was duly adopted as the emblem of the Nazi Party.

Once the Nazis were in power, they banned several traditional Christian holidays and substituted others more appropriate to the 'new Germany'. South-west of Hanover stand the Externsteine rocks, also known as 'the Rocks of the Sun'. Four weathered limestone pillars dating back 70 million years rise a hundred feet, topped by a Neolithic observatory whose round window lines up precisely with dawn at the summer solstice. Hitler, cherishing anything that might glorify the Teutonic race's ancient forebears, was enthusiastic about the place, and the SS promoted it as the old Germanics' most sacred site, holding feasts, weddings and Hitler Youth ceremonies there at equinoxes and solstices.

All these beliefs required intellectual buttressing, and the National Socialists cast about for suitable candidates. In the First World War, 150,000 copies of *Thus Spake Zarathustra* had been printed for the Kaiser's army; the Nazis now distributed it to the Hitler Youth, and in 1934 laid a de luxe copy in the Tannenberg Memorial (which honoured German soldiers who had fallen at the great victory there in 1914). To their delight, they discovered that the glories of 'natural' paganism had also been propounded by Goethe himself, who just eleven days before he died had written:

---

*The swastika (from the Sanskrit 'su' meaning 'good' and 'asti' meaning 'to be', with the suffix 'ka') has been used in cultures around the world (excepting those south of the Sahara and in Australia). The ancient peoples of the Indus Valley believed the Sun to be square, and conceived of the swastika as its symbol, its bent arms representing the solar wheel on its yearly course through the sky. Madame Blavatsky adopted it as part of the insignia of the Theosophical Society and of her personal crest.[22] In Ireland, farmers affixed swastikas, called 'Brigid's crosses', on their doors. During the First World War swastikas were sewn on to the shoulder patches of the American 45th Division and even after the Second could be found on Finnish air force uniforms. All these societies saw the right-handed (clockwise) swastika as an expression of life and good fortune. The left-handed swastika, on the other hand, represents Kali, the terrifying dark goddess of death, and also has black-magical significance. The Nazis adopted both clockwise and anticlockwise versions, although Hitler is said to have favoured the latter. Its popularity in Germany had begun in the 1870s, when the archaeologist Heinrich Schliemann found many swastikas on his digs at ancient

If I am asked whether it be in my nature to give [Christ] adoring reverence, then I say, 'Completely!' I bow before him as the divine revelation of the highest principle of morality. And if you ask me whether it be in my nature to worship the Sun, then I say again, 'Completely!' For it, likewise, is a revelation of the most high, and to be sure the most powerful ever granted us mortals to perceive.[24]*

Not only Goethe and Nietzsche were at hand. From 1901 Carl Jung had expressed enthusiasm for *völkisch* ideas, his interest in them roused by a patient who believed that the Sun had a gigantic phallus that could control the weather. Jung noted similarities between his patient's delusions and pagan sun worship, and offered them as proofs of the existence of what he called a collective unconscious. During late 1909 and 1910 he became fascinated by the literature on Zoroastrianism and Mithraism and their roots in ancient Iranian solar worship, and developed a 'Sun as God' theme, adding 'libido' and 'hero' to the existing list of solar correspondences (light, God, father, fire, heat).[26]

In 1912 Jung published *Wandlungen und Symbole der Libido* (*The Psychology of the Unconscious*), which Noll calls 'the *Völkisch* Liturgy . . . A modern mystical contribution to the solar mythology of Müller'.[27] For in it, what should the great Swiss analyst do but resurrect the writings of that solar faithful Max Müller, whose many books and articles could now be seen to fit conveniently with Jung's own preoccupations. 'Is everything the Dawn? Is everything the Sun?' Müller had pondered.

Troy and Mycenae and wrote about them in two best-selling accounts of his discoveries. Many anti-Semitic and militaristic groups had started using the swastika by the time that it was adopted by the National Socialists at Nuremberg in 1935. Usually termed *Hakenkreuz* – hooked cross – or Thorshamarr, it became the official emblem of the German Gymnasts League, while solar disc symbols began to appear on posters, armbands, banners, book and magazine covers and elsewhere throughout Central Europe. (A German Jew, buying toiletries in 1933, noted that even the toothpaste tube was so decorated.)[23]

*While still a young man, Goethe decided that the greatest symbol of God was the Sun, and set himself to perform a fire sacrifice, whose flames must be kindled directly from the star. To make the ritual more fundamental, he chose not to use wood or coal, selecting instead incense sticks, because, he said, 'this gentle burning and evaporating seemed much better to express what is happening in the mind than an open flame'. He fastened some sticks to a small table, kindling them with a magnifying glass as soon as the Sun appeared over the roofs of the neighbouring houses. In his absorption he failed to notice that the incense sticks had burned down, to destroy the fine flowers painted on the family table.[25]

'This question I had asked myself many times before it was addressed to me by others . . .'[28] And here was Jung writing of exactly these concerns – of the hero as Sun, the Sun as god, the god as self-immolating deity. Müller was resurrected, to become the poster professor of the new Germany.*

Müller's writings presented the Nazis with much more than an opportunity to build up solar mythology, because, of course, there were *two* prongs to his thought: the centrality of the Sun in language and myth, and the supreme place occupied by Aryan civilization. He himself may never have been a racist, but he was the perfect advocate for the new dispensation, and his work on Aryan origins was soon being quoted to bolster Nazi claims for a *scientific* basis to anti-Semitism.

The appropriation of the Sun manifested itself in several ways, but the most malevolent and far-fetched was the fantasy created by Heinrich Himmler, *Reichsführer SS*. It was he who spearheaded the quest for the ancestral heritage and pagan culture that had been suppressed by the Christian Church. It was also he who embraced wholeheartedly the doctrine of the Black Sun† – *Schwarze Sonne* – which reaches back to ancient Sumer and Akkadia, and holds that the Sun manifests in two powers, one 'white', at the centre of our planetary system, the other 'black', hidden yet spiritually illuminating, 'the strongest and most visible expression of God'. This unseen or burned-out Sun, a source of mystical energy capable of regenerating the Aryan race, has obvious affinities to the Sun in Madame Blavatsky's Theosophy (*The Secret Doctrine* was first published in German in 1901), to Jung's 'inner Sun' and to the contrasting concepts of Noumenon and Phenomenon posited by Schopenhauer. The term *Schwarze Sonne* was also to become a central

---

*In contrast, Rudolf Steiner, whose schools were shut down by the Nazis, referred to Jews as *Der Sonnenwiese* – the sun beings – not a racist slur but part of a personal philosophy, 'anthroposophy', that saw Christ as the ultimate sun being. Steiner's system, dating from 1913, was based on the premise that the soul can contact the spiritual world. The concepts of reincarnation and karma were central to it, as was the concept that Christ was a cosmic force, a 'Sun Being' who became incarnate at the turning point of man's spiritual evolution. Anthroposophy was condemned by the Catholic Church in 1919.

†The Black Sun has recently been revived as a rallying point for right-wing extremists. In Austria and Germany, a former SS man, Wilhelm Landig (1909–97), appropriated the Black Sun symbol, swastikas and other images from the *völkisch* movement to become by the late 1980s 'a major political icon of opposition to democracy and liberalism in the West'.[29]

When Charlie Chaplin satirized Adolf Hitler in *The Great Dictator*, playing the demented Adenoid Hynkel of Tomania, he held court behind a vast desk, above which was an even larger image of two black crosses in a circle, representing the party of the 'Sons and Daughters of the Double Cross'. From this central symbol burst forth the rays of the Sun. Released in 1940, the film was remarkably prescient

element in the secret initiation of Hitler's senior SS generals. For the rank and file of the SS (about 50,000 men), the initials 'SS' stood for *Schutzstaffel,* meaning a special staff or military unit; but for initiates it signified *die Schwarze Sonne.*

☼

Japan did not need the insights of a Jung or a Goethe to see the Sun's political possibilities. From earliest times its myths had rested on its founding by the Sun goddess Amaterasu and the divinity of its *Tenno* (or heavenly emperor), believed to be a direct descendant of the Sun: 'even as the Sun is the centre of the universe so is the Imperial Dynasty the centre of Japanese race life'.[30] Japan's flag, which has for over a thousand years displayed the rising Sun, affirms the nation's sense of its solar ancestry, as does Shinto, the state religion. With the outbreak of the Second World War, the solar image became the nation's most visible symbol: soldiers departed for battle wearing the national flag tied diagonally across their chests and headbands emblazoned with the rising Sun. Kamikaze pilots were similarly inspired by the god-emperor's bond with the Sun.

510 CHASING THE SUN

The link between sun goddess, emperor and Japan's predatory ambitions of the 1930s was eloquently described by the famous American reporter John Gunther, who wrote just after the war: 'The legend of [the emperor's] lineal descent from the Sun goddess and consequent "divinity" . . . made the emperor sacrosanct and united the nation in a blindly intense feeling not merely of loyalty but of actual kinship to the imperial house.' Following Japan's surrender in 1945, the commander of the Allied occupying forces, General Douglas MacArthur, set on demystifying the dynasty, told Emperor Hirohito that he must declare to his people that he was not a god. This was no idle request – as Gunther comments:

> Think back. Before the war a traffic cop gave the wrong signal in an imperial procession; he killed himself in shame. The court tailor could not measure the emperor's clothes except at a distance, because touching the imperial person was forbidden. When the emperor traveled, all blinds along the entire route had to be drawn, because of the theory that direct view of the Son of Heaven might cause blindness.[31]

Knowing that there were many in the West who wanted to see him tried as a war criminal, Hirohito apparently submitted to MacArthur's demands. In a historic 'Rescript to Promote the National Destiny', delivered on New Year's Day 1946, this longest-living monarch in Japan's history (1901–89) repudiated his divine status.

Or did he? This statement, the first made by the emperor to his subjects since he commanded them to surrender the previous August, has never been easily available in Japanese. Originally drafted by the American occupation authorities, it was ingeniously edited by Hirohito's courtiers to appease their conquerors rather than to convey meaning to the emperor's subjects; and while in the West Hirohito's declaration was hailed as a momentous concession (a *New York Times* editorial said the announcement dealt the 'jungle religion' of Shinto 'a blow from which it can never recover'), even today many Japanese are unaware that Hirohito ever set aside his divinity. The statement, known as the emperor's 'declaration of humanity' (*Ningen-sengen*), has never been discussed or analysed in Japan's national media. MacArthur wanted the renunciation to emphasize the break with the pre-war system; for

Hirohito and his advisers, the object was to minimize the document's impact. The crucial three sentences run:

> The ties between Us and Our people have always been formed by mutual trust and affection. They do not depend on mere legends and myths. Nor are they predicated on the false concept that the Emperor is divine and that the Japanese people are superior to other races and destined to rule the world.[32]

These words were printed on one occasion only, in the 1946 New Year's Day newspapers; unlike the emperor's earlier speech (of 14 August 1945) admitting defeat, they were never broadcast. Moreover, the whole statement is deliberately vague and equivocal, the subject of each of the clauses being not Hirohito himself but 'the ties between Us and Our people'. The premise that the emperor is renouncing his divinity rests on a single noun qualifier: 'false'. Further to confuse the issue, the language was the archaic formal Japanese employed by the imperial household. The phrase 'concept that the Emperor is divine' is embodied in the single obscure word *Akitsumikami*, rendered by three characters, or ideographs, that even many educated Japanese would not have recognized: when a near-final version of the script was submitted to the Japanese Cabinet on 30 December, a phonetic reading was written alongside the word so that ministers could grasp the reference. No wonder Hirohito's advisers likened making sense of the directive's import to 'cutting smoke with scissors'.[33] In this contest of words it was the imperial myth that had secretly triumphed. Hirohito was allowed to retain the three sacred items of sovereignty: the mirror, the curved jewel and the sword, which together form the proofs of divine descent. Any effective renunciation would have seen these handed over to the occupying powers, or at least to a recognized museum. The divine wind had blown again. That June, Hirohito was formally exonerated of charges of aggression and war crimes.*

---

*Hirohito's struggle was the subject of a compelling 2005 film, *The Sun*, produced by the Russian director Aleksandr Sokurov as the third in a series about the downfall of powerful leaders. In it, the forty-four-year-old emperor wanders around under virtual palace arrest, turning over in his mind how he will respond to MacArthur's ultimatum. 'I am no longer a god,' he

In recent years traditionalists have set about taking advantage of Hirohito's sleight of hand. His son, the current emperor, Akihito (b. 1933) is reported to have resumed the rituals of homage to the rising Sun in the privacy of his palace grounds. Across the country, conservative politicians have toiled to bring patriotism back into the schools. The nationalist movement has been enthusiastically promoting the imperial mystique, while offering a version of its country's past that shows no remorse for the Second World War. The imperial family, said a former minister at a rally organized by Nippon Kaigi, one of the country's largest nationalist groups, 'is the precious treasure of the Japanese race, as well as a world treasure'.[34]

Although historians trace the start of Japan's imperial system to the fourth or fifth century, according to myth the first emperor, Jimmu, the son of Amaterasu, began his reign 2,665 years ago. Politicians are now asserting this myth as fact, all the while promoting the corollary myth that the Japanese are an exceptional race and pressing for a revisionist account of the country's wartime past. 'Japanese patriots,' wrote one observer in autumn 2006, after the country had voted in a nationalist prime minister, 'are sensing their Sun is rising after decades of shame.'[35]*

☼

Over the last century the rulers of China have also made full use of the symbolic authority of the Sun. Mao Zedong (1893–1976) came to full power only in 1949, but as early as 1936, following his reign of terror in Yenan (the destination of the 8,000-mile Long March), he was planning

---

openly tells his manservants at one point; 'I have renounced my divine nature' – then later he says, with conscious irony: 'A deity in this imperfect world can speak only in Japanese.' The sound engineer ordered to work on the radio broadcast commits hara-kiri, and Hirohito, a bathetic Chaplinesque figure in top hat and dress coat, ends by declaring: 'The Sun will go before the people completely in shadow.' Since there was no actual broadcast, one wonders what other liberties have been taken; not surprisingly the film, for all the denunciations that greeted its Japanese opening, played to standing-room-only audiences.

*In an often harshly male-dominated society, female voices have occasionally brought the Sun directly into politics. The writer and activist Hiratsuka Raicho (1886–1971) – and founder of the fledgling women's movement in Japan – sought to reclaim the Sun as a feminist symbol. Her declaration that 'In the earliest days of our history, the Sun was a woman' became nationally famous, and in her poem 'The Hidden Sun' she developed this theme, insisting that 'Originally woman was an authentic person; but now woman is the Moon'. By this she meant that at one time women defined themselves, but now they had allowed themselves to be defined by men. Asian women's struggle for liberation was 'the struggle to be the Sun again'.[36]

'The Communist Party is like the Sun', runs the anthem of the Cultural Revolution,
'Wherever it shines, it is bright.' But Mao was using solar imagery long before he
came to power

a personality cult, which included associating himself with solar imagery.
In 1948 Liu Shaoqi (1898–1969), his original party vice-chairman,
referred to Mao's thought being 'like the rays of the Eastern Sun'.

The Chairman encouraged depictions of himself framed within the
solar orb, and was consistently represented as 'the Sun that never sets'.
By the 1960s this imagery was showing up nationwide in songs, photographs, woodcuts and paintings, as well as on such artefacts as
teacups, money and swimming trunks. A typical graphic from the
Cultural Revolution displays a red Sun over vast grainfields – Mao as the
source of prosperity. His famous Little Red Book (first published in
April 1964, with more than 900 million copies printed) shows him in a
circle, golden rays spreading outwards, both sun figure and god. The de

facto anthem of the Cultural Revolution – 'The East Is Red' – lends its own special force to the identification of Mao and his party with the Sun:

> The East is red, the Sun is rising.
> China has brought forth a Mao Zedong.
> He amasses fortune for the people,
> Hurrah, he is the people's great saviour.
> Chairman Mao loves the people,
> He is our guide, to build a new China,
> Hurrah, he leads us forward!
> The Communist Party is like the Sun,
> Wherever it shines, it is bright.

The lyrics were attributed to one Li Youyuan, a farmer from northern Shanxi. Throughout Mao's decades-long supremacy, they blared over PA systems in every city and village from dawn to dusk. 'The East Is Red' became the title of a 'song and dance epic' promoting Communism, while a film version was released in 1965. The song's title is also the name of a series of Chinese satellites, the first of which, *Dong Fang Hong 1*, included a transmitter broadcasting the famous lyrics across a somewhat unaware planet.

Because of the anthem's associations with the Cultural Revolution, it was rarely heard after the rise of Deng Xiaoping in the late 1970s, and today it is viewed as an unseemly reminder of the cult of personality. So too with visual representations of the Great Leader. In the new modern China, it may seem bizarre to look back at the promotional films of a 'radiating Mao', his image set against a blazing red background, with rays shooting outwards like so many tracer bullets: but that was the everyday reality for hundreds of millions of Chinese for nearly thirty years.*

Despite its political eclipse, the Sun is in no danger of disappearing

---

*Of course, solar imagery had been employed in Chinese politics long before Mao. The activist and artist Lu Hao dong (1868–96), one of the leaders of the Kuomintang (the revived China Society, which became the National People's Party), designed two flags, the first for the Kuomintang, showing a twelve-ray white sun against a blue sky. Later, when the United League was established, a red background was added and the design thereafter adopted as the flag of the Republic of China. Hao dong became 'the first martyr of the people's revolution' when he was arrested by the Manchu government, tortured and put to death.

from the network of Chinese symbols, for it has long been part of traditional medicine, neatly summed up in the statement: 'Chinese doctors believed the human body was built up, like all the universe, from *Yin* and *Yang*, etymologically, dark coiling clouds and bright sunshine.'[37] While written in the past tense, this describes a tradition that is alive and well in China today.*

☼

Not only leaders of state have sought to align themselves with the Sun. Many a religion has adopted solar imagery, none more effectively through the millennia than the Roman Catholic Church. The year 2000 saw the publication of a book written by Pope John Paul II's most powerful aide – Joseph Cardinal Ratzinger (b. 1927) – who in 2005 became Pope Benedict XVI. Ratzinger made his reputation as the foremost theologian at the Second Vatican Council and wrote *The Spirit of the Liturgy* while serving as the Prefect of the Congregation for the Doctrine of the Faith, a position second only to the Pope's in determining what Catholics must believe.

In the course of his propagation of the liturgy, which he considers to have been handed down by God, Ratzinger discusses the range of solar imagery in Christianity – its source, its meaning, its importance and, most significantly, its identification with Christ, which in his view is free of any pagan influences. This allows him to argue that, since Christ is 'represented by the Sun' and 'we find Christ in the symbol of the rising Sun', we should change the direction in which we pray, even the orientation of our church buildings, so that 'the symbolism of the Cross merges with that of the East'. St Peter's Basilica itself 'because of topographical circumstances' faces west, but theologically this is a misfortune. He understands that current theologians think 'that turning to the east,

---

*In the spring of 2005 my daughter Mary spent several months before university working in two hospitals in Shanghai, one of which had returned to using the old physic. She showed me a book that had been recommended to her, *Basic Theories of Traditional Chinese Medicine*, which early on explains: 'The original meaning of the concept of yin and yang is very simple. Initially, the terms referred to whether or not a place faced sunlight. The place that faced sunlight or that was filled with sun was called yang, while the place that faced away from the Sun was called yin.'[38] Methods based on this doctrine are used side by side with modern, Westernized medicine.

towards the rising Sun, is something that nowadays we just cannot bring into the liturgy. Is that really the case? Are we not interested in the cosmos any more?' He argues that we in the West must be, and should change our ways to face east to pray.[39]

In the following chapter, 'Sacred Time', Ratzinger considers how 'the first day of the week was regarded as the day of the Sun', and by the same cosmic symbolism 'The Sun proclaims Christ'. He goes on: 'We have seen how deeply Christianity is marked by the symbolism of the Sun', which becomes 'the messenger of Christ'. With each New Year, the feasts of Christmas and Epiphany mark 'the dawning of the new light, the true Sun, of history'. He adds, with breathtaking disingenuousness: 'The complicated and somewhat disputed details of the development of the two feasts need not detain us in this little book . . . The claim used to be made that December 25 developed in opposition to the Mithras myth, or as a Christian response to the cult of the unconquered Sun promoted by Roman emperors in the third century in their efforts to establish a new imperial religion. However, these old theories can no longer be sustained.' And for him that assertion is argument enough.

Joseph Ratzinger, now a beleaguered Pope Benedict XVI

The same confident rejection of any connection between a pagan winter solstice and the celebration of Christmas also applies to the observation of the Resurrection at Easter, the time of the spring equinox: 'the close interweaving of incarnation and resurrection can be seen precisely in the relation, both proper and common, that each has to the rhythm of the Sun and its symbolism.' By chapter's end the Sun has become 'the image of Christ'. Our old friend M. de Volney would have laughed (or, more likely, been enraged) had he read such sophistry: two hundred years on, and Mother Church is still appropriating the ancient pagan ways while insisting the contrary. But the old religions did not own the Sun any more than those that followed them. It can be harnessed by anyone, and ever will be. Politics, of religion or state, is no more than the art of government: and what better to call on for help, symbolic or otherwise, than that great star that governs us? All leaders, in their own fashion, find ways to land on the Sun.

PART SIX

THE SUN AND THE FUTURE

This whimsical sketch graced the title page of a seventeenth-century manuscript,
De thermis – Of Temperatures. The Sun looks suitably reflective

OVER THE HORIZON

Some think that solar work is pretty well played out. In reality,
it is only beginning.

George Ellery Hale, 1893[1]

If you can't explain it simply, you don't understand it well
enough.

Albert Einstein[2]

In the autumn of 1930 Lord Rothschild gave what was, even for him, an
unusually grand dinner at the Savoy Hotel to benefit Eastern European
Jewish refugees flocking to Britain to escape their increasingly unsafe
homelands. The master of ceremonies, George Bernard Shaw, introducing the guest of honour, wound up: 'Ptolemy made a universe which
lasted fourteen hundred years. Newton also made a universe, which
lasted three hundred years. Einstein has made a universe, which I suppose you want me to say will never stop, but I don't know how long it will
last.' The guest in question laughed out loud, and when his turn to speak
came chided Shaw for invoking his 'mythical namesake, who makes life
so difficult for me'.[3]

The universe Einstein unveiled has endured more than a hundred
years so far. It began in 1905, when in less than twelve months he wrote
four papers that transformed the scientific landscape. The first, published
three days after his twenty-sixth birthday, helped lay the foundations of
quantum physics. The second redirected the course of atomic theory and

statistical mechanics. The other two initiated what came to be known as the special theory of relativity, to be followed by the general theory of 1915. These papers revolutionized our understanding of the nature of gravity, of the passage of light and our concepts of space and time. How much longer Einstein's universe will prevail is indeed unclear: *The New York Times* in 2005 quoted a leading physicist as forecasting that 'The smart money says that something is going to happen: general relativity won't last another two hundred years'.[4] The beginning of the end may have come with a concept that surfaced even before Einstein's papers of 1905 and 1915: quantum theory, which, among so much else, ushered in a totally new approach to understanding solar energy.

The new way of thinking had taken form five years before Einstein's annus mirabilis, when the great German physicist Max Planck came up with the idea that all energy is emitted in discrete units, which he called 'quanta' (from the Latin, 'how much'). It was as if, in the words of George Gamow, 'one could drink a pint of beer or no beer at all, but not any amount in between'. Other physicists, among them Niels Bohr, Erwin Schrödinger, Wolfgang Pauli, Max Born and Werner Heisenberg, helped extend this disconcerting insight, which redefined the nature of energy.

Just as Einstein's general theory applied to interactions between the largest objects in the universe, quantum physics described what goes on at the atomic and subatomic levels, where events occur that are completely at odds with our everyday experience. For example, a quantum particle (a photon, say, a massless and chargeless discrete particle) is so insubstantial that it will move from one place to another without going through any intervening space: it ceases to exist at one point while simultaneously appearing at another – a 'quantum leap'. This seems to confound common sense: surely to travel from A to a non-adjacent C involves passing *through* something. As Niels Bohr famously said to one of his Copenhagen students, who complained that quantum mechanics made him giddy: 'If anybody says he can think about quantum problems *without* getting giddy, that only shows he has not understood the first thing about them.'[5] Brian Cathcart explains further in his book on atomic physics:

> Planck's technique had a significant drawback: it worked only if you
> set aside an essential component of the classical laws, the principle of

continuity. Continuity has an analogy in the kitchen: in baking, milk is 'continuous' in the sense that any given amount may be measured out and added to the mixture, while eggs tend to be 'discontinuous' – it is a perverse cookbook that asks you to separate one-quarter of an egg . . . Here was the first inkling of a sensational possibility: that the laws of physics that applied in the observable world might not, after all, be valid at the atomic level.[6]

Such laws were not easy to accept. Even as recently as 1999, at a conference of physicists, a poll was taken about quantum mechanics. Of the ninety respondents, only four said that they accepted the standard interpretation which follows the Bohr/Planck understanding, while fifty ticked the box labelled 'None of the above or undecided.'[7] But – then as now – quantum theory worked: it explained and predicted phenomena for which there was no other explanation. Classical physics is deterministic: If A, then B – the bullet fired at the window shatters the glass. On the quantum scale this is only *usually* true. Quantum physics shows that from time to time quanta, behaving not like particles but like waves, break through the force field 'as if a cannonball were to fly untouched through a fortress wall',[8] a phenomenon that was to become known as 'quantum tunnelling'. This was demonstrated by a key experiment conducted in 1909, in which certain radioactive elements were discharged at a wafer-thin sheet of gold foil, and a small number bounced back, a phenomenon that Ernest Rutherford was to describe as 'almost as incredible as if you fired a fifteen-inch shell at a piece of tissue paper and it came back to hit you'.[9] Over the next forty years scientists went from a minimal knowledge of the protons and neutrons that make up the nuclei of atoms to understanding the primary thermonuclear fusion process that powers the Sun. This is a process that only quantum physics could explain, because it involves the fusing of nuclei, which classical physics says is impossible – because all nuclei bear a positive charge, and positively charged particles repel each other. As one leading physicist put it: 'According to classical physics, two particles with the same sign of electrical charge will repel each other, as if they were repulsed by a mutual recognition of bad breath.'[10]

Classical theory denies that two protons in a star can go fast enough to break the walls of their electromagnetic force fields and merge into a single nucleus. But quantum tunnelling permits protons to traverse the barrier created by electromagnetic repulsion. Under intense heat and density – the result of the gravitational energy generated by the solar mass contracting – protons overcome the usual forces of repulsion and collide, to form stable helium nuclei; this mass is released as energy. So here is the Sun, blazing away, proving that classical physics has it wrong.

After quantum theory was generally accepted, so many different types of particles were discovered that physicists had to consult a booklet, the *Particle Properties Data Handbook*, to keep track. (The outside world eventually reacted by producing T-shirts with the legend *Protons have mass? I didn't even know they were Catholic*.) A table was drawn up: sixteen basic particles, twelve of matter, called fermions, and four quanta, called bosons, which mediated the interactions between particles. Fermions are the basic building blocks of matter, divided into either leptons (from the Greek for 'thin') or quarks.*

Quarks are never found in isolation but only in groups of three (when together they are called baryons) or in doublets (as mesons): bound states of quarks are collectively known as hadrons. More complex things tend to be made out of fermions – protons, neutrons, atoms, molecules, buildings, people – 'If I could remember the names of all these particles I would have been a botanist,' complained Enrico Fermi, for all that this largest group of particles bears his name. Fermions also comprise neutrinos, which interact so weakly with other particles as to make them exceptionally difficult to detect, and indeed trillions stream through our bodies every second. As John Bahcall, most renowned of neutrino researchers, observed: 'a solar neutrino passing through the entire earth has less than one chance in a thousand billion of being stopped by terrestrial matter . . . About a hundred billion solar neutrinos pass through your thumbnail every second and you don't notice them.'[11] No wonder John Updike felt compelled to write,

*A term borrowed from *Finnegans Wake* by their hypothesizer, the Caltech physicist Murray Gell-Mann, from a line as spoken in a dream by a drunken seagull. Instead of asking for 'three quarts for Mister Mark', the inebriated bird demands 'three quarks for Muster Mark'; since the theory originally posited three quarks, the name made some sense. Thus were quarks immortalized, but their namesakes' lives are incredibly brief, their most enduring estimated to last no more than 1×10^{-24} seconds.

Neutrinos, they are very small,
And do not interact at all.[12]

The last particles identified have been the W and Z bosons in 1983, the top quark in 1995 and the tau neutrino in 2000, so this subatomic world is still very much in the process of being discovered.*

From Galileo's day on, we have been learning about the Sun's shape, size, rotation and spots; its mass and density, and how it moves. We went on to assess its age, record its infrared and ultraviolet radiation, dark-line spectrum, radioactive output, activity cycle, prominences and chromosphere, corona, chemical composition, bright-line spectrum, radio transmissions, X-ray emissions, neutrino output, coronal holes and whole-body oscillations. Despite such a roster, it is humbling that many of the Sun's basic processes are only now beginning to be understood: how its magnetic field is generated, for instance, or its atmosphere heated; why, since the Sun is not itself on fire, flames burst from it. And some are even further beyond our grasp. What creates the corona, and how is it heated to such extremely high temperatures? What switches the solar magnetic poles? Why does the atmosphere's temperature suddenly soar in the thin transition region between the lower corona and the chromosphere? Where is the solar wind produced, and how far out does it spread? What kind of shielding do we get from solar magnetic clusters? Even, why do sunspots exist? We have a long way to go.

Three 'levels' of research seek to answer these questions – literally 'levels', as they take place on the ground, in the skies above, and in the earth beneath; together they form the tripartite structure of this chapter.

---

*Indeed, in this bizarre new world it has been shown that, at the micro-microscopic level of photons (the minute entities that make up a beam of light), our usual expectations of how our globe works once again fall apart. Thus it was no surprise when in the 1970s there was hypothesized a medium called 'slow glass' through which light might take months or even years to pass. If light were to take a year in its passage, then everything that had been seen in front of the glass would become visible only a year later on the far side. 'Slow glass', which would be created from a complex form of plasma, is not yet a reality, but, even though it would delay light by a factor of a quintillion or more, it could be, and soon: in 1999 the Rowland Institute for Science at Harvard reduced light speed to below a metre per second. As the historian of science Brian Clegg says: 'If the speed of light can be controlled, then so can reality itself. Extraordinary scientific work is now under way, work that will make technical miracles such as slow glass possible . . . and that will, in a few short years, make light the most exciting area of study in all science.'[13]

☼

Scientists generally agree that the Sun is fired by thermonuclear reactions that fuse light elements into heavier ones, converting mass into energy. To demonstrate this truth, however, has been difficult, since the nuclear furnace lies deep in the Sun's interior, and conventional instruments can record only those particles emitted by the outermost layers.[14] Ninety-seven per cent of this energy is made up of photons, but about three per cent sweeps outwards in the form of neutrinos – 200 trillion trillion trillion of them every second, outnumbering all ordinary particles by a billion to one; and analysing them provides our best means of learning how this fusion works.

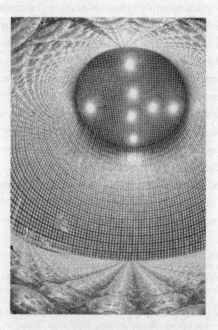

A neutrino detector built a kilometre below ground in a disused Japanese mine. The huge stainless-steel vessel, forty metres in height and width, has been filled with 50,000 tons of highly purified water, and its walls lined with 13,000 light sensors, to pick up a flash generated by electrons recoiling from neutrino collision in the water – a necessarily complex way of discovering how the Sun works

However, it is intensely difficult to register neutrinos, which have minimal mass (they were long thought to have none), travel at the speed of
light and change their nature as they go. As Wolfgang Pauli confessed: 'I
have done a terrible thing. I have postulated a particle that cannot be
detected.' Updike again:

> The Earth is just a silly ball
> To them, through which they simply pass,
> Like dustmaids down a drafty hall.

Neutrinos pass through ordinary matter (not just human bodies but the
great globe itself) as if it were entirely transparent: to them, everything is
a vacuum. Not until the late 1960s was the ideal apparatus developed,
when Raymond Davis and John Bahcall made a detector out of a disused
mine outside the town of Lead, South Dakota: a tank located 4,850 feet
underground and filled with 100,000 gallons of tetrachloroethylene – a
simple cleaning fluid, yet highly sensitive to registering neutrinos, each
one of which, should it react with the chlorine, may produce a radioactive isotope of argon. Such exposure is useful only underground, since
the detector must be shielded from the endless shower of other subnu-
clear particles, many from beyond the solar system.

About twenty times a day, a neutrino crashes into a neutron in this dark
sphere, creating a faint flash. The network of some 9,600 photomultipliers
that coat the vessel detects the flash, which is then analysed for data about
the neutrino that set it off. It was here that Bahcall made his discovery,
dubbed 'the Neutrino Problem'. Simply, not nearly enough neutrinos were
reaching Earth – between a third and a half of the predicted number. It was
an astounding shortfall: where were the rest of them? Had the Sun's scientific acolytes miscalculated? Bahcall wrote: 'The most imaginative idea
[is that] neutrinos have a kind of double identity . . . It will not be easy to
test this unusual hypothesis, but it cannot be lightly dismissed.'[15] He went
on to say that the most imaginative solution suggested was Stephen
Hawking's: that the Sun's core might contain a small black hole.

In fact, the question preoccupied solar physicists, cosmologists and
astrophysicists for the next thirty years. (To complicate the problem, neutrinos had only one disposition: i.e. they spun as they travelled, but only

backwards along the direction of their flight, like left-handed corkscrews. As the long-time editor of *Nature*, John Maddox, asked: 'What is it about our world that caters only to left-handed neutrinos?')[16] A number of answers were put forward, including that solar models were indeed wrong, and that the temperature and pressure in the Sun's interior were substantially different from those posited in current theory; or that the nuclear processes at the Sun's core might temporarily have shut down: since it takes thousands of years for heat energy to move from the core to the outer layers, such a disconnection would not be apparent for thousands of years.*

An answer was eventually posited in 2001. Neutrinos in the course of their journey to Earth may change from the type expected to occur in the Sun's interior into two other forms undetectable by the old machines. Scientists pondered the three different kinds: electrons – the ones produced in the Sun – muons and taus. (When the muon was discovered, a physicist famously asked, 'Who ordered that?') After extensive statistical analysis, it was found that about thirty-five per cent of the neutrinos reaching Earth were electron-neutrinos, the rest being made up of muons and taus. Once all three categories could be accurately tracked, the total number lay well within the range of earlier predictions. Solar physics was vindicated. As one of its practitioners told me, 'It used to be believed that we were being dumb because we were doing something wrong. We knew the neutrino particles' reactions, we knew everything, so, if we weren't capable of tracing their journey from the Sun, well, it wasn't a *physics* problem, it was a *solar* physics problem – "These guys can't get these things right!" Well, it turns out it's the other way around: we *were.*'[17]

But the solution of 2001 is still a 'best theory', and who is to say when or if it will be confirmed? One further problem is that the materials being examined are *so* minute. As a character in Tom Stoppard's *Hapgood* says, 'When things get really small, they get truly crazy, and you don't know how small things can be, you think you know but you don't know . . . Every atom is a cathedral.'[18] Another visitor from the world outside science, Bill Bryson,

---

*Modern neutrino detectors grow ever more sophisticated. They register their quarry in real time, and thus, for instance, examine the day–night effect: neutrinos passing through the Earth (detected at night) are slightly different from those that don't (detected during the day). When in 1934 Enrico Fermi submitted to *Nature* a paper on neutrinos, postulating that they escaped the Sun due to certain radioactive decays, it was rejected because 'it contained speculations too remote from reality to be of interest to the reader'.

begins *A Short History of Nearly Everything* by complaining that a proton 'is just way too small . . . A little dib of ink like the dot on this *i* can hold something in the region of 500,000,000,000 of them, rather more than the number of seconds contained in half a million years.'[19]* It is daunting to know that the smallest things in nature hold the clue to one of the largest.

☼

On to what is done up in the air. The progress of solar astronomy since the Second World War has required overcoming distances so great that the challenge involved is at least as daunting as coming to grips with the subatomic realm. To date, rockets have been our means of getting closest to the Sun. In the 1920s the prototypes created by Robert Goddard (1882–1945) were derided, *The New York Times* notoriously declaring that Goddard seemed to 'lack the knowledge ladled out daily in high schools', and mocking his experiment as 'Goddard's Folly'. Any such attempt on the heavens was regarded as the stuff of fantasy, best kept for cinemas and comic strips: it was impossible to move in a vacuum, as there was nothing to push against. However, on 15 November 1936 a group of students at Caltech, known as 'The Suicide Squad', scraped together enough cheap engine parts to make worthwhile a trip to Arroyo Seco, an isolated canyon wash at the foot of the San Gabriel Mountains, and test-fire a small rocket. As they huddled behind sandbags, the motor burned for about three seconds before an oxygen hose tore loose, bathing the testing area in flame. Over the coming weeks they tried again, and on their fourth attempt, on 16 January 1937, the motor fired long enough – forty-four seconds – to heat its metal nozzle red. Lift-off.

The adventurers' explosive and noisy experiments in time proved too dangerous for the campus, but by now people who could finance such activities were interested. In 1938 Caltech's aeronautical lab received a surprise visit from the chief of the US Army Air Corps, and within two years a facility was built in the Pasadena foothills. Three test stands and tarpaper shacks marked the first buildings of what would become the Jet Propulsion Laboratory. The first substantial financing came from the Air

*A rare Bryson miscalculation. He should have written 15,000, not half a million: 60 × 60 × 24 × 365 × 15,000, totalling 473,040,000,000.

15 November 1936: students from Caltech, known as 'The Suicide Squad' from
their early mishaps, on the day they tried to send a rocket into space.
Thanks to them, we are now sending rockets to the Sun

Corps, which wanted small rockets to help lift heavily loaded planes from
short runways. After their successful development, and Pearl Harbor, the
army called for other models, and JPL began to develop guided missiles.
But these were early days, with no vision yet of solar research.*

The Second World War hampered solar physicists in countless ways,
but it also prepared for a later flourishing. British and American scientists
working on radar discovered that the Sun emitted radio waves (first mistaken for German jamming), while both Axis and Allied solar physicists
participated in weather-forecasting, some Germans trying to observe the
Sun from above the atmosphere with 'Retribution Weapon 2' – V-2 rockets. They failed, but their initiative paved the way for much that followed.

*JPL's official history records that in those days most people's exposure to rockets was lim-
ited to Saturday matinees, where Flash Gordon blasted from planet to planet. According
to an eminent founder-member of JPL, the great Theodore von Kármán, a US official told
him bluntly: 'I don't understand how a serious scientist or engineer can play around with
rockets.' Accordingly, as Kármán writes in his autobiography: 'The word "rocket" was in
such bad repute that for practical reasons we decided to drop it from our early reports and
even our vocabulary.' The codeword for 'rockets' became 'jets', and by 1959 the lab was
almost entirely out of the propulsion business. JPL started out as part of the 'Guggenheim
Aeronautical Laboratory at the California Institute of Technology' (after the mining mag-
nate Daniel Guggenheim), being renamed the Jet Propulsion Laboratory on 1 July 1944.[20]

In the spring of 1942 Wernher von Braun (1912–77), technical director for the V-2, seeking to give the programme a scientific dimension, asked the physicist Erich Regener to develop a special payload and nose cone.[21] The most sophisticated instrument that Regener considered was an ultraviolet spectrograph, aimed at correlating the amount of atmospheric ozone to altitude. All the while, hundreds of missiles were being readied for their primary, destructive purpose. The first of more than 3,000 blasted off in September 1944, and the British War Cabinet discussed, as it never had throughout the Blitz, evacuating London. (The rockets, each carrying about 1,700 pounds of explosive, were particularly feared since, travelling faster than sound, they arrived in advance of audible warning.)

That December, paying no heed to the imminent ruin closing in on the Reich, von Braun scheduled the launch of a research prototype for the following month. By mid-January the instruments were moved up for take-off. It never took place. The last V-2 roared towards England in late March; six weeks later the Red Army took Berlin. As Germany collapsed, a special US intelligence unit rounded up von Braun's 118 scientists and all available V-2 components – enough for a hundred rockets. The importance of this find was not lost on Western leaders. Leo Goldberg, later NASA's principal investigator, would write to a colleague:

> If anyone asked you what technological development could, at one stroke, make obsolete almost all of the textbooks written in astronomy, I am sure your answer and mine would be the same, namely, the spectroscopy of the Sun outside of the Earth's atmosphere . . . [The] V-2 Rocket has attained a height of 60m[iles], and with the host of control mechanisms that have come out of the war it should be possible to point the rocket at the Sun . . .[22]

In 1944 the Jet Propulsion Lab (JPL) formally became an army facility, operated under contract by Caltech. Between 1945 and 1957 several small devices were indeed sent outside the atmosphere to observe the Sun's extreme-ultraviolet and X-ray spectra (all getting considerably scorched in the process). A V-2 panel was set up, an informal committee that initiated experiments in such subjects as variation in atmospheric pressure, radio propagation, cosmic rays, temperature and solar radiation. In 1946–7 alone, physicists managed to mount instruments to photograph the Sun's ultraviolet spectrum on to eleven of the twenty-eight V-2s launched; from these beginnings was inaugurated ultraviolet astronomy, and the entire field became known as solar physics.[23]

British and American scientists had meanwhile been drawing on solar data to predict both magnetic storms and the behaviour of the ionosphere (i.e. that part of the Earth's atmosphere where the Sun affects the transmission of radio waves). The US navy employed such forecasts to anticipate radio frequencies used in its surveillance of Soviet submarines. Neither activity played a decisive part in the Cold War, but they were significant enough to establish solar physics as a military asset.

By 1953 fourteen of the fifty observatories in the world conducting visual studies of the Sun were equipped with coronagraphs (the special telescopes that, by blocking out certain solar light, enable scientists to see the Sun more clearly); between 1945 and 1951 about seventy per cent of all publications in the emerging discipline of radio astronomy were devoted to its solar aspects. In 1955 the US government announced plans to launch a satellite during an 'International Geophysical Year', the eighteen months from July 1957 to December 1958, and ninety-five observatories and stations around the world were set to participate. Then a thunderbolt struck.

On 4 October 1957 the Soviet Union launched Sputnik (in Russian, 'attendant' or 'fellow traveller') – a simple sphere weighing 184 pounds and not quite two feet wide that, in the words of The New York Times, 'changed everything: history, geopolitics, the scientific world'.[24] The 'beep-beep-beep heard round the world'[25] plunged the United States into a crisis of self-doubt, precipitated a furious missile build-up and stimulated prodigious amounts of funding for scientific and engineering research, all aimed at maintaining America's security and prestige. After Sputnik, politicians and military men set the agenda for man's explorations in space.*

---

*In 1961 Russia's pioneer cosmonaut, Yuri Alekseyevich Gagarin, the first person ever to see Earth in its entirety, set out to describe the experience. During his 108-minute flight he had been in orbit for less than an hour and a half, but even that long a view dazzled him: 'The Earth has a very characteristic, very beautiful blue halo, which is seen well when you observe the horizon. There is a smooth colour transition from tender blue, to blue, to dark blue and purple, and then to the completely black colour of the sky.' After so much discussion of how the Sun appears to us on Earth, suddenly such visions were reversed.[26]

In January 1958 President Eisenhower created the National Aeronautics and Space Administration (NASA), and the JPL was put under its control. That same month, *Explorer 1* went into orbit, to 'the collective sigh of relief from an anxious American public'.[27] Circling the Earth every 113 minutes, the craft radioed back data about temperatures, meteorites and radiation. Even more instruments for studying the Sun were fixed to stratospheric balloons, high-altitude aircraft and rockets, while solar scientists obtained new telescopes or substantially improved existing ones at ground level.[28] Between 1957 and 1975 the solar physics community roughly doubled; *Solar Physics: A Journal for Solar Research and the Study of Solar Terrestrial Physics* appeared in 1967 and was soon publishing more than 200 articles a year. From 1959 on, US and Soviet programmes sent ever larger and more complex spacecraft on ever longer journeys.*

On 7 March 1962 NASA launched *Orbiting Solar Observatory 1*, which was powered by solar photovoltaic cells and which had a special midsection stabilized to point directly at the Sun. It was an unqualified success; but by early 1964 the space programme was torn apart by arguments over funding, by the disastrous malfunction of a satellite at Kennedy Space Flight Center which killed three engineers, then by a third observatory's failure to make orbit when a rocket fired prematurely. At the end of 1965 the Advanced Orbiting Solar Satellite was cancelled, and within four years the Nixon administration had terminated the programme that had built the Saturn V Moon rocket – in one stroke depriving mankind of 'a ubiquitous presence throughout the solar system'.[29]

Fortunately, the case for using spacecraft for solar observation was compelling enough that NASA soon recovered, over the next decade launching *Solar Observatory 3*, as well as embarking on its most ambitious project, a crewed observatory in space, the Skylab Apollo Telescope Mount. An unmanned Skylab was launched in May 1973, with four main

---

**Explorer 1 was kept secret at the Jet Propulsion Lab, where it was nicknamed 'Project Deal'. As its manager Jack Froehlich, a poker player, explained after Sputnik: 'When a big pot is won, the winner sits around and cracks bad jokes, and the loser cries, "Deal!"' Names are not always appropriate; in later years, when America was attempting to land men on the Moon, it called the programme 'Apollo'.

solar instruments – an ultraviolet spectrograph and spectroheliograph (cost \$40.9 million), an ultraviolet spectroheliometer (\$34.6 million), a white-light coronagraph (\$14.7 million) and an X-ray telescope (a comparatively cheap \$8.3 million). Expenditures at this level mean that virtually all the material wealth that has gone into astronomy has been invested over the last forty years.

However, over that very decade of 1971–1980 national priorities shifted away from science, for two prime reasons. First, economic: the cost of the Vietnam War, OPEC's vast increase in the price of oil, and the funds required for much-needed social and environmental programmes raised doubts about the returns on the stream of federal money flowing into extraterrestrial technology. Second, political: scientists were often at the forefront of opposition to America's overseas and military policies, which raised questions about whether money spent on this community was fostering unpatriotic elements. Solar physics was particularly hard hit: an official report noted that by the end of the 1970s it appeared to be an isolated field of research, underfunded and under-studied. Astronomers as a whole were 'largely unaware of and maybe indifferent' to it.[30] Financial problems and cancellations continued, a situation exacerbated in January 1986 by the loss off Florida of the space shuttle *Challenger* with her entire company. NASA postponed for four years the launch of *Ulysses*, the spacecraft for observing the solar output from above the Sun's poles, and abandoned altogether the development of the major solar telescope it had been planning. While America was foremost in reducing its programmes, Europe was cutting back too: even as early as 1976, Greenwich had discontinued its sunspot photography after 102 years of continuous operation, and four years later the Swiss government closed its centre for sunspot statistics. Researchers refused to be bowed, however, and from the 1970s onwards came up with some remarkable discoveries. Most notable of all was what came to be known as 'solar wind'.

This requires going back a bit in time. In 1951, building on the hypothesis that the Sun discharged not just energy but matter, the German astronomer Ludwig Biermann (1907–86) had suggested that the tails of comets were blown persistently away from the Sun, not, as had been assumed, by the known pressure of light but by an outward flux of particles, which deflected the ions in each comet's tail. Building on Biermann's insight, Eugene N. Parker (b. 1927), a distinguished theorist from Chicago, argued that such an efflux originated in a continuous, million-mph expansion of the corona, which he dubbed 'solar wind', to emphasize its dynamic character. This wind travels inwards from the Sun's north pole and outwards from its south pole, argued Parker, the Sun's surface brightness being particularly sensitive to vertical currents. The storm's influence on the Earth might be feeble compared with that of solar radiation (it would not ruffle the hair on one's head) but, Parker went on, it affected the magnetic fields in inner-planetary space, drawing lines of force out of the corona deep into the solar system. Among Parker's hypotheses was that the Earth's own magnetic field was 'elongated' by the winds, which shaped it into something closer to a teardrop than a sphere.

Towards the end of the 1950s the German geophysicist Julius Bartel ascribed the source of such wind to the Sun's 'M regions' – 'M' standing appropriately for 'Mystery' – probably areas of unusually weak coronal emissions. Scientists at MIT decided to measure them – with a probe in *Explorer 1*, launched on 25 February 1961, which measured flux, speed and direction. *Mariner 2*, launched a year later, established that the Sun discharged plasma continuously, at the speed of 400–700 kps, occasionally mounting to 1,250 kps, temperatures increasing or decreasing along with speed. The corona indeed had sufficient energy to escape into space.

There followed a gap in research, possibly because of cutbacks, certainly because many scientists still did not accept that Parker was right. Then in August 1977 *Voyager 2* was launched, which discovered yet stronger winds as they crossed the orbits of planets ever further from the Sun. It became clear that a constant surge of particles was indeed spilling relentlessly from the solar upper atmosphere. Parker was vindicated: matter was leaving the Sun 'like streams of water whipping out from the whirling head of a lawn sprinkler'.[31]*

In 1978, in a report to the National Academy's Space Science Board,

---

*In the summer of 2005 I visited the main Beijing observatory and spoke to a lively young researcher, Chen Jie, who was completing a PhD on coronal loops that cross the solar equator. She spoke reverently of Dr Parker: 'When they give a Nobel Prize to solar physics, it should go to him.' As yet, it hasn't, possibly because he has not produced 'one extraordinary piece of work', possibly because solar physics seems perpetually out of favour with the wise men of Stockholm.

Parker declared: 'Our daytime star is sufficiently near [to reveal] a variety of phenomena that at first sight defy rational explanation . . . but ultimately stimulate the theoretical understanding of new effects . . . elsewhere in physics and astrophysics.' Solar physics, he concluded in a later essay, was the 'mother of astrophysics'.[32] Further,

> the temperature, luminosity, mass and radius of the Sun had to be known before there was a firm basis for arguing that the distant stars are suns. The gaseous nature of the Sun had to be established before the concept of the self-gravitating, self-supportive, luminous gaseous sphere could be developed quantitatively to give an idea on conditions in the deep interior of the Sun . . .[33]

In short, there would seem to be a natural progression in how we learn about our star. That progression has been hugely accelerated by half a century of discovery. By the time Parker penned these words, researchers had created three-dimensional models, mapped interplanetary magnetic realms and begun to explore the dynamics of solar gales. *Voyager 1* (launched two weeks *after* its successor, on 5 September 1977) is still operating even as it draws away from the Sun, now 9.6 billion miles distant, having entered the 'heliopause', where the solar wind is in balance with the surrounding interstellar medium. Around 2015 it will have left the solar system altogether, to plunge on in interstellar space.*

Nearly every year now sees another craft sent aloft. Arguably the most important of all solar missions has been SOHO (December 1995), among its many services being early warning of mass ejections that can affect astronauts, and giving up to three days' notice of Earth-directed disturbances.[34] *TRACE* (Transition Region and Coronal Explorer: 1998) images the solar photosphere, transition region and corona.

---

*As *Voyager* spacecraft leave the solar system (*Voyager 2* will follow *Voyager 1* in 2017), they carry a message for any intelligent alien life that may find them. At JPL, I was shown a copy of the golden record, no larger than a dinner plate, fixed to each craft's side. It contains eighty-seven images representing humanity: our bodies, our ways of life and our interaction with our planet, as well as instructions on how to build a record player. Delegates from the UN have recorded greetings in almost every known national language, from 'Please come to visit when you have time' to 'May the honours of the morning be upon your heads'. Music, from Beethoven to rock 'n' roll (Chuck Berry) and Australian Aboriginal chants to a chorus of crickets and frogs, finds its place – as well as the sound of a kiss.

Stardust (1999) returned to Earth in early 2006 with a million motes – weighing in total no more than a few grains of salt – picked up in three circuits of the Sun, while in 2000 a joint NASA/European project, *Cluster 2*, put into space four craft to analyse both how sunspots come into being and the interaction between solar winds and other magnetized particles.

In August 2003 the Spitzer Space Telescope blasted into orbit, after twenty-three years of preparation. The fourth and last of NASA's great observatories, it monitors the heat radiated by heavenly objects and records how primordial matter formed into galaxies: one galaxy, the Sombrero, is said to contain 800 billion suns.[34] Then, in the last months of 2006 alone, both *STEREO* (Solar Terrestrial Relations Observatory: two complementary NASA spacecraft that track massive solar electromagnetic storms) and Japan's now-named *Hinode* ('Sunrise': it is a Japanese custom not to name their satellites until they are up and working) joined them in the heavens. *Hinode*'s telescope was soon sending back spectacular images of sunspots and solar storms, some of the most beautiful ever seen. More is yet to come: in 2010 NASA hopes to launch, on the back of its *Atlas 5* rocket, a Solar Dynamic Observatory, designed to study the causes of the Sun's variability and its impact on Earth, and the first mission of NASA's 'Living With a Star' programme.

We do not know – even approximately – just how many objects (say of at least modest house size) circle the Sun, as we still lack the observational powers. Currently, some forty man-made research bodies are in orbit, fewer than six still functional, the rest useless hulks. TV and other satellites are legion, and there is already enough equipment out there to make controllers careful. For example, in September 2003 the fourteen-year-old *Galileo* probe, which brought off the exploration of Jupiter and its (then) sixty-three known moons as well as sending back crucial information about solar winds, was deliberately plunged into the giant planet's seething atmosphere so that it would not contaminate Europa, a Jovian moon believed capable of harbouring life.[36]*

*But it is not humanity that has cluttered the skies: billions of 'stones in heaven' circle the Sun, from the mass of Jupiter to the uncountable populations of meteor swarms. Trillions of comets and about 14,000 known asteroids join them: thus, the once-tidy community of Sun, Moon and five planets has fragmented into a cosmos formidably crowded.

Since 2004 the world has sent out at least eighty commercial satellites, and the next decade will see at least twenty more. A new generation of super-rich entrepreneurs who grew up fascinated by space is pouring money into rockets of all types and sizes. Peter Diamandis, a founder of the Ansari X Prize, a $10 million incentive to put a pilot in space without government financing, says that given the energy and minerals to be found in the heavens 'the first trillionaires are going to be made in space'. Economic considerations will play a bigger role than political ones in advancing the interplanetary frontier.[37] Paul G. Allen, a founder of Microsoft, paid for *SpaceShipOne*, the tiny craft that won the X Prize in 2004. Elon Musk, a founder of PayPal, is developing rockets through his company Space Exploration Technologies; Jeffrey P. Bezos, founder of Amazon.com, is developing rockets at a site in western Texas.

For the moment, however, NASA still holds the lead in research. When visiting the Jet Propulsion Lab in 2007, I was invited to sit in on a committee considering a solar monitoring device on the Moon. Plans are currently under consideration in Congress for a solar-powered, human-tended, continuously inhabited research outpost there, rising from either its north or south pole, where sunlight endures and water may exist.[38] But an even more dramatic venture, and one of the main reasons for my visit to Pasadena, is perhaps the most ambitious unmanned satellite project ever.

☼

'Jet Propulsion Lab' is a proudly carried misnomer – the employees just like the title: 'We're the only place like us that's not named after some person in the past,' one staff member told me, counting off the Goddard, Johnson, Kennedy, George C. Marshall and John C. Stennis space centres. 'That's kinda cool.' The whole place is kinda cool: perched on the edge of a national forest fourteen miles from downtown Los Angeles, it covers 177 acres and overlaps the town boundaries of Pasadena and La Cañada Flintridge, which vie with one another for primacy of address. Some 5,000 full-time employees work in the facility, plus a few thousand contractors on any given day, so the place has the feel of a small town, with its own peculiar traditions (these once included the annual crowning of a 'Miss Guided Missile', later 'Queen of Outer

Space', before the contest was discontinued without explanation in 1970) and an almost comically strict security code (although, to be fair, a drug-crazed young employee did sell an intelligence satellite programme to the Soviets, an episode tracked in the book and film *The Falcon and the Snowman*).

I was to interview two of the scientists working on *Solar Probe Plus*: Dr Neil Murphy, the group's supervisor, a Manchester-born astrophysicist in his mid-forties who, after working on the *Galileo* project, went out to Pasadena in 1999; and the man Murphy himself describes as 'the smartest guy at JPL', Marco Velli, for the last four years their designated principal scientist, while remaining a tenured professor at the University of Florence. Murphy and I were set up in a small, windowless conference room deep within JPL's research area. 'For hundreds of years we've been saying that history is much longer than we think,' he told me,

Then we said: how did the Sun get all its energy? There's a huge amount, and it's tied up in the magnetic field; it turns out that all the exciting stuff on the Sun is. So the next question is: We know the source of the Sun's energy: can we find out how it functions?

What we're really looking for are trapped waves inside the Sun. We've been developing *Solar Probe* for about thirty years now, because the only way to understand the Sun's heating process is to get really close to it – there's no simple formula that describes how the dynamo inside the Sun works.[39]

Solar Probe has had its ups and downs, and has always been considered the thing that was just *beyond* what NASA could bring off; but Neil Murphy thinks their current blueprints will result in a funded mission to look at the plasma as it evolves. Astronomers use a 'solar radius' to measure the size of planets and stars. Ten solar radii, he explained, are a little less than 7 million kilometres, or 4,360,000 miles – close enough that one would expect anything at that proximity to be fritzed in a millisecond. Somehow *Solar Probe Plus* has been designed to cope.

I was beginning to get out of my depth, and for a moment he relented. Plasma, he explained, is the 'fourth state of matter', solids, liquids and gases being the more familiar three. Formed from ionized atoms, it is the most common form of matter in the universe, making up the Sun, the other active stars and interstellar gas. 'The solar cycle is basically a process of stretching all this stuff around until it bursts through the surface. It's like twisting a rubber band: if you twist and twist, eventually this stuff will pop out, and that's what provides the energy for a sunspot.'

By this time there was no stopping Neil, his hands weaving about to illustrate what his satellite would be looking for. 'One of the things that has evolved in the last ten years,' he continued,

is our view of the Sun and the solar system as being all one large system. We've seen the effect of geomagnetic storms on sunlight and on the sunspot number. The main thrust of our group over that time is to have studied that solar wind, and how it interacts with the planets, and the effect of solar disturbances on geomagnetic environments – or what we call geospace.

He went on to describe the 'tachocline' (from *tachos*, fast, and *clene*, break), the solid layer of the Sun that sits between the outer convection zone and that region of the Sun's interior where rotation is uniform. Within the tachocline everything also rotates at the same rate. This, Neil explained, was where the plasma motions that generate the Sun's power come into being, via the process of ionization. 'If you take an ionized gas with a magnetic field in it, and push the gas, the field moves with it. Ionized gases are almost perfect conductors, and if the magnetic field is really strong, they'll move the plasma; but plasma is highly energetic, and can push the magnetic field – which is particular to the Sun, and, in physics, very unusual. What we are dealing with is how all these particles interact once they've been magnetized like that.'

So what was he hoping to discover? 'Over the next fifty years we'll go through these stages of understanding what the processes are that drive the dynamo and so on. Hopefully, we'll be able to gain some predictive power, to say: "Given the situation, ten days from now we may see a sunspot appear."' Nodding, I asked about solar winds. Neil leaned forward, grinning: 'Outside the Sun's photosphere we've found some weird things happen.'

You have this transport of energy uphill, apparently because the temperature is going up, but we're not sure how the balances work yet. Two processes could take energy out of this magnetic field and accelerate the wind. One is when you take two magnetic fields pointing in the opposite direction to a plasma and you push them together and they break and re-form and liberate some energy doing that. And that's one of the processes responsible for explosive leaks of energy – putting energy into the solar wind.

The second one, to make a crude analogy, is like a microwave oven. The material flowing out from the Sun starts subsonically to go upwards, but as it gets further out it becomes supersonic – as well as being heated it's being *accelerated* – and by the time it gets to Earth, it can be travelling from anything between three to four hundred kilometres per second. So there's this process we don't understand; but we're now at the point of putting it all together.

Neil was about to leave for the Antarctic, to analyse the external structure of solar wind gusting there (later that morning I would watch footage with another JPL scientist, Paulett Liewer, of a comet being blown off course by the wind, both the comet's tail and the wind as clearly visible as the wisps of cloud in a medieval painting). But at this point came a knock on the door, and in walked a slight, well-groomed man a few years younger than his team leader. Marco Velli.

Marco told me that he had been involved with trying to reach the Sun for over thirty years, 'and it finally looks as if we're getting there'. The original plan, he explained, had been for *Solar Probe Plus* to get as near as three radii from the Sun's surface: approximately 1,297,356 miles. 'Now we're talking about ten, but that still means the radiation will be unbearable.' The satellite, he said, should be ready for launch in 2015. He handed over a hologram showing how *Solar Probe Plus* will appear in orbit: part of the great red furnace taking up two-thirds of the picture, with, bottom left, a cone of laser light, shining forth from a thick protective disc, behind which stood a yellow metal tower containing the recording devices. As one moved the card, the beam of light moved even deeper into the Sun's depths, while the star itself billowed out its winds and particles. Even as a mock-up, it made me catch my breath.*

What would photos from the satellite likely show? The Sun's mottled, grainy texture grows progressively denser as one moves towards its core, he said. The consistency around that core has been likened to thick yoghurt; some of the magnetic fields found there are 6,000 times more powerful than Earth's.

Solar Probe is designed to go into the Sun as far as we can, to measure particles and also the structure of the solar atmosphere. We'll be asking: where is all its energy stored? How can it be released so fast? Why is the wind the way it is? There is a seeming contradiction in the universe being so far away, requiring such complex techniques to explore it, while the Sun is right there, and we're looking at it all the time. In the next twenty years we will come to know how the solar corona is powered, how solar wind is formed and we will begin to learn the limits of the climate system, and of meteorology, and how solar reactions are generated by the wind.[40]

That wind, he continued, does not blow uniformly, but in fast and slow streams: around the Sun's equator it billows out at three to four hundred kilometres a second, decreasing from around the region of sunspots, while at the polar regions it travels much faster, about eight hundred kilometres a second. Marco, now as fired up as Neil had been, explained about plasma, how to define the photosphere and what might cause the Poles to switch magnetic forces, until it was time for a final question: 'Could you have chosen a more complicated subject to have immersed yourself in?' From out of Marco's small frame rolls a great basso profundo 'Ho-ho-ho-ho', like a department store Santa Claus. 'I doubt it.'

☼

*At the same time as the $750 million *Solar Probe Plus*, the European Space Agency is developing *Solar Orbiter*, to be launched in 2017. After travelling for three and a half years, covering some 75 million miles, it will enter into orbit at 20 million miles distance, and is programmed to take the first-ever photos of the Sun from above each of its poles. The satellite's leading edge will rise to a temperature of 600°C, whereas the spacecraft's interior will remain at room temperature.

By their very nature, satellites orbiting high in the heavens are dramatic, but they pose major problems. For example, one carrying a large telescope needs special protection once it comes within ten radii of the Sun, and will be hugely expensive. In comparison, ground-based telescopes are achieving what satellites may only occasionally attain. Consequently, larger and larger telescopes are being developed for ground use: monsters, such as the Quest camera and Samuel Oschin telescope on Mount Palomar in southern California, enabling astronomers, in the words of one writer, to 'peel away cosmic history like the layers of an onion'.[41]*

These devices can pick up the most distant known quasar, 13.5 billion light years away – and correspondingly many years ago. As of December 2009 some 347 planets had been detected outside our solar system, the smallest of which (unpoetically named 'Gliese 581') is five times as massive as the Earth. Staring up through our atmosphere used to be like examining a passing plane from the bottom of a swimming pool, but a new technique called adaptive optics, developed by the US military and now used extensively in astronomy, compensates for the fluctuations of air, so that, rather than having their view blurred by turbulence, stargazers get hard-edged pictures.

The latest examples of this breed are the Advanced Technology Solar Telescope currently under construction at Sunspot, New Mexico, to be sited finally at Haleakalā, in Hawaii, and Giant Magellan T, with a main lens 83.3 feet in diameter, being built in Chile for operation in 2013, which will have four times the light-gathering power of any previous telescope, and should be able to view extra-solar planets directly. While such grand instruments will harvest information far beyond our solar system, they are also crucial to learning about the Sun. The ATST, for instance (brainchild of the Global Oscillation Network Group – GONG – a consortium of twenty-two institutions worldwide), is set to reveal the nature of the small fibrils of flux tubes within the chromosphere, believed to be the fundamental building blocks of its magnetic structure. The great

---

*While high science will be employing ever larger instruments, technological developments in the 1990s have made digital imaging of the sky possible for amateurs. Now, rather than use telescopes, many stargazers employ cameras that transfer the images to computer screens.

fields tend to break up into separate compressed tubes of flux; the new telescope should explain why.

Such innovations are not the only ground-based technologies interrogating the Sun. Computers can model key parts of our system, from moons to solar winds, with increasing realism. In yet another new field, laboratory astrophysics, researchers create laboratory versions of plasmas whose behaviour mimics phenomena such as the explosion of stars, the formation of galaxies, coronal ejections and solar prominences – huge gaseous arches extending outwards from the Sun's surface. Meanwhile, GONG has helped construct six stations of velocity imagers around the world (Mauna Loa, in Hawaii, Big Bear Observatory, in southern California, Idaipur, on an island in Rajasthan in north-west India, Learmouth, on the north-west cape of Australia, Cerro Tololo, in Chile, 310 miles north of Santiago, and El Teide, in the Canary Islands), so spaced that at least two of them can monitor the Sun at any given time.

☼

Though an accounting of all the current solar research ventures may sound a little like the chairman's annual report, beyond the dry facts is the reality that we are on the verge of huge breakthroughs. In 1952 the astrophysicist G. P. Kuiper could write confidently that 'The golden age of solar physics has not yet been experienced'.[42] Sixty years on, it is upon us. As early as 1979 Carl Sagan too was referring to a coming golden age:

> In all the history of mankind, there will be only one generation that
> will be first to explore the solar system, one generation for which, in
> childhood, the planets are distant and indistinct disks moving through
> the night sky, and for which, in old age, the planets are places, diverse
> new worlds in the course of exploration.[43]

'We are positioned for a new era of discovery about the Sun and the subtle ways it affects life on Earth,' states GONG authoritatively.[44] Carlos Frenk of Durham University told *National Geographic*: 'It's no exaggeration to say that we're going through a period of change analogous to the Copernican revolution.'[45] Then there were Neil Murphy's final words to me: 'We're now in a position to pick up information beyond all previous imagining, data so complex that at last we should be able to deduce the nature of this intolerably inaccessible object! We're trying to find all the pieces of a puzzle that go together; and I think we will.' 'Indeed,' concludes Eugene Parker, with a certain welcome acidity, 'the activity of the Sun provides so many effects outside the realm of conventional laboratory physics that its contemplation is a humbling experience for the serious physicist, repeatedly demonstrating the incorrect nature of our best ideas and explanations.'[46]

UNDER THE WEATHER

'I've read somewhere that the Sun's getting hotter every year,' said Tom [Buchanan] genially. 'It seems that pretty soon the Earth's going to fall into the Sun – or, wait a minute – it's just the exact opposite. The Sun's getting colder every year.'

F. Scott Fitzgerald, *The Great Gatsby*[1]

Some say the world will end in fire,
Some say in ice.
From what I've tasted of desire
I hold with those who favor fire.

Robert Frost, 'Fire and Ice'[2]

Over the centuries it has been established that the odds are heavily against our world surviving for ever. In *A Choice of Catastrophes* (1978), Isaac Asimov examines the range of disasters that threaten us,[3] defining five classes of peril: that the entire universe might so change that our world becomes uninhabitable; that the Earth might undergo a convulsion making life impossible; that something (perhaps man-made) might destroy us; that civilization as we know it might be undermined, condemning what was left of humanity to a primitive existence; and that something might happen to the Sun. In all, he lists sixty-six different kinds of catastrophe, from expanding and contracting universes, black holes and quasars through comets, meteorites and asteroids on to volcanoes, earthquakes, cosmic rays, pandemics, nuclear bombs, pollution

Lightning pierces the volcanic ash of the Icelandic volcano Eyjafjallajökull on 17 April 2010. On average, over the entire Earth's surface, lightning strikes approximately 100 times per second – 8,640,000 times a day

and depletion of resources. By the end one becomes almost light-headed, and it is a relief to turn to such age-old worries as wars and population explosions.

What is notable thirty-odd years on is that 'hole in the ozone layer' and 'global warming' are virtually absent from Asimov's list, the former appearing just once, the latter not at all. While both involve noxious gases, greatly increased heat and catastrophic denouements, they are very different. The ozone layer has been, in the view of most scientists, to a large extent repaired; not so global warming.*

Ozone (from the Greek *ozein*, 'to reek') is an unstable form of oxygen created when sunlight splits a standard oxygen molecule (O_2) into two single oxygen atoms that in turn combine with two more molecules of oxygen to form two molecules of ozone (O_3). It is found in the thin layer that encases the atmosphere some twelve to sixteen miles above the

---

*On 3 March 2006 *The Times* published a letter in its 'Q&A' column from a Mr Thornton of the Scottish border town of Galashiels, which asked: 'When was the term "global warming" first used?' Unlike nearly all other entries, this question was never answered. The first use I can find is in *Science* (8 August 1975) by Wallace S. Broecker, a long-time researcher at Columbia University, who titled his article, 'Climate Change – Are We on the Brink of a Pronounced Global Warming?' – and the phrase caught on.

Earth – and on Earth itself, where it has a strong metallic odour that becomes evident whenever an electric discharge (such as lightning) passes through oxygen. Perhaps because of ozone's chlorine-like smell, it was for many years viewed as having cleansing properties; the Victorians spoke of 'taking the ozone' at the seaside for their health (though in fact what they smelled was rotten seaweed, which is similarly pungent). Tubercular patients were often sent to mountain resorts, with their supposedly higher ozone levels; to attract customers such places would advertise themselves as 'ozoniferous'. In 1865 a letter in *The Times* called the gas 'Nature's Great Cleanser' and extolled its 'bactericidal capabilities'.

Ozone is found at ground level, but not in any bracing way. In city conditions, it causes smog, irritates the eyes, nose and lungs and has been linked to asthma. It also harms plants, insects and fungi, and reduces by as much as half the capacity of trees to sop up greenhouse gases through photosynthesis. This is what one science writer calls its 'Mr Hyde side'.[4] On the other hand, the 'ozone layer' – first detected in 1913 by the French physicists Charles Fabry and Henri Buisson – is the Dr Jekyll side: it shields us from all kinds of ultraviolet light, and has done so for a billion years. Yet the ozone layer, which encircles the Earth far less densely – comparatively – than a film of mud on a tennis ball, is on average just three miles deep. It turns out to be fragile too: in May 1985 scientists with the British Antarctic Survey detected a hole in the layer, and the following year American research confirmed the bad news. The polar ozonosphere thins seasonally, but the size of these gaps was something new – indeed, between 1978 and 1985 the Earth's whole ozone shield eroded by 2.5 per cent. It was soon predicted that before long these depletions would spread to over thirty per cent of the globe's polar regions, allowing solar radiation to do incalculable damage.

It now appears that chlorofluorocarbons (CFCs), for decades hailed as one of the century's great inventions, may be the main cause of the hole. CFCs were the brainchild of an American mechanical engineer-turned-chemist and inventor, Thomas Midgley, Junior (1889–1944), who ended up with over a hundred patents. (It has been remarked that Midgley 'had more impact on the atmosphere than any other single organism in Earth history'.)[5] In 1930 he developed a non-toxic refrigerant for household appliances, dichlorodifluoromethane, a chlorinated fluorocarbon (CFC) that he dubbed 'Freon'. CFCs replaced the various toxic or explosive substances previously used in refrigerator heat pumps. This wonder gas, odourless, colourless and non-reactive, served not only as a coolant in refrigerators but also as a foam-blowing agent in fire extinguishers and a propellant in aerosols.

For over four decades chlorofluorocarbons seemed wholly beneficial. Their lack of reactivity confers on them a lifespan of more than a hundred years, allowing them time to diffuse into the stratosphere. In the 1970s three chemists (in 1995 to share the Nobel Prize), the Dutchman Paul Crutzen and the Americans Mario Molina and F. Sherwood Rowland, began a joint research project, simply as an academic exercise, to find out what happened to CFCs as they drifted on upwards, and discovered that, although they were stable near the Earth's surface, once in the stratosphere ultraviolet radiation broke off their chlorine atoms, which on their own are highly reactive and have the effect of decomposing the ozone layer. This occurs with particular severity around the South Pole, where extreme weather patterns, super-cold winters and icy clouds concentrate the effect; but diffusion spreads the thinning as far equatorward as the middle latitudes.[6]

Although the first popular study of the damage was published in 1978, the same year as Asimov's book, governments long ignored the evidence.[7] At one point President Reagan's Interior Secretary suggested that, if CFCs were indeed so destructive, anyone worried should just wear sunglasses and buy a hat: 'People who don't stand out in the Sun – it doesn't affect them,' he claimed.

Eventually, a United Nations protocol, signed in Montreal in 1987, severely restricted the production of CFCs. The agreement has been acclaimed as 'far and away the most important piece of regulatory legislation on any environmental issue';[8] but given the chemical's long lifetime, the situation had to get worse before it got better. By January 1993 the Earth's average ozone layer concentration was the lowest on record (this remains the case); by March 1999 researchers were claiming that fish in Antarctica would soon be at risk of sunburn. Away from the Poles, solar radiation was said to be causing other problems, visiting extreme deformities, such as extra hind legs, on more than sixty species of frog and toad.[9]

Only in recent years, as the UN restrictions have begun to take effect, has the ozone layer shown signs of returning to its presumed natural state. Even so, many scientists maintain that it will take years before all CFCs are processed out of the atmosphere (several countries, indeed, still use them), and the layer is still about ten per cent thinner than it should be. Full recovery is not expected at the middle latitudes until about 2050, in the Antarctic until 2080. In 2002 satellite observations indicated that the Antarctic hole had shrunk from 9 million square miles to 6 million, the first dramatic reduction since the 1987 ban; but the hole evidently fluctuates in size, and as recently as September 2006 it matched its 2003 extent – exposing an area larger than the whole of North America. The patient is still convalescent.

Of course, this presumes that the diagnosis is correct in the first place. For the Sun seems to play a role here too. In 2004 I went to talk to scientists at the Kitt Peak solar centre in Arizona, and spent a morning in the library of the local paper, the *Tucson Citizen*, which I discovered had, back in 1986, carried an AP report of satellite observations suggesting that it was not only CFCs but the Sun too that had been depleting the ozone. The article did not deny that CFCs carried particular responsibility – they plainly did – but it pointed out that an unusually intense climax of solar activity in late 1979 and early 1980 had set off a cascade of chemical changes that had reduced ozone worldwide, with a pronounced depletion over Antarctica, where ozone levels rose again as solar activity subsided. In other words, solar variations were affecting the ozone layer, and will continue to do so – a useful corrective to bear in mind as we turn to that other matter in which Isaac Asimov had displayed so little interest back in 1978: global warming.

☼

People have long suspected that human activity could alter the climate (which word experts usefully define as 'thirty years of weather'). But to what extent? And how does that relate to the various natural factors that have been identified – fluctuations in solar output, outpourings of smoke, dust and sulphur from volcanoes, the effects of clouds ('still the climate expert's Achilles heel')[10] and water vapour; over a longer term, the rise and erosion of mountain ranges (which divert wind patterns and ocean currents) – and even changes in the composition of the air itself?

The first human-related warming effect may have been set in motion some 10,000 years ago, produced by the fires, clearing of vegetation and soil disturbance of early farmers. That gases in the atmosphere might warm up the Earth was recognized in 1827 by Jean-Baptiste Fourier (1768–1830), a mathematician who had served under Napoleon in Egypt. Fourier pointed out the similarity between what happens in the atmosphere and in a greenhouse, as incoming sunshine warms up plants and soil faster than heat can escape either planet or glass-walled building. Were there no greenhouse gases, the Earth's temperature would average about –40°F (–40°C; the one temperature where the two scales coincide). However, up until the end of the nineteenth century it remained unclear to what extent humans were responsible for these gases.[11] Then in 1896 the great Swedish chemist Svante Arrhenius (1859–1927) described how the burning of coal during the Industrial Revolution, which was releasing the trapped carbon from millions of years' worth of vegetation, might alter the carbon balance – and the temperature. The increase in CO_2 (which he called 'carbonic acid gas', its name at the time) would trap solar infrared energy that would otherwise radiate back into space, thereby raising the planet's average temperature. This was what his much older contemporary John Tyndall (the man who explained why we see the sky as blue), picking up on Fourier, had a few years before dubbed the 'greenhouse effect'. 'We are evaporating our coal mines,' wrote Arrhenius – coal being about seventy per cent carbon by weight. But such warnings found few supporters, most scientists arguing that whatever humans did was not significant enough to have any effect. Anyway, they added, most carbon dioxide came from volcanic emissions and other natural sources, not from human activity.

Over the next three decades the argument did not greatly develop; but then, in the 1930s, scientists reported that the North Atlantic region and the US had warmed significantly during the previous half-century, although they supposed this was just a phase in some natural cycle. The only voice suggesting otherwise was that of a professional coal engineer and amateur climatologist, Guy Stewart Callendar, who, standing before the Royal Meteorological Society in London in 1938, laid out the previous hundred years' worth of carbon dioxide measurements and warned that greenhouse heating should be a concern. Again, there was a period of silence. Then in the 1950s researchers brought improved skills to bear on the question. In 1959 a California geochemist, Charles Keeling (1928–2005), showed that carbon dioxide levels rise and fall with the seasons, following the ebb and flow of land vegetation in the northern hemisphere (the southern has far less land, hence less variability); but he also demonstrated that carbon levels were rising yearly, a trend that came to be known as the Keeling Curve.*

Over the next ten years the study of ancient pollens and fossil shells confirmed that grave changes could take place within as little as a few centuries or even decades. In 1967 Roger Revelle (1909–91), both figuratively and literally a giant of science – he was six foot four – and the dynamic researcher who brought Keeling to join him at the Scripps Institution of Oceanography near San Diego, became the first person to make concerted measurements of carbon dioxide in the Earth's atmosphere, sending up weather balloons over the Pacific and plunging bottles down into it, to measure radiocarbon in air and ocean respectively. Up until that time, scientists believed the vast seas could absorb any increases in CO_2, but Keeling calculated that the amount being added by humans would take millennia to absorb. Within months, other scientists, assessing the gases released by burning tropical forests, by livestock and rice cultivation, and from town dumps and pipeline losses, were suggesting that average temperatures might rise significantly within the next century. However, the warming seemed a long way off, not a clear and present danger.

There was not even a consensus on what was actually happening over the relatively short term. Analysis of northern-hemisphere weather statistics, for example, showed that a cooling trend had begun in the 1940s. The one point most scientists agreed on was that they scarcely understood this intricate system, which appeared to respond to so many different inputs. It was apparently so delicately balanced that almost any small perturbation might set off a great shift – improved computer

*Keeling employed a gas analyser, which worked by passing an infrared beam through an air sample and measuring how much of the beam got through. The more carbon dioxide in the sample, the more it blocked the beam. Each analyser cost around $20,000 then, about $145,000 now.

models began to suggest how such jumps might come about, for example through a change in the circulation of ocean currents. However, the modellers who suggested this had to make many arbitrary assumptions, and some reputable scientists disputed the value of their results; others, raising questions about the effects of agriculture and deforestation on CO_2 levels, pointed out how little was known about the way ecosystems interact with climate. One unexpected discovery was that the levels of certain other gases were also rising, some of these (those ubiquitous CFCs) in addition degrading the ozone layer.*

In the 1970s, as temperatures began to rise again, international scientific panels warned for the first time that humanity faced a serious threat: a molecule of CO_2 would in the course of its lifetime trap 100,000 times more heat than was released in producing it (indeed, the byproducts of our industrial civilization trap almost 100 times more energy than we actually use).[12] More than half the atmospheric CO_2 was coming from human activity, each of us on average shovelling an entire ton of carbon into the air every year. Al Gore summed up the coming crisis in his 2006 documentary *An Inconvenient Truth*: 'The most vulnerable part of the world's ecological system is the atmosphere, because it's so thin. And we're in danger of changing its basic composition.'†

That agenda-setting film was still a long way off at the time that scientists' warnings first caught the attention of a wider audience – during the

*In *Emma*, Jane Austen describes an apple tree blooming at an odd time of year, which has always been viewed as a mistake on her part. Even Austen's brother Edward questioned her reputation for meticulous observation, writing: 'I wish you would tell me where you get those apple-trees of yours that come into bloom in July.' However, the weather for 1814–15, when Austen was writing *Emma*, was unusually chill, and H. H. Lamb in *Climate, History and the Modern World* cites the decade as the coldest since the 1690s. It was within the period known as the 'Little Ice Age' (1350–1850), further chilled by volcanic dust.

†Although these figures have never been satisfactorily refuted by global warming deniers, a comment made by Oliver Morton remains telling: 'It strikes me how unfortunate it is that English gives only a technical name to something as basic to life as carbon dioxide, something as fundamental as blood and breath. Because it is indiscernible without instruments and has thus never furnished our perceptual world, I'm forced to write about it under a name that, even if it doesn't alienate, certainly can't carry any freight of emotion. "Water" is as rich in imagery as a word can be; "oxygen", coined just over two hundred years ago, has far fewer associations, but still has some general aura of necessity and energy and freshness. "Carbon dioxide" is just a chemical . . . It's salutary to remember that language itself can contrive to mask the richness and relevance of the world that science reveals.'[13]

A weather balloon goes up during hail-prevention operations in Georgia

stifling heatwave of 1988, then the hottest summer on record in the northern hemisphere, when the climatologist James E. Hansen (b. 1941) informed a sweating US Senate committee that global warming threatened mankind. But the continuing uncertainties, and the sheer complexity of climate, made for vehement debate, with Hansen tagged alarmist in chief. Even so, governments were sufficiently worried to establish, that same year, a special Intergovernmental Panel on Climate Change (IPCC) under UN auspices, issuing reports in 1990, 1995, 2001 and 2007.

The first three of these surveys were all phrased with great caution, so that there might be some consensus among members: it was *much more likely than not*, they said, that our planet faced severe warming; but its cause could as easily be natural as man-made. The year following the 1990 initial appraisal, a US National Academy of Sciences report separately declared that there was 'no evidence yet' of dangerous climate change, a question-begging formulation: it was rumoured that the Clinton administration had asked for much stronger wording. The 2001 IPCC report declared that sixty-six per cent of warming was *probably* due to humans. The last time the world had been this warm was

50 million years ago, when, as Elizabeth Kolbert wrote in *The New Yorker* in 2005, 'crocodiles roamed Colorado and sea levels were nearly three hundred feet higher'.[14] By the time of the next survey, issued in 2007 and consisting of 1,572 pages that drew on the work of more than 2,000 scientists from 154 countries, both the tone and the figures had changed: the probability that human activities were responsible was assessed at ninety per cent, and the panel was forthright in asserting that the late-twentieth-century phase had been the result of greenhouse gases, the effect of which was estimated to outweigh the Sun's influence by a factor of thirteen to one.* By the end of the year the panel had been awarded a Nobel Peace Prize, which it shared with Al Gore.

The tide of opinion had clearly turned. Speaking of the high levels of CO_2, as well as of other greenhouse gases such as methane and nitrous oxide (the Kyoto Protocol of 1997 covers no fewer than twenty-four such compounds), Mark Lynas, Environmental Correspondent of the *New Statesman*, declared: 'That greenhouse gases have a warming effect, rather like an extra blanket around the globe, is indisputable, and has been established physics for over a hundred years.'[15]† Gernot Klepper of the Kiel Institute for World Economy, a member of the IPCC, has said bluntly: 'The world is already at or above the worst-case scenarios in terms of emissions.'[17] Another expert, George Philander of Princeton University, recently told *National Geographic*: 'We're now

---

*CO_2 doesn't have to be released into the atmosphere. Carbon capture and storage ('CCS') procedures absorb it, say from a power plant or similar large emitter, then inject it at high pressure into such handy spaces as depleted oilfields. The world's longest-existing CCS effort, maintained by the state-owned Norwegian energy company Statoil at a natural-gas field in the North Sea, began in 1997. Columbia University is offering a $200,000 prize for ideas on how to keep CO_2 below an acceptable level, while in 2007 Virgin's Richard Branson offered $25 million for a workable plan to remove a billion tons of carbon dioxide per year. Meanwhile, a mystery exists: roughly that weight of carbon dioxide pours into the atmosphere every year, but nearly half, forty-three per cent, disappears – nobody knows where.
†In January 2008 Lynas became involved in an acrimonious debate over whether global warming had stopped. The *New Statesman* had run a website article by David Whitehouse (author of *The Sun: A Biography*) which had argued that for the last ten years the Earth's average temperature had evened out and was no longer rising. Lynas contributed a fierce rebuttal castigating his fellow contributor. But the resultant debate on the Internet (more than 1,200 replies on the *New Statesman*'s site, while 12,000 continued on another website, making it the Internet's longest common thread) excoriated Lynas for jumbled reasoning and misapplied statistics, and by some margin upheld Whitehouse's view.[16]

geological agents, capable of affecting the processes that determine climate.'[18] Personal opinions all – but by far the majority of experts share them.

How can we quantify such gas increases? 'In the 1780s,' wrote Elizabeth Kolbert, 'ice-core records show, carbon dioxide levels stood at about 280 parts per million. Give or take ten parts per million, this was the same level that they had been at two thousand years earlier, in the era of Julius Caesar, and two thousand years before that, at the time of Stonehenge, and two thousand years before that.'[19] When industrialization began to drive up these levels, they rose gradually at first, then much more steeply. When we started measuring, in the late 1950s, the ratio had reached 315 ppm. In May 2005 the figure stood at 378; *Scientific American* of April 2007 gave it as 379. By February 2008 it was 383. It has been speculated that the maximum tolerable proportion is 445. To keep concentrations even to that figure, the world must reduce emissions by eighty per cent within forty years – an almost impossible aim, given the costs involved. In June 2009 James Hansen declared that CO_2 levels had already reached 385 ppm, and that CO_2 was being pumped into the air some 10,000 times faster than natural processes can remove it.[20] 'So humans now are in charge of atmospheric composition.' And that is just one greenhouse gas, albeit 'the most general contaminant of all'.[21] Other heat traps include water vapour (pre-eminently the single most common greenhouse gas on Earth), methane, chlorofluorocarbons, nitrous oxide and ozone.

Even if these emissions stopped overnight, the world would warm a further 0.2° by the century's end, due to the release of energy already absorbed by the oceans: actual warming over the past 150 years has amounted to 0.6°C (just over 1°F), to around 59°. Removal of excess CO_2 will become less efficient as the planet heats up. The IPCC predicts that the Earth could be 3.5° to 8° warmer by 2100, while in May 2009 researchers at MIT put the figure as high as 13.3°F.[22] One thinks of Kurt Vonnegut's despairing cry in *Armageddon in Retrospect*, 'What can we do about global warming? We could turn out the lights, I guess, but please don't. I can't think of any way to repair the atmosphere. It's way too late' – an outburst that carries all the more weight when one learns that his brother was one of America's leading meteorologists.[23] No wonder that the idea of constructing some huge planetary sunshade has recently migrated from the fringe to the mainstream.*

All of which brings us back to the Sun. How much is this 'mildly variable star' responsible?[25] At a rough average, its rays strike the Earth's surface at about 288 watts per square metre – the power emitted by three good-sized incandescent light bulbs. The Sun is about fifteen per cent broader than it was about 4.5 billion years ago, and thus is radiating about twenty-five per cent more heat than at its birth. But in what is known as the 'faint sun paradox', the Earth had a warmer surface early in its life – when the atmosphere consisted mainly of carbon dioxide and water. The present-day 'global dimming' is caused by various pollutants that prevent the energy from reaching us, either by reflecting radiation or by condensing more water droplets out of the air, to generate thicker, darker, more obscuring clouds.

From about 1960 to the early 1990s the average amount of sunlight, both in duration and strength, reaching the Earth's surface has decreased by as much as ten per cent. In some regions like Asia, the US and Europe, the drop has been even steeper: sunlight in Hong Kong has decreased thirty-seven per cent.[26] The amount reaching us has diminished everywhere, not just over cities. Air travel may have exacerbated this effect: when commercial traffic was grounded for a couple of days after 11 September 2001, local temperature swings, from day to night, widened by several degrees, due to the absence of the jet contrails that normally absorbed sunlight and at night acted like a blanket (that metaphor again). So it might seem that the Sun was not responsible for global warming, and – as Al Gore put it – 'the science is all settled'.

☼

*According to the Soviet climate expert Mikhail Budyko, all we would have to do is dump gaseous sulphur dioxide into the stratosphere; it would form drops of sulphuric acid, which within months would be whirled around the globe by stratospheric winds, wrapping the planet in a white shroud. Since 2000 US government sources have advocated such techniques as giant mirrors in space, reflective dust pumped into the atmosphere, enormous numbers of tiny, shiny balloons or tinkering with clouds to make them more reflective. The IPCC has labelled such ideas 'speculative, uncosted and with potential unknown side effects'. But in July 2009 the US Department of Energy issued a report analysing the possibility of using aerosol particles to reflect short-wave solar radiation back into space. Geoengineering seems to be the way of the future.[24]

I was willing to accept that it was, but then I heard about Piers Corbyn, a climatologist who specializes in thirty-day, forty-five-day and twelve-month forecasting. Corbyn is director of Weather Action, and the vigorous foe of the British Meteorological Office, which he attacks without respite as being in the hands of those who mistakenly believe that the world is getting warmer solely because of greenhouse gases. The Met Office would probably like him to shut up or go away, but they cannot dismiss him as a freak, because too often, to their embarrassment, his forecasts have trumped their own.

Corbyn is in his mid-sixties, with a luxuriant beard and mop of black hair, both speckled with grey, atop the kind of wiry frame one sees on a long-distance runner (which he used to be). One recent interviewer likened him to Doctor Who. His office, on Borough High Street, the long, slightly down-at-heel arterial road through south-east London, is one of the smallest I have ever squeezed into, hardly affording room for two chairs amid all the papers, teacups, books and printouts. On the wall hangs a framed *Guardian* profile headlined, 'All This Man's Clouds Have Silver Linings'. While still at secondary school, he published three papers on aspects of astronomy and meteorology, as well as building his own weather station. After winning a scholarship to Imperial College, London, he received a first-class degree in physics and moved on to research work – successively on superconductivity, the mean matter density of the universe, galaxy formation, and solar activity.

He had for some time been interested in climate history, and by 1982 had begun to move into forecasting. Two years later came the miners' strike against the Thatcher government, and he was asked to predict whether it would be a cold winter. Yes, he reckoned, the turn of the year would be bitter – both his scientific view and his political hope, as he believed a cold winter would help the strike succeed (one of his brothers is Jeremy Corbyn, the fierce far-left Member of Parliament). His political foresight fell short – the strike failed – but his weather predictions were spot-on: the end of 1984 saw a mild Christmas but the New Year began very cold. 'By 12 January it was dire – my breath was freezing into icicles on my beard – a transcendental moment,' Corbyn says, laughing.

By the summer of 1988 he felt confident enough to place bets on the weather with William Hill, one of the big three bookmakers and the only betting set-up that would take them. They were offering odds of 10–1 against that July being one of the ten wettest of the century. Corbyn put down his money, and won. Soon he was making more than £2,500 a year, but in 2000 William Hill stopped taking his wagers. Now, says Corbyn, they accept bets on anything except the death of the monarch – which they are forbidden to do under law, since it gives the punter an interest in the sovereign's demise – and weather bets placed by Piers Corbyn, MSc, FRAS.

Plainly an evangelist for integrity in weather prediction, Corbyn is affronted by what he sees as the bad science behind most forecasts. 'The global warmers' claim that CO_2 is or has been the main controller of climate fails when past data are examined,' he explained, while energetically searching for a teabag. His research has suggested that water vapour, volcanoes and upper-level clouds all play a greater role than CO_2.[27]*

Also of far more importance, he says, has been the motion of the magnetic towards the geographic North Pole, driving temperatures down sharply enough to cancel out much of their net rise in the last century. In fact, some scientists have been predicting that, far from being threatened by global warming, we may instead be facing an ice age, perhaps only a few hundred years from now. In 2005 the Russian astronomer Khabibullo Abdusamatov predicted that the Sun would reach a peak of sunspot activity in 2011, which would cause 'dramatic changes' in temperatures – downwards, not upwards.[30]

During the last 2 million years, the planet has undergone more than twenty glacial advances and retreats, regions of the world having warmed and cooled differently under the influences of such factors as oceans, mountains and winds. The last ice age began to break up about eleven

---

*I am aware that I have given clouds – those ponds floating above our heads – short shrift, yet they play a vital role in modulating Earth's basic radiation balance. 'All clouds have a split personality,' scientist-reporter Richard Monastersky has observed. 'They feature some characteristics that cool the Earth and others that warm it.'[28] Or as the dying monarch in Ionesco's *Exit the King* complains: 'They don't listen – they do what they like.' Relatively few data have been established about the ways in which clouds respond to other variables, but recently two Danish scientists revealed that the Earth's cloud cover varied in lockstep with a phenomenon known to be tied to the solar cycle. They called their discovery 'a missing link in solar-climate relationships'.[29]

thousand years ago, at which point the glaciers covering much of North America, Scandinavia and northern Asia began to retreat to approximately their current positions.

One possible scenario for the onset of a new ice age would involve the shutting down of the Gulf Stream – which, unexpectedly, would be one consequence of global warming. This is because, as it flows, the North Atlantic Current, an extension of the Gulf Stream that cuts north-east across the Atlantic Ocean, surrounds the lands of the high latitudes of Europe with warm equatorial water and releases heat equivalent to the annual output of a million medium-sized power stations into the overlying air. Together, warm water and warm winds are what keep Europe's climate so mild. One of the factors that drives this system involves vast tracts of seawater that freeze over each winter to form ice shelves, coincidentally releasing salt, which makes the surrounding water heavy enough to sink to the seabed, whence it flows south-west towards the Caribbean before warming again and rising. However, since 1977 the shelves have stopped forming as before (probably because of global warming), and the water has stopped sinking. By 2005 scientists were saying there was an almost one in two chance that the Gulf Stream would cease to flow. Europe's temperatures would plummet, with winter figures in Britain, for instance, dipping to $-30°C$.[31] (Of course, as with almost every view on the subject of global warming, there is a counter view: in September 2009 the journal *Science* published fresh evidence that human activity was not only warming the globe, particularly the Arctic, but might even be delaying the inevitable descent into an ice age.)[32]

Another global-warming-related factor that could, paradoxically, help precipitate a new ice age involves the great volumes of frozen water in Antarctica, which measure several kilometres deep. As the ice pushes slowly oceanward, its progress is delayed by the sheets at the edges. A rise in sea level could disturb these edge sheets, however, and release vast glaciers into the sea, which could drive temperatures down dramatically. There are yet further scenarios for an ice age that do not involve global warming – by humans or otherwise. 'Glacial cycles are initiated by slight, periodic variations in the Earth's orbit,' writes Kolbert, which variations 'alter the distribution of sunlight at different latitudes during different seasons and occur according to a complex cycle that takes 100,000 years to complete.'[33]

Whether we should prepare ourselves for global warming or for a new ice age indirectly caused by a hotter Earth or by some other factor is almost impossible to forecast, given the current state of our technology. (Until the sixteenth century 'weather' and 'whether' were interchangeable spellings.) The various computer models for global climates are – to date – insufficiently accurate. As one modeller complains: 'We are starved for measurements . . . When natural variability is so great, it is dangerous to attribute any particular change to any particular cause.'[34] Nigel Calder, author of *The Weather Machine*, calls climatic change 'one of the untidiest areas of modern science'.[35] Or, as Joyce's Leopold Bloom so charmingly puts it, weather is 'as uncertain as a child's bottom'.[36]

☼

This is where Piers Corbyn, for all his eccentricity (the Met Office early on dubbed him 'the mad scientist'), comes into his own. Unlike the Met Office, he believes that the Sun is central to climate change. While he is secretive about his methods, his forecasts are predicated on the assumption that solar particles and the magnetic links between Sun and Earth, control our climate on all timescales, from days to hundreds of thousands of years. Solar particles are dramatically affected by alterations in the Earth's magnetic field, particularly shifts of the magnetic poles and the particles' excursions nearer the equator. Corbyn charts solar flares and other coronal mass ejections, eagerly ticking off each further influence. 'There is the twenty-two-year cycle, the Moon, the Sun's magnetic field – these all affect the weather. One should also mention the shock waves in solar winds, known as "red spikes", and changes in the tilt of the Earth.' The stronger the Earth's magnetic field, the more particles it collects from the solar wind; and the more particles it collects, the warmer the Earth becomes. The Sun's magnetic activity has more than doubled since 1901, Corbyn points out, reinforcing the field by a factor of 1.4 since 1964: and there is growing evidence that solar magnetic activity is reaching an eight-thousand-year high.

He certainly has made his fair share of incorrect forecasts, such as a white Easter in 1989 and 'raging weather' in September 1997: both periods of serene warmth. But the record of the Met Office is considerably worse: since 1923 it has provided the BBC with forecasts, but the franchise has now been opened up to bidders after a particularly bad year in 2009 – the Met predicted an 'odds-on barbecue summer', which turned out so cold and wet the organization had to apologize – and then predicted a mild winter: cue the coldest January for twenty-three years. A recent poll claimed that seventy-four per cent of people believe the Met Office forecasts are generally inaccurate.[37] Weather forecasters worldwide have hardly a better history.

In this increasingly contentious debate, one point in Corbyn's favour – to my mind – is that the IPCC's first report ignored solar variability, and its recent 'summary for policymakers' still mentions sunlight only in passing, even then omitting the Sun's particle and magnetic effect.[38] As for the Copenhagen conference of December 2009, the emphasis was again on human-caused warming, and solar factors received hardly a mention – just that recent output has been at a low ebb.[39]

For Corbyn, however, everything the Sun does is vital in establishing what we can expect on Earth. The last time I saw him, in March 2009, he had embarked on two fresh areas of research about which he was very excited: first, the effects of the Moon on solar winds and on the Earth's magnetic field, which are sometimes powerful enough to disrupt the magnetic pathways from the Sun to the Earth; and second, what he dubs 'virtual particle heating', in which charged solar particles change the poleward heat flow from the equator, and so affect the circulation of heat worldwide.

Corbyn's record in forecasting (for 2010 he predicts his success rate will be as much as eighty per cent; as ABC News put it, 'he's not perfect, but he's pretty good') suggests that he has it right, and that the IPCC has indeed underestimated the Sun's influence. Just how seriously is hard to gauge. Sami Solanki, the director of the prestigious Max Planck Institute in Katlenburg-Lindau, Germany, sits on the fence: 'The brighter Sun and higher levels of so-called "greenhouse gases" have both contributed to the change in the Earth's atmosphere, but it is impossible to say which has had the greater impact.' Timothy Patterson, Director of the Ottawa-Carleton Geoscience Center of Canada's Carleton University, comes down more squarely on Corbyn's side: 'Solar activity has overpowered any effect that CO_2 has had before, and it most likely will again. If we were to have even a medium-sized solar minimum,' he goes on, using the term for a period of minimal sunspot activity, 'we could be looking at a lot more bad effects than global warming would ever have had.'[40] To round off this list of heavy hitters, Willie Soon, a solar and climate scientist at the Harvard-Smithsonian Center for Astrophysics, wrote to me: 'Bill Clinton used to sum up politics by saying, "It's the economy, stupid!" Now we can fairly sum up climate change by saying, "It's the Sun, stupid!"'[41]

Which side, then, to believe – those who sign on with the IPCC, for whom human-mediated global warming is 'a fact like a rock you could hold in your hand',[42] or those who talk of such alarms as scams? Are we due for another ice age or a planetary roasting? Or both? One recalls the great mathematical philosopher Alfred North Whitehead's reflection: 'There are no whole truths. All truths are half-truths. It is trying to treat them as whole truths that plays the devil.' Two extreme points of view are evident. One paints the bleak picture implicit in Roger Revelle's 'we're at a hinge of history'; the other is close to total denial. One side rallies scientists who believe in the human-caused reasons for climatic disaster, bristling with charts, graphs and research data, while the other side musters its own roster of experts, who deploy a different set of charts, graphs and research to argue their way to a different conclusion – both sides growling that the other is direly wrong.*

What we can say with certainty is that the science is *not* all settled – how can it be, when there is still so much we do not know? Given all I have learned over the last eight years from the sun-watchers in Tucson and on Mount Wilson, the Jesuit astronomer at Castel Gandolfo and the weather analysts in East Anglia, the physicists in Pasadena and scientists in China, Japan, India, Western Europe, Russia and South Africa, it seems plain that nowhere amid this flux of information does there yet exist an adequate theory. We are still very much in the process of understanding climate. I do believe that we fail to control greenhouse gases at our peril,

*In November 2009 a hacker released thousands of emails and documents from the Climatic Research Unit at the University of East Anglia showing that researchers there had been covering up or tampering with evidence, but the outcry was ridiculously overheated: there is plenty of other good evidence to consider.

and that the opening decade of our century was the warmest for 150 years is cause for alarm; although, as Robert Kunzig points out in his study of the world's oceans: 'Maybe a century from now, our fretting about carbon pollution will cause our great-grandchildren to chuckle, as we do now when we read the forecasts, made before the rise of the motor car, of city streets buried in horse manure.'[43] Maybe it is even a form of self-importance to attribute such phenomena as climate shifts to primarily human causes.

Yet one fact in all this debate rings clear and undeniable: the Sun so obviously embraces us, its atmosphere and wind surrounding us, its *matter* washing on our shores, that we must accept that our presiding star remains by far the major influence on our lives, and on our climate. Beyond that, we are all still searching.

31

THE IMPOSSIBLE AND BEYOND

Law 1. When a distinguished but elderly scientist says that something is possible, he is almost certainly right. When he says that something is impossible, he is very probably wrong.

Law 2. The only way of discovering the limits of the possible is to venture a little way past them into the impossible.

Law 3. Any sufficiently advanced technology is indistinguishable from magic.

Arthur C. Clarke[1]

One cannot choose but wonder.

H. G. Wells, *The Time Machine*[2]

For nearly two thousand years, writers of what we now call science fiction have been conjuring up scenarios involving the Sun. In the earliest days of the genre their stories tended to reflect the values and concerns of their age, but had little in common with anything that we would today think of as science. As time passed and man's knowledge of his universe grew, writers of science fiction (never 'sci fi', they insist) began to anticipate – sometimes by hundreds of years – the discoveries of scientists, so that academic discipline and literary endeavour seemed to leapfrog one another in their speculations, on occasion building on each other in remarkable ways.

The first science fiction of which we know was written by Lucian of Samosata (AD 120–c. 180), a Syrian-Roman rhetorician whose tale offers interstellar wars, kidnapping by extraterrestrials and a trip to the Moon, in a pastiche of some of the more outlandish tales in the *Odyssey*. It is called 'True History', because the whole story is nothing but lies. A character named Endymion describes how he had been carried away to the Moon while he slept, but has since escaped. He plans to make war on the People of the Sun, whose King Phaethon has refused to allow him to colonize Venus. In the ensuing titanic struggle, the People triumph and Phaethon builds a high wall that prevents the light from his domain from ever reaching the Moon, thus casting it into permanent darkness.

Those other literary lions of ancient times, both of whom preceded Lucian – Cicero, in *The Dream of Scipio*, and Plutarch, in *Of the Face in the Moon Disc* – had also written works of speculation about celestial bodies, but these were non-fiction. To find anyone else among these early imaginers who wrote science fiction (at about the same time as Lucian, so far as we can judge) we have to travel around the globe to Zhāng Hêng (AD 78–139), the great Chinese court astronomer, who in his story 'Meditation on the Mystery' described a journey beyond the Sun.[3] Many centuries later, Kepler would pick up again on the idea of space travel when he wrote the beginnings of a story called 'The Dream', involving a boy's trip to the Moon.

The seventeenth century saw a proliferation of epic tales of interplanetary adventure,[4] a fad that, fuelled by the explosion in scientific enquiry and the attention given it by writers of all kinds, continued full tilt into the eighteenth. In 1705 Daniel Defoe published *The Consolidator, or Memoirs of Sundry Transactions from the World in the Moon*, in which he anticipates a machine powerful enough to reach our then-distant satellite.

Twenty years on, Jonathan Swift published *Gulliver's Travels*, in which, perhaps for the first time, even if only in jest, a writer contemplates the idea that both the Earth and the Sun are mortal. (One of the fears of the people of Lilliput, for instance, is that 'the face of the Sun will by degrees be covered in its own effluvia, and give no more light to the world'.) When Gulliver visits the great academy at Lagado, he finds that its members live in dread of a comet with a tail 'ten hundred thousand and fourteen miles long' travelling too close to the Sun and acquiring 'ten thousand times the heat of red-hot iron' – enough to burn up the Earth. 'When they meet an acquaintance in the morning,' Gulliver reports, 'the first Question is about the Sun's Health: how he looked at his Setting and Rising, and what Hopes they have to avoid the Stroak of the approaching Comet . . .'[5] But that is not the only troubling scenario; Swift goes on to satirize their fears about the possible demise of the Sun from sheer exhaustion of fuel, an extraordinarily prescient idea, albeit one Swift himself does not seem to take too seriously.

By the nineteenth century speculative writers had started to consider the Sun a heavenly body like any other. William Herschel (1738–1822), although not meaning to write science fiction, left science well behind when he suggested that the Sun might be inhabited. E. Walter Maunder, of the Maunder Minimum, summarized what Herschel had proposed:

> He conceived that it was possible that its stores of light and heat might be confined to a relatively thin shell in its upper atmosphere, and that below this shell a screen of clouds might so check radiation downward that it would be possible for an inner nucleus to exist which should be cool and solid. This fancied inner globe would then necessarily enjoy perpetual daylight, and a climate which knew no variation from Pole to Pole. To its inhabitants the entire heavens would be generally luminous, the light not being concentrated into any one part of the vault; and it was supposed that, ignorant of time, a happy race might flourish, cultivating the far-spread solar fields, in perpetual daylight . . . but we now know that it corresponds in not a single detail to the actual facts.[6]

Accurate or not, this might seem a promising area for fiction, but writers left it alone. Jules Verne, whom one might have expected to have been curious about the Sun, actually had little to say about it, though he does mention it (if somewhat slightingly) in *20,000 Leagues Under the Sea* and its sequel *The Mysterious Island* (1874), a romance rather than a work of science fiction: Verne writes about the effect of equinoxes, the contrasting natures of solar and terrestrial mean time, the rotation of the Earth and using a gnomon to tell the time; but imaginative leaps about the star were outside his orbit.

A host of other nineteenth-century novelists tried their hands at science fiction at least once – Kipling ('With the Night Mail' and 'As Easy as ABC'), Poe, Fenimore Cooper, Twain, Melville, even Henry James – but for a major imaginative work that draws on the Sun one has to wait until 1895, when a London publisher paid £100 (no small sum then for a young man just out of college) for a story of some 38,000 words – one of the most memorable works of science fiction ever written: *The Time Machine*.

H. G. Wells (1866–1946) had penned at least three previous versions of the story for various magazines, but in book form it mesmerized the reading public. No one had written a tale quite like it before. A product of its time, and of the distinct sensibility of its author, already a propagandist for his conviction that the future will ever be the direct outcome of existing social forces, it is as much dystopian socialist outpouring as science fiction. Wells imagines an inventor who discovers how to transport himself into the future and return, starting off in the seemingly stable days of the 1890s and ultimately travelling 30 million years forward in time. In this far-off world, the never-named Time Traveller sees the last living things on a dying Earth, which 'has come to rest with one face to the Sun', one hemisphere for ever seared, the other lost to eternal cold and darkness. In the former, menacing crab-like creatures wander blood-red beaches, amid 'intensely green vegetation'. The Time Traveller jumps yet further forward, to see the Sun growing ever dimmer and more crimson:

At last a steady twilight brooded over the Earth, a twilight only broken now and then when a comet glared across the darkling sky. The band of light that had indicated the Sun had long since disappeared; for the Sun had ceased to set – it simply rose and fell in the west, and grew ever broader and more red . . . At last, some time before I stopped, the Sun, red and very large, halted motionless upon the horizon, a vast dome glowing with a dull heat, and now and then suffering a momentary extinction. At one time it had for a little while glowed more brilliantly again, but it speedily reverted to its sullen red heat. I perceived by this slowing down of its rising and setting that the work of the tidal drag was done. The Earth had come to rest with one face to the Sun, even as in our own time the Moon faces the Earth.

> So I travelled, stopping ever and again, in great strides of a thousand years or more, drawn on by the mystery of the Earth's fate, watching with a strange fascination the Sun grow larger and duller in the west-ward sky, and the life of the old Earth ebb away.[7]

Wells was in fact expanding on a theme that has fascinated writers at least since Swift: the question of how long the Sun (and hence our planet) might last. Several works of Wells's time – such as the eminent astronomer Camille Flammarion's *Omega* (1893–4), George C. Wallis's 'The Last Days of Earth' (1901) and William Hope Hodgson's *The House on the Borderland* (1908) – rest on the assumptions that the Sun is sustained by chemical combustion, and that there will come a time, within the foreseeable future of civilization, when it will burn itself out.[8]

Winston Churchill, in his one novel, *Savrola* (1899), also envisioned a depleted Sun, but he posited this gradual process of extinction as an opportunity, a challenge mankind could meet, given certain moral and psychological qualities. In one episode his protagonist (a highly idealized version of himself) gazes at Jupiter through a telescope and thinks of 'the incomprehensible periods of time that would elapse before the cooling process would render life possible on its surface', which could conceivably lead to some kind of extraterrestrial Utopia. However, in the end he concludes that even if this came about, inexorably 'the perfect development of life would end in death; the whole solar system, the whole universe itself, would one day be cold and lifeless as a burned-out firework'.[9] This meditation might almost be a précis of the end of Wells's story – both melancholic conclusions written by ambitious depressives, two great men dedicated to improving the common lot.*

By the 1930s it was known that the Sun produced heat by thermonuclear fusion, and that it would one day die out. Clare Ashton, one of the few women then in the genre, produced *Phoenix*, a poignant story about a journey through space to rekindle the fading star – an idea used yet

---

*Churchill was a great fan of Wells's work, borrowing occasional phrases from it such as 'the gathering storm', and in 1931 declaring that he could even 'pass an exam' on his hero. In 1947 he paid an amusing tribute to Wells as a 'seer. His *Time Machine* is a wonderful book, in the same class as *Gulliver's Travels*. It is one of the books I would like to take with me to Purgatory.' Wells for his part created a caricature of Churchill in *Men Like Gods* (1923).

Danny Boyle's 2007 film *Sunshine* imagined a dying Sun, and a manned rocket
sent to revivify it by blasting it with a thermonuclear bomb

again in Gene Wolfe's series *Book of the New Sun* (1980–3).* Although the
solar surface temperature had been established in the 1890s, John
Mastin also describes such a voyage in *Through the Sun in an Airship*
(1909), and H. Kaner set *The Sun Queen* (1946) on a sunspot.

From the 1930s on, science-fiction practitioners were beginning to be
recognized as something more than just enthusiastic amateurs: in the
November 1945 editorial of the influential magazine *Astounding Stories
of Super-Science*, the great science-fiction editor John W. Campbell roundly
declared of the previous decade: 'The science-practitioners were suddenly recognized by their neighbors as not quite such wild-eyed
dreamers as they had been thought, and in many soul-satisfying cases
became the neighborhood experts.'[10] Clifford D. Simak's 'Sunspot
Purge' (1940) and Philip Latham's 'Disturbing Sun' (1959) correlated
events on Earth with sunspot cycles. Robert Heinlein's six connected

---

*Film and television have shared an interest in this idea. There has been little interest in the
Sun per se (although in *Star Wars* Luke Skywalker sees across the vast desert landscape of
his home planet *two* setting suns), films most understandably concentrating on space travel
and transspacial rivalries. The premise of the 2007 movie *Sunshine*, directed by Danny Boyle
of *Slumdog Millionaire* fame, is that the Sun is rapidly losing power, so a manned rocket is
sent into space to set off a rejuvenating thermonuclear explosion: 'Our Sun is dying; our
purpose – to create a star within a star,' as one scientist puts it. The scriptwriter Alex
Garland kindly took me around the set, a few acres of warehouses in south-east London;
every detail, even the ship's oxygen gardens, had been checked with astronomers: a possible first.

stories, *The Man Who Sold the Moon*, of 1950 ('daring adventures of the bold men of tomorrow . . . a world where energy from the Sun will be converted directly into power') ends with a chart of 'Future History 1951–2600', forecasting weather control by 2075, interstellar space travel by 2100, followed by consolidation of the solar system.[11]

The 1950s began a second golden age for science fiction, and both Cordwainer Smith (the pseudonym of the great counter-Chinese psychological warrior Paul Linebarger, 1913–66) and Poul Anderson (1926–2001) belong to this period. In Smith's story 'The Lady Who Sailed the *Soul*' (1960) he depicts voyagers crossing the heavens on the wings of the solar wind.* Anderson, equally happy writing within classic science fiction or in the new world of science fantasy, interested himself in plausible non-Earth-like planets, the benefits of space exploration and faster-than-light travel beyond a solar system he regarded as 'a hostile barren mess'. In his tales 'A Sun Invisible' (1966), 'World Without Stars' (1966) and 'Day of Burning' (1967) his space travellers visit other systems, while in *The Long Way Home* (1955) he envisions the possibility of a solar system that has become the seat of an interstellar empire, Earth turned into an Eden by global warming, with the Sun as an innocent bystander; but then the Sun is hardly ever a guilty participant.

A similar confidence in human ingenuity appears in Theodore L. Thomas's *The Weather Man* (1962), which features technicians who skim the solar surface in 'sessile boats' to modify its radiation – a characteristically optimistic vision by this chemical engineer-turned-patent lawyer. In Philip E. High's *Prodigal Sun* (1964), a sudden outpouring of intense solar radiation heads towards Earth, and scientists have to create a shield of gas in the upper atmosphere.

Other writers have imagined the Sun heating up rather than cooling down or disappearing. David Brin's *Sundiver* (1980), featuring a spectacularly close encounter between man and star, shows an exceptional

---

*As recently as November 2009, the Science Correspondent of *The New York Times* wrote: 'About a year from now, if all goes well, a box about the size of a loaf of bread will pop out of a rocket some 500 miles above the Earth. There in the vacuum it will unfurl four triangular sails as shiny as moonlight and only barely more substantial. Then it will slowly rise on a sunbeam and move across the star . . . [It] will mark a milestone for a dream that is almost as old as the rocket age itself, and as romantic: to navigate the cosmos on winds of starlight . . . ' By the time you read this, science fiction should have become science fact.[12]

grasp of modern science. In *One in 300* (1954), John D. MacDonald follows ten people who take off for Mars, shortly after it has been discovered that the world is doomed: 'Earth by this time was dead, boiled sterile.' (More fiction has been written about Mars than about any other heavenly body.) Robert Silverberg's 'Thomas the Proclaimer', one of three separately authored novellas that make up *The Day the Sun Stood Still* (1972), has the Sun stopped in its path, to become a potential flytrap for spaceships.

None of these authors had much interest in exploring the Sun for its own sake, or making it the centre of a story because of what was happening within it – this is true even of Hugh Kingsmill's 'The End of the World' (1924) and Larry Niven's 'Inconstant Moon' (1971), both based on the idea that the Sun might go nova. One exception is Ray Bradbury, who in his 1953 tale 'The Golden Apples of the Sun' sends another Icarus thither 'to touch it and steal part of it away for ever'.[13] Left with the job of bringing this venture to life on the page, Bradbury makes a game if not wholly convincing attempt, depicting a huge metallic cup, held by a strangely anthropomorphic hand, that reaches out and scoops the required amount of 'precious gas'. It proves equal to the task, and the story ends with the rocket heading home, to 'take to Earth a gift of

On 19 May 2005, NASA's Mars Exploration Rover Spirit captured this extraordinary view as the Sun sank below the rim of the red planet's Gusev crater. As Mars is further from the Sun than Earth, the Sun appears only about two-thirds as large

fire that might burn for ever'. Why, the captain asks, this whole exercise of journeying to the Sun, 'playing tag, hitting and running'? This time Bradbury's prose gives us vivid answer: 'Because the atoms we work with our hands, on Earth, are pitiful; the atomic bomb is pitiful and small, and only the Sun knows what we want to know . . . ' This is not the only time that the Sun has been credited with consciousness, but it rates as one of the more compelling.

Somewhere between wild-eyed dreaming and scientific forecast lies the notion of igniting new suns to replace the current one. In 1957–8 Frederik Pohl and C. M. Kornbluth co-authored *Wolfbane*. The story is set in 2203, after a rogue planet, populated by machines known as Pyramids, has kidnapped the Earth and the Moon and taken them off into interstellar space. 'The Sun grew too distant to be of use, and out of the old Moon the Pyramid-aliens built a new small sun in the sky – a five-year sun, that burned out and was replaced, again and again and endlessly again.'[14] The novel begins with a character ruminating on whether the latest sun will be regenerated:

In a week astronomers knew something was happening. In a month the Moon sprang into flame and became a new sun – beginning to be needed, for already the parent Sol was visibly more distant, and in a few years it was only one other star among many.

When the inferior little sun was burned to a clinker *they* . . . would hang a new one in the sky; it happened every five clock-years, more or less. It was the same old moon-turned-sun; but it burned out, and the fires needed to be rekindled. The first of these suns had looked down on an Earthly population of ten billion. As the sequence of suns waxed and waned there were changes; climactic fluctuation; all but immeasurable differences in the quantity and kind of radiation from the new source.

This is as far as Pohl and Kornbluth carry their notion, and the story soon takes more conventional turns; but it was an interesting variation on the theme of suns burning down – a subject that would be taken up and dealt with more ambitiously in works by Isaac Asimov and Arthur C. Clarke, the two 'kings' of twentieth-century science fiction.

Asimov (1920–92) was born in the Petrovichi *shtetl* of Smolensk Oblast, in the Soviet Union, but in 1923 his family moved to the US, where his parents bought a succession of corner shops, most of which sold sci fi in the heyday of the old-fashioned pulps. By the time he was eleven Asimov was composing his own stories (in all, he would write or edit more than 500 books and an estimated 9,000 letters and postcards, publishing works in every major category of the Dewey Decimal System except philosophy). By the age of twenty-one he had written thirty-one stories. 'My status . . . was as nothing more than a steady and (perhaps) hopeful third-rater.'[15] Then on 17 March 1941 he visited the offices of *Astounding*, whose editor, John Campbell, had an idea he wanted to put to his fledgling author:

> He had come across a quotation from an eight-chapter work by Ralph Waldo Emerson called *Nature*. In the first chapter, Emerson said: 'If the stars should appear one night in a thousand years, how would men believe and adore; and preserve for many generations the remembrance of the city of God . . .' Campbell asked me to read it and said, 'What do you think would happen, Asimov, if men were to see the stars for the first time in a thousand years?'
>
> I thought, and drew a blank. I said, 'I don't know.'
>
> Campbell said, 'I think they would go mad. I want you to write a story about that.'[16]

The next day Asimov started 'Nightfall', a 28-pager (13,300 words) that, 23 years later, the Science Fiction Writers of America would vote (by a considerable margin) the best of its kind ever written, and which has been so frequently reprinted that, as Asimov himself recognized, 'it is now so well known that nothing like it can be published again' (though Asimov later wrote a novel-length treatment of the same story, fleshing out many details, including the final catastrophe). I confess to finding the characters tiresome and the plot lacklustre, but along the way it does pose some interesting ideas, particularly about the possibility of the existence of planets that have multiple suns.

The action takes place in Saro City, on the planet Lagash, which over the ages has been bathed in perpetual light from six suns. These are in constant motion, so that at least one is always shining; but when the story opens, one by one five of the six have passed out of sight. When Beta, the final sun, disappears, the whole planet will face mind-breaking night. The people of Lagash totally lack the mental equipment to deal with such a moment, and are poised to become the latest in a series of nine civilizations that have foundered at the height of their culture.[17]

Lagash's multiple suns have real-life counterparts, as we now know. In 2006 a professor of astrophysics reported that researchers continue to discover systems outside our own with at least one planet that has more than one star: one of the most recent, called HD 188753, lies 149 light years away in the constellation Cygnus, and is heated by three suns.[18]

The theme of replenishing – or replacing – a dying sun, indeed the idea of making a moon into a sun (as occurs in *Wolfbane*), had earlier been taken up by Arthur C. Clarke (1917–2008) in his 1951 novel *The Sands of Mars*. In Clarke's story scientists are at work on 'Project Dawn', which involves the ignition of the moon Phobos to burn for at least a thousand years and serve as a second sun for Mars, in the hope that the extra heat, together with the mass growth of oxygen-generating plants, will eventually make the Martian atmosphere breathable. The science of *The Sands of Mars* is in some ways ahead of *Wolfbane*'s, which is not surprising, given its author's stature.

Arthur C. Clarke was born in Somerset and, like Asimov, grew up on a diet of American sci-fi pulp. He served as a radar specialist involved in the early-warning defence system during the Battle of Britain, and he continued to be fascinated by space flight. His early stories appeared in fanzines between 1937 and 1945, his first professional sale coming in 1946, and from 1951 on he devoted himself full time to writing. Many of his later books (he wrote or edited over seventy) feature a technologically advanced but blinkered mankind confronted by a superior alien intelligence, although overall his writing is characterized by the optimistic view that science will enable humankind to explore the universe.*

*Around 1972, after sharing a cab in Manhattan, Asimov and Clarke together published their 'Treaty of Park Avenue', which proclaimed, tongue in cheek, that Asimov was required to insist that Clarke was the best *science-fiction* writer in the world (reserving second place for himself), while Clarke had to insist that Asimov was the best *science* writer (reserving second place for himself). Thus, the dedication in Clarke's book *Report on Planet Three* (1972): 'In accordance with the terms of the Clarke–Asimov treaty, the second-best science writer dedicates this book to the second-best science-fiction writer.' But the implicit acknowledgement that Clarke was the better fiction writer is justified.

UFO enthusiasts believe that flying saucers are particularly visible during solar eclipses. The two circular blobs in this photograph are cited in evidence – as photographed by Miss Mildred Maier of Chicago on 30 June 1954

Seven of his stories directly concern the Sun. One early tale, 'The Star', was published in November 1955 after being entered (and altogether discounted) in a *Observer* short story competition. The action begins on a spaceship ('No other survey ship has been so far from Earth'), whose mission is to visit the remnants of a supernova, and the crew finds 'a single small world circling the star at an immense distance', which had once been host to intelligent life. The long-perished inhabitants have left behind a huge vault, evidence of 'a civilization that in many ways must have been superior to our own'. As he leaves the supernova, the chief astrophysicist, a Jesuit priest, reflects:

I know the answers that my colleagues will give when they get back to Earth. They will say that the universe has no purpose and no plan, that since a hundred suns explode every year in our galaxy, at this very moment some race is dying in the depths of space. Whether that race has done good or evil during its lifetime will make no difference in the end: there is no divine justice, for there is no God. Yet, of course, what we have seen proves nothing of the sort.[19]

Armed with a ship full of computer-generated information, he calculates the exact date of the explosion that caused the supernova, which must have outshone 'all the massed stars of the galaxy', and the story ends with the revelation that it must be the very same one that the shepherds and wise men saw above Bethlehem. That central conceit is clever enough, but what is most interesting is Clarke's early fascination with yet-to-be-mapped worlds, and with the possibility of other suns.

In 1958 he published 'A Slight Case of Sunstroke' – a minor offering, but one that uses the huge power of reflective sunlight as a plot device. In the capital of the Latin republic of Perivia, a crucial football match is in progress and passions are running high. Then the referee, obviously paid off by the visiting side, disallows a Perivian goal. A mistake: the 'handsome souvenir programme' bought up by the home side's supporters is made of tinfoil, and in seconds 50,000 makeshift mirrors are trained on him:

> I never knew, until then, just how much energy there is in sunlight; it's well over a horsepower on every square yard facing the Sun. Most of the heat falling on one side of that enormous stadium had been diverted into the single small area occupied by the late ref . . . He must have intercepted at least a thousand horsepower of raw heat . . . They play football for keeps in Perivia.[20]

Clarke's other solar-oriented stories are interspersed with his *2001* sagas – *2001: A Space Odyssey, 2010, 2060: Odyssey Three* and *3001: the Final Odyssey* – all of them written between 1968 and 1997. The 1960 story 'Summertime on Icarus' has a more serious tone, building on the original fable* to tell of a spaceship, the *Prometheus*, that tries to get close to the Sun by riding the near-orbiting asteroid Icarus, 'the hottest piece of real estate in the solar system':

---

*Daedalus, imprisoned with his son Icarus in a high tower by a vengeful King Minos, constructs two sets of wings, securing them to his own and his son's shoulders with wax. Daedalus warns Icarus that should he fly too high the wax will melt. Father and son lift away. Exulting in his new-won freedom, the boy soars upwards: sure enough, the blazing Sun softens the wax, and Icarus plunges headlong into the sea.

Here was a unique opportunity for a research ship to get within a mere 17 million miles of the Sun, protected from its fury by a two-mile-thick shield of rock and iron. In the shadow of Icarus, the ship could ride safely round the central fire.[21]

Such overreaching seems destined to fail, but Clarke must have been in forgiving mood, for even when Things Go Wrong he allows his main character, Astronaut Sherrard, to survive (with the help of a large metal-foil radiation screen) – though not before he comes terrifyingly close to being grilled alive, and we the readers have been given a lesson on the dangers of getting so close to ultimate power.

Clarke's next solar adventure, 'The Wind from the Sun' (1964), described 'a yacht race with a difference', in which half-square-mile sails run before the storm of radiation blowing from the Sun. Again, Clarke was ahead of most solar scientists:

Deep within the surface of the Sun, enormous forces were gathering. At any moment the energies of a million hydrogen bombs might burst forth in the awesome explosion known as a solar flare. Climbing at millions of miles an hour, an invisible fireball many times the size of Earth would leap from the Sun and head out across space . . .[22]

Throughout his adult life, Clarke urged that humanity's destiny lay beyond the confines of Earth (hence the *2001* books). His work, such as his detailed forecast of telecommunication satellites, published in *Wireless World* magazine in 1945, two decades before one ever orbited the Earth, was often prophetic. Borrowing a phrase from the nineteenth-century psychologist-philosopher William James, he once suggested that exploring the solar system could serve as the 'moral equivalent of war', giving an outlet to energies that might otherwise bring nuclear holocaust.[23] If this seems far-fetched, consider that in 2004 George W. Bush proposed sending humans to Mars, and that Richard Branson is currently proposing space flights for paying passengers.[24]*

*Virgin Galactic, an offshoot of Virgin Atlantic, is due to launch VSS *Enterprise* (in homage to Captain James Kirk's fictional starship), a six-passenger hybrid rocket, some time in 2011. The Sun is some way off, however: the rocket will blast to just sixty-eight miles above the

For Clarke's most memorable solar story we have to go back to 1958, and 'Out of the Sun'. The narrator is aboard a spaceship within the (then believed) twilight zone of Mercury, part of a team exploring what goes on within the Sun at the height of a sunspot cycle: 'From the far X-rays to the longest of radio waves, we had set our traps and snares; as soon as the Sun thought of something new, we were ready for it. So we imagined . . . ' What follows is less a story than a meditation on man's attitude towards the Sun, and is still worth listening to:

> These great clouds of ionized gas moving far out from the Sun are completely invisible to the eye and even to the most sensitive of photographic plates. They are ghosts that briefly haunt the solar system during the few hours of their existence; if they did not reflect our radar waves or disturb our magnetometers, we should never know that they were there . . .[25]

He picks up on his radar scanner a 'tight little echo . . . outside all previous records of solar phenomena', which he reckons to be a gas cloud about 500 miles long and half that figure in width – and seeming to exhibit the characteristics of an intelligent being:

> Today . . . the idea no longer seems so strange to me. For what is life but organized energy? Does it matter *what* form that energy takes – whether it is chemical, as we know it on Earth, or purely electrical? . . . Only the pattern is important: the substance itself is of no significance. But at the time I did not think of this; I was conscious only of a vast and overwhelming wonder as I watched this creature of the Sun live out the final moments of its existence.

As soon as the cloud hits Mercury in the centre of its daylight side it will be dispersed – destroyed for ever:

> Was it intelligent? Could it understand the strange doom that had

---

Earth, and passengers will enjoy four and a half minutes of weightlessness – and stupendous views – over a two-and-a-half-hour journey that will cost them $200,000 per head. Branson promises that within ten years he will get the cost down to $40,000; already 340 seats have been booked. Another company, Bigelow Aerospace, plans to launch an inflatable 'space hotel' by 2015.

befallen it? Perhaps, in those last few seconds, it knew that something strange was ahead of it . . . for it had begun to change . . . It may be that I was looking into the brain of a mindless beast in its last convulsion of fear – or of a godlike being making its peace with the universe.

Whatever one makes of Clarke's flight of imagination, this is science fiction of the highest order – thrilling as a story (yet in total less than six pages), scientifically accurate within its limits and addressing itself to questions that astronomers would find pertinent today. J. G. Ballard has asserted that sooner or later all the predictions of science fiction will come true,[26] a pardonable exaggeration; more accurate is Clarke's observation that 'most technological achievements were preceded by people writing and imagining them'. What we come to do to the Sun – or to escape from it – may well play out scenarios that science fiction has already described.

For instance, in one of his early short stories, Isaac Asimov imagines a technology that could collect the Sun's rays in space and transmit that energy back to Earth. Space-based power stations would be able to capture solar energy twenty-four hours a day and send it anywhere on our planet. At the time of the story – 1941 – launching such a system was not commercially viable; but in 2010 the California-based Solaren Corporation contracted with Pacific Gas and Electric to provide 200 megawatts of power (about half the output of an average coal-fired power plant) from a station in space. Solaren's deadline is 2016.

THE DEATH OF THE SUN

If the Sun should desert the day, What would life be?

Cole Porter, 'Do I Love You, Do I?'

Suns sink on suns, and systems systems crush
Headlong, extinct, to one dark centre fall.
And dark, and night, and chaos mingle all!

Erasmus Darwin[1]

In the final volume of C. S. Lewis's *The Chronicles of Narnia*, that enchanted kingdom perishes when its sun explodes: 'At last the Sun came up . . . They knew at once that this sun also was dying. It was three times – twenty times – as big as it ought to be, and very dark red . . . and in the reflection of that sun the whole waste of shoreless waters looked like blood.'[2] So it must be for all worlds, and will be for our sun too. Before such a time, we may have to endure, if we can, several other horrors. Isaac Asimov, for instance, in his *A Choice of Catastrophes*, considers whether, with both Sun and Earth slightly on the move from their current orbit, they might collide. He dismisses that possibility; whether the Sun might crash into another star is another matter.

The Sun is 32,000 light years from the centre of its galaxy of a hundred billion stars, which it orbits at about 155 miles a second, taking about 200 million years to complete a revolution, which means that it has completed 24 or 25 passages in its existence. In the 13.73 billion years of the universe's history, the stars of our galaxy seem to have arranged

themselves into orbits that avoid any further collision; yet there remains
the possibility of a maverick star or globular cluster of stars changing
direction and, if not colliding with the Sun, passing by close enough to
alter its orbit, which would surely affect life on Earth.[3]

The chance of any such calamity Asimov admits is 'very small indeed',
pointing out that anyway we would probably have a million years of warning. However, a drifting black hole (his next category of possible
catastrophes) might give only a few years' notice: in 2005 scientists
detected a hole the size of our solar system that has so far swallowed up
the mass equivalent of 300 million stars the size of our Sun, although
thankfully it is about 26,000 light years (153,400 billion miles) away, or
about 63,000 times the distance between the Earth and the Sun. Should
a mini black hole strike the Sun, our star could probably ingest it without major repercussion; on the other hand, there is at least a chance that
it could bring the Sun to terminal collapse or set off an explosion: not
good news for Earthlings. Asimov goes on to consider clusters of anti-
matter and non-shining bodies (or 'free planets'), both of which might
wander into the Sun's path, to disastrous effect: but then in theory
Britney Spears could win an Oscar. One feels he is ticking off anything
at least physically possible, however unlikely.

In the decades since Asimov's book appeared, the possibility of Earth's
being hit by asteroids has been endlessly rehearsed. There are enough in
our system to worry about: between 1.1 and 1.9 million, with 5,000 new
ones being discovered every month.[4] In October 2010, it was reported
that an asteroid almost thirty feet across flew 28,000 miles above
Singapore: a near miss. Scientists have predicted that a 1,000-foot asteroid will miss Earth in 2029 by between 15,000 and 25,000 miles, then
re-orbit, just possibly to score a direct hit a few years later, probably
around 2034.[5] The Earth's atmosphere is torn by various objects, ranging
from those of basketball-size (several a day) to car-size (twice a year).
Currently, NASA's Large Synoptic Survey Telescope is monitoring the
heavens for such missiles, and it is here that the Sun plays an unusual role:
the telescope has a blind spot, since bodies that are either just ahead or
just behind Earth are easily lost in the solar glare. Sunlight also affects the
rotation of asteroids less than 1.2 miles in diameter at roughly average distance from the Sun: as they absorb its rays, their spin will alter, and they

accelerate on their axes 'like pinballs in a breeze'.[6] Many break up under
the stress, but were an asteroid travelling towards Earth, to divert it we
would need to change its velocity – according to one recent estimate – by
a millimetre per second a decade in advance.*

☼

One way or another, the question of the solar system's stability has fascinated and tormented astronomers for more than 200 years. Somewhat
to the embarrassment of contemporary experts, it remains one of the
most unresolvably complicated calculations in celestial mechanics. But
long before such calculations could be attempted, superstition and ignorance were advancing their own dire predictions. In Scandinavian myth,
for instance, the Sun will eventually fail, to be followed by the three terrible years of 'Fimbulwinter', when the Earth is caught in the grip of a
bone-chilling freeze, after which comes 'Ragnarok', when eternal darkness settles, and Sun and Moon are both devoured, heralding the end
of all things. Such a belief touches a raw nerve: Charles Darwin, writing
to a friend in 1865, voiced 'his own pet horror . . . that of the Sun some
day cooling and we all freezing'.[8] The world's passing is often linked to
the Sun's demise, only, of course, in Darwin's day the source of stellar
energy was not known, since nuclear fusion had yet to be discovered.
Now that we know how the Sun gets its power, and that its stores of
hydrogen are finite, we can plot how the end may come.

The Sun's life is held in exquisite balance: while it is contracting
under its gravitational field, the violence of thermonuclear reaction at
its core exerts an expansive force, the two cancelling each other out. It
can retain its size only if these opposing forces are in sync, which requires
an ever more rapid consumption of its hydrogen. At some point that fuel
will run out. Our star's photosphere is made up of about 90 per cent
hydrogen, 9.9 per cent helium and a mix of 67 elements, such as iron,

*Meteors hit the Earth every year at the rate of one every 7,700 square miles. In February
2005 *The Sun* magazine ran the front-cover headline: 'End Times Shocker – Killer Meteor
Heads for US.' Due to impact by 2012, the meteor 'could easily fulfil the End Times
prophecies of the Four Horsemen of Revelation'. According to this powerful authority, 'the
5,000-foot-wide ball of rock and ice is not only massive enough to destroy the planet, it's
wrapped in an unusual, four-layered cloud of dust and debris thousands of miles across'.
The magazine, however, was equal to the threat, advising its readers: 'If you can afford it,
install an air purification system.'[7]

calcium and sodium, as well as 18 chemical compounds, reflecting the make-up from which the Sun was originally formed (although nuclear reactions will steadily alter the mixture at the Sun's core, where plasma is compacted to such pressure and temperatures as to induce fusion). The larger any star is, the hotter it must be, and the more rapidly it must consume its hydrogen. More massive stars than the Sun run through their nuclear fuel far more rapidly than does our star, because they produce far higher temperatures.

This is all part of a process that has been going on for billions of years, and will continue for many billions more. By at least 2.5 billion years ago the Sun had contracted to a diameter of 186 million miles; only after that did it shrink to a size smaller than the Earth's orbit. At some point thereafter the Earth was formed (by coincidence, at just about the same time that the Sun began to radiate light). The Sun has continued its contraction, but not at a rate to signal its end any time soon: it loses several million tons of matter every second, the mass equivalent of the energy it produces through thermonuclear reactions. Another few million tons are lost as solar wind and other particle emissions.

Although the Sun will eventually cool, it has first to go through a long period during which it actually grows hotter. As hydrogen converts into helium at its core, the resulting contraction will intensify the Sun's gravitational field and raise its inner temperature so that it will slightly expand. Eventually, it will get so hot that it will initiate new thermonuclear reactions, the core's helium nuclei combining to form nuclei of the next higher elements, such as carbon, oxygen, magnesium and silicon; but by now its essential balance will be lost to expansion. As the Sun swells, becoming at its maximum some 256 times as wide across as it is today, and 2,730 times as luminous, it will radiate more intensely from its extending surfaces: it will have become a red giant.

For the next few billion years, the Sun will fuse atoms quietly and steadily – only, rather than burning hydrogen, once its principal nuclear fuel starts to become depleted it will reorganize its structure so it can burn the next most plentiful fuel source: helium. To do this it needs an even higher core temperature, so its inner regions will shrink and get hotter still. However, at the same time its outer regions will expand greatly. During its last billion years, it will get ten per cent brighter still, searing the Earth's surface to 2,000°C, hot enough to start melting our planet. As Dennis Overbye paints the scene:

> Skimming over the flame tops of this giant, the bare, burned Earth would produce a bulge in the Sun. But friction would cause the bulge to lag as it tried to follow the Earth. The gravitational tug from the bulge would slow the Earth and would cause it to spiral inward, where friction from gases in the Sun's expanded atmosphere would slow it even more. Then it would go down [into the Sun].[9]

According to the scientific consensus, we can look forward to about 5.7 billion years of some kind of life on our planet – but not human life, which will have a far shorter span. Our long-range enemies will be heat and lack of carbon dioxide – half a billion years from now, the carbon dioxide concentration will be too low for most plants to continue photosynthesizing, and the Earth's average temperature will be about 120°F. Over the next billion years, after the extinction of plants, the atmosphere will fill with steam, as it did at the Earth's birth, and be bombarded by intense sunlight. Two and a half billion years from now, the last water molecule will have disappeared, and the Earth will be a hot, dead world, covered by molten rock.

It could conceivably escape being engulfed – unlike Mercury and Venus, both of which will be swallowed up by the Sun once it reaches the greatest extension of its red gianthood. The most recent research suggests that although the Earth is at the borderline between being engulfed and surviving, the former fate is the more likely. But if it did remain outside the Sun's great swollen bulk, the enormous heat it would receive is likely to vaporize it.

Should the Earth actually escape being both engulfed and vaporized, what would that mean for us? Would human life even be an option? One scenario is suggested by what happened to a planet in a neighbouring galaxy. About twice a month we discover another extrasolar world; about 750 have been documented. In 2007 astronomers came across an Earth-like planet known as V 391 Pegasi,* a gas giant at least three times

---

*The bookmakers William Hill, on learning that such a planet had been discovered, cut its odds on the existence of extraterrestrial life from 1,000–1 to 100–1. For them to pay out, the Prime Minister has to confirm the proof of such life officially, within a year of the bet's being placed.

Computer artwork of the Sun, around five billion years hence, heating a dying Earth.
The oceans have evaporated, leaving salt-encrusted rocks. The Moon passes in front of
the massively-swollen Sun, already in its first red giant stage

as massive as Jupiter, orbiting about 150 million miles from a faint star
in the constellation Pegasus. That star blew up as a red giant and lost half
its mass, but did not destroy the planet; maybe, if V 391 Pegasi can survive, so might the Earth.

As for other possibilities, as Isaac Asimov suggests, 'There is, at least,
ample warning. If humanity survives those billions of years, it will know
for all those billions that it will have to plan an escape somehow. As its
technological competence increases . . . an escape may become possible.'
So not only will we need to construct timetables for the various stages of
the Sun's demise, but at the same time we must look for ways in which to
leave our planet or to move it away from an expanding Sun. A particularly
dangerous time for mankind would be at the end of the first red giant
phase, when the Sun may become so luminous that its helium fuses in a
gigantic explosive flash. Another period of jeopardy will occur as the Sun
sheds its upper layers, thus diminishing its mass and weakening its grav-

itational field, which will loosen its grip on the remaining planets (including ours). No longer so closely tethered to their orbital paths, they might intersect each other's orbits, or even collide with the Sun. At this point I am tempted to seek refuge in Woody Allen's take on forecasting the future: 'I thought to myself that if our great golden star suddenly exploded this planet would fly out of orbit and hurtle through infinity for ever – another good reason to always carry a cell phone.'[10]

But all is not necessarily lost. Human ingenuity is almost endless, and we are already researching the practicalities of moving to another planet in our solar system, or to one of its satellites; or moving outside the solar system altogether. One possibility, remote as it may be, is to use a series of small nuclear bombs to propel a starship: such charges would be exploded in sequence, so that the spacecraft would 'ride' on the nuclear flash either into permanent orbit away from the Earth or somewhere more habitable. President Obama has already announced a goal of landing astronauts on an asteroid by 2025 as a precursor to a human mission to Mars. Another recent idea, outlined in a 2001 paper by Don Korycansky and Gregory Laughlin of the University of California, Santa Cruz, and Fred Adams of the University of Michigan, is that in the same way that space probes can boost their trajectory by playing gravitational pinball with Venus or Jupiter in order to be flung outwards, so we, or some version of us, could bring about regular encounters between the Earth and a comet or asteroid, thus widening our orbit and propelling us further from the Sun.[11] All that is required is for an asteroid about 62 miles in diameter to fly past us and transfer some of its orbital energy. The asteroid would then move out to encounter Jupiter, acquire more energy and on a subsequent encounter pass it on to us. But this would be about as likely as moving an aeroplane by having a bumblebee go past it every ten years: even the scientists busy developing these scenarios say, 'We are not advocating them as policy.' One wonders what Arthur C. Clarke would have made of such ideas.

At least we can take solace from the fact that, sceptical as the scientists themselves are, we are talking of so many millions of years hence that the technologies that will be available then are impossible to predict. While the Earth may not survive indefinitely, mankind may outlive its destruction. If we assume for the moment that we do manage to leave Earth on a nuclear-propelled starship, or some other vehicle of the future, where would we go? In March 2009 the Kepler space telescope was launched, to detect planets outside the solar system that have roughly the same size, conditions and distance from their stars as Earth. By December of that year a planet, GJ 1214b, was found that was 2.7 times the size of Earth and was orbiting a star smaller and less luminous than our Sun; it was also comparatively near to our solar system – about forty light years away – and was water-rich.[12]

Our means of making such a planet humanly habitable already has a name – 'terraforming'. And it may even be possible, since by the time the Sun has begun to expand sufficiently to start cooking the Earth, enough aeons will have passed that we should have the technology to have established ourselves on hundreds of such extrasolar worlds. As H. G. Wells wrote: 'There is no way back into the past. The choice is the universe – or nothing.' His fellow novelist Tom Wolfe, writing on the fortieth anniversary of man's landing on the Moon, asked: 'When do we start building that bridge to the stars? We begin as soon as we are able, and this is that time. We must not fail in this obligation; we have to keep alive the only meaningful life we know of.[13]*

Not that we should underestimate the extraordinary challenges involved. Should we wish to settle on another planet in our solar system (highly unlikely, as none seems capable of supporting life), we know that Jupiter, Saturn, Uranus and Neptune will remain circling the Sun, even after it has collapsed; and that although their orbits will expand, it will not be by much (less than double), so they will not be moving conveniently nearer to us. Even if such planet-hopping were to turn out to be viable, it is far more likely that by such a time we would have established structures in space capable of supporting large numbers of people, each community ecologically self-sustaining and independent. As the Sun grew hotter, they would adjust their orbits and slowly spiral away.

Such scenarios are not so much science fantasy as science forecast.

*A recent astronomy paper has concluded that 'the probability of finding an exoplanet with a habitat nearly identical to that of Earth's by 2264 is 95 percent.'[14] Astronomers currently estimate that there are an astonishing 2×10^{11} planets within our galaxy that could support human life – over 200 billion (although some experts more pessimistically believe that the Earth may be the only true example). But then the number of possible other universes is said to be 10^{100}.

An artist's impression of a manned base on Mars – modelled on similar bases in the
Antarctic; only the artist has set the camp at the heart of a raging
Martian dust storm, so casting an eerie and lurid light

The astronomer Carolyn Porco voiced the current orthodoxy when she
wrote: 'Humanity's future need not be confined to mere survival on our
home planet. Other worlds beckon, we know how to reach them and we
will once more be outward bound . . . This won't be a space race so
much as a global exodus undertaken by an international community . . .
There could be no better way to say: the future is boundless, and it
belongs to us.'[15] And if that doesn't seem visionary enough, Michio
Kaku, in his book *Physics of the Impossible,* tells us that 'physicists have now
demonstrated that a law that prevents time travel is beyond our present-
day mathematics . . . Statements about what is possible and impossible
have to take into account technologies that are millennia to millions of
years ahead of ours.'[16] In other words, we cannot rule out anything that
we cannot prove impossible.

Though we may be able to dodge catastrophe – in what form, who
can say, because the length of time involved is so great that the human
race will likely have been succeeded (or will have replaced itself) by

an entirely different species – the Sun will not. Eventually, its surface will cool to a point where it glows red-hot rather than white-hot, as in its early days, and enters a second red gianthood. Once hydrogen fusion is no longer the main source of its energy, it can maintain itself for only a comparatively short additional period. It enters a brief unstable phase of diminishing energy returns as it depletes its helium and goes on to burn progressively heavier and scarcer elements – beryllium, boron, carbon, nitrogen and oxygen – each one for a shorter period. The ability of this new giant to keep itself distended against the pull of gravity will falter, and it will begin to collapse, shrinking and expanding before shrinking for one last time. Over the span of a couple of hundred million years (a very brief period on the scale of most stellar lifetimes) the Sun will become lighter, its outer layers blowing into space in the form of a much denser solar wind. It will not be large enough to explode violently, so there will be no danger that in a few hours of fury the entire solar system will be scoured of life. It will simply contract into a white dwarf, a tiny shrunken cinder probably smaller than Mars, but with a density of two tons or more per cubic centimetre – where a teaspoonful of matter would weigh as much as a Rolls-Royce – leaving behind at most a thin film of its outer layer, making of itself a planetary nebula.

Once a white dwarf, it will be a dense ball of carbon and oxygen, no more than a dot of light even within its own system. Seen from the satellites of Jupiter (supposing mankind has pitched its tents there), it will be only 1/4,000 as bright as the Sun appears to us now, and will deliver but a fraction of that energy, as it fades away over the rest of time. Conceivably, long before the moment of white dwarfdom comes, future space settlers will have developed some form of hydrogen-fusion power stations, and so would be independent of the Sun; more likely they will have left the solar system altogether.

As for the Sun itself, as one writer has envisioned its last days: 'At 100,000°C, this pinprick of white light . . . will continue to shine for billions of years, illuminating the giant gas clouds from the Sun's expelled outer layers to form a glorious, multicoloured nebula – our Sun's tombstone.'[17]

Not quite: a final stage is yet to come in that long, drawn-out death. As the Sun cools, it will become a black dwarf, with the detritus of the planets circling round it. It will not detonate into a supernova – again, it is too small; nor will it become a black hole through implosion or any other means. Our great, terrible, even lovable companion will simply dwindle away, a dark cinder of degenerate matter drifting in the vacuum, its life-giving journey done.

SUNSET ON THE GANGES

When I, sitting, heard the astronomer, where he lectured with
much applause in the lecture-room,
How soon, unaccountable, I became tired and sick,
Till rising and gliding out, I wander'd off by myself,
In the mystical moist night-air, and from time to time
Look'd up in perfect silence at the stars.

Walt Whitman[1]

Let us obtain the adorable splendour
of the Sun; may He arouse our minds.

Rig-Veda[2]

Soles occidere et redire possunt, wrote Catullus: Suns can set and rise again. In English the word 'sunset' goes back at least to the 1440s. Shakespeare, as one would expect, wrote of it; so too Milton, Byron, Browning, Longfellow, Frost, Ginsberg and Emerson, as well as many writers of less fame. Different languages summon up different associations – the French *soleil couchant*, for instance, or the German *Abendröte* (meaning the red glow of the evening sky); but in nearly all cultures the splendours that light up the horizon at the Sun's going down are celebrated for their beauty and exploited for their symbolic potential.

Films, and sometimes books, end with hero and heroine departing into the sunset. Deathbed sayings are often sunset references, from Tiberius's contemptuous dismissal of his attendants ('You know when to abandon the setting and hasten to the rising Sun!') to Jung's 'Help me out of bed; I want to see the sunset'.[3] Stephen Vincent Benét (1898–1943) won a Pulitzer Prize for his book-length poem *John Brown's Body*, in which he describes Brown's majestically plain address to the court on being sentenced to death:

> *Here is the peace un-begged, here is the end,*
> *Here is the insolence of the Sun cast off,*
> *Here is the voice already fixed with night.*

Whether seen over an ocean or across a vast prairie, from a great height or simply from inside a city flat, sunsets are vivid experiences that move people ever and everywhere, sometimes to melancholy, elsetimes to rapture. Winston Churchill, in his latter years, would travel to the Atlas Mountains to paint, because sunsets at the desert's edge are so exquisite. Thomas Hardy, when a boy, would sit on the stairs at his parents' home and watch as the evening Sun awoke a special intensity in the Venetian-red walls of the staircase.[4] Andy Warhol once filmed an entire sunset, with no action except the dying light and a jet's contrail drawing across the sky. During the imperial entertainments he put on in British East Africa between the wars, the prodigiously rich Sir Philip Sassoon, cousin of the poet, had the Union Jack hauled down from his mansion's roof because the colours clashed with the evening sky. Rangers in the Sierra Nevada National Park are said to be 'paid in sunsets', while every evening at the eighteenth hole of Pebble Beach golf course, in California, a bagpiper laments the close of day.[5] Perhaps no one on earth, however, has seen a sunset as spectacular as those viewed from space. Astronauts have spoken of being overwhelmed by the prismatic radiances spreading across the face of the Earth.

Artists, of course, are particularly responsive. Van Gogh wrote of sunsets during the mistral:

> Towards sunset it generally grows a little calmer, then there are superb
> sky effects of pale citron, and the mournful pines with their silhouettes
> standing out in relief against it with exquisite black lace effects.
> Sometimes the sky is red, sometimes of an extremely delicate neutral
> tone, and again pale citron, but neutralized by a delicate lilac.[6]

In 2007 a group of scientists produced a study of 554 paintings of sunsets, including many of Turner's 115 renderings, as well as works by Claude, Rubens, Rembrandt, Gainsborough, Hogarth, Copley, Degas, Caspar David Freidrich, Alexander Cozens and Gustav Klimt, to help understand changes in climate. They found 181 artists who had painted sunsets between 1500 and 1900, then used a computer to work out the proportions of red and green along each picture's horizon. As we now know, the magnificence of sunsets depends on cloud formation, or the level of particles in the air. The more airborne particles, the more sunlight scatters towards the red; so the dustier the sky, the redder the sunset.[7] Most of the paintings with the highest red–green ratios were painted in the three years following a documented eruption, there being fifty-four such 'volcanic sunsets'. Turner (all unknowing) recorded the effects of three: those of Tambora, in what is now Indonesia, in 1815; Babuyan, in the Philippines, in 1831; and of Cosiguina, in Nicaragua, in 1835.[8] Though painters are not savants attempting precise depictions of nature, it is interesting to realize that such visions as *The Fighting Téméraire* (1838), once sharply criticized for its outrageous colouring, may be truer representations than once thought.

☼

One of several sunset sites on the Internet publishes a list of the 'world's top ten': 10: Key West, Florida; 9: Ipanema Beach, Rio de Janeiro; 8: The Maldives; 7: Paradise Island, Bahamas; 6: Natadola Beach, Fiji; 5: Grand Canyon, Arizona; 4: Great Pyramids, Giza, Egypt; 3: Anchorage, Alaska; 2: Kaunoao Beach, Hawaii; and, in first place, Oia, a village on Santorini, the shattered volcanic Greek island. I don't begrudge each its appeal, but I wanted to witness a sunset that *meant* something in addition to being visually spectacular. In October 2006 I set off, not for Hawaii or Los Angeles, but for the city of Varanasi, on the banks of the Ganges, taking a few sun-related side-trips along the way.

After landing in New Delhi, and visiting Jaipur and its famous ancient observatory, I flew on to Udaipur, about 250 miles further south. October is the month of Diwālī, the Hindu festival of light, and 21 October, the day I arrived in Udaipur, their New Year. I had arranged that night to see the maharana, Arvind Singh Mewar, known as Sriji (noble leader), and received at my hotel a printed invitation for my summons, the envelope embossed with a sun sporting a mighty whirly moustache, accepted emblem of virility.

At 6.50 p.m. I presented myself at Shambhu Niwas Palace, the buildings ablaze with Diwālī lights, and was led on to a large patio overlooking the famous lake of Udaipur, where Sriji was waiting. He was little more than five foot three, his outsize head adorned by a fittingly luxuriant moustache and full beard beneath lazily watchful eyes. Sixty-two years old (as he told me), he was wearing light-coloured trousers, an untucked striped shirt and loafers, as if holidaying at some Caribbean resort. He asked a manservant for a gimlet (gin and lime), and looked disappointed when I opted to stay non-alcoholic.

'Now: what would you like to know?' Having learned that he had an interest in solar energy, I asked about his fourteen sun-powered vehicles for hire, from seven-geared pedal rickshaws with two seventy-five-watt panels to scooters with twelve-volt lead-acid batteries – and, on the river, a Surya-powered water taxi. Udaipur was already known as 'India's solar state'.

Sriji's response was brisk. 'Oh, yes, all that. It's really not important.' He turned instead to his family tree, whose roots go back to 569 and are ultimately 'descended from the Sun', which is why his family crest shows the star so prominently. 'For me the Sun is a human being,' he said easily. Really? A mass of gases, human? Without missing a skip, he replied, 'Well, maybe not a human *being*, but in terms of spirit, energy, yes, in what he gives to us, I *do* mean that the Sun connects in a special way. He is a god, a part of us. I'm not an intellectual, but it's what I feel, and I suppose what I'm saying is about how we *relate* to the Sun. People in the West seem to have lost that connection, to have lost their way.

'I think we need to redefine "divinity",' he went on. 'Because the Sun *is* divine – only not in the way the word is used in your part of the world. It doesn't so much bring us light as take away the darkness. Each of us needs to find the Sun in ourselves.'

The star had certainly endowed him with more than average energy. In his youth he had been a fine cricketer, an able bat who represented Rajasthan. More recently he had overseen the building of a dozen museums and libraries, put together a vintage car collection, formed a polo complex, managed a successful stud farm, pioneered local music initiatives and, helped by a post-university course in hotel management, had doubled his chain of palaces into luxury hotels. He was chairman of three gem companies and several religious and educational trusts. In an era when most Indian princes are looking for a role and worrying about money (Indira Gandhi abolished the princedoms in 1971), he has both.

By the time I left the palace, Sriji had given me more than two hours of his time. Three years later, in July 2009, the Prime Minister of India dedicated his country to the most ambitious solar energy development programme in the world. All government buildings are to have solar panels by 2012, and 20 million households are to have installed solar lighting by 2020; to achieve these and similar goals, the government plans to invest 920 billion rupees (about $20 billion). But Sriji was the pioneer.

☼

Over the next four days I visited Modhera, in Gujarat, to see one of the great Sun temples, thence to Ahmedabad, Delhi, and finally, my ultimate destination, Varanasi, where I was met by my guide, Ravid ('It means "sun",' he told me). Aged thirty-eight, he looked like a lead from a Bollywood movie – except he was too slim and un-jowly, and his hair was too obviously treated with henna. One of the taller Indians I had met, he strode, gangly and loose-limbed in sockless shoes, a silver bracelet on his right arm. Over the next three days he would oversee my education.

Varanasi is Hinduism's holiest city and pre-dates the modern metropolises of India – Chennai (Madras), Mumbai (Bombay), Kolkata (Calcutta), and New Delhi – by at least 2,000 years, making it one of the most ancient in the world, as old as Jerusalem. It has a permanent population of 1.1 million. Here the faithful come to die or their bodies are brought to be cremated (as was Mrs Gandhi's), so absolving them of the burden of *samsara*, or reincarnation. No city is more symbolic of Indian culture. Called Banāras (or Benares, the ruling British attempts at the Mughal version of the ancient Hindu name) until independence, it is also known as Kāshī – from the Sanskrit root *to shine*, thus 'the city of light'.[9] The wordplay underlines the relation of the name to enlightenment. It was here that India's most famous seeker, Siddhārtha Gautama, putting aside his life of luxury, preached his first sermon. Before long, he was acclaimed as Buddha.

All the gods in the Hindu pantheon are manifest in Kāshī – no need to go anywhere else, say the locals; everything the soul requires is present. But while primarily a holy city host to 1,500 temples, it is also renowned for its universities (foreign visitors have named it 'the Athens of India') as well as for its courtesans, poets, thugs, brass shops, silk palaces, mangoes, *pān* leaves – even its sweets.

That first day, Ravid guided me to Lolārka Kund ('Trembling Sun'), one of the two shrines mentioned in that supreme Sanskrit epic of ancient India, the Mahabharata. The shrine was set in a small square similar to those behind the main piazzas in Venice, but from this simple rectangle thirty-five startlingly steep steps descend to a bathing pool fifty feet down. The shrine was built up to its current size by a local ruler in the eighteenth century, after his bathing here had cured his leprosy. During the month of Bhadrapada, explained Ravid – some time in August or September – the faithful flock in homage and hope. Infertile women pray for their barrenness to be lifted, the pregnant for the good fortune of a son. To effect the latter, married couples must bathe together, the wife's sari tied to her husband's dhoti. As she bathes, the woman is expected to release some vegetables (a symbol of fertility) into the water. Those suffering skin diseases leave their clothes behind. How many people come to so small a spot? 'Twenty thousand, all in one day, at most in two.'*

Ravid was delighted to display his learning, as he led me through the city's most cherished buildings. Hindu tradition is more than a lifetime's study – its complex myths, its manifold divine identities, its elaborate rituals and its teachings about life and death. Those who lead a good life have brightness in their faces – *Dhejas*. The Sun is compassionate, healing, friendly, Ravid continued, but Hindus believe that you must not

*The spiritual pull of Varanasi is felt throughout India. At some Hindu weddings, even today, the groom will appear at the doorway of the bride's home shortly before the extensive marriage rites and declare to his future father-in-law that he is leaving for Varanasi to renounce the world. This is but a ritual threat and initiates a round of pleading and bargaining in which the bride's family inevitably dissuades him, with gifts and promises, from his noble goal – although one wonders whether some unimpressed future in-law might choose *not* to wheedle the young man back . . .

An 1890s photograph of Dasaswamedh Ghat (literally, 'the river front of ten sacrificed horses'), one of the oldest and holiest landings in Varanasi

follow a shadow, for it entails misfortune, nor tread on another's shadow, for it will drive him mad.

By now it was nearly six in the evening, and few people were about, just some children playing hopscotch, the odd mangy dog and some monkeys. After sunset, the better class of person stays at home, while Untouchables scavenge for leftovers from the public wells or collect the warm droppings of cows to sell for fuel. It was time to return to my hotel.

The next morning I was up by 5.40 and was soon joined by Ravid, whose forehead was marked by a *roli* of red turmeric: different sects have different colours and signs, and red is for sun worshippers. We wound our way through the crumbling stone alleys, many no wider than footpaths, then walked along the row of ghats, or landings, whose staircases extend root-like into the river. Some of the landings (eighty-four in all, stretching along just one side of the river) were almost deserted, others thronged with people, as we wended our way past

waking fakirs and stallholders putting out their wares – bright orange
garlands, ghee for lamps, costume jewellery, brass pots, trays and bowls,
wooden toys, plastic watercans and bottles of water. The evening before,
I had read a late-nineteenth-century traveller's description:

> Up and down the ghats, all day long, but especially in the early morn-
> ing, stream the endless course of pilgrims, ragged tramps, aged crones,
> horrible beggars, hawkers, Brahmin priests, sacred bulls and cows,
> Hindu preachers, wealthy *rajas* or bankers in gay palanquins, fakirs,
> pariah dogs and scoffing globetrotters . . .[10]

Not much seemed to have changed in the 110 years since.

At last we came in sight of the skiff Ravid had hired, and trod carefully
over the hardened mud of the sandbank to board it. Our oarsman was
about forty, small but sinewy and seemingly impervious to my presence:
for him, just another day at the office. Although it was not yet dawn,
there was light enough, and as we began to move slowly down the river
I could see some people already at prayer, a few having their hair or
beard cut by a street barber, others preparing themselves for immersion
in the great Ganga. Before sunrise, the water is said to be not just warm
but *hot*; after sunrise, cooler. Generally, the women come to bathe first,
as early as 4 a.m., and will go to any of the ghats that line the water's
edge.

Varanasi sits between two streams, the Varanā ('Averter'), which flows
into the Ganges on the north, and the lesser Asi ('Sword'), which joins
the river on the south. Ravid was defensive about the Ganga's reputation
for being heavily polluted. The great river had amoeba-like bacteria in
it, he said, and these eat harmful bacteria. *All* harmful bacteria? *Much*
harmful bacteria.

We followed the river, its ghats alternating with the pavilions, temples
and terraces that crowd the banks, most of them hypnotically beautiful in
the morning light, which at sunrise tinted them a uniform coral red. The
local cactus, *Religios perseca*, seemed to sprout everywhere. As we passed
Digpatiya Ghat, a chant rose from a gathering of priests. We slipped by an
imposing twelfth-century ruin, overrun by vagrants and goats. A wide
expanse of riverbank had been covered with saris laid out to dry – a feast

of colour, the full length of each nearly twenty feet. Inevitably, Mark Twain had come by here, and for once wrote without aiming at comic effect:

> The Ganges front is the supreme showplace of Benares. Its tall bluffs are solidly caked from water to summit, along a stretch of three miles, with a splendid jumble of massive and picturesque masonry, a bewildering and beautiful confusion of stone platforms, temples, stair-flights, rich and stately palaces – nowhere a break, nowhere a glimpse of the bluff itself; all the long face of it is compactly walled from sight by this crammed perspective of platforms, soaring stairways, sculptured temples, majestic palaces, softening away into the distances; and there is movement, motion, human life everywhere, and brilliantly costumed – streaming in rainbows up and down the lofty stairways, and massed in metaphorical flower gardens on the miles of great platforms at the river's edge.[11]

As the Sun rose in the sky, I could see men and women of all ages practising breath control, meditation disciplines or the yogic sunrise ceremony, *Sūrya Namaskar*, which requires one to bow 108 times – in the late 1950s the longest name in the world belonged to an Indian prince, who required *Sūrya* (the Sun) to be included exactly that number of times. At sunset a similar ceremony is enacted, with a minimum of nine bows.*

Some men were already out hosing down the silt piled up by the seasonal flooding, while others raised hands cupped to hold the sacred water in supplication, or clutched the ceremonial *kusha* grass used in performing their rites. An elder of seventy, completely bald, the three deep furrows on his forehead etched in yellow, gazed enrapt towards the horizon. Everyone seemed caught up in his or her own world. Most of the men were burly, certainly not slim: following my eyes, Ravid explained

---

*Since my visit, the question of whether schoolchildren should be required by law to take up the sun salutation has become a serious topic. In January 2007 the Hindu Nationalist-led government of the state of Madhya Pradesh put forward a measure that required public school students to salute the Sun and recite certain chants in Sanskrit, but Muslims and Christians took issue, saying the chants were essentially a Hindu rite, violating the constitutional separation of faith and state. A court later ruled that neither the chants nor the salutation should be compulsory, but the issue remains contentious.[12]

that Varanasi is famous for its wrestlers. Three women were busy washing one another using a polythene bag as a cloth, two girls splashing about joyfully. An older woman was doing her best to preserve her modesty while bathing in a minimal sari. A beautiful mother of about eighteen carried a baby, turning seven times round, as the holy books prescribe, to cleanse herself and her child before the gods. Several bathers, up to their waists, were scouring their teeth with twigs from a holy tree. After their ablutions, most of those I watched crowded into temples along the river, their hands full of jasmine and sweets destined for the lap of a favoured god, or else they carried the waters home in polished pots – made from the brass for which the city is celebrated.

One of the greatest Indian poets, Kabīr (1440?–1518), although he lived in Varanasi, couldn't resist heaping disrespect on his hometown:

> *Going on endless pilgrimages, the world died,*
> *Exhausted by so much bathing.*[13]

Maybe so, but those who had braved the waters were not the only ones making use of the river. As we glided along, Ravid pointed out Manikarnika Ghat, the larger of the two main places of cremation – the only points along the river where photography is forbidden, although bodies are immolated almost everywhere. This burning ground has traditionally gone by yet another name: Jalasāī Ghat, the 'Sleeper on the Waters', and hosts a sacred fire said to have burned constantly for all history. Further along, at the other main cremation centre, Harishchandra Ghat, there now stands an electric crematorium where funeral pyres once blazed. But most prefer the old way.

All may be cremated, Ravid said, except snakebite victims, pregnant women, children or the victims of smallpox. Cremation by flame costs 551 rupees, the mechanized version 151. The burnings go on without cease, each taking three to four hours: the great stacks of wood that line the shore, higher than houses, are for ever being replenished and depleted. As we drifted slowly by, smoke wafted up from the low-lying pyres to curl about the temples that surround the sites. I could smell charring flesh.

When a body arrives – dead women draped in orange, men in white, explained Ravid – it is dipped into the Ganges, then taken to the funeral pyre, where it is sprinkled with sandalwood oil and garlanded. Tall bamboo poles dot the scene, each hung with a small basket filled with oil lamps to honour the ancestors of those passing through the fire. Wherever possible, the eldest or youngest son of the dead will take the central part in the ceremony. He will have had his head and eyebrows freshly shaven and will be clad in a seamless long white shirt, with the sacred thread, which usually hangs from the left shoulder, cast over the right. He will lead the mourners five times around the unlit pyre, anti-clockwise (because, in death, everything is reversed), then he will light the pyre. Once the body is near-consumed, the chief mourner performs the 'rite of the skull', cracking the head with a bamboo stick, so releasing the soul from entrapment. Then he turns to cast a pot of Ganges water over his left shoulder, dousing the embers, and strides away.

I wanted to venture beyond the formal line of ghats, so Ravid and I disembarked and walked on through the city. Before he left me, I asked Ravid to join me that evening so that during the rituals he could explain anything I didn't understand. He grinned. 'Of course.' Back at my hotel, I returned to Diana Eck's classic history *Banāras: City of Light.* A distinguished visitor to Varanasi, she writes, 'described the dawning riverfront as "one vast Sun temple", and yet it is not the Sun alone which is honored. It is the Divine, of many names and forms, here physically manifest, present in warmth and light . . .'[14] How much of this devotion centres on the Sun, and how much on other powers?

I had a chance to ask these questions a few hours later, when Ravid accompanied me to a meeting he had arranged with Dr Sudharka Mishra, an eminent scholar of Sanskrit, at the professor's house. My host was a friendly if earnest figure, as intent as Ravid on enlightening my ignorance, and began by explaining something of the Indian attitude towards eclipses (as related in Chapter 4), breaking off only to admonish one of his many children, who scampered about in delight at having such an unusual visitor.

Soon it was time for me to be instructed in the three 'levels' of sun worship: the mythic, where the Sun is part of a ritualistic framework; the astronomical, where it is the king of stars (in southern India, for instance, the basis of the calendar); and the everyday, where it is an

intrinsic part of normal life. 'Without sun worship,' Dr Mishra concluded, 'nothing is going to happen in Benares.' The Sun has a soul, he
said – *sutrama*. Inside each of us is a small map of the universe (somewhere near the bridge of the nose): proper meditation releases this
knowledge, and the trillions of galaxies within us. Hinduism recognizes
330 million gods . . .

I interjected that that was a crazy number – perhaps a metaphor for
the many facets of God . . . 'No!' he interrupted, suddenly leaning
towards me. There really are that many, and Hinduism honours each
one. In the vastness of the universe, is that so surprising? The West likes
its cleanliness outside, not inside, he added, so does not understand. 'We
have no solution for you. What is God? That is what you ask. Just as useful
to demand, "What is the colour of pain?" There is no answer.' Nor could
I come up with one, although my mind returned to the white shrouds of
the funeral pyres.

By the time Ravid and I took our leave and strolled together back into
town, the Sun was about to set, and two of the main ghats were filling up,
although the honouring service would not start until at least half an hour
later, around 6.30. The largest landing, Dashashwamedh Ghat, already
hosted about 400 people, many of them having taken off their shoes,
including the women, the majority dressed in their finest saris. Rows of
pilgrim-priests huddled under bamboo umbrellas on low wooden
chaukīs.

Seven small altars, draped in fine orange silk (ruby silk being reserved
for the central altar), were lined up at the river's edge, and acolytes
passed from one to another, checking that all was in order: flowers,
incense, joss sticks, the special fan each priest would use, conches, bells,
combustibles, the *prasada* (consecrated food). I wanted to ask Ravid
about all these, but he had wandered off into the melee.

I found a space directly behind the main altar, giving an unimpaired
view. A large portrait had been placed on the altar, framed with white-
and-red flowers: Arsalan Bapu, a local holy man. Beside me, a
whippet-thin devotee of about fifty was already at prayer; I felt awkward
stealing glances at him. Above the altars a large sign flapped in the wind:
Gaya Nidhi Seva, the major corporation that subsidizes the nightly ceremony. The stone steps afforded easy sitting, but many would still have

to stand. Behind the altars, scores of tourist-laden boats had started to jockey for position, while minute leaf coracles bobbed their lamps of ghee and camphor, a long galaxy of lights under a perfect crescent Moon. From somewhere along the river rose a rhythmic popping of fire-crackers. The atmosphere was as charged as it is in a theatre just before the curtain goes up. At last seven young priests, clad in orange shirts, light orange shoulder sashes, white trousers and strings of beads, took up their places at the altars.

Suddenly Ravid appeared by my side, evidently agitated. 'You must come with me,' he said sharply. I rose and followed. He took me to a small boat manned by the same oarsman we had had that morning. For a moment I thought he just wanted to ensure a good vantage point, but after a bit he calmed down enough to explain: the year before, a bomb had been set off almost exactly where I had been sitting. Twenty worshippers had died. The government had whitewashed the atrocity as a simple matter of 'some gas cylinders exploding', but locals were sure it was the work of provocateurs planted by Pakistani intelligence. No evidence was offered either way. The bomb was said to have been set for 6.45, when it would have killed several hundred people, but exploded at 16.45.

Soon after I boarded our skiff, the seven priests, their black hair glistening under the lights, began to intone a great rumbling 'Ohhhmmm',* and soon I began to lose myself in the extraordinary tableau around me. Huge, vividly coloured umbrellas, each a good eight feet across, lined the shores – maybe twenty on one side of the ghat, thirty on the other. More than a thousand spectators had gathered on shore by now, on the river a further five hundred, the boats so hugger-mugger that a young girl selling lamps skipped shoeless from one to another.

The priests eventually ceased their chanting and began to sound conch-shells, turning east, north and south (for other gods), and finally west (to honour the Sun). The conches were meant to echo the 'Om' resonance, Ravid explained. Next they raised seven-branched cande-labra, symbolizing the world's hierarchy of species, and began to chant again, to a new rhythm. Once this part of the service was completed, they

---

*'Om', sometimes written 'Aum', is 'the verbal symbol of the Supreme Reality', and is said to consist of five separate sounds, 'A', 'U' and 'M', plus the nasalization (bindu) and the res-onance (nāda). One temple in Varanasi consists of five shrines, one for each of the notes.

raised their outsize fans – like peacock feathers, only tasselled and made from cow hair, which Ravid explained repels small bees and flies, as well as disease and bacteria. Then once more they made their four turns, as they invoked Vishnu, then Shiva.

On another surge of ohhhmmm, less synchronized than before, there arose a clamour of bells, the priests chanted for peace in the world – *Ohm Shanti* – then the high priest cast flowers over the dark current as a final homage to his many gods, the Sun, the Moon, the planets and the stars, and the ceremony was over. The great assembly began to break up, and we pushed off, first heading upriver where the cremation flames still smouldered, the rich reds of the pyres yielding new colours to the evening. Near the bank, the water was strewn with human ashes and bits of bone. By now it was nearly eight o'clock and the neighbouring ghats were for the most part deserted, except for the occasional stray animal prowling for scraps.

In Geoff Dyer's recent novel set in Varanasi, his narrator offers a dyspeptic view of the rites I had just watched: 'You didn't have to be a particularly discerning tourist to see that this was an exhausted pageant, drummed up for tourists, a son et lumière with a cast of hundreds. Any significance it was supposed to have had had been drained, possibly a long time ago or maybe just yesterday, or even now, right before our eyes.'[15] While I do not deny that the tourists may have gone off to their beds happy but completely ignorant of the meaning of what they had just seen, I sensed a genuine reverence among the celebrants themselves. As with the Cuzco celebrations for their summer solstice, or the noisy crowds circling Stonehenge, or even that small group of travellers who watched the Moon pass across the Sun over the Antarctic ice, a feeling of awe, even something like fear, had filled the air, as tribute was paid to forces that hold the power of life and death.

After a while our oarsman turned us about and pulled hard against the current towards the steps closest to my hotel. The air was chill. Above one of the main buildings, a banner read, in English, 'Banaras is Banaras. Banaras was Banaras. Banaras will be Banaras.' We passed by Dashashwamedh Ghat, where the young priests and their helpers were clearing the fittings and instruments away, like orderlies in an operating theatre. They would perform the same ceremony the following sunset, and the next. In their language, *kal* means both 'yesterday' and 'tomorrow'.

AFTERWORD TO THE FIRST
PAPERBACK EDITION

The Sun, for all our rich imaginings at summer and winter solstice, is never motionless. Nor do we keep still in our attempts to understand our star or conjure new ways to take it over. Since this book was completed at the beginning of July 2010, almost every category of our relations with the Sun has been enlivened by new discoveries; the following are from just the four months September–December 2010.

It was in September that a Spanish woman, Ángeles Durán, went to her local notary in Galicia and registered the Sun as being her property. Her proclamation became worldwide news. There is an international agreement, she informed the press, that states that no country may claim ownership of a planet or of a star, but it says nothing about individuals. She was taking advantage of a loophole. Señora Durán, who is fifty, added that she intended to charge a fee for anyone who uses the Sun and give half the proceeds to the Spanish government, forty per cent to good causes and retain ten per cent for herself. Her announcement was met by an outcry across the Internet from people who said that if her claim stood they would sue her for skin cancers or for damage to their property as a result of solar storms.

A more practical benefit has been secured by those inmates of British jails who declare themselves pagan. They are to be given four days off prison work a year to celebrate their festivals, including those devoted to the Sun – which certainly puts into perspective the cartoon I include on page 618. Paganism has been recognized by the Prison Service as a religion for more than a decade, and the number of those incarcerated declaring themselves its adherents has tripled in the last few years to 366. As the festival of Beltane, which falls on Mayday, is said to involve celebrating the Sun 'with unabashed sexuality and promiscuity', prison governors may be a little apprehensive.

Even ancient stories are being reinterpreted. The veracity of Archimedes using mirrors to reflect the Sun's rays onto the Roman fleet has been questioned by mechanical engineers at the University of Naples, who argue that the Greek astronomer and mathematician is far more likely to have used steam cannons and fiery cannonballs. Several accounts, including one by Leonardo da Vinci, who sketched a steam cannon in the late fifteenth century, credit such a weapon to Archimedes: so this most famous of legends about the power of the Sun may be a chimera. Or maybe the cannons were operated by Sun power (which would make sense), and the two stories got confused.

In December 2010, astronomers at Yale and Harvard reported that the number of stars in the universe had been seriously undercounted, and that there could be three times as many as previously thought. This could seriously undermine our understanding of how galaxies formed and grew over the eons, and thus affect our knowledge of how our Sun came into being.[1]

Another story, not immediately solar-based but dealing with two important figures in the Sun's history, concerns the death of Tycho Brahe, said to have died from failing to relieve himself while drinking profusely at a banquet, so ruining his bladder. It was always an unlikely way of passing, and last November Danish and Czech scientists exhumed the astronomer's remains and advanced the theory that Brahe died from mercury poisoning – thus possibly by human hand. Could his assistant, Johannes Kepler, be the culprit? He had the means, the motive and the opportunity . . .

More central to the Sun's influence on Earth is the report, in December's *New Scientist*,[2] that animals, like plants, may be able to turn light into food. Might, one day, photosynthetic creatures help feed the world? A research assistant at Harvard Medical School injected photosynthetic bacteria into the eggs of zebrafish (a popular aquarium fish native to the southeastern Himalayas) – initially, just to see whether the bacteria could survive. As the magazine points out, the ability to turn light into food evolved in cyanobacteria, and plants developed when more sophisticated cells stole the technology by enslaving cyanobacteria within themselves. Such bacteria and fish do not mix, but two weeks after the fish hatched, the bacteria were still alive, raising tantalizing possibilities.

Our ideas about time and the Sun are being recast too. In a remote part of Australia, Aborigines believe that time rises in the east and sets in the west, in accord with the Sun's daily arc across the sky. Thus the day after tomorrow would be considered as two days to the west, two days past as being two days to the east. For these particular tribesmen, the Pormpuraawans, time flows from left to right when facing south, from right to left when facing north. They rarely use terms for right and left but use absolute directions – thus: 'Move your cup to the north-north-west a little bit.' In the United States and Europe, of course, time is generally thought of as running from left to right. How we use the Sun thus depends on our special knowledge.[3]

The Sun's power can win tributes from unlikely sources. In his newly published second volume of autobiography, the actor Michael Caine relates the occasion, around 1980, he and his mother were having tea in Los Angeles with David Hockney and *his* mother, a native of industrial Yorkshire. Caine asked Mrs Hockney how she liked Beverly Hills. 'It's lovely, dear,' she replied. 'But it's a terrible waste – all that sunshine, and no one's got the washing out.'[4] Much the same point was made by Neil MacGregor, Director of the British Museum, whose BBC radio series and book, *A History of the World in 100 Objects,* became major hits in Britain at Christmas. The final object in his survey was a portable solar-energy panel, made in China and sold for about £15. 'The power of the Sun seems a good place to end this global history,' said MacGregor. 'Solar energy is an energy source that has been with us since the world began. Everybody has dreamed of harvesting the Sun, to have its light and power available at will. One of the great things about solar power is its independence from the grid . . . It really is empowering, and may ultimately help to save the planet.' The final programme ended with music created by researchers at Sheffield University from vibrations in the Sun's atmosphere – 'It's the noise of a new day.'

MacGregor's book was published at the same time as Keith Richards's autobiography, *Life.* An unlikely pairing; but here, as a final homage to the Sun's ubiquity, is the Stones lead guitarist on his reputation as a drugtaker: 'It's thirty years since I gave up the dope! Image is like a long shadow. Even when the sun goes down, you can see it.' So while I have attempted to keep *Chasing the Sun* as up to date as possible, I realize that one can never actually catch up. It races on ahead.

Richard Cohen, New York, January 2011

ACKNOWLEDGEMENTS

Chasing the Sun has taken almost eight years to research and write, during which time I have visited eighteen countries. That such trips were possible – vastly expanding what I first envisaged for the book – is due to the generous grant I received from the Alfred P. Sloan Foundation. To the Foundation, and particularly to its director, Doron Weber, I am extremely grateful.

Such an undertaking also required the help of an extraordinary range of people. To particularize: material on Japan was checked with Yoshiko Chikubu and by Junzo Sawa and Hamish Macaskill (of the English Agency, Japan); I benefited from early discussions on solar myths with Dr Sue Benson of Murray Edwards College and Stephen Hugh-Jones, professor of anthropology at King's College, both of Cambridge; the details on Peru were checked by Marie Arana of *The Washington Post*; solstice customs by Professor Ronald Hutton of Bristol University and North American dance rituals by Susan Gardner, my colleague at Kingston University, and by Wade Davis, Explorer in Residence at *National Geographic*. The chapters on eclipses were read by Thomas Crump, whose own book was so useful to me, and by Jay Pasachoff, Field Memorial Professor of Astronomy at Williams University.

For the chapters on ancient Babylonian, Egyptian and Greek solar astronomy I received early advice and encouragement from Richard Parkinson, assistant curator of Ancient Egyptian Pharaonic Culture at the British Museum, Christopher Walker and Michael Wright, two other BM alumni, Sir Geoffrey Lloyd, Master of St Catherine's College, Cambridge, and Dennis Rawlins, Editor of *Dio*. I would single out for special thanks James Allen, ex-Curator of Egyptian Art at the Metropolitan Museum in New York: we came to an agreement that he would explain Egyptian culture to me if afterwards I would give him a fencing lesson, so that well after the Museum was closed to normal visitors the mummies of ancient Egypt would look benignly down on us as our sabres clashed.

Chinese solar history was read for me by Nathan Sivin, Professor of Chinese Culture and the History of Science at the University of Pennsylvania (though we did not always agree); Simon Winchester, recent biographer of Joseph Needham; and Christopher Cullen, head of the Needham Institute in Cambridge; and the chapters from Copernicus to Newton by Owen Gingerich, Professor Emeritus of Astronomy and of the History of Science at Harvard University. Brian Cathcart, Professor of Journalism at Kingston University and author of *The Fly in the Cathedral*, provided some invaluable advice on the physicists who helped build the atomic bomb.

On tree-ring details I was advised by Dr Mike Baillie, of Queen's University, Belfast. The chapter on how the Sun affects our bodies was read by Dr Seth Orlow, chairman of dermatology at NYU Langone, while Linda Prestgaard at the Norwegian embassy in New York checked details about Norway and other matters to do with SAD. David Davidar of Penguin Books told me about the Indian skin-bleaching techniques he employed for his novel, *The House of Blue Mangoes*. The material on photosynthesis and other matters of natural history was read by Dr Oliver Crimmen, Head Curator at London's Natural History Museum, and by Dr Olivia Judson, the author and journalist. The details about animal migration were checked by Dr André L. Martel and Pierre Poirier, of the Canadian Museum of Nature, and Henry Bouchard, of the US Fish and Wildlife Service.

Details about solar navigation were corrected by Robert Ferguson, author of *The Vikings*, Dr Richard Dunn, Curator of the History of Navigation at the National Maritime Museum in Greenwich, London, and Gloria Clifton, Head of the Royal Observatory. The section on sundials benefited from the expertise of Dr Frank King of Churchill College, Cambridge, and my visit to Freiburg and the details on solar energy research generally were checked by Larisa Kazantseva of Nitol Solar and by Thomas Dresel of the Solar Energy Centre in Freiburg; and my Spanish trip by José Soler of the Plataforma Solar de Almeria.

The chapter on the Sun in art was read through by Turner's recent biographer James Hamilton (who 'teased the text a bit'), and the passages on cinema by Kenneth Turan, film critic of the *LA Times*. For much of the material on classical music I am indebted to the late Gabrielle Roetke, professor of opera at the New School. My musings on Nabokov were read by Azar Nafisi, author of *Reading Lolita in Tehran*, and on literature generally by two author friends, Clare Asquith and Betsy Carter. The chapter on politics and the Sun was read by Richard Bernstein, the author and sinologist, while Alex Cook of King's College, Cambridge, first alerted me to *The Ruins*, and Don Cohn of Columbia University was helpful on Mao and the Sun.

The labyrinths of quantum physics were explained to me by Andrew Blake, research associate at the Cavendish Laboratory in Cambridge, and by Dr Alan Walton, a senior Cambridge University physicist; Katie Yurkewicz, at CERN, was also an invaluable help. David Agle and Curtis D. Montano at the Jet Propulsion Lab gave me useful material on current solar projects and also on the early days of rocketry. Piers Corbyn, of Weather Action, gave me hours of his time, and the chapter on global warming was read by Dr Willie Soon of Harvard University and by Dr Robert Carter, Professor at the Marine Geophysical Laboratory, James Cook University (both of whom, I fear, will not like all my conclusions). The history of science fiction and the Sun was checked by David Compton, editor and renowned science-fiction author, and I was also helped by the screenwriter and novelist Alex Garland. The epilogue was read by Louise Nicholson, writer and authority on India, who in addition arranged my time in that country with such a sure hand.

Silke Ackermann of the Department of Prehistory and Europe was my 'control' at the British Museum, leading me to expert after expert. During my researches I used almost a dozen libraries, from the one at Castel Gandolfo to that of the Munch Museum in Oslo. In particular, the Society Library in New York City and the London Library never let me down, while the New York Public Library was a home away from home, be it the Allen Room, the Wertheim Study or the General Reading Rooms. David Smith at the NYPL has even won a profile in the pages of *The New York Times* for the quality of help he gives to struggling authors; for me, his business-card byline – 'librarian to the stars' – seems especially appropriate.

Others who helped on one matter or another include: Professor Babette Babich of Fordham University's Department of Philosophy, Kai

Cai, Dr Nicholas Campion of Bath Spa University (on matters astrological), Fr. Juan Casanova, Lesley Chamberlain, Al Clement, Margaret Cook, Peter d'Epiro, Timothy Ferris, Emeritus Professor at the University of California, Dr Valerie Fomin in St Petersburg, Manuel Perez Garcia at the University of Almeria, Carol Gaxiola, John Gerrard (who shared with me his *Watchful Portrait (Caroline)*), Professor Ai Guoxiang of the Chinese Academy of Science, Bob Hay, Jeff Hester, professor of physics and astronomy at Arizona State University, Miles Hilton-Barber (for telling me what it is like to fly a plane when you are blind), Chen Jie, who showed me round Beijing Observatory, Professor Phil Jones of the University of East Anglia, Chow Kii, Adam Kuper, longtime professor at Brunel University, Bill Livingston and John Leibacher of the National Solar Observatory in Tucson, Ved Mehta, Oliver Morton, Don Nicholson at the Mount Wilson Observatory in Pasadena (the one person in all my researches who called one of my questions 'stupid': I asked how far his telescope could see), Colin Pearce, Alberto Righini, Mr Tetsuya of the Watanabe National Astronomical Observatory in Tokyo and Ingebjorg Ydstie of the Munch Museum in Oslo. I am also grateful to the astronomers at the observatory in Shanghai who spent time with me explaining their work, and the Mullard Space Science Laboratory in Dorking, Surrey.

The script was read by Nick Webb, fresh from his first-class degree in astronomy, Jonathan Weiner, author of *Beak of the Finch* and *Long for This World*, and by Peter Petre, one-time science editor of *Fortune*. Two years after starting work on the book, I became aware that David Whitehouse, a professor of astronomy at Jodrell Bank and an award-winning science correspondent for the BBC, had published *The Sun – A Biography*. We went on to become friends, and one day he arrived at lunch with a large paper bag stuffed with books – the library he had used for his Sun book. 'They're for you,' he said, then handed over a DVD – 'It's got all my research,' he added. It needs hardly adding how generous a gift this was, or how useful. He also kindly read an early complete draft of the book.

Friends and relations all did their bit – in many ways: John and Nina Darnton, Liza Darnton (for early research work on Carl Jung), Jamie Darnton (on some tricky Chinese etymology), Darrell McLeod, Sara Wheeler, Hilary Spurling, Elaine Shocas, Linden Stafford, Brian Brivati, my old university friend Colin Gleadell, Richard Oldcorn, Paul Pickering, Michael Johnson, Allen Kurzweil, David Bodanis (when not preoccupied reinterpreting the Ten Commandments), Bill and Linda Rich (for that insight about pirates), John de Stefano, my cello teacher Laura Usiskin (for her research into matters ranging from Freemasonry to bird migration), Harry Hotz (for his knowledge of temperature ratios), Andrew Di Rienzo (for befriending me on a bus, and teaching me some crucial physics thereafter), Ron Rosenbaum (for some Shakespeare tuition), Kevin Jackson, Jim Landis (not least, for the foul-mouthed quotation from August Strindberg), Elizabeth Sifton (who told me of the Sun Museum in Riga), Ben Cheever (whom I forgot to mention last time), Peregrine Hodson (for making me climb Mount Fuji),Woodrow Campbell (my accomplice and photographer in Antarctica), Nancy Campbell (no relation), Mary Cunnane in Australia, my old tutor Dr Michael Tanner of Corpus Christi College, Cambridge, Jackie Albers (for sharing her homework) and – not least – Toby, Mary and Guy Cohen; it was Toby who one day put his hand on my shoulder and asked, 'Daddy, why don't you make some money and write *a nice, short book*?' One day I will.

The head doorman of my apartment building, Bill Albers, made sure I was kept up to the mark with a constant flow of magazine items which he would slip under my door, along with pertinent questions about progress. Ann Godoff and Vanessa Mobley were both kind and helpful in the book's early days. At Random House, my editor Beth Rashbaum enormously improved the book with her powers of organization and her determination that nothing should appear that would not be clear to everyone; I am also extremely grateful for the skill and help of many others at Random, including Tim Bartlett, Tom Perry, Barbara Bachman, Susan Kamil, London King, Meghan Cassidy, Angela Polidoro and Gina Centrello, making my publication there feel like a family affair. The same is true of my British publishers, led by my long-time friend Ian Chapman, along with Mike Jones, Katherine Stanton, Rory Scarfe, Anna Robinson, Suzanne Baboneau and Sally Partington. In Germany I am lucky to have as my friend and publisher Niko Hansen, whose patience and enthusiasm for the project never wavered (not in front of me, anyhow!). Catherine Talese has been the ideal illustration researcher, as well as a wonderful cheerleader and an indispensable guide to New York swimming pools.

At the Robbins Office, David Halpern, Kate Rizzo Munson, Rachelle Bergstein, Karen Close, Mike Gillespie, Coralie Hunter, Katie Hut and Ian King each read part or all the script, often more than once, and made valuable comments. Having them around and in support has been fun. Tim Dickinson, as he was on *By the Sword*, has been companion, sounding board, fount of knowledge both pertinent and not-too-pertinent (he would insist on the hyphens) but always highly interesting; I cannot thank him enough for all he did.

Last and supremely there is my literary agent, Kathy Robbins, aka Mrs Cohen. Over the years I have become used to other authors thanking her in their books for her friendship, her counsel, her unrivalled humour, her ability to buoy up and push forward, her negotiating powers, her tenacity, her empathy, her peculiar ability to time both good and not-so-good news to maximum effect. I have experienced all these and more, but at least I can add one unique claim: it is she who, when woken in the small hours and told yet another extraordinary way in which chapter four, or maybe fourteen, might begin, can say to me without note of rancour, 'That's *really* interesting, sweetheart,' then turn over and at once return to a well-earned sleep.

Despite this roster of helpers, any mistakes in the book are of course my responsibility. I am comforted in this by a sentence I came across in Charles Darwin's autobiography (1882), where he writes: 'Whenever I have found out that I have blundered, or that my work has been imperfect, and when I have been contemptuously criticized, and even when I have been over-praised, so that I have felt mortified, it has been my greatest comfort to say hundreds of times to myself that "I have worked as hard and as well as I could, and no man could do more than this".'

SOURCES

PREFACE

1. *Mother Earth News*, Issue No. 41, September/October 1976
2. *Encyclopedia of Astronomy and Astrophysics*, New York: Nature Publishing, 2001, Vol. 4, 'Sun'. Astronomers classify stars based on colour, which relates to temperature, size and longevity. Throughout the book, 'Sun' appears thus, as astronomers and scientists prefer; when referred to in a non-specific way, and often when mentioned in quoted matter, it is 'sun'. It also appears in lower case when part of a word – 'sun-glazed' or 'sun-worshipper', for instance
3. Doyle, Arthur Conan, *A Study in Scarlet*, 1897
4. Eddy, John, 'Climate and the Role of the Sun', in *Journal of Interdisciplinary History*, x:4 (Spring 1980), pp. 725–47
5. Cf. US Scouting Service Project at http://www.usscouts.org 'Confidence and Team-Building Games'. John Keats recalled a game at his school in which the boys whirled round the playground in a huge choreographed dance, trying to imitate the solar system: there is truly no new thing under the Sun

SUNRISE: MOUNT FUJI

1. Stevens, Wallace, 'Notes Toward a Supreme Fiction', *The Collected Poems of Wallace Stevens*, New York: Random, 1990
2. Pink Floyd, 'Time', from the album *Dark Side of the Moon*, Capitol Records, 1973
3. Gateways are associated with the Sun in many cultures, Japanese gateways probably deriving from Indian *torana*, or 'celestial gates'; the palace of the god-king at Kyoto is entered by 'the Gate of the Sun'. Cf. Lethaby, William, *Architecture, Mysticism and Myth*, New York: Braziller, 1975, pp. 186–9

CHAPTER 1 TELLING STORIES

1. Müller, Max, *Lectures on the Science of Language*, 1864, reprinted New York: Kessinger, 2007
2. Donne, John, *Ignatius His Conclave*, 1611
3. Cf. the myth of Taios in Cook, Arthur, *Zeus, A Study in Ancient Religion*, 3 vols, Cambridge: CUP, 1914–40, pp. 633–5 and p. 719 ff.

4. Cf. Chatwin, Bruce, 'Art and the Image-breaker', *The Morality of Things*, London: Cape, 1973, p. 179

5. Cf. Lévi-Strauss, Claude, *The Origin of Table Manners*, New York: Harper and Row, 1978, p. 395

6. Genesis XI, 16–17. Cf. also Olcott, William Tyler, *Sun Lore of All Ages, a collection of myths and legends concerning the Sun and its worship*, New York, London: Putnam, 1914, pp. 38–9

7. Cf. Crow, John A., *The Epic of Latin America*, Berkeley: University of California Press, 1992, p. 33

8. *The Essential Max Müller*, Ed. Jon R. Stone, New York and Basingstoke: Palgrave, 2002, pp. 155–6

9. Müller, Max, *India: What Can It Teach Us?*, New York, 1883, p. 216. This work, written a decade or more after *On the Philosophy of Mythology*, contains the more pithy of Müller's opinions, hence my quoting it here

10. Perry, W. J., and Smith, Grafton Elliot, *The Children of the Sun, A Study in the Early History of Civilization*, London, 1930, p. 164

11. Cf. Kesser, Armin, 'Solar Symbolism among Ancient Peoples', *Graphics*, No. 100, 1962, p. 115

12. Perry, W. J., *The Children of the Sun, op. cit.*, p. 141

13. *Ibid*, p. 160

14. Jung, Carl Gustav, *Memories, Dreams, Reflections*, New York: Pantheon, 1961, pp. 250–1

15. *Ibid*, p. 252. Mountain Lake was the tribe's spokesman, since he spoke the best English. He also told Jung that if the whites did not stop interfering with their religion, the Pueblos would make them regret their actions bitterly – they would stop helping Father Sun on his daily journey across the sky

16. Cf. Storr, Anthony, *Jung*, New York: Routledge, 1973, p. 26

17. Blair, Deirdre, *Jung*, New York: Little, Brown, 2004, p. 354

18. Eliade, Mircea, *Patterns in Comparative Religion*, New York: Bison, 1996, pp. 124–51

19. George Gordon, Lord Byron, *Don Juan*, Canto the Second, Verse XCIII (1819)

"They're keeping me from practicing my religion. I'm a sun-worshipper."

CHAPTER 2 CELEBRATING THE SEASONS

1. Donne, John, *A Nocturnal Upon St Lucy's Day, Being the Shortest Day*, c. 1611–12? Donne believed 13 December to be the shortest day; in the northern hemisphere it is 21 or 22 December

2. Furst, Alan, *Dark Star*, New York: Random, 1991, p. 124

3. Twain, Mark, *The Adventures of Huckleberry Finn*, New York: Oxford, 1999, p. 93

4. Lopez, Barry, *Arctic Dreams*, New York: Scribner, 1986, p. 20

5. Macrobius, *Saturnalia*, Columbia University Press, 1969, Bk. 1, Section 17, p. 114

6. Cf. *Cambridge Medieval History*, Vol. IV, Pt. 1, Cambridge: CUP, 1966, p. 43

7. Cf. Golby, J. M., and Perdue, A. W., *The Making of Modern Christmas*, Athens, Georgia: University of Georgia Press, 1986, p. 123, p. 24

8. Cf. Harline, Craig, *Sunday:A History of the First Day from Babylonian Times to the Super Bowl*, New York: Doubleday, 2007, pp. 10, 17–24

9. Hutton, Ronald, *The Pagan Religions of the Ancient British Isles*, Oxford: OUP, 1991, p. 36

10. Brand, John, *Observations on Popular Antiquities*, Ed. Henry Ellis, London, 1813, Vol. 1, pp. 238 ff

11. Cf. *Encyclopedia of Religion*, New York: The Free Press, 1987, Vol. 14, pp. 139–40

12. Hagan, Rune, private paper on Midsummer Witches, University of Trømso

13. Brand, John, *Observations on Popular Antiquities*, *op. cit.*, pp. 238–9

14. Cf. Cressy, David, *Myths of the English*, Ed. Porter, Roy, Cambridge: CUP, 1992, p. 78

15. Hardy, Thomas, *The Return of the Native*, New York: Signet Classics, 1959, p. 23

16. Ladurie, Emmanuel Le Roy, *Carnival in Romans*, New York: Braziller, 1979

17. Ozouf, Mona, *Festivals and the French Revolution*, Cambridge, Mass: Harvard, 1988, pp. 158–167

18. Broughton, Jack M., and Buckskin, Floyd, 'Racing Simloki's Shadow: The Ajumawi Interconnection of Power, Shadow, Equinox and Solstice', in *Prehistoric Cosmology in Mesoamerica and South America*, Smithsonian Occasional Publications in Mesoamerican Anthropology, 1988, pp. 184–9

19. Holler, Clyde, *Black Elk's Religion: The Sun Dance and Lakota Catholicism*, New York: Syracuse University Press, 1995, p. 60

20. Cf. Lowie, Robert, *The Crow Indians*, Lincoln and London: University of Nebraska Press, reprint 1983, p. 215: 'Most characteristic was the intertwining of war and religion. The Sun Dance, being a prayer for revenge, was naturally saturated with military episodes'

21. Lévi-Strauss, Claude, *The Origin of Table Manners*, New York: Harper and Row, 1978, pp. 222, 212

22. Brown, Joseph Epes, *Black Elk's Religion*, 1978, p. 12; quoted in Goodman, Ronald, 'Lakota Stellar Theology'. For the importance of scalping and its link with solar myths, see Claude Lévi-Strauss, *op. cit.*, pp. 404–5

23. Davis, Wade, *One River*, New York: Simon and Schuster, 1996, pp. 79–82; cf. also Matthiessen, Peter, *In the Spirit of Crazy Horse*, New York: Viking, 1980, p. 47

24. Davis, Wade, *ibid.*, p. 82

25. Holler, Clyde, *Black Elk's Religion*, *op. cit.*, p. 201

26. Cf. Nietzsche, Friedrich, *The Birth of Tragedy*, New York: Anchor, 1956, p. 22

CHAPTER 3 THE THREE THOUSAND WITNESSES

1. This poem was actually written in Italy in May 1948 and first published in *Horizon* that July; it then appeared in *Nones* (1951) and, revised, in the last chronological section of Auden's *Collected Shorter Poems, 1922–1957* (1966)

2. Cf. Alexander, Caroline, 'If The Stones Could Speak', *National Geographic*, June 2008, pp. 34–59

3. Deakin, Roger, *Wildwood: A Journey Through Trees*, London: Hamish Hamilton, 2007, p. 120

4. Stacey, C. P., *The American Historical Review*, Vol. 56, No. 1 (October 1950), pp. 1–18

5. Hutton, Ronald, *The Pagan Religions of the Ancient British Isles*, Oxford: OUP, 1991, p. 36

6. *Ibid*, p. 112

7. Cf. Barthal, Thomas, *The Eighth Land*, Honolulu: University of Hawaii, 1974, pp. 248–9

8. Chatwin, Bruce, *The Songlines*, London: Penguin, 1988, p. 136

9. Delaney, Frank, *Ireland*, New York: HarperCollins, 2004, pp. 68 ff

10. Cf. Dumas, Alexandre, *The Man in the Iron Mask*, Oxford: OUP, 1998, 'The Grotto of Locmaria'

11. Cf. *In Search of Ancient Astronomies*, Eddy, John A., 'Medicine Wheels and Plains Indian Astronomy', p. 147 ff

12. Cf. Atkinson, Jay, 'America's Stonehenge: A Classic Whodunnit and Whydunnit', *The New York Times*, 11 December 2009, C38

13. Cf. Aveni, Anthony, *Skywatchers*, Austin: University of Texas Press, 2001, p. 318; Urton, Gary, *At the Crossroads of the Earth and the Sky*, Austin: University of Texas, 1981, p. 6; and Zuidema, R. T., 'The Inca Calendar', in Aveni, A. F., *Native American Astronomy*, Austin: University of Texas Press, 1998, p. 219

14. *Science and Stonehenge*, Ed. Cunliffe, Barry, and Renfrew, Colin, Oxford: OUP, 1997, p. 572. The Great Wall had no major solar connections (although China is the other country to boast pyramids, such as those near Ch'ang, capital of the Western Han Dynasty of 206 BC–AD 220)

15. Isler, Martin, *Sticks, Stones and Shadows*, Oklahoma: University of Oklahoma Press, 2001, p. 111. Even so, as the orientalist Kate Spence writes, 'The Egyptians were almost certainly unaware that their orientation method was precession-dependent and that it was therefore initially inexact, otherwise they would have to have used a more accurate bisection method.' (*Nature*, Vol. 412, 16 August 2001, p. 700). Isler's book has been particularly useful to the second half of this chapter

16. Cf. Grinsell, Leslie V., *Folklore of Prehistoric Sites*, London: David and Charles, 1976, p. 25

17. Cf. Heilbron, J. L., 'Churches as Scientific Instruments', annual invitation lecture to the Scientific Instrument Society, Royal Institution, 6 December 1995

18. Cf. *New York Review of Books*, 15 December 2005, p. 28

19. Cf. *The New York Times*, Rao, Joe, 'Sky Watch', 9 July 2006, p. 26

CHAPTER 4 TERRORS OF THE SKY

1. Clark, Stuart, *The Sun Kings: The Unexpected Tragedy of Richard Carrington and the Tale of How Modern Astronomy Began*, Princeton: Princeton University Press, 2007, p. 15
2. Ray, Dorothy Jean, 'Legends of the Northern Lights', *Alaska Sportsman*, April 1958
3. Henning, Robert A., quoted in Akasofu, S.-I., *Aurora Borealis: The Amazing Northern Lights*, Alaska: Alaska Geographic, 1979, p. 5
4. Huxley, Leonard, *Scott's Last Expedition, The Personal Journals of Captain R. F. Scott, R.N, CVO, on His Journey to the South Pole*, London: Murray, 1941, p. 257
5. Hawkes, Ernest W., *The Labrador Eskimo*, Ottawa: Government Printing Bureau, 1916, p. 153
6. The explorer Samuel Hearne, quoted in Lopez, Barry, *Arctic Dreams, op. cit.*, p. 235
7. 2 Maccabees 5, 1–4, written about 176 BC
8. Scott, Sir Walter, *The Lay of the Last Minstrel*
9. Devens, R. M., *Our First Century, being a Popular Descriptive Portraiture of the One Hundred Great and Memorable Events of Perpetual Interest in the History of Our Country, Political, Military, Mechanical, Social, Scientific, and Commercial: embracing also Delineations of all the great Historical Characters Celebrated in the Annals of the Republic; Men of Heroism, Statesmanship, Genius, Oratory, Adventure and Philanthropy*, Chicago: C. A. Nichols, 1878, p. 379 ff
10. Cf. Leary, Warren E., 'Five New Satellites With a Mission of Finding a Source of Color in Space', *The New York Times*, 23 January 2007
11. Quoted by http://hyperphysics.phy-astr.gsu.edu/hbase/atmos/redsun/html
12. Cf. *The Times*, 8 March 2006, and also http://epod.usra.edu for 23 February 2006
13. De Meis, Salvo, *Eclipses: An Astronomical Introduction for Humanists*, Rome: Istituto Italiano per L'Africa e L'Oriente, 2002
14. Cf. *The Eclipse Chasers*, a three-part series originally broadcast on BBC 4 on 31 March, 9 April and 16 April 2006
15. The NASA claim may seem far-fetched, but cf. Stockdale, Cally, 'Agnihotra Ancient Healing Fire Practice', www.rainbownews.co.nz, August/September 2006; Priyadarshini, Dr Shantala, 'Homa Fire Ceremony', AHC magazine, 3 February 2005; and Internet, 'NASA and Cow Dung'
16. Cf. Lévi-Strauss, Claude, *Introduction to a Science of Mythology, op. cit.*, p. 42
17. Cf. McEvoy, J. P., *Eclipse: The Science and History of Nature's Most Spectacular Phenomenon*, London: Fourth Estate, 1999
18. Cf. Sayce, Rev. A. H., *The Astronomy and Astrology of the Babylonians*, San Diego: Wizards Bookshelf, 1981, p. 29
19. Dicks, D. R., *Early Greek Astronomy to Aristotle*, Ithaca: Cornell University Press, 1985, p. 10
20. Stephenson, F. R., *Historical Eclipses and Earth's Rotation*, Cambridge: CUP, 1997, p. 226
21. Needham, Joseph, *Science and Civilization in China*, Cambridge: CUP, 1959, Vol. 3, p. 188

22. Niccolò Machiavelli, *Discorsi*, 1:56

23. Dio Cassius, *History of Rome*, Bk. lx, Ch. 26. He does not say how Claudius acquired his knowledge

24. Cf. BBC Radio 4 broadcast on Lawrence, narrated by John Simpson, 16 October 2005

25. Nakayama, Shigeru, *A History of Japanese Astronomy*, Boston: Harvard University Press, 1969, p. 51

26. *Iliad*, Bk. xvii, ll. 266–8: 'Nor would you say that the Sun was safe, or the Moon, for they were wrapped in dark haze in the course of the combat'; and *Odyssey*, Bk. xx, ll. 356–7: 'The Sun has utterly perished from heaven and an evil gloom is overspread.' Salvo de Meis makes the point in his *Eclipses:An Astronomical Introduction for Humanists* (p. 21) that such lines may be poetical licence. 'They contain no faithful descriptions of an eclipse, neither information which could lead to an acceptable dating'

27. Cf. Ferguson, Charles W., *Naked to Mine Enemies*, Boston: Little, Brown, 1958, p. 31

28. Urmy, William Smith, *The King of Day*, New York: Nelson and Phillips, 1874

29. Luke 23, 44

30. Cf. Stephenson, F. R., *Historical Eclipses and Earth's Rotation, op. cit.*, p. 344

31. Cabaniss, J. A., *Agobard of Lyons, Churchman and Critic*, Syracuse: Syracuse University Press, 1953, p. 18

32. Lewalski, Barbara K., *The Life of John Milton*, Oxford: Blackwell, 2000, p. 210, p. 259 and p. 278

33. Quoted in Appleby, Derek, and McCann, Maurice, *Eclipses:The Power Points of Astrology*, Northampton: Aquarian Press, 1989, p. 11

34. *Diary of Samuel Pepys*, Ed. M. Bright, 1879, Vol. vi, p. 208

35. Crump, Thomas, *Solar Eclipse*, London: Constable, 1995, pp. 115–7. Halley invited observations from those members of the public 'with a pendulum clock of which many people are furnished'

CHAPTER 5 THE FIRST ASTRONOMERS

1. Emerson, Ralph Waldo, *Journals and Miscellaneous Notebooks*, 14 November 1865, Harvard: Belknap Press, 1982

2. Flaubert, Gustave, *Dictionary of Accepted Ideas*, New York: New Directions, 1954

3. Lockyer, J. Norman, *The Dawn of Astronomy, A Study of Temple Worship and Mythology of the Ancient Egyptians*, New York: Dover, 2006 (first published 1894)

4. Cf. Sayce, Rev. A. H., *The Astronomy and Astrology of the Babylonians*, San Diego: Wizards Bookshelf, 1981, p. 145, reprinted from Vol. 3, Pt. I of *Transactions of the Society of Biblical Archaeology*, 1874

5. Cf. Britton, John, and Walker, Christopher, 'Astronomy and Astrology in Mesopotamia', in *Astronomy Before the Telescope*, London: British Museum Press, 1996, p. 42

6. Cf. Macqueen, J. G., *Babylon*, New York: Praegar, 1964, p. 212

7. Cf. *The Cambridge History of Islam*, Vol. 3, p. 534, and Swerdlow, N. M., *The*

Babylonian Theory of the Planets, Princeton: Princeton University Press, 1998, p. 26

8. Cf. *Cambridge Ancient History,* Vol. 3, Pt. 2, Cambridge: CUP, p. 285

9. Swerdlow, Noel M., *The Babylonian Theory of the Planets,* Princeton: Princeton University Press, 1998. I found this book very useful, but it received a critical mauling on publication. The world of Babylonian studies is a fierce one. It is worth recalling what was told me by Christopher Walker, the great expert on Babylonian astronomy: 'We have just the first few filaments of the spider's web. To my mind, the debate as to what the Babylonians knew is still wide open; we are all operating in a vacuum'

10. McEvoy, J. P., *Eclipse, op. cit.,* pp. 63–4

11. Cf. Wells, Ronald A., 'Astronomy in Egypt', *Astronomy Before the Telescope,* p. 29

12. Cf. Hawkes, Jacquetta Hopkins, *The First Great Civilizations: Life in Mesopotamia, the Indus Valley and Egypt,* London: Hutchinson, 1973, pp. 230–1; cf. also Neugebauer, Otto E., and Parker, Richard A., *Egyptian Astronomical Texts, Vol. III: Decans, Planets, Constellations and Zodiacs,* Providence: Brown Egyptological Studies 3, 1969

13. Cf. Clemens Alexandrinus, *Stomata,* Lib. VI. Also cf. *Encyclopedia of Religion, op. cit.,* p. 145

14. Cf. Allen, James P., *Genesis in Egypt: the Philosophy of Ancient Egyptian Creation Accounts,* New Haven, Conn.: Yale University Press, 1988, p. 8 and p. 12

15. *Ibid,* pp. 13–14. Atum's seed becomes all life: 'Atum is the one who developed growing ithyphallic, in Heliopolis. /He put his penis in his grasp /That he might make orgasm with it.' Among the Greeks, the Stoic Zeno too chose a biological model. At first there is nothing except God, then God creates a difference within himself, such that he is 'contained' in moisture. This is the living 'sperm' that produces the cosmos

CHAPTER 6 ENTER THE GREEKS

1. Cf. *Greek Thought: A Guide to Classical Knowledge*, Ed. Brunschwig, Jacques, and Lloyd, Geoffrey E. R., Boston: Harvard University Press, 2000, p. 269

2. Cf. Linton, Christopher M., *From Eudoxus to Einstein*, Cambridge: CUP, 2004, p. 15

3. Dicks, D. R., *Early Greek Astronomy to Aristotle*, New York: Norton, 1970, p. 29

4. Plato, *Apology*, 26D

5. Cf. Durant, Will, *The Life of Greece*, New York: Simon and Schuster, 1939, pp. 337–8, p. 627 and p. 161

6. Cf. Heath, Thomas, *A History of Greek Mathematics*, New York: Dover, 1981, pp. 176, 178. Sir Thomas Heath was Permanent Secretary for Administration at the British Treasury in the 1920s and in his spare time a formidable historian of mathematics

7. Cf. Tarn, William, *Hellenistic Civilization*, London: Arnold, 1966, p. 299, and Fowler, Michael, 'How the Greeks Used Geometry to Understand the Stars', http://galileoandeinstein.physics.Virginia.edu/lectures/greek_astro.htm.

8. Aristotle, *On the Heavens*, trans. J. L. Stock, Charleston: Forgotten Books, 2007, pp. 15, 23

9. Durant, Will, *The Life of Greece, op. cit.*, p. 337

10. Cf. Neugebauer, Otto E., *Studies in Civilization, op. cit.*, p. 25

11. Sivin, Nathan, and Lloyd, Geoffrey, *The Way and the Word: Science and Medicine in Early China and Greece*, New Haven and London: Yale, 2002, p. 101

12. Ferris, Timothy, *Coming of Age in the Milky Way*, New York: Anchor, 1989, p. 41. In a later footnote (p. 65) Ferris writes: 'One could write a plausible intellectual history in which the decline of sun worship, the religion abandoned by the Roman emperor Constantine when he converted to Christianity, was said to have produced the Dark Ages, while its subsequent resurrection gave rise to the Renaissance'

13. Rawlins, Dennis, 'Astronomy v. Astrology: The Ancient Conflict', *Queen's Quarterly 91/4* (winter 1984), pp. 969–89

14. Cf. Newton, Robert R., *The Crime of Claudius Ptolemy*, Baltimore: John Hopkins, 1997, p. 76 and ff

15. Ronan, Colin A., *The Astronomers*, London: Evans, 1964, p. 91

16. *Ibid.*, p. 95

17. Cf. 'The Universe of Aristotle and Ptolemy', http://csep10.phys.utk.edu/astr161/lect/retrograde/aristotle/html.

18. Cumont, Franz, *The Oriental Religions*, New York: Dover, 1956 (first published 1911), p. 134

CHAPTER 7 GIFTS OF THE YELLOW EMPEROR

1. Quoted by Joseph Needham in *Science and Civilization in China, op. cit.*, p. 171

2. Zou Yuanbiao, 'Da Xiguo Li Madou', in *Yuan xue ji*, 3/39. Quoted in Spence, Jonathan D., *The Memory Palace of Matteo Ricci*, London: Faber, 1985, p. 151

3. Cf. Winchester, Simon, *Joseph Needham: The Man Who Loved China*, New York:

SOURCES 625

HarperCollins, 2008, pp. 17–21. Winchester adds: 'Also a chain-smoker and devastating womanizer'
4. Cf. *The New York Times*, 27 March 1995
5. Cf. Judson, Horace Freeland, 'China's Drive for World-Class Science', MIT's *Technology Review*, January 2006. When Needham's first volume appeared, eminent American academics published indignant reviews, rubbishing the notion that Chinese science and technology played any consequential part in history, and warning that the book was part of the Red Menace
6. Cf. Ronan, C., 'Astronomy in China, Korea and Japan', *Astronomy Before the Telescope, op. cit.*, pp. 245–68
7. Quoted by Needham, *Science and Civilization in China, op. cit.*, p. 193
8. Cf., for example, Article 110 in *The T'ang Code: Vol. II, Specific Articles*, trans. Wallace Johnson, 1997
9. Needham, *op. cit.*, p. 406
10. *Ibid*, p. 227.
11. Cf. 'Ancient Chinese Cosmology', www.astronomy.pomona.edu/archeo/china/china3.html.
12. Cronin, Vincent, *The Wise Man from the West*, New York: Dutton, 1955, p. 232
13. Needham, *op. cit.*, p. xlv
14. Cullen, Christopher, 'Joseph Needham on Chinese Astronomy', *Past and Present*, No. 87 (May 1980), p. 42. Cullen ends his essay: 'Compilers of works of popularization and "cross-disciplinary syntheses" who try to treat Needham's work as a bran-tub will find a few mousetraps awaiting their fingers as they rummage.' Point taken
15. Cronin, Vincent, *The Wise Man from the West, op. cit.*, p. 11. Needham lists 40 outstanding inventions over 30,000 years up to 1500 BC, and 287 inventions from then until AD 1700
16. Xue Shenwei, *Journal of Ancient Civilizations*, Vol. 2, IHAC North-East Normal University, Changchun, Jilin Province, 1987, 'Brief Note on the Bone Cuneiform Inscriptions', pp. 131–4
17. Conversation with Christopher Walker, Department of Ancient Near East, British Museum, 12 January 2006
18. Cf. Hadingham, Ivan, 'The Mummies of Xinjiang', *Discover*, 15.68, April 1994; cf. also Drauenhelm, 'Ulghur-I Ulgur Mummies Say "No" to China', *East Bay Monthly*, Vol. XXIX, No. 3, December 1998; Wong, Edward, 'The Dead Tell a Tale China Doesn't Care to Listen To', *The New York Times*, 19 November 2008, A6; Barber, Elizabeth Wayland, *The Mummies of Urumchi*, New York: Norton, 1999; and Wade, Nicholas, 'A Host of Mummies, A Forest of Secrets', *The New York Times*, 16 March 2010, D1
19. Cronin, Vincent, *The Wise Man from the West, op. cit.*, p. 100
20. *Ibid*, pp. 192–3
21. Needham, *op. cit.*, p. 240
22. Cf. Rothstein, Edward, 'A Big Map That Shrank The World', *The New York Times*, 20 January 2010, C1 and C7
23. Cf. Internet blog entry, 'Eunuchs, Sleuths and Otherwise', 21 April 2008

24. Cronin, Vincent, *The Wise Man from the West*, *op. cit.*, pp. 140–2; this is largely Cronin's account, with just small changes

25. Fan Tsen-Chung, *Dr Johnson and Chinese Culture*, London: The China Society, 1945, p. 9

CHAPTER 8 THE SULTAN'S TURRET

1. Qur'an, Surah 6, 76–8

2. Koestler, Arthur, *The Sleepwalkers*, London: Hutchinson, 1959, p. 105

3. Cf. Internet article on Arab astronomy by J. J. O'Connor and E. F. Robertson

4. Cf. Ilyas, Mohammad, *Islamic Astronomy and Science Development: Glorious Past, Challenging Future*, Petaling Jaya, Selangor Darul Ehsan, Malaysia: Pelanduk Publications, 1996, p. 22

5. Khan, M. A. R., *Muslim Contributions to Science and Culture*, Ashraf, Lahore, Internet edition, p. 14. Cf. also King, David A., *Astronomy in the Service of Islam*, Brookfield, Vt.: Variorum, 1993, p. 246

6. Smith, Zadie, *White Teeth*, London: Penguin Books, 2001, p. 397

7. Ilyas, Mohammad, *Islamic Astronomy and Science Development, op. cit.*, p. 253. Cf. also 'Aspects of Applied Science in Mosques and Monasteries' in *Science and Theology in Medieval Islam, Judaism and Christianity*, Madison, Wisconsin, 15–17 April 1993

8. King, David A., 'Islamic Astronomy', *Astronomy Before the Telescope, op. cit.*, p. 160

9. Ilyas, Mohammad, *Islamic Astronomy and Science Development, op. cit.*, pp. 1–2. The office of the *muwaqqit*, or mosque astronomer, was developed during the thirteenth century, in Egypt. Most Egyptian and Syrian astronomers of consequence for the next two centuries were *muwaqqits*

10. Cf. Neugebauer, Otto, 'The Early History of the Astrolabe', *Astronomy and History: Selected Essays*, Berlin: Springer, 1983, p. 279. A nice description of the astrolabe also appears in Edward Rutherfurd's novel, *London*, London: Century, 1997, p. 340

11. Maalouf, Amin, *The Crusades Through Arab Eyes*, New York: Schocken, 1989, pp. 98–105

12. Ṣā'id ibn Ahmad, Abūl-Qāsim, *The Categories of Nations*, quoted in Neugebauer, 'The Transmission of Planetary Theories in Ancient and Medieval Astronomy', *Astronomy and History, op. cit.*, p. 3/129

13. Cf. Linton, C. M., *From Eudoxus to Einstein, op. cit.*, p. 85

14. Neugebauer, Otto, *A History of Ancient Astronomy*, Berlin: Springer, 1975, 3 Vols, p. 127

15. Cf. Pingree, David, 'Astronomy in India', *Astronomy Before the Telescope, op. cit.*, pp. 123–42

16. Cf. Henry, John, *Moving Heaven and Earth*, Cambridge: Icon Books, 2001, p. 41

17. St Augustine, *Works*. Quoted by Linton, Christopher, *From Eudoxus to Einstein, op. cit.*, p. 113

18. North, John David, *Stars, Minds and Fate: Essays in Ancient and Medieval Cosmology*, London: Hambledon Press, 1989, p. 403

19. Cf. Neugebauer, Otto E., *A History of Mathematical Astronomy*, Boston: Birkhäuser, 1975, p. 943

20. Cf. Schott, Ben, 'On the Division of Our Three Score and Ten', *The New York Times*, 19 October 2009, A27

21. Cf. Green, V. H. H., *Renaissance and Reformation*, London: Arnold, 1952, p. 50

22. Cf. Koestler, *op. cit.*, p. 110. Much of this paragraph paraphrases his analysis

NARCISSISM: THINKS WORLD REVOLVES AROUND IT

CHAPTER 9 THE EARTH MOVES

1. Luther, Martin, *Works*, Vol. 22, c. 1543

2. Marx, Groucho, *The Groucho Letters*, London: Sphere, 1967, p. 10

3. Bailey, Thomas A., *The American Pageant: A History of the Republic*, Boston: Houghton Mifflin, 13th Edition 2006, Ch. 1

4. Manchester, William, *A World Lit By Fire*, Boston: Little, Brown, 1992, p. 89

5. Cf. Linton, C. M., *From Eudoxus to Einstein: A History of Mathematical Astronomy*, *op. cit.*, p. 116

6. Sobel, Dava, *Galileo's Daughter*, New York: Penguin, 2000, p. 59

7. Cf. Heath, Thomas, *A History of Greek Mathematics*, *op. cit.*, p. 301. The omitted passage runs: 'Philolaus believed in the mobility of the Earth, and some even say that Aristarchos of Samos was of that opinion.' One also wonders what Copernicus would have made of the assertion by Leonardo da Vinci (1452–1519) that 'The Sun does not move . . . The Earth is not in the centre of the circle of the Sun, nor in the centre of the universe.' (*Notebooks*, I, 310, 298)

8. Cf. Joshua X, 12–13, and Ecclesiastes I, 4–5

9. Moore, Patrick, *Astronomy Before the Telescope*, *op. cit.*, p. 10

10. Cf. Hardin, Garrett, 'Rewards of Pejoristic Thinking', Internet essay

11. Neugebauer, Otto, *Astronomy and History*, *op. cit.*, 'On the Planetary Theory of Copernicus', p. 103. This essay gives in detail the many ways Copernicus adhered to Ptolemaic calculations. Cf. also an interesting discussion of the extent to which Copernicus may have plagiarized Arab theorems in Teresi, Dick, *Lost Discoveries*, New York: Simon and Schuster, 2002, pp. 3–5

12. Koestler, Arthur, *The Sleepwalkers, op. cit.*, p. 202

13. Brahe, Tycho, 'An Oration on Mathematical Principles'. Cf. Linton, C. M., *From Eudoxus to Einstein, op. cit.*, p. 155

14. Christianson, John Robert, *On Tycho's Island: Tycho Brahe and His Assistants, 1570-1601*, New York: CUP, 2000, p. 82

15. Koestler, Arthur, *The Sleepwalkers, op. cit.*, p. 302

16. Kepler, Johannes, *The New Astronomy*, Cambridge: CUP, 1992, cf. introduction. Its subtitle, 'Celestial Physics', implies a programme of investigation that struck Kepler's contemporaries as paradoxical: when the book was first published it was almost universally ignored

17. Gingerich, Owen, *The Book Nobody Read*, New York: Walker, 2004, p. 168, p. 47. In his entertaining account, Gingerich not only writes off what he calls 'the legend of epicycles upon epicycles' but is dismissive of Koestler's theory of scientific advance: 'Some critics said his book should have been entitled *The Sleepwalker*, because he had only one good example of a scientist groping in the dark, Kepler himself. A few even more perceptive critics said he had zero examples, an opinion borne out by recent scholarship' (p. 49). Maybe some scientists bridle at being likened to sleepwalkers, but the fact remains that many important discoveries come out of left field – and that was Koestler's point

18. Brecht, Bertholt, *Galileo*, New York: Grove, 1966, p. 51

19. Clegg, Brian, *Light Years, op. cit.*, p. 74

20. Cf. Overbye, Dennis, 'A Telescope to the Past As Galileo Visits U.S.', *The New York Times*, 28 March 2009, A8. But Galileo's colleague Cesare Cremonini, a professor of natural philosophy, was paid 2,000 florins a year: mathematicians were still regarded as inferior to philosophers

21. Ferris, Timothy, *Coming of Age in the Milky Way, op. cit.*, p. 84

22. Clegg, Brian, *Light Years, op. cit.*, pp. 54–5

23. Swerdlow, Noel M., 'Galileo's Discoveries with the Telescope and Their Evidence for the Copernican Theory', *The Cambridge Companion to Galileo*, Ed. Peter Machamer, Cambridge: CUP, 1998, pp. 244–5

24. Cf. *Discoveries and Opinions of Galileo*, trans. and ed. Drake, Stillman, New York: Doubleday, 1957, pp. 87–144

25. Burton, Robert, *The Anatomy of Melancholy*, New York: New York Review of Classics Books, 2001, Pt. 2, pp. 57–8

26. Mencken, H. L., *A Treatise on Right and Wrong*, New York: Knopf, 1934, p.279

27. Koestler, Arthur, *The Sleepwalkers, op. cit.*, p. 483

28. Geymonat, Ludovico, *Galileo Galilei: A Biography and Enquiry Into His Philosophy of Science*, trans. Drake, Stillman, New York: McGraw-Hill, 1965, p. 73

29. Milton, John, *Paradise Lost*, Bk. VIII, ll. 167 ff

30. Cf. Johnston, George Sim, article on the Internet, 'The Galileo Affair', available from Scepter Press, PO Box 1270, Princeton, NJ, 08542

31. Cf. Westfall, Richard S., *Essays on the Trial of Galileo*, Vatican Observatory Foundation, 1989, p. v; also Owen, Richard, and Delaney, Sarah, 'Vatican Recants with a Statue of Galileo', *The Times*, 4 March 2008, p. 11, which quotes the current Pope, while still a cardinal, making a speech in 1990 in which he described Galileo's trial as fair. The Vatican later said he had been misquoted

CHAPTER 10 STRANGE SEAS OF THOUGHT

1. Brewster, David, *Life of Sir Isaac Newton*, London: John Murray, 1831, Vol. 2, Ch. 27
2. Tolstoy, Leo, *War and Peace*, trans. Richard Pevear and Larissa Volokhonsky, New York: Vintage, 2008, p. 606
3. Westfall, Richard S., *Never at Rest, A Biography of Isaac Newton*, Cambridge: CUP, 1983, p. 10
4. *Ibid*, p. 581
5. Einstein, Albert, 'What is the Theory of Relativity?', *The Times*, 28 November 1919
6. Herivel, John, *Background to Newton's Principia*, Oxford: Clarendon Press, 1965, p. 67
7. Voltaire, François-Marie Arouet de, *An essay upon the civil wars of France [. . .] and also upon the epick poetry of the European nations*, reprinted in *Letters on England*, New York: Prentice Hall, 2000, p. 75
8. Stukeley, William, *Memoirs of Sir Isaac Newton's Life*, London: Taylor & Francis, (reprint) 1936, p. 20
9. Quoted in Whitehouse, David, *The Sun: A Biography*, West Sussex: Wiley, 2005, pp. 71–2
10. Cf. Gregory, R. L., *Eye and Brain: The Psychology of Seeing*, London: Weidenfeld, 1966, p. 16
11. Eliade, Mircea, *Patterns in Comparative Religion*, op. cit., p. 128
12. Cf. Needham, Joseph, *Science and Civilization in China*, op. cit., p. 219
13. Dizikes, Peter, 'Word for Word on the Web, Isaac Newton's Secret Musings', *The New York Times*, 12 June 2003
14. Keynes, J. M., 'Newton, The Man', *The Royal Society Newton Tercentenary Celebrations*, Cambridge: CUP, 1947, p. 27
15. Milton, John, *Paradise Lost*, Bk. III, ll. 606–9
16. Gleick, James, *Isaac Newton*, New York: Pantheon, 2003. Gleik's concise study is the source of many of the biographical details in this chapter
17. Fernie, Donald, *The Whisper and the Vision:The Voyages of the Astronomers*, Toronto: Clarke, Irwin, 1976, p. 10
18. Cf. Holmes, Richard, *The Age of Wonder*, London: Harper, 2008, p. 17
19. Cf. Horowitz, Tony, 'A Week Before the Mast', *The New Yorker*, 22 July 2002, p. 42
20. Kooser, Ted, *Delights and Shadows*, Port Townsend, WA: Copper Canyon Press, 2004, p. 62
21. Cf. Hufbauer, Karl, *Exploring the Sun*, op. cit., p. 33
22. Burney, Fanny, *Diary*, September 1786, pp. 169–70
23. Cf. Clark, Stuart, *The Sun Kings, The Unexpected Tragedy of Richard Carrington and the Tale of How Modern Astronomy Began*, Princeton: Princeton University Press, 2007, p. 35
24. Linton, C. M., *From Eudoxus to Einstein*, op. cit., p. 360
25. Laplace, Pierre-Simon, *Oeuvres Complètes*, Paris, 1884, VI, p. 478, and VII, p. 121

CHAPTER 11 ECLIPSES AND ENLIGHTENMENT

1. Joyce, James, *Ulysses, The Corrected Text*, New York: Random, 1986, p. 136, ll. 364–370
2. Beaglehole, J. C., *The Life of Captain James Cook*, Stanford: Stanford University Press, 1974, p. 87
3. *Sky and Telescope*, April 2000, pp. 63–5
4. Cf. Clerke, Agnes, M. *A Popular History of Astronomy during the Nineteenth Century*, London: A & C Black, 1893, p. 74
5. Todd, D. P., *Total Eclipses of the Sun*, Boston, 1894, p. 110
6. Cf. *The Nat Turner Rebellion*, Ed. John B. Duff and Peter M. Mitchell, New York: Harper and Row, 1971, p. 19
7. Styron, William, *The Confessions of Nat Turner*, New York: Random House, 1967, p. 348
8. *Month. Not.*, R. A. S., Vol. xx, p. 301, May 1850. Cf. Chambers, *The Story of Eclipses, op. cit.*, p. 143
9. Bailey, Paul D., *Wovoka, the Indian Messiah*, Los Angeles: Westerlore Press, pp. 93–4
10. Woolf, Virginia, *The Diary of Virginia Woolf, Vol. III, 1925–30*, Ed. Anne Olivier Bell, New York: Harcourt, 1980, pp. 142–3
11. Woolf, Virginia, *The Waves*, London: Penguin, 1951, p. 244 and p. 246
12. Asimov, Isaac, *The Relativity of Wrong*, 'New Stars', New York: Doubleday, 1988, p. 138
13. Gould, Stephen Jay, *Dinosaur in a Haystack*, New York: Harmony Books, 1985, 'Happy Thoughts on a Sunny Day in New York City', p. 3
14. Updike, John, *The Early Stories, 1953–75*, New York: Knopf, 2003, pp. 645–6
15. Cf. Pasachoff, Jay M., 'Solar-Eclipse Science: Still Going Strong', *Sky and Telescope*, February 2001, pp. 40–5
16. Einstein, Albert, '*Grundlage der allgemeinen Relativitätstheorie*' ('The foundations of the theory of general relativity'), *Annalen der Physik*, 1916, Vol. 49, pp. 769–822
17. Cf. Linton, C. M., *From Eudoxus to Einstein: A History of Mathematical Astronomy, op. cit.*, p. 450
18. Eddington, A. S., cf. *Obituary Notices, Fellows of the Royal Society*, No. 8, Vol. 3, January 1940, p. 167
19. Mosley, Nicholas, *Hopeful Monsters*, New York: Vintage, 2000, p. 26
20. Cf. Clark, Ronald W., *Einstein: The Life and Times*, London: Hodder, 1973, p. 225; see also Miller, Arthur I., *Empire of the Stars: Obsession, Friendship and Betrayal in the Quest for Black Holes*, Boston: Houghton Mifflin, 2005, pp. 51–2
21. Crump, T., *Solar Eclipse, op. cit.*, p. 134; I have drawn in some detail from his description of the expedition

CHAPTER 12 THE SUN DETHRONED

1. Tommaso Campanella (1568–1639), *A Dialogue between a Grand Master of the Knights Hospitallers and a Genoese Sea Captain: The City of the Sun*

2. Cf. Baggott, Jim, *The First War of Physics: The Secret History of the Atomic Bomb 1939 – 1949*, New York: Pegasus Books, 2010

3. Cf. *Catholic Encyclopedia* on CD-ROM, entry for Secchi

4. Cf. *The New York Times*, Science section, 11 July 2006

5. Haddon, Mark, *The Curious Incident of the Dog in the Night-Time*, London: Cape, 2003, p. 12

6. Cowen, Ron, 'The Galaxy Hunters', *National Geographic*, February 2003, p. 16

7. Odenwald, Sten F., *The 23rd Cycle, Learning to Live with a Stormy Star*, New York: Columbia University Press, 2001, pp.xii–xiii

8. Housman, A. E., *Selected Prose*, Ed. Carter, J., and Sparrow, J., 1961

9. Gorst, Martin, *Measuring Eternity: The Search for the Beginning of Time*, New York: Broadway, 2001, p. 11

10. Hawking, Stephen W., *A Brief History of Time*, New York: Bantam, 1988, p. 7

11. Gorst, *Measuring Eternity, op. cit.*, p. 95, of whose excellent account I have made significant use in this section

12. Ferris, Timothy, *Coming of Age in the Milky Way, op. cit.*, p. 246

13. Gorst, *Measuring Eternity, op. cit.*, p. 187

14. *Ibid*, p. 188

15. Herschel, J. F. W., *Treatise on Astronomy*, London: Longman, 1833, p. 212

16. Gamow, George, *The Birth and Death of the Sun: Stellar Evolution and Subatomic Energy*, New York: Viking, 1949, p. 12

17. For a discussion of how 'outsiders' contributed to solar physics from 1910 on, cf. Hufbauer, Karl, *Exploring the Sun, Solar Science since Galileo*, Baltimore: John Hopkins University Press, 1993, p. 81 ff

18. Payne, Cecilia H., *Stellar Atmospheres* (Harvard Observatory Monograph No.1), Cambridge, 1925, p. 185

19. Cf. Gingerich, Owen, 'The Most Brilliant PhD Thesis Ever Written in Astronomy', Harvard Center for Astrophysics, p. 16

20. Cf. Gamow, George, *The Birth and Death of the Sun, op. cit.*, p. v

21. Cf. Aaserud, Finn, *Redirecting Science: Niels Bohr, Philanthropy and the Rise of Nuclear Physics*, Cambridge: CUP, 1990, p. 2

22. Kaiser, David, 'A x B # B x A', *London Review of Books*, 26 February 2009, p. 21

23. Letter to Bohr, 21 April 1932, quoted in Aaserud, *Redirecting Science*, p. 55

24. Aaserud, Finn, *Redirecting Science, op. cit.*, p. 7

25. Cf. seedmagazine.com, 'The Elements'. Freeman Dyson, one of his students, has called Bethe 'the supreme problem solver of the twentieth century'

26. Cf. Kaku, Michio, *Physics of the Impossible, A Scientific Exploration into the World of Phasers, Force Fields, Teleportation, and Time Travel*, New York: Doubleday, 2008, p. xv

27. Cf. Rhodes, Richard, *Dark Sun*, New York: Simon and Schuster, 1995, p. 222

28. A full account of how Einstein's letter was drafted and how it reached the White House is given in Isaacson, Walter, *American Sketches*, New York: Simon and Schuster, 2009, pp. 149–155

29. Cf. also Bernstein, Jeremy, 'The Secrets of the Bomb', *New York Review of Books*, 25 May 2006, p. 41. The wags at the Project were not the only ones to make play with nicknames. Klaus Fuchs, who was to hand over so many vital secrets to the Russians, was tagged by his masters 'Santa Klaus'

30. Cf. Dobbs, Michael, 'If You Build It . . . ', *The New York Times Book Review*, 9 May 2010, p. 18
31. Cf. Hersey, John, *Hiroshima*, Harmondsworth: Penguin, 1946, Publisher's Note, p. viii

CHAPTER 13 SUNSPOTS

1. Urmy, William, *The King of Day*, op. cit., pp. 82–3
2. Cf. Needham, Joseph, *Science and Civilization in China*, op. cit., p. 435. Needham mentions several Arab observations, including those in 840 (mistaken for a transit of Venus), 1196 and 1457
3. Casanovas, Father Juan, 'Early Observations of Sunspots', San Francisco: Astronomical Society of the Pacific Conference Series, Proceedings of a Meeting held in Puerto de la Cruz, Tenerife, Spain, 2–6 October 1996, Vol. 118, pp. 3 and 19
4. McClure, Ellen M., *Sunspots and the Sun King*, Chicago: University of Illinois Press, 2006, p. 1
5. Schwabe, Heinrich, *Astronomische Nachrichten*, Vol. 20, No. 495, 1843
6. Mitton, Simon, *Daytime Star: The Story of Our Sun*, New York: Scribner, 1981, p. 122
7. Odenwald, Sten F., *The 23rd Cycle, Learning to Live with a Stormy Star*, New York: Columbia University Press, 2001, p. 54
8. Cf. Internet essay on Schwabe and Wolf by Dr David P. Stern, email address: education@phy6.org
9. Clark, Stuart, *The Sun Kings*, op. cit., p. 23
10. Cf. *Encyclopedia of Astronomy and Astrophysics*, op. cit., p. 3203
11. Livingston, W., Harvey, J. W., Malenchenko, O. V. and Webster, L., 'Sunspots with the Strongest Magnetic Fields', *Solar Physics*, Vol. 239, Nos 1–2, December 2006, pp. 41–68
12. Cf. Mitton, op. cit., p. 130. Cf. also: Livingston, William, and Bhatnagar, Arvind, *Fundamentals of Solar Astronomy*, New Jersey: World Scientific Publishing, 2005
13. Cf. Clark, Stuart, *The Sun Kings*, op. cit., p. 142
14. Cf. Zeuner, F. E., *Dating The Past*, London: Sutton, 1952, p. 19
15. Stetson, Harlan True, *Sunspots and Their Effects*, New York: Whittesley House (McGraw-Hill), 1937, p. 43
16. Brody, Judit, *The Enigma of Sunspots*, London: Floris Books, 2002, pp. 102–3
17. Twain, Mark, *The American Claimant*, New York: Kessinger, 2004
18. *Ibid.*, p. 272
19. Johnson, Samuel, *Rasselas, Poems, and Selected Prose*, ed. Bronson, Bernard H., New York: Holt, 1965, pp. 592–3
20. Clark, Stuart, 'Quiet Sun Puts Europe on Ice', *New Scientist*, 14 April 2010
21. Cf. Eddy, J. A., *Climate and the Role of the Sun*; and Eddy, J. A., *The New Solar Physics*, pp. 11–34, Westview Press, Boulder; entry on Douglass, Andrew E., *Dictionary of American Biography*; and *Nature*, Vol. 431, p. 1047
22. Cf. Weart, Spencer R., *The Discovery of Global Warming*, op. cit., p. 131

23. Calder, Nigel, *The Weather Machine*, London: BBC, 1974, p. 131

24. Cf. Ruddiman, William, F., 'How Did Humans First Alter Global Climate?', *Scientific American*, March 2005, pp. 46–53; also Goldsmith, Donald, 'Ice Cycles', *Natural History*, March 2007, pp. 14–18

25. Cf. Sidis, William James, 'A Remark on the Occurrence of Revolutions', *Journal of Abnormal Psychology*, 1918, 13, pp. 213–228. Cf. also Gardner, Martin, *Mathematical Carnival*, New York: Knopf, 1977

26. Cf. *Citizen Cohn*, scripted by Nicholas von Hoffman, HBO film for television, Breakheart Films/Spring Creek Productions, 1992; von Hoffman's book of the same title, New York: Doubleday, 1988, p. 180; and Phillips, Wayne, 'Harassing Feared by "Voice" Suicide', *The New York Times*, 7 March 1953, p. 10

27. Cf. Brody, Judit, *The Enigma of Sunspots, op. cit.*, pp. 162–3

28. Urmy, William, *The King of Day, op. cit.*, p. 79

29. Cf. Odenwald, Sten F., *The 23rd Cycle, op. cit.*, p. 8 ff and p. 70

30. Cf. Guesnerie, Roger, *Assessing Rational Expectations: Sunspot Multiplicity and Economic Fluctuations*, Cambridge, Mass.: MIT Press, 2001, p. 1

31. Cf. Wrigley, E. A., 'The Quest for the Industrial Revolution', read at the British Academy 12 November 2002

32. La Marche, Jr., Valmore, C., and Fritts, Harold C., 'Tree-Rings and Sunspot Numbers', *Tree-Ring Bulletin*, Vol. 32, 1972, pp. 19–33

33. Cf. Baillie, Mike, 'Tree-rings focus attention on global environmental events that could be caused by extraterrestrial forcing', School of Archaeology and Palaeoecology, The Queen's University, Belfast. Cf. also Jones, P. D., and Mann, E., 'Climate Over Past Millennia', *Review of Geophysics*, 42, RG2002, May 2004, p. 21 of 42

34. Cf. Simons, Paul, 'Summer 1816', *The Times*, 1 October 2008, p. 9

35. Stuart Clark, 'Absence of sunspots make scientists wonder if they're seeing a calm before a storm of energy', *Washington Post*, 22 June 2010; and 'What's wrong with the Sun?', *New Scientist*, 14 June 2010

36. Simons, Paul, 'Weather Eye', *The Times*, 10 April 2010, p. 85

'Call it "Astronomy Beyond the Invisible–Light Spectrum".'

CHAPTER 14 THE QUALITIES OF LIGHT

1. Quoted in Sims, Michael, *Apollo's Fire, op. cit.*, p. 29

2. Greene, Graham, *The Power and the Glory*, London: Penguin, 1940, p. 70

3. Galilei, Galileo, *Dialogue Concerning Two New Sciences*, Philadelphia: Running Press, 2005

4. Cf. McCullough, Katie, 'Speed of Light Formal Report (The Foucault Method)', Internet entry

5. Kaku, Michio, *Physics of the Impossible*, *op. cit.*, p. 19

6. Cf. *The Master of Light*, Dorothy Michelson Livingston, Chicago, 1973; and Achenbach, Joel, 'The Power of Light', *National Geographic*, October 2001, p. 15

7. Lloyd, John, and Mitchinson, John, *The Book of General Ignorance*, London: Harmony, 2007, p. 57

8. Cf. Attenborough, David, *The Private Life of Plants*, Princeton: Princeton University Press, 1995, p. 102

9. Clark, Stuart, *The Sun Kings*, *op. cit.*, p. 94

10. Cf. Steel, Duncan, *Marking Time*, *op. cit.*, pp. 145–50. Cf. also *Why Don't Penguins' Feet Freeze? And 114 Other Questions*, Ed. Mick O'Hare/New Scientist, London: Profile, 2006, pp. 154–5

11. Clark, Stuart, *The Sun Kings*, *op. cit.*, p. 96

12. Cf. Bucke, Richard M., *Cosmic Consciousness*, Secaucus, NJ: University Books, 1961, quoted in Matthiessen, Peter, *The Snow Leopard*, New York: Viking, 1978, p. 98

13. Wood, Elizabeth, *Science for the Airplane Passenger*, Boston: Houghton Mifflin, 1968, p. 60

14. *Encyclopedia Britannica*, 10th Edition, 'Meteorology', pp. 278–9

15. *National Geographic*, *op. cit.*, p. 33

16. *Natural History*, October 2005, p. 11

17. Holden, Anthony, *Big Deal*, London: Abacus, 2002, pp. 49 and 120. In Charles Bock's breakout first novel *Beautiful Children* (New York: Random, 2008) Las Vegas is: 'The neon. The halogen. The viscous liquid light. Thousands and millions of watts, flowing through letters of looping cursive and semi-cursive, filling then emptying, then starting over again. Waves of electricity, emanating from pop-art facades, actually transforming the nature of the atmosphere, creating a mutation of night, a night that is not night – daytime at night.'

18. Woolf, Leonard, *Beginning Again*, London: Hogarth Press, 1964, p. 19

19. Cf. Hibbert, Christopher, *The Great Mutiny: India, 1857*, London: Penguin, 1980, p. 147; and Cornwell, Bernard, *Sharpe's Rifles, Richard Sharpe and the French Invasion of Galicia January 1809*, London: Penguin, 1989, p. 86

20. Nesbit, Edith, *The Wouldbegoods*, London: Puffin, 1996, Ch. 3, 'Bill's Tombstone'

21. Cf. Air Vice Marshal J. E. Johnson, *The Story of Air Fighting*, London: Chatto, 1964, pp. 24–5

22. Cf. Bennett, Geoffrey, *Coronel and the Falklands*, Edinburgh: Birlinn, 2000, pp. 13, 28

23. Shepard, Sam, 'A Short Life of Trouble', originally published in *Esquire*, 1987, reprinted in *Bob Dylan, The Essential Interviews*, Ed. Jonathan Cott, New York: Wenner Books, 2006, p. 365

24. For a discussion of how shadow and glare have affected baseball in the US, cf. Branch, John, 'Postseason's Afternoon Start Times Put Shadows in Play',

The New York Times, 7 October 2009, B17

25. Snow, C. P., *Variety of Men*, New York: Scribner's, 1967, p. 22

26. Hazlitt, William, *Essays*, 'The Fight', London: Penguin, 2000. Hazlitt was himself a good amateur boxer, and in this essay introduced what has become a renowned sporting cliché: 'In the first round everyone thought it was all over'

27. Hopman, Harry, *Lobbing into the Sun*, New York: Bobbs Merrill, 1975, pp. 43–5

28. Cf. also Marjoribanks, Edward, *For the Defence, The Life of Sir Edward Marshall Hall*, New York: Macmillan, 1929, p. 147, pp. 408–12

29. Cf. *Encyclopedia Britannica*, 1973 Edition, Vol. 5, p. 25

30. Ekrich, A. Roger, *At Day's Close: Night in Times Past*, New York: Norton, 2005; cf. also *The New Yorker*, 30 May 2005, pp. 89, 59

31. Anisha Vranckx, in an interview with the author, October 2006. Ms Vranckx, of Greek extraction but now based in Udaipur, is a Romany, and has twice represented gypsy interests at the United Nations

32. Sangster, W., *Umbrellas and Their History*, London: Cassell, 1871, p. 15

33. Cf. St Joseph, J. K., 'Air Photography and Archaeology', *The Geographical Journal*, Vol. 105, No. 1/2 (Jan–Feb 1945), pp. 47–59

34. Hull, Edmund C. P., *The European in India, or Anglo-Indian's Vade Mecum*, London, 1871, p. 61

35. Cf. Baring-Gould, S., *Historic Oddities and Strange Events*, London: Geoffrey Bles, 1945, pp. 132–8

36. Stevenson, Robert Louis, *The Strange Case of Doctor Jekyll and Mr Hyde and Other Tales of Terror*, Ed. Mighall, Robert, London: Penguin, 2002, p. 23

37. Sherriff, R. C., *Journey's End*, London: Penguin Classics, 2000, p. 43

38. Hegel, G. W. F., *Philosophy of Right*, trans. T. M. Knox, New York: OUP, 1967, first published 1820

39. Nelson, Jill, *Sexual Healing*, Chicago: Agate Publishing, 2003; cf. review of this first novel by Beverly Lowry, *The New York Times Book Review*, 22 June 2003, p. 20

40. Nabokov, Vladimir, *Speak Memory*, New York: Putnam, 1960, p. 81; cf. also Wertheim, Margaret, 'The Shadow Goes', *The New York Times*, 20 June 2007; and Champlin, Charles, and Tseng, Derrick, *Woody Allen at Work: the Photographs of Brian Hamill*, New York: Abrams, 1995, p. 29, photo caption showing Allen on a beach in Southampton, Long Island, during the shooting of *Interiors* (1978)

41. A belief promulgated by Rabbi Jacob Emden (1697–1776), who cited the biblical verse (Josh. I, 8), 'You will study [the Torah] day and night', leaving room for secular studies during hours that are neither truly day nor truly night
42. Cf. Coldstream, John, *Dirk Bogarde*, London: Weidenfeld, 2005, p. 517
43. Jung, Carl Gustav, *Memories, Dreams, Reflections, op. cit.*, p. 269

CHAPTER 15 BENEATH THE BEATING SUN

1. Kipling, Rudyard, 'Dear Auntie, Your Parboiled Nephew', Holograph letter to Edith Macdonald, 12 June 1883
2. Clarke, Thurston, *Equator*, New York: Avon, 1988, pp. 11, 137 and 143
3. Spence, Jonathan, *The Memory Palace of Matteo Ricci, op. cit.*, pp. 37, 84
4. Tolstoy, Leo, *War and Peace, op. cit.*, p. 1013
5. Jhabvala, Ruth Prawer, *Heat and Dust*, London: John Murray, 1975, p. 71
6. Coward, Noël, *The Complete Illustrated Lyrics*, New York: Overlook, 1998
7. Kipling, Rudyard, *Kim*, London: Penguin, 1987, p. 110 and p. 330
8. Cf. Grann, David, *The Lost City of Z*, New York: Doubleday, 2009, p. 56
9. Cf. Wheeler, Sara, *Too Close to the Sun: The Audacious Life and Times of Denys Finch Hatton*, New York: Random, 2007, pp. 55, 69 and 95
10. Collier, Richard, *The Sound of Fury, An Account of the Indian Mutiny*, London: Collins, 1963, pp. 120 and 14–15
11. Weiner, Jonathan, *The Next One Hundred Years*, New York: Bantam, 1990, p. 154
12. Obama, Barack, *Dreams from My Father*, New York: Three Rivers Press, 2004
13. Babich, Babette, 'Reflections on Greek Bronze and "The Statue of Humanity"', *Existentia*, Vol. XVII, 2007, p. 436
14. Shakespeare, William, Sonnets, 62, l. 10; there are 253 references to the Sun in his writing as a whole
15. Cf. 'Questions Answered', *The Times*, 4 and 5 October 2004
16. Seymour, Miranda, 'White Shoulders', her review of Deborah Davis's biography *Strapless, John Singer Sargent and the Fall of Madame X*, *The New York Times Book Review*, 28 September 2003, p. 19
17. Fussell, Paul, *Abroad, British Literary Travelling Between the Wars*, Cambridge: CUP, 1980, p. 137. Several of the details on these pages come from this study, notably the chapter 'The New Heliophily', pp. 137–41
18. Du Maurier, Daphne, *Myself When Young*, New York: Doubleday, 1977, p. 61
19. Hamilton-Paterson, *Three Miles Down, A First-Hand Account of Deep-Sea Exploration and A Hunt for Sunken World War II Treasure*, London: Cape, 1998, p. 99
20. Fitzgerald, Scott, *Tender Is the Night*, London: Grey Walls Press, 1953, p. 83. Sun worship also appears in *The Beautiful and the Damned*. He called the beach at Antibes a bright tan prayer mat
21. Cf. the review of *Dark Lover, The Life and Death of Rudolf Valentino*, by Emily W. Leider, 'We Lost It at the Movies', by Barry Gewen, *The New York Times Book Review*, 11 May 2003, p. 15
22. Cf. Jules-Rosette, Nennetta, *Josephine Baker in Art and Life*, Chicago: University of Illinois, 2007

23. Blume, Mary, 'Vive the Vacation: The French Look Back', *International Herald Tribune*, 20 July 2006, p. 20

24. Potter, Stephen, *Lifemanship*, New York: Holt, 1950, pp. 18–20

25. Clarke, Thurston, *Ask Not: The Inauguration of John F. Kennedy and the Speech That Changed America*, New York: Holt, 2004, pp. 41–2. Clarke may have chosen not to stress the effects of the Sun in his travel book *Equator* but he makes his point here

26. Vidal, Gore, *Palimpsest: A Memoir*, New York: Random, 1995, p. 355

27. Templeton, Sarah-Kate, 'Harley Street sells "risk-free" skin whitener', *The Sunday Times*, 8 July 2007

28. Davidar, David, *The House of Blue Mangoes*, London: Weidenfeld, 2002, p. 158

29. Asim, Jabari, *Washington Post*, 16 October 2007

30. Fowles, John, *Wormholes, Essays and Occasional Writings*, New York: Holt, 1998, pp. 284–7

31. Austen, Jane, *Persuasion*, Cambridge: CUP, 2006, p. 110

32. Madden, Thomas More, *On Change of Climate*, London, 1874, p. 32

33. Cf. Hobday, R. A., 'Sunlight Therapy and Solar Architecture', *Medical History*, 1997, 42: 455–72. Cf. also Hobday, Richard, *The Light Revolution: Health, Architecture and the Sun*, Forres: Findhorn Press, 2006, p. 12

34. Ransome, A., *The Principles of 'Open-Air' Treatment of Phthisis and of Sanatorium Construction*, London: Smith, Elder, 1903, pp. 72–3

35. Cf. Stuttaford, Thomas, *The Times*, 2 February 2006, Section 2, p. 9

36. Cf. Green, Martin Burgess, *Children of the Sun*, New York: Stein and Day, 1976, p. 6. Green openly acknowledges taking his title from W. J. Perry and Grafton Elliot Smith's 1930 study

37. MacCarthy, Fiona, *The Times*, 19 December 1991

38. 'The Solar Revolution' is a phrase of the British critic John Weightman (1915–2004)

39. Cf. Mughall, Robert, *Sunshine: One Man's Search for Happiness*, London: Murray, 2008, p. 46

40. Spender, Stephen, *World Within World*, London: Hamish Hamilton, 1951; reissued by Faber 1997

41. Fowles, John, *Wormholes, op. cit.*, p. 285

42. Quoted in Fussell, Paul, *Abroad, op. cit.*, 1980, p. 136

43. Cf. Mighall, Robert, *Sunshine, op. cit.*, p. 39

44. Cf. Child, Lee, *Gone Tomorrow*, New York: Delacourt, 2009, p. 1, and the author's Amazon essay on his research of suicide bombers

"Listen, pal, it's a free country. You worship the sun god your way, and I'll worship the sun god my way."

45. Camus, Albert, *Lyrical and Critical Essays*, New York: Knopf, 1968, p. 352, pp. 80–83. After his death, his long-time publisher Blanche Knopf wrote an appreciation in the *Atlantic* (February 1961, pp. 77 ff) entitled 'Albert Camus in the Sun': 'Camus was born of the Sun and always had a yearning to be in it'

46. Cf. *The Tucson Star*, 19 August 1986

CHAPTER 16 SKIN DEEP

1. Cf. Swerdlow, Joel L., 'Unmasking Skin', *National Geographic*, 13 November 2002, pp. 36–63

2. Hart-Davis, Alice, 'Take More Cover', *Evening Standard*, 20 May 2009, p. 29

3. Cf. Jones, Steve, *In the Blood*, London: HarperCollins, 1996, p. 194

4. Gettleman, Jeffrey, 'Albinos, Long-Shunned, Face Deadly Threat in Tanzania', *The New York Times*, 6 June 2008, pp. 8, 20; cf. also McNeil, Donald, 'Bid to Stop the Killing of Albinos', *The New York Times*, 17 February 2009, D5; 'The Enduring Curse of African Witchcraft', *The Week*, 11 April 2009, p. 16

5. Quoted by Fussell, Paul, *Abroad, op. cit.*, p. 185

6. Melville, Herman, *Moby Dick, op. cit.*, pp. 166–7. Cf. also Miller, Walter M., *A Canticle for Leibowitz*, New York: Bantam, 1959, p. 44

7. Cf. *The New Yorker*, 6 December 2004, p. 52

8. Chesterton, G. K., *The Annotated Innocence of Father Brown*, Ed. Gardner, Martin, New York: Dover, 1997, p. 215

9. *Why Don't Penguins' Feet Freeze? And 114 Other Questions, op. cit.*, pp. 4–5

10. Brody, Jane E., 'A Second Opinion on Sunshine: It Can Be Good Medicine After All', *The New York Times*, 17 June 2003, F7. Cf. also *Science News*, 24 April 2010, p. 9

11. Cf. Holick, Michael, and Jenkins, Mark, *The UV Advantage: New Medical Breakthroughs Reveal Powerful Health Benefits from Sun Exposure and Tanning*, New York: iBooks, 2003. In 2007 the *Independent* gave up its whole front page to tell its readers: 'A new study reveals that vitamin D, unlocked by sunlight on our skin, can ward off colds and flu. It also halves the risk of cancer, combats heart disease and fights diabetes. Now scientists are calling it the "wonder vitamin".' (14 April)

12. Jones, Steve, *In the Blood, op. cit.*, pp. 177–97

13. Sanders, Lisa, 'Perplexing Pain', *The New York Times Magazine*, 1 November 2009, pp. 24–5

14. Cf. Cloudsley-Thompson, J. L., 'Time Sense of Animals', *The Voices of Time*, Ed. Fraser, J., New York: Braziller, 1966, p. 299

15. Cf. Kostova, Elizabeth, *The Historian*, New York: Little, Brown, 2005, p. 55; also p. 148, 'He is not going to jump on you in broad daylight, Paul' – discussing a modern-day vampire

16. Cf. Lane, Nick, 'New Light on Medicine', *Scientific American*, January 2003, Vol. 288, No. 1, p. 38; and 'The Straight Dope', *Washington City Paper*, 'Did Vampires Suffer from the Disease Porphyria – or Not?'

17. Cf. Denver, Jones, *An Encyclopedia of Obscure Medicine*, New York: University Books, 1959

18. Cf. Rohl, John C. G., Warren, Martin, and Hunt, David, *Purple Secret: Genes, Madness and the Royal Houses of Europe*, London: Bantam, 1998, p. 212. Cf. also Macalpine, Ida, and Hunter, Richard, 'The Insanity of King George III: A Classic Case of Porphyria', *British Medical Journal*, 1966, pp. 65–71; and Macalpine, Ida, Hunter, Richard, and Rimington, Claude, 'Porphyria in the Royal Houses of Stuart, Hanover and Prussia', *British Medical Journal*, 1968, pp. 7–18; 'King George III and Porphyria: an elemental hypothesis and investigation', *The Lancet*, Vol. 366, Issue 9482, article by T. Cox, N. Jack, S. Lofthouse, J. Watling, J. Haines, M. Warren; and Arnold, W. N., *The Lancet*, 29 June 1996, 347, 1811–3

19. Cf. Mel Sinclair, http://www.eczemasite.com

20. Updike, John, *Self-consciousness: Memoirs*, New York: Knopf, 1989, pp. 42–78

21. Cf. 'Winter Depression', *Harvard Mental Health Letter*, November 2004, p. 4

22. Cf. Zuger, Abigail, 'Nighttime, and Fevers are Rising', *The New York Times*, 28 September 2004, F5

23. Hawkes, Nigel, 'Why We Get So Gloomy in Winter', *The Times*, 17 February 2004, p. 5. Cf. also Brody, Jane, 'Getting a Grip on the Winter Blues', *The New York Times*, 5 December 2006, F7. Ms Brody had articles on Sun-related matters for four consecutive months that year, none of which questioned the SAD lobby's findings

24. Chowder, Ken, 'Copenhagen', *The New York Times Magazine*, 21 November 2004, p. 92

25. Lingjaerde, O., and others, *Journal of Affective Disorders 33*, (1995), pp. 39–45

26. Reichborn-Kjennerud, Ted, *Patients with Seasonal Affective Disorder: a Study of the Clinical Picture, Personality Disorders, and Biological Aspects*, Oslo: University of Oslo, 1997, p. 119

27. Perry, Judith A., Silvera, David H., Rosenvinge, Jan H., Neilands, Tor, and Holte, Arne, 'Seasonal Eating Patterns in Norway: A Non-clinical Population Study', *Scandinavian Journal of Psychology*, 2001, 42, pp. 307–12

28. Magnússon, Andrés, and Axelsson, Jóhann, 'The Prevalence of Seasonal Affective Disorder Is Low Among Descendants of Icelandic Emigrants in Canada', *Archives of General Psychiatry*, December 1993, Vol. 50, p. 947 ff

29. Brennen, Tim, Martinussen, Monica, Hansen, Bernt Ole, and Hjemdal, Odin, 'Arctic Cognition: A Study of Cognitive Performance in Summer and Winter at 69° N', *Applied Cognitive Psychology*, 13: 561–580 (1999)

CHAPTER 17 THE BREATH OF LIFE

1. Quoted in *The New York Times*, 16 November 2002

2. Galilei, Galileo, *Dialogue Concerning the Two Chief World Systems*, New York: Modern Library, 2001

3. Bryson, Bill, *A Short History of Nearly Everything*, New York: Broadway, 2003, p. 298

4. Morton, Oliver, *Eating the Sun: How Plants Power the Planet*, London: Fourth

Estate, 2007, pp. xvii, 56. Cf. also Raven, Peter H., *Biology*, New York: McGraw-Hill, 7th Edition, 2007; Internet essay on Priestley by David Williams

5. Cf. Hartwig, Georg, *The Subterranean World*, New York: Scribner, 1871, p. 88, and Twain, Mark, *The Innocents Abroad, Or The New Pilgrim's Progress*, London: Hotten, 1870. 'Our author makes a long, fatiguing journey to the Grotto del Cane on purpose to test its poisoning powers on a dog – got elaborately ready for the experiment, and then discovered that he had no dog'

6. Kuhn, Thomas S., *The Structure of Scientific Revolutions*, Chicago: University of Chicago Press, 1962, pp. 53–6

7. Ingenhousz, Jan, *Experiments upon Vegetables, Discovering their great Power of purifying the Common Air in the Sun-shine, and of Injuring it in the Shade and at Night*, London, 1779

8. Kiang, Nancy Y., 'The Color of Plants on Other Planets', *Scientific American*, April 2008, pp. 48–55

9. Cf. Maddox, John, *What Remains to be Discovered*, New York: Free Press, 1998, p. 147

10. Cf. Morton, Oliver, *Eating the Sun, op. cit.*, pp. 168–75

11. Cf. Sawicki, Mikolaj, 'Myths about Gravity and Tides', *The Physics Teacher*, 37, October 1999, pp. 438–41

12. *Excursions, The Writings of Henry David Thoreau*, Vol. 9, Boston: Houghton Mifflin, 1893, p. 292

13. Attenborough, David, *The Private Life of Plants*, Princeton: Princeton University Press, 1995, pp. 45 ff

14. Morton, Oliver, *Eating the Sun, op. cit.*, p. 222

15. Attenborough, David, *The Private Life of Plants, op. cit.*, p. 1

16. Deakin, Roger, *Wildwood, A Journey Through Trees*, London: Hamish Hamilton, 2007, p. 109

17. Quinn, Vernon, *Stories and Legends of Garden Flowers*, New York: Frederick Stokes, 1939, pp. 116–17. In alchemy, both the lily and the holm oak are associated with the Sun, the lily embodying purity, the holm oak strength and glory

18. Cf. Tompkins, Peter, and Bird, Christopher, *The Secret Life of Plants*, New York: Harper, 1973, pp. 58 and 170–1

19. Morgan, Dan, *Merchants of Grain*, New York: Viking, 1979, p. 16

20. Cf. Broad, William J., 'CIA Revives Data Sharing on Environment', *The New York Times*, 5 January 2010, A1

21. Cf. 'Night schools for Fish', *Washington Post*, 30 March 2009, p. A5

22. Cf. Ray, C. Claiborne, 'Birds of a Feather', *The New York Times*, 3 October 2006, F2, and Angier, Natalie, 'Some Blend In, Others Dazzle', *op. cit.*, F1

23. Cf. Alford, D. V., *Bumble Bees*, London: Davis-Poynter, 1975, p. 75

24. Frisch's account is authoritative but a little leaden; here I have paraphrased the excellent description in Attenborough, David, *Life on Earth*, Boston: Little, Brown, 1979, p. 103

25. Cf. Dacke, M., and others, 'Built-in Polarizers Form Part of a Compass Organ in Spiders', *Nature*, 30 September 1999, p. 470 ff

26. Cf. Wilson, Edward O., *The Insect Societies*, Boston: Harvard University Press, 1971, p. 216

27. Cf. Maier, N. H. R., and Schneirla, T. C., *Principles of Animal Psychology*, New York: McGraw-Hill, 1935, p. 188

28. Cf. Crafts, Leland, Schneirla, Theodore C., Robinson, Elsa E., and Gilbert, Ralph W., *Principles of Animal Psychology, op. cit.*, 'Migration and the "Instinct" Problem', pp. 25–39

29. Cf. *Gathering of Angels: Migrating Birds and Their Ecology*, Ed. Able, Kenneth P., Ithaca, New York: Cornstock Books, 1999, pp. 14–15, and Attenborough, David, *The Life of Birds*, Princeton: Princeton University Press, 1998, p. 62

30. Cf. *Nature Neuroscience Reviews*, February 2005, and *The New York Times*, 1 February 2005, F1

31. Cf. Gill, Frank P., *Ornithology*, New York: W. H. Freeman, 1995, Ch. 13, and Berthold, P., *Bird Migration: A General Survey*, New York: OUP, 1993, pp. 156–7

32. Cf. Attenborough, David, *Life on Earth*, CD 7, Victors of the Dry Land, and *Life on Earth, op. cit.*, Boston: Little, Brown, 1979, p. 152

33. Cf. Attenborough, David, *Trials of Life: A Natural History of Behavior*, Vol. 5, Homemaking

34. Cf. Lovelock, James, *Gaia: A New Look at Life*, Oxford: OUP, 1979, and *The Ages of Gaia*, Oxford: OUP, 1988

35. Morton, Oliver, *Eating the Sun: How Plants Power the Planet, op. cit.*, p. 256

CHAPTER 18 THE DARK BIOSPHERE

1. Kunzig, Robert, *Mapping the Deep, The Extraordinary Story of Ocean Science*, New York: Norton, 2000, p. 207

2. Hamilton-Paterson, James, *The Great Deep, The Sea and Its Thresholds*, New York: Holt, 1992, p. 165

3. Wallis Budge, E. A., trans. of Pseudo-Callisthenes, 1933, cf. Hamilton-Paterson, James, *The Great Deep, op. cit.*, p. 167

4. Cf. Grann, David, 'The Squid Hunter', *The New Yorker*, 24 May 2004, pp. 56–71

5. Broad, W. J., 'Deep Under the Sea, Boiling Founts of Life Itself', *The New York Times*, 9 September 2003, F1–4

6. Maddox, John, *What Remains to be Discovered, op. cit.*, p. 150

7. Cf. Huber, Julie A., Butterfield, David A., and Baross, John A., *Microbiology Ecology*, April 2003

8. Cf. Kunzig, Robert, *Mapping the Deep, op. cit.*, p. 53. In *The Ecological Theater and the Evolutionary Play* (1965), the famous Yale ecologist G. Evelyn Hutchinson suggested that it might be possible in theory for organisms to make a living off the internal heat of the Earth rather than off sunlight, but he had no idea how they might do it

9. Attenborough, David, *Life on Earth, op. cit.*, p. 22

10. Cf. Whitehead, Peter, *How Fishes Live*, London: Phaidon, 1975, pp. 111–12

11. Carson, Rachel, 'The Sunless Sea', *Great Essays in Science*, Ed. Gardner, Martin, New York: Prometheus, 1994, pp. 287–304

12. Cf. Hamilton-Paterson, James, *The Great Deep, op. cit.*, pp. 112–13

13. Cf. Herring, Peter, *The Biology of the Deep Ocean*, Oxford: OUP, 2002, Ch. 8, 'Seeing in the Dark'
14. Hamilton-Paterson, James, *The Great Deep, op. cit.*, p. 203
15. Choi, Charles Q., 'Thousands of Weird Creatures Found', LiveScience.com, 22 November 2009
16. Cf. Moody, M. F., and Parris, 'Discrimination of polarized light by octopus', *Nature*, 186, 839–840 (1960); Shashar, Nadav, Rutledge, Phillip S., and Cronin, Thomas W., 'Polarization vision in cuttlefish – A concealed communication channel?', *Journal of Experimental Biology*, 199, 2077–2084 (1996)
17. Cf. Robinson, Bruce, 'Life in the Ocean's Midwaters', *Scientific American*, July 1995, Vol. 273, p. 60
18. 'Deep-sea Fish Sees Red', *BBC Wildlife*, August 1998, p. 59
19. Cf. *The Voices of Time, op. cit.*, p. 309
20. Cf. Angier, N., 'Out of Sight, and Out of Predator's Stomach', *The New York Times*, 20 July 2004, F4. Cf. also Johnsen, Sonke, 'Transparent Animals', *Scientific American*, February 2000, Vol. 282, Issue 2, pp. 80–90, and Madin, K., and Kovacs, D., *Beneath Blue Waters: Meetings with Remarkable Deep-Sea Creatures*, New York: Viking, 1996
21. Kunzig, Robert, *Mapping the Deep, op. cit.*, p. 8
22. Cf. Chapman, Piers, 'Ocean Currents', www.waterencyclopedia.com/Mi-Oc/Ocean-Currents.html
23. Cf. Broad, William J., 'Rogue Giants at Sea', *The New York Times*, 11 July 2006, F1 and F4
24. Cf. *The Encyclopedia of Climate and Weather*, Ed. Schneider, S. H., Oxford: OUP, 1996, Vol. 2, p. 734
25. Whewell, William, *The Philosophy of the Inductive Sciences*, 1847
26. Broad, William, 'Rogue Giants at Sea', *op. cit.*, F4
27. Cf. Sawicki, Mikolaj, 'Myths about Gravity and Tides', *The Physics Teacher*, 37, October 1999, pp. 438–41
28. Cf. Chang, Kenneth, 'Findings', *The New York Times*, 8 August 2006
29. Cf. *Dictionary of National Biography*, entry on Darwin, Sir George Howard, 1901–11 supplement to main dictionary
30. *Encyclopedia Britannica*, 11th Edition, p. 944, 'Tides'. This is a very Darwin-family deprecation: Charles Darwin is constantly saying that subjects are too difficult to tackle

CHAPTER 19 THE HEAVENLY GUIDE

1. Melville, Herman, *Moby Dick*, New York: R. R. Donnelley, 1930, p. 716
2. Letter to Thomas Wentworth Higginson, 1862, *Selected Letters*, Ed. Johnson, T., Cambridge, Mass.: Harvard, 1971
3. Boorstin, Daniel, *The Discoverers, op. cit.*, p. 46
4. Brown, Dan, *The Da Vinci Code*, New York: Doubleday, 2003, p. 106
5. Boorstin, Daniel, *The Discoverers, op. cit.*, p. 219

6. *Konungs Skuggsja*, or *The King's Mirror*, trans. Larsen, Laurence Marcellus, New York: Twayne Pil., 1917. In another place and time, Samuel Pepys would set up an examination for naval officers which included navigation, and naval schoolmasters were put on board ships to instruct the crew in mathematics: cf. also *The Voices of Time: A Cooperative Survey of Man's Views of Time As Expressed by the Sciences and by the Humanities*, Ed. Fraser, J. T., New York: George Braziller, 1966, pp. 216–17

7. Ferguson, Robert, *The Hammer and the Cross: A New History of the Vikings*, London: Allen Lane, 2009, p. 62

8. Cf. the Internet essay, 'The Viking Sunstone: Is the Legend of the Sun-Stone True?', and Schaefer, Bradley E., 'Vikings and Polarization Sundials', *Sky & Telescope*, May 1997, p. 91

9. Cf. Chang, Kenneth, 'Etched in Lava, Diary of the Earth's Magnetic Field Shows a Temporary Calm', *The New York Times*, 16 May 2006, F3

10. Ferris, Timothy, *Coming of Age in the Milky Way*, *op. cit.*, p. 128

11. Cervantes, Miguel de, *Don Quixote*, trans. Cohen, J. M., Harmondsworth: Penguin, 1950, Pt. II, Ch. 29, p. 58

12. Cf. Spence, Jonathan, *The Memory Palace of Matteo Ricci*, *op. cit.*, pp. 66–9

13. Braudel, Fernand, *The Mediterranean and the Mediterranean World in the Age of Philip II*, trans. Reynolds, Sian, London: Collins, 1972, Vol. 1, p. 104

14. Cf. Grant, Michael, ed., *The Ancient Mediterranean*, New York: Penguin, 1988, p. 146

15. Braudel, F., *The Mediterranean and the Mediterranean World in the Age of Philip II*, *op. cit.*, p. 138

16. Cf. Taylor, E. G. R., *The Haven-Finding Art*, London: Hollis and Carter, 1971, pp. x–xi. Cf. also Hay, David and Joan, *No Star at the Pole, A History of Navigation from the Stone Age to the Twentieth Century and For All Who Enjoy the Sea*, London: Charles Knight, 1972, p. 125

17. *Encyclopedia Britannica*, 11th Edition, p. 284, 'Navigation'

18. Cf. Childe, V. Gordon, *What Happened in History*, Harmondsworth: Penguin, 1946, pp. 246–7

19. Cf. Walton, Steven A., Michigan Technological University, Internet essay; also Howse, Derek, *Greenwich Time and the Discovery of Longitude*, London: OUP, 1980, p. 12, and Jardine, Lisa, *Ingenious Pursuits*, New York: Random, 1999, pp. 137–8 and 159

20. Grann, David, *The Lost City of Z*, *op. cit.*, p. 51

21. Cf. Shirley, Rodney W., *The Mapping of the World*, London: New Holland, 1993, p. xii

22. Cf. Grann, David, 'The Map Thief', *The New Yorker*, 17 October 2005, p. 68

23. Cf. Burt, John, *History of the Solar Compass*, Detroit: O. S. Gulley's presses, 1878

24. Melville, Herman, *Moby Dick*, *op. cit.*, p. 717

25. Cf. *The Definitive Time Machine*, Ed. Geduld, Harry M., Bloomington: Indiana UP, 1987

26. Melville, Herman, *Moby Dick*, *op. cit.*, pp. 740–3

27. From time to time in the writing of this book I have consulted Wikipedia,

"Now that we can tell time, I'd like to suggest that we begin imposing deadlines."

but with considerable trepidation. Its summary of the longitude problem, however, is hard to improve on, and I am duly grateful

28. Watson, William, *Ode on the Coronation of King Edward VII*, 1902

CHAPTER 20 OF CALENDARS AND DIALS

1. Bondi, Sir Hermann, quoted in Davies, Paul, *About Time, op. cit.*, p. 1
2. Bickerman, E. J., *The Ancient History of Western Civilization*, New York: Harper & Row, 1976
3. Eco, Umberto, 'Times', *The Story of Time*, Ed. Lippincott, Kristen, London: Merrell, 1999, p. 11
4. Cf. *Encyclopedia Brittanica*, 11th Edition
5. Allan R. Holmberg, quoted by Mann, Charles C., *1491: New Revelations of the Americas Before Columbus* (New York: Knopf, 2005) and by Grann, David, *The Lost City of Z, op. cit.*, p. 30
6. Cf. *The Cambridge History of Islam, op. cit.*, p. 98; Ilyas, Mohammad, *Islamic Astronomy and Science, op. cit.*, pp. 97–107; and Carlebach, Elisheva, *Palaces of Time: Jewish Calendar and Culture in Early Modern Europe*, Mass.: Harvard, 2011
7. Needham, Joseph, *Science and Civilization in China, op. cit.*, p. 189, and Cullen, Christopher, *Astronomy and Mathematics in Ancient China: the Zhou Bi Suan Jing*, New York: CUP, 1996, p. 6. Cullen, who currently oversees the Needham Research Institute in Cambridge, reckons that Needham badly underestimated

the importance of calendrical astronomy in China; his own writing makes only a partial case

8. Cf. Weiner, Tim, 'Hailing the Solstice and Telling Time, Mayan Style', *The New York Times International Edition*, 23 December 2003

9. Cf. Zuidema, R. T., 'The Inca Calendar', *Native American Astronomy*, Ed. Aveni, Anthony F., London: University of Texas Press, 1977

10. Steel, Duncan, *Marking Time: The Epic Quest to Invent the Perfect Calendar*, New York: Wiley, 1999, p. 10

11. Cf. Hirst, Christopher, 'A Thousand Years of Tinkering with Time', *The Week*, 8 March 2008, pp. 44–5

12. Tolkien, J. R. R., *The Lord of the Rings*, op. cit., p. 1140 ff

13. Hazlitt, William, 'On a Sun-Dial', first published in *New Monthly Magazine*, October 1827; reprinted in *Selected Essays*, Ed. Keynes, G., New York: Random House, Nonesuch Press, 1930, p. 345

14. Cf. Chaucer, Geoffrey, *The Parson's Prologue*; also the Introduction to *The Man of Law's Tale*, when he calculates the hour as ten o'clock by the same method. The original runs:

> Foure of the clokke it was tho, as Igesse,
> For ellevene foot, or litel moore or lesse,
> My shawe was at thilké tyme, as there,
> Of swiche feet as my lengthé parted were
> In sixe feet equal of proporcioun'

15. Needham, Joseph, 'Time and Knowledge in China and the West', *The Voices of Time*, op. cit., p. 106

16. Cf. Waugh, Albert E., *Sundials, Their Theory and Construction*, New York: Dover, 1973, pp. 4–5

17. Field, J. V., 'European Astronomy in the First Millennium: the Archaeological Record', *Astronomy Before the Telescope*, op. cit., p. 121

18. Sobel, Dava, 'The Shadow Knows', *Smithsonian*, January 2007, p. 91. There are two distinctive terms for sundials: the style, a straight edge whose shadow is a straight line on a plane dial, which shadow normally identifies a *direction* on the dial; and the nodus, a small ball or disc whose shadow identifies a *point* on the dial. Cf. Dohrn-van Rossum, Gerhard, *History of the Hour: Clocks and Modern Temporal Orders*, Chicago: Chicago University Press. The gnomon as a whole serves as a style and its tip as a nodus

19. '*Horam non possum certam tibi dicere: facilis inter philosophos quam inter horologia convenient.*' *Apocolocyntosis*, 2:2

20. *Julius Caesar*, II, i, ll. 4 ff; *Cymbeline*, II, ii; *Richard II*, V, v. Cf. also *As You Like It*, II, vii, ll. 20–2

21. Cousins, Frank W., *Sundials: A Simplified Approach by Means of the Equatorial Dial*, London: John Baker, 1972, p. 9

22. Sobel, Dava, 'The Shadow Knows', op. cit., p. 91

23. *Henry VI Pt. III*, II, v, ll. 21–30. As so often with Shakespeare, these lines bear a second reading, based on 'quaint' also meaning female genitalia

24. Hazlitt, William, 'On a Sun-Dial', op. cit., p. 336

25. Cf. Friend, Tad, 'The Sun on Mars', *The New Yorker*, 5 January 2004, pp. 27–8

CHAPTER 21 HOW TIME GOES BY

1. Joyce, James, *Ulysses, The Corrected Text, op. cit.*, p. 47
2. *The Taming of the Shrew*, IV, ii, l. 197
3. Quoted by Jackson, Kevin, *The Book of Hours*, London: Duckworth, 2007, pp. 164–5
4. Burgess, Anthony, *One Man's Chorus*, New York: Carroll and Graf, 1998, pp. 120–3
5. Jackson, Kevin, *The Book of Hours, op. cit.*, p. 15
6. Boorstin, Daniel, *The Discoverers, op. cit.*, p. 40
7. *History Magazine*, February/March 2008, p. 12
8. Jardine, Lisa, *Ingenious Pursuits: Bulding the Scientific Revolution, op. cit.*, p. 133
9. Woolf, Virginia, *Mrs Dalloway*, London: The Hogarth Press, 1925, reprinted Granada, 1979, p. 91. The novel was originally entitled *The Hours*; in a book where so much is given a precise time and place, critics disagree about the hour that it opens – 9 a.m. or 10 a.m. – while the hour of 11 a.m. to noon fills nearly a quarter of the novel
10. Low, Valentine, *The Week*, 'The King of Clocks', 10 January 2009, p. 37
11. Cf. James, Peter, *Dead Simple*, London: Macmillan, 2005, p. 318
12. Cf. Blaise, Clark, *Time Lord, Sir Sandford Fleming and the Creation of Standard Time*, New York: Pantheon, 2000, p. 170
13. Lawrence, T. E., *Seven Pillars of Wisdom, A Triumph*, London: Cape, 1973, p. 573
14. Blaise, Clark, *Time Lord, op. cit.*, pp. 69, 129 and 35
15. Burgess, Anthony, *One Man's Chorus, op. cit.*, p. 123
16. Wheeler, Sara, *Too Close to the Sun, op. cit.*, p. 136
17. Blaise, Clark, *Time Lord, op. cit.*, p. 142 ff
18. Smith, Mark M., *Mastered by the Clock: Time, Slavery and Freedom in the American South*, Chapel Hill: University of North Carolina Press, 1997

19. Cf. 'Q+a', *The New York Times*, 8 May 2007, F2
20. Cf. Jackson, Kevin, *The Book of Hours, op. cit.*, pp. 188–9
21. Cf. Prerau, David, *Seize the Daylight: The Curious and Contentious Story of Daylight Savings Time*, New York: Thunder's Mouth Press, 2005, and Downing, Michael, *Spring Forward: The Annual Madness of Daylight Saving Time*, Emeryville, California: Shoemaker & Hoard, 2005
22. O'Connor, Anahad, 'Really?', *The New York Times*, 10 March 2009, D5
23. Collins, Gail, 'The Great Clock Plot', *The New York Times*, 23 August 2007, A21
24. Cf. Adams, Cecil, *The Straight Dope*, 'Why Is India 30 Minutes Out of Step with Everybody Else?', *Washington City Paper*, 5 June 1981
25. Adams, Cecil, *The Straight Dope, Washington City Paper*, 11 May 2007, p. 18
26. Cf. Simons, Marlise, 'Synchronizing the Present and Past in a Timeless Place', *The New York Times International Edition*, 12 September 2005
27. Cf. www.yeswatch.com
28. Cf. *The Week*, 21 February 2009, p. 17
29. Hirst, Christopher, 'A Thousand Years of Tinkering With Time', *op. cit.*, p. 44

CHAPTER 22 THE SUN IN OUR POCKET

1. Swift, Jonathan, *Gulliver's Travels*, Part III, A Voyage to Laputa
2. More recently, researchers at MIT and the University of Arizona have reached similar conclusions: cf. Sample, Ian, 'Doubts Cast on Archimedes' Killer Mirrors', *Guardian*, 24 October 2005
3. Kryza, Frank T., *The Power of Light, The Epic Story of Man's Quest to Harness the Sun*, New York: McGraw-Hill, 2003, p. 53
4. *Ibid*, p. 30
5. Clark, Stuart, *The Sun Kings, op. cit.*, p. 60
6. Cf. www.californiasolarcenter.org/history_solarthermal.html
7. Cf. Smith, Charles, 'Revisiting Solar Power's Past', *Technology Review*, July 1995
8. Gamow, George, *The Birth and Death of the Sun, op. cit.*, p. 1
9. Cf. Archer, Mary, 'Hello Sunshine', Royal Institute Proceedings No. 66, p. 10
10. Ishikawa, Eisuke, *O-edo ecology jijo* (*The Edo Period Had a Recycling Society*), Tokyo: Kodansha, 1994: cf. also 'Japan's Sustainable Society in the Edo Period (1603–1867)', Japan For Sustainability Newsletter, Internet essay, 6 April 2005
11. Murphy, Mary Jo, 'Becoming the Big New Idea: First, Look the Part', *The New York Times*, 24 August 2008, p. 4. As recently as 1976 *Nature* thought it necessary to inform its readers, 'The Sun can be used for a variety of energy-related purposes', in an editorial, 'Is the Sun being oversold?' (20 May 1976)
12. Clarke, Arthur C., *Astounding Days*, New York: Wiley, 1984, p. 203
13. *The New York Times*, 3 August 2006, C1. Tidal and wave projects are currently under way in Rhode Island, Cantabria (Spain), Daishan (China), northern Portugal and at least three sites in the United Kingdom
14. Cf. Kolbert, Elizabeth, 'The Car of Tomorrow', *The New Yorker*, 11 August 2003, p. 40

15. Cf. Alford, Henry, 'Solar Chic', *The New Yorker*, 24 September 2007, p. 128

16. *Gourmet*, January 2005, p. 8; *The New York Times Magazine*, 21 August 2005, p. 51

17. Spudis, Paul D., 'Why We're Going Back to the Moon', *Washington Post*, 27 December 2005

18. Cf. Kolbert, Elizabeth, 'The Climate of Man – Part III', *The New Yorker*, 9 May 2005, p. 57; and Gibbs, W. Wayt, 'Plan B for Energy', *Scientific American*, September 2006, p. 84 – an issue devoted to alternative sources of power; but then the magazine devoted an issue to energy problems as far back as 1913. *Plus ça change*. Cf. also Smyth, Chris, '"Sailing" on sunlight may help polar observations', *The Times*, 11 September 2009, p. 26

19. Cf. Revkin, Andrew C., 'Budgets Falling in Race to Fight Global Warming', *The New York Times*, 30 October 2006, A1, A14, and Woody, Todd, 'Desert Vistas vs. Solar Power', *The New York Times*, 22 December 2009, B1, B5

20. Cf. Goodell, Jeff, 'Black Gold or Black Death?', *The New York Times*, 4 January 2006, A15

21. Cf. Whitlock, Craig, 'Cloudy Germany a Powerhouse in Solar Energy', *Washington Post*, 5 May 2007, A1, A14

22. Bradsher, Keith, 'China Tries A New Tack To Go Solar', *The New York Times*, 9 January 2010, B1 and B4; and Bradsher, Keith, 'China Leading Race to Make Clean Energy', *The New York Times*, 31 January 2010, p. 1

23. Areddy, James T., 'Heat for the Tubes of China', *Wall Street Journal*, 'Marketplace', 31 March 2006, p. 1. The article's title is a play on that of a famous 1930s book (and later film), *Oil for the Lamps of China*, by Alice Tisdale Hobart

24. *Mother Earth News*, Issue No. 41, September/October 1976

25. Cf. *The New York Times*, 1 September 2003

26. '(Solar) Power to the People Is Not So Easily Achieved', *The New York Times*, 23 January 2008, B2, and Zweibel, K., Mason, J., and Fthenakis, V., 'Solar Grand Plan', *Scientific American*, January 2008, p. 66

27. Cf. 'As Earth Warms, the Hottest Issue is Rethinking Energy', *The New York Times*, 4 November 2003

28. Cf. Tyson, Neil de Grasse, 'Energy to Burn', *Natural History*, October 2005, pp. 17–20

29. Morton, Oliver, *Eating the Sun: How Plants Power the Planet*, op. cit., p. 395

30. Since I wrote this chapter these words were quoted back at me – by Al Gore, addressing the Democratic Convention in Denver in August 2008

CHAPTER 23 THE VITAL SYMBOL

1. Dostoyevsky, Fyodor, *The Brothers Karamazov*, London: Penguin, 1967, p. 628. The great atomic physicist Paul Dirac read *Crime and Punishment* and judged it 'nice', complaining that in one place the Sun rises two times in one day

2. Blake, William, *A Vision of the Last Judgement*, c. 1810

3. Cf. 'The Netherlands: Anne Frank's Chestnut Tree to be Cut Down', Associated Press, 15 November 2006

4. Mantel, Hilary, *Wolf Hall*, New York: Holt, 2009, p. 78. Edward's words come

from a different, contemporary source

5. Cf. Weir, Alison, *Lancaster and York: The Wars of the Roses*, London: Pimlico, 1998, pp. 261–2

6. Cf. Thompson, Ian, *The Sun King's Garden, Louis XIV, Andre le Nôtre and the Creation of the Gardens of Versailles*, New York: Bloomsbury, 2006, p. 48

7. From the records of André Félibien, quoted by Ranum, Orest, 'Islands of the Self', in *Sun King, The Ascendancy of French Culture During the Reign of Louis XIV*, Ed. Rubin, David Lee, London: Folger, 1992, pp. 30–1

8. Thompson, Ian, *The Sun King's Garden, op. cit.*, p. 751

9. *Souvenirs and Prophecies: the Young Wallace Stevens*, New York: Knopf, 1977, Ed. Stevens, Holly: journal entry for 20 April 1920

10. Cf. McDonald, William B., *Christian Sun Symbols, op. cit.*, pp. 135–7

11. Cf. Haffert, John, *The Peacemaker Who Went to War*, New York: Scapular Press, 1945, p. 196

12. Cf. Brown, Andrew, 'Resurrection Shuffles', *Guardian*, 18 June 2009

13. Cf. Mitchell, John, and Richard, Robert, *Phenomena: A Book of Wonders*, New York: Pantheon, 1977

14. Smith, Craig, 'Nearly 100, LSD's Father Ponders His Problem Child', *The New York Times*, 7 January 2006

15. Cf. Eliade, Mircea, *Occultism, Witchcraft, and Cultural Fashions, Essays in Comparative Religions*, Chicago: University of Chicago Press, 1978, Ch. entitled 'A South American'

16. Cf. Isil, Olivia A., *When a Loose Cannon Flogs a Dead Horse There's the Devil to Pay*, New York: Ragged Mountain Press, 1996

17. Pitman, Joanna, *On Blondes*, London: Bloomsbury, 2003, p. 12

18. Chandler, Raymond, *Farewell, My Lovely*, London: Penguin, 1940, pp. 76–7

19. Aldersey-Williams, Hugh, *Periodic Tales: The Curious Lives of the Elements*, London: Viking, 2011, p. 327

20. Pitman, Joanna, *On Blondes, op. cit.*, p. 13

21. Joyce, James, *Ulysses, op. cit.*, p. 50, ll. 240–2. Joyce famously makes reference to the Sun in Episode 14: The Oxen of the Sun (ll. 383–428), when the introductory Latin verses, the maternity hospital, the nurses and Dr Horn are all symbols of the fertility that the oxen of the Sun embody

22. Cf. Brooks, R. R., *Noble Metals and Biological Systems*, New York: CRC Press, 1992, pp. 13, 99, 111, 116, 121, 199 and 297

23. Abraham, Lyndy, *A Dictionary of Alchemical Imagery*, New York: Cambridge, 1998, p. 130

24. Hugo, Victor, *The Hunchback of Notre-Dame*, New York: OUP, 1999, p. 235

25. Cf. Arnott, Sarah, 'What is so special about Gold, and should we all be investing in it?', *Independent*, 10 September 2009, p. 34

26. Cf. Grann, David, *The Lost City of Z, op. cit.*, pp. 148–9

27. Keynes, J. M., 'The Return to the Gold Standard', *Essays in Persuasion*, New York: Harcourt, 1932, p. 182

28. Cf. Healey, Matthew, 'Century Later, Gold Coin Reflects Sculptor's Vision', *The New York Times*, 25 November 2008

29. Cf. Schwarz, Ted, *T. H. Paget of the Royal Mint*, New York: Arco, 1976, pp. 190–1

30. Cf. Goldberry, Ben, *The Mirror and Man*, Charlottesville: University Press of Virginia, 1985, p. 37

31. Cf. Melchior-Bonnet, Sabine, *The Mirror, A History*, New York: Routledge, 1994, p. 46

CHAPTER 24 DRAWING ON THE SUN

1. Cf. Simons, Paul, 'Windmills foretell a thaw in the air', *The Times*, 14 March 2006, p. 64

2. Cf *Encyclopedia Britannica*, 15th Edition, 1974, p. 695, column 1

3. Reynolds, Graham, *Turner*, London: Thames and Hudson, 1969, p. 12

4. Cf. Bailey, Anthony, *Standing in the Sun*, London: Sinclair-Stevenson, 1997, p. 102

5. *Ibid*, p. 248

6. Hamilton, James, *Turner, A Life*, London: Hodder, 1997, p. 216

7. Cf. Bailey, Anthony, *Standing in the Sun, op. cit.*, p. 385

8. *Ibid*, pp. 343–4

9. Hamilton, James, *Turner, A Life, op. cit.*, p. 283. A letter to *The Times* of 24 August 2005 pointed out that the *Téméraire* was being towed upstream, so Turner, in sketching the scene, must have been facing east as the ship was tugged west. Thus the painting may not portray the nostalgic sunset of the age of sail at all, but symbolize instead the dawn of the exciting age of steam

10. Dickens, Charles, *Our Mutual Friend*, London: Penguin Books, 1997, pp. 80–1

11. Cf. *American Light: the American Luminist Movement 1850-75*, Washington, DC: Camera Three Productions Inc., video recording; and Mansthorne, Katherine, and Mitchell, Mark, *Luminist Horizons: The Art and Collection of James A. Suydam*, New York: National Academy Museum, 2006

12. *The Times*, 5 December 1926. Cf. also Fletcher, Valerie J., 'The Light of Art', *Fire of Life: The Smithsonian Book of the Sun*, New York: Norton, 1981, p. 192

13. Cf. Hobl, Reinhold D., 'The Sun in Contemporary Painting and Sculpture', *Fire of Life, op. cit.*, p. 228

14. *The Complete Letters of Vincent Van Gogh to His Brother*, New York: New York Graphic Society (1958), Vol. 3, p. 374; letter No. 519, 1888, p. 3; p. 128; Nos 531 and 533; p. 257; p. 202; and p. 205

15. Mighall, Robert, *Sunshine, op. cit.*, p. 235; cf. also Updike, John, 'The Purest of Styles', *New York Review of Books*, 22 November 2007, p. 16

16. Van Gogh, Vincent, *The Complete Letters, op. cit.*, p. 298, letter Nos 444 and 524; p. 204

17. For this recollection of a Paul Nash painting I am indebted to my friend, Nicola Bennett. She also reminds me that in *Guernica*, Picasso paints a light bulb shedding jagged light on to horses and bulls, a scene that one might expect to be placed outdoors – only it is too awful to be in the healing Sun

18. Cf. *Fontana Economic History of Europe*, Ed. Cipolla, Carlo M., New York: Barnes and Noble, 1976, Vol. 3, p. 79

19. Spurling, Hilary, *Matisse the Master*, London: Hamish Hamilton, 2005: the quotations here are drawn from the second volume of the biography, and my discussion of Matisse draws considerably on her observations

20. Cf. Kimmelmann, Michael, 'The Sun Sets at the Tate Modern', *The New York Times*, 21 March 2004

21. Cf. Spurling, Hilary, *Matisse the Master, op. cit.*, pp. 216–7. Spurling dates their first meeting as 16 January 1918

22. Gabriel, Trip, 'David Hockney; Acquainted with the Light', *The New York Times*, 21 January 1993

23. Adams, Tim, 'David Hockney: Portrait of the Old Master', *Observer*, 1 November 2009

24. Sims, Michael, *Apollo's Fire: A Day on Earth in Nature and Imagination*, New York: Viking, 2007, p. 79

CHAPTER 25 NEGATIVE CAPABILITIES

1. Le Corbusier, *Towards a New Architecture*, New York: Praeger, 1960, p. 16

2. Cf. Jardine, Lisa, *Ingenious Pursuits, op. cit.*, pp. 104–5

3. Leggat, Robert, *A History of Photography*, Internet website, 15–32

4. Cf. Galassi, Peter, *Before Photography: Painting and the Invention of Photography*, New York: Museum of Modern Art, 1981, p. 11

5. Barthes, Roland, *Camera Lucida: Reflections on Photography*, New York: Hill and Wang, 1982, p. 15

6. Frank, Robert, *The Americans*, University of Michigan: Scalo, 1959. A French edition had appeared the previous year

7. Cf. Olshaker, Mark, *The Instant Image; Edwin Land and the Polaroid Experience*, New York: Stein and Day, 1978, p. 11

8. This photograph, and the insight about how it captures light, is taken from Dyer, Geoff, *The Ongoing Moment*, New York: Pantheon, 2005, pp. 131–2

9. Quoted in Delany, Paul, *Bill Brandt: A Life*, Stanford: Stanford University Press, 2004, p. 281

10. Di Bello, Patrizia, 'From the Album to the Computer Screen: Collecting Photographs at Home', *Multimedia Histories: From The Magic Lantern to the Internet*, Ed. Lyons, James, and Plunkett, John, Exeter: University of Exeter Press, 2007, p. 58

11. Cf. Thomas, Ronald R., 'Making Darkness Visible', *Victorian Literature and the Victorian Visual Imagination*, Ed. Christ, Carol, and Jordan, John O., Berkeley: University of California Press, pp. 134–56

12. Cf. Bowen, Thomas David, *The Origins of the Motion Picture*, London: HMSO, 1964, pp. 7–21. Cf. also Russo, Vito, 'Adventures in Cybersound': A History of Motion Pictures, http://www.acmi.net.au/AIC/ENC-CINEMA.html; and *Edison: The Invention of the Movies*, 4-disc boxed set, Museum of Modern Art/Kino International, 2005; the first film kiss occurred in 1896, taking up fifty feet of film and nineteen seconds

13. Thomson, David, *A Biographical Dictionary of Film*, London: Deutsch, 1975, p. 602, entry on Vincent Price

14. Sims, Michael, *Apollo's Fire*, New York: Viking, 2007, p. 27

15. *Ibid.*, p. 140

16. Cotter, Holland, 'Full Constant Light', *The New York Times*, 26 December 2008, C39

17. Tauranac, John, *The Empire State Building*, New York: St Martin's, 1995, p. 51

18. Quoted in *The Ideas of Le Corbusier*, Ed. Guitton, Jacques, New York: Braziller, 1981, p. 104

CHAPTER 26 TALK OF THE DAY

1. Rimsky-Korsakov, Nikolay Andreyevich, *My Musical Life*, New York: Knopf, 1923, pp. 141–2

2. Cf. Donington, Richard, *Opera and Its Symbols*, New Haven: Yale University Press, 1990, p. 125 ff

3. Cf. Magee, Bryan, *Wagner and Philosophy*, New York: Holt, 2000, p. 216–21

4. Cf. Tanner, Michael, *Wagner*, London: HarperCollins, 1996, p. 101

5. Mann, William, *The Operas of Mozart*, New York: Oxford University Press, 1977, p. 597

6. Mackey, A. G., *An Encyclopedia of Freemasonry and Its Kindred Sciences:*

Comprising the Whole Range of Arts, Sciences and Literature as Connected with the Institution, New York: Masonic History Company, 1912, p. 765

7. Ross, Alex, 'Apparition in the Woods', *The New Yorker*, 9 and 16 July 2007, p. 55

8. *The Beatles Anthology*, New York: Chronicle Books, 2000

9. Cf. interview with Tony Burrell, 'Best of Times, Worst of Times', *The Sunday Times Magazine*, 11 June 2006. Cf. also Anthony, Ted, *Chasing the Rising Sun*, New York: Simon and Schuster, 2007, p. 230

10. These can be heard on the Internet: cf. The Singing Sun

CHAPTER 27 BUSY OLD FOOL . . .

1. Longfellow, Henry Wadsworth, *Evangeline*, Pt. I, Section IV

2. Francis, Dick, *Smokescreen*, New York: Harper, 1972, p. 6

3. Nabokov, Vladimir, *Ada: A Family Chronicle*, New York: Vintage, 1990, p. 45

4. Cf., for example, Nabokov, Vladimir, *Pale Fire*, in *Novels 1955–62*, New York: Library of America, 1996, p. 460: *The pen stops in mid-air, then swoops to bar /A cancelled sunset or restore a star, /And thus it physically guides the phrase /Toward faint daylight through the inky haze*

5. Nabokov, Vladimir, *Speak, Memory*, New York: Putnam, 1966. The many references in *Speak, Memory* are often delightful, without either the perverse sensual element or the doom-laden significance they possess in *Lolita*. For instance, as a boy Nabokov plays with some wax, 'using a candle flame (diluted to a deceptive pallor by the sunshine that invaded the stone slabs on which I was kneeling)' (p. 58), and on another occasion he notes the lurid gleam cast by the Sun at the end of a rainy day on the fresh-picked mushrooms lying on a garden table (p. 44)

6. Boyd, Brian, *Vladimir Nabokov, The Russian Years*, Princeton: Princeton University Press, 1990, pp. 434–7; p. 44

7. Boyd, Brian, *Vladimir Nabokov, The American Years*, Princeton: Princeton University Press, 1990, p. 159; p. 297; p. 306

8. Nabokov, Vladimir, *Lolita*, London: Penguin, 2006. Cf. p. 7; p. 8; p. 182; p. 42; p. 68; p. 44 p. 184; p. 48; p. 60; p. 63; p. 65; p. 66; p. 79; p. 86; p. 107; p. 109; p. 176; p. 125; p. 101; p. 215; p. 141; p. 142; p. 241; p. 244; p. 250; p. 269; p. 270; p. 300; p. 292; a 'green Sun', being over in an exquisite instant, may be a Nabokovian reference to orgasm; p. 324; p. 320; p. 326; p. 334 (the phrase 'burning like a man' is terrifying as an image, but not easy to explain); and p. 348

9. Wood, Florence, with Wood, Kenneth, *Homer's Secret Iliad: The Epic of the Night Skies Decoded*, London: John Murray, 1999, p. 5

10. *Iliad*, 1. 472. Cf. also Dicks, D. R., *Early Greek Astronomy to Aristotle, op. cit.*, pp. 29–32

11. Fact came to mirror fiction when in 1783 the French Academy of Science sent up a balloon, its pilot becoming the first person known to history to see the Sun set twice – and finding it terrifying. Cf. Longyard, William, *Who's Who of Aviation History*, Novato, CA: Presidio Press, 1994, p. 41

12. Cullen, Dolores L., *Chaucer's Pilgrims: the Allegory*, Santa Barbara: Fithian Press, 2000, p. 15. Cf. also the Internet essay 'Astronomy and Astrology in the Works of Chaucer'; and Gingerich, Owen, 'Transdisciplinary Intersections: Astronomy and Three Early English Poets', *New Directions for Teaching and Learning*, Vol. 1981, Issue 8, pp. 67–75. Chaucer's name, incidentally, means 'shoemaker'

13. Milton, John, 'Ode on the Morning of Christ's Nativity', ll. 229–33

14. Marvell, Andrew, 'Upon Appleton House', ll. 661–4

15. Herrick, Robert, 'To the Virgins, to Make Much of Time', ll. 5–8

16. Cf. Milton, John, *Paradise Lost*, Bk. III, ll. 571–87. Other notable mentions in *Paradise Lost* are Bk. I, ll. 594–9, when the Sun in eclipse is used as a simile for Satan's blasted appearance; Bk. III, ll. 608–12, a reference to the Sun generating precious stones in the ground; Bk. IV, ll. 32–9, when Satan addresses the Sun; Bk. IX, l. 739, when he tempts Eve at high noon; and in *Paradise Regained* (II: 292) he tempts Christ at the same ominous time of day. There are also references in *Paradise Lost*, Bk. VI, ll. 479–81, and *Comus*, ll. 732–6, to gems that are said to shine in the dark because they store solar fire within them

17. Yates, Frances Amelia, *The Art of Memory*, London: Routledge, 1966, p. 153

18. Cf. Manquel, Alberto, and Guadalupi, Gianni, *Dictionary of Imaginary Places*, New York: Harcourt, 2000, p. 632

19. *Romeo and Juliet*, III, v, ll. 12–24; another favourite poem about lovers and the Sun is from Catullus, *Ode to Lesbia*, V, the lines beginning 'Though the Sun each night expires, Morning renovates its fires . . .'

20. More, Sir George, *A Demonstration of God in His Workes: Agaynst all such as eyther in word or life deny there is a God*, 1597, p. 24, quoted in Stubbs, John, *Donne, The Reformed Soul*, London: Viking, 2006, p. 178

21. Donne, John, 'An Anatomy of the World, The First Anniversary'. For a longer discussion of Donne's anti-Copernican views, cf. Koestler, Arthur, *The Sleepwalkers, op. cit.*, p. 214 ff

22. Coleridge, Samuel Taylor, letter to Humphry Davy, 15 July 1800, *Collected Letters*, Vol. 1, London: OUP, 2002, p. 339

23. Housman, Alfred Edward, letter to *The Times Literary Supplement*, 12 December 1928. For his own poetry mentioning the Sun, see *Collected Poems*, 'The West', and *Last Poems*, p. 97, 'The Sun Is Down'; 'The Welsh Marches', No. 28 of *A Shropshire Lad* (published in 1896), p. 44, 'The vanquished eve, as night prevails'; and No. 10, p. 23, 'The . . . golden wool of the ram'. Best of all is *Revolution*, p. 143, 'the golden deluge of the morn'

24. Lawrence, David Herbert, *Sons and Lovers* (1913), Ed. Helen and Carl Baron, Cambridge: CUP, 1992, p. 464

25. Lawrence, D. H., *Collected Stories*, New York: Knopf, 1994, p. 981; p. 983; p. 990; p. 996, p. 998. Cf. also *The Letters of D. H. Lawrence*, Ed. Zytaruk, George J., and Boulton, James T., Cambridge: CUP, 1982, p. 481; and Reeve, N. H., 'Liberty in a Tantrum: D. H. Lawrence's *Sun*', *Cambridge Quarterly*, Vol. 24, No. 3, pp. 209–20

26. Lawrence, D. H., *Apocalypse*, London: Penguin, 1996, p. 27 and ff

27. Brunner, Edward, 'Harry Crosby: A Biographical Essay', *Modern American Poetry*, 2001

28. Fussell, Paul, *Abroad, op. cit.*, p. 139

29. Brunner, Edward, 'Harry Crosby: A Biographical Essay', *op. cit.*, p. 2

30. The details about the Black Sun in Mandelstam are taken from an Internet essay of 30 April 2003 by 'language hat'. Later contributors to the site quoted Nerval's writing about a Black Sun in 'El Desdichado', 'the Black Sun of melancholy', and Blake in 'Marriage of Heaven and Hell': 'By degrees we beheld the infinite Abyss, fiery as the smoke of a burning city; beneath us at an immense distance was the Sun, black but shining . . .'

31. Cf. Jenkins, Simon, 'Betjeman's discreet, dignified muse makes today's look like mere groupies', *Guardian*, 18 April 2008, p. 34

32. Melville, Herman, *Moby Dick, op. cit.*, p. 711; cf. also p. 774

"Can Hiram call you back? He's adjusting our solar panels."

33. Cf. footnote 2 to the Introduction to *Paul Verlaine: Selected Poems*, trans. Martin Sorrell, Oxford World's Classics, 1999

34. Cf. Nicholl, Charles, *Somebody Else: Arthur Rimbaud in Africa 1880–91*, London: Cape, 1999, p. 356; p. 439

35. Cf. Robb, Graham, *Rimbaud*, New York: Norton, 2000, p. 93; Robb makes the point (p. 386) that Rimbaud exaggerated the heat of the Sun

36. Gide, André, *The Immoralist*, trans. Bussy, Dorothy, Harmondsworth: Penguin, 1960, pp. 25, 51, 26. Gide's sun worship is discussed at length in

Mighall, Robert, *Sunshine*, *op. cit.*, pp. 86–8. For the remarks on Gide and Camus I have drawn on an unpublished essay by Lesley Chamberlain

37. Strout, Elizabeth, *Olive Kitteridge*, New York: Random, 2008, pp. 176, 3, 80, 8, 103, 202–3, 270

CHAPTER 28 THE RISING STAR OF POLITICS

1. Morgan, Peter, *Frost/Nixon*, London: Faber, 2006, p. 66
2. 'Mr. Bush received one D in four years [during his freshman year at Yale], a 69 in astronomy', *The New York Times*, 8 June 2005, A10. By comparison, John Kerry had four Ds in his freshman year, but in this area comedians prefer to target Republicans. Thus, during Ronald Reagan's presidency, the *David Letterman Show* of 12 August 1987 had 'Top Ten Surprises in the President's Speech' and at No. 1: 'Hysterical shouts of "We're hurtling towards the Sun!" made poor closing statement'
3. Steadman, Ralph, *The Joke's Over: Bruised Memories, Gonzo, Hunter S. Thompson, and Me*, London: Harvest Books, 2007, p. 36
4. Cf. Campion, Nicholas, 'Prophecy, Cosmology, and the New Age Movement', private MS, Ch. 3, p. 52
5. Cf. *Cambridge Ancient History*, *op. cit.*, p. 110
6. Nietzsche, Friedrich, *Ecce Homo: How One Becomes What One Is*, London: Penguin Books, 1992, p. 3
7. Nietzsche, Friedrich, *The Gay Science*, 1882, republished Cambridge: CUP, 2001. Its first translated title was *The Joyous Wisdom*, but 'the Gay Science' was a well-known phrase at the time, derived from a Provençal expression for the technical skill required for poetry writing. In Bk. 3, Para. 108, Nietzsche first proclaims that 'God is dead'
8. From an unpublished article, 'Philosophy Under the Sun', by the novelist Lesley Chamberlain
9. Translated from the draft of a letter in the Munch Museum, quoted in Berman, Patricia Gray, 'Monumentality and Historicism in Edvard Munch's University of Oslo Festival Hall Paintings', a dissertation submitted to New York University, 1989, p. 62
10. Quoted to me in a letter from James Landis, author of a novel about Strindberg; no source
11. Munch, Edvard, Notebook OKK Reg. No. N55, Munch Museum
12. Prideaux, Sue, *Edvard Munch: Behind the Scream*, London: Yale University Press, 2005, p. 276
13. This is one of at least eight alternative texts Munch wrote in Norwegian and in German, as well as a version in French. The texts are spread from 1895 to 1930, and are reproduced in Reinhold Heller, *Edvard Munch: The Scream*, London: Allen Lane, Penguin, 1973, pp. 105–6
14. Cf. Meade, Marion, quoted by Grann, David, *The Lost City of Z*, *op. cit.*, p. 41
15. Blavatsky, H. P., *Isis Unveiled, A Master Key to the Mysteries of Ancient and Modern Science and Theology*, Pasadena: Theosophical University Press, 1877,

reissued 1931; l. 302; l. 502; l. 271; l. 258

16. Tacitus, Cornelius, *Germania*, Ch. 4, pp. 135–7, quoted by Schama, Simon, *Landscape and Memory*, New York: Knopf, 1995, p. 82

17. *Ibid.*, p. 117

18. Noll, Richard, *The Jung Cult: Origins of a Charismatic Movement*, Princeton, NJ: Princeton University Press, 1994, p. 104. He attacks Jung again in *The Aryan Christ, The Secret Life of Carl Jung* (New York: Random, 1997), branding him an ambitious charlatan. The case is well argued, to the point that Noll has become a demon figure among Jungians

19. Cf. Goodrick-Clarke, Nicholas, *The Occult Roots of Nazism*, New York: New York University Press, 1985, p. 5

20. Mosse, George L., *The Crisis of German Ideology*, New York: Howard Fertig, 1964, p. 59

21. Noll, Richard, *The Jung Cult, op. cit.*, pp. 80–1

22. Cf. Servando González, *The Riddle of the Swastika: A Study in Symbolism*, privately published by the author, Box 9555, Oakland, CA 94613, USA. Cf. also the section on 'Psyche and Swastika' in Cocks, Geoffrey, *Psychotherapy in the Third Reich*, New York: OUP, 1985, pp. 50–86

23. Cf. Evans, Richard J., *The Third Reich in Power, 1933–1939*, New York: Penguin, 2005

24. Goethe, Johann Wolfgang von, *Gespräche mit Eckermann*, 5 vols, Leipzig, 1909, 4: 441–2

25. Cf. Schrâder, Dr Otto F., 'The Religion of Goethe', Adyar Pamphlets No. 38, Theosophical Publishing House, Adyar, Chennai (Madras), February 1914

26. Cf. Jung, Carl Gustav, *Symbols of Transformation*, New York: Bolingen Foundation, 1976, first published in 1912, pp. 121–31, 171–206

27. Noll, Richard, *The Jung Cult, op. cit.*, p. 133

28. Müller, Max, *Lectures on the Science of Language, delivered at the Royal Institution of Great Britain February–May 1863*, second series, New York: Scribner, 1869, p. 520

29. Goodrick-Clarke, *Black Sun: Aryan Cults, Esoteric Nazism, and the Politics of Identity*, New York: New York University Press, 2002, p. 4. Cf also Landig's lurid and racially repellent trilogy of novels beginning with *Götzen gegen Tule* (1971); and *The Occult Roots of Nazism: The Aseosophists of Austria and Germany 1890–1935*, Wellingborough: Aquarian Press, 1985

30. James, David H., *The Rise and Fall of the Japanese Empire*, London: Allen and Unwin, 1952, p. 50 and p. 181

31. Gunther, John, *The Riddle of MacArthur; Japan, Korea, and the Far East*, New York: Harper, 1950, pp. 107–8

32. Cf. Dower, John W., *Embracing Defeat, Japan in the Wake of World War II*, New York: Norton, 1999, p. 310. For a fuller discussion, see pp. 308–18; cf. also Bix, Herbert P., *Hirohito and the Making of Modern Japan*, New York: HarperCollins, 2000, pp. 560–2. The official translation appears in the US Department of State, *Foreign Relations of the United States 1946*, 8: 134–5

33. Cf. Dower, John W., *op. cit.*, p. 307

34. Cf. Onishi, Normitsu, *The New York Times*, 12 March 2006, p. 4

35. Sheridan, Michael, *The Sunday Times*, 17 September 2006, p. 31

36. Cf. Tomida, Hiroko, *Hiratsuka Raicho and Early Japanese Feminism*, Leiden: Brill, 2004; and Chung Hyun Kyung, *Struggle To Be the Sun Again*, London: Orbis, 1991

37. Cronin, Vincent, *The Wise Man from the West, op. cit.*, p. 97

38. *Basic Theories of Traditional Chinese Medicine*, Beijing: Academy Press, 1998, p. 23

39. Ratzinger, Joseph, *The Spirit of the Liturgy*, San Francisco: Ignatius Press, 2000, pp. 24; 54; 42; 68; 69; 82; 96; 101; 103; 107; 109; and 128–9

CHAPTER 29 OVER THE HORIZON

1. Hale, George Ellery, letter to H. M.Goodman, 5 March 1893, quoted in Wright, H., *Explorer of the Universe, a Biography of George Ellery Hale*, New York: Dutton, 1966, p. 102

2. Quoted in *The Week*, 10 April 2010. A whole industry exists, busily making up fake Einstein quotations, but this one appears genuine: on his fiftieth birthday, 14 March 1929, Einstein was showered with gifts, and as a way of thanking everybody composed a humorous doggerel. These are its (translated) final two lines

3. Cf. Levenson, Thomas, 'Einstein's Gift for Simplicity', Internet essay posted 30 September 2004

4. Overbye, Dennis, 'The Next Einstein? Applicants Would Be Welcome', *The New York Times*, 1 March 2005, F4

5. Moore, Ruth, *Niels Bohr*, Cambridge, Mass.: MIT Press, 1985, p. 127

6. Cathcart, Brian, *The Fly in the Cathedral: How a Group of Cambridge Scientists Won the International Race to Split the Atom*, New York: Farrer, Straus, 2005, p. 70. Boris Kachka, Roger Straus's current biographer, incidentally, told me: 'I've lost count of the number of people who refer to him as the Sun King'

7. Cf. Kumar, Manjit, Quantum: *Einstein, Bohr, and the Great Debate about the Nature of Reality*, New York: Norton, 2010

8. Ferris, Timothy, *Coming of Age in the Milky Way, op. cit.*, p. 262

9. Cf. Kolbert, Elizabeth, 'Crash Course', *The New Yorker*, 14 May 2007

10. Bahcall, John N., 'How the Sun Shines', published in the Nobel e-Museum, June 2000, p. 8: http://www.nobel.se/phsics/articles

11. *Ibid*, p. 11. Ironically, of course, the accepted solution to the neutrino problem was announced within the year

12. Quoted in Kolbert, Elizabeth, 'Crash Course', *op. cit.*

13. Clegg, Brian, *Light Years*, London: Piatkus, 2001, p. 3

14. Cf. Bahcall, John N., 'Neutrinos from the Sun', *Scientific American*, Vol. 221, No. 1, July 1969

15. *Ibid*, pp. 28–37

16. Maddox, John, *What Remains to be Discovered*, New York: Free Press, 1998, p. 86

17. Professor Marco Velli, interview at the Jet Propulsion Lab, 6 December 2007. Velli added, with some feeling: 'Solar physics has always been "in between", a subgroup of a subgroup. Some of the famous physicists of history haven't always given the right consideration to solar research'

18. Stoppard, Tom, 'Playing with Science', *Engineering and Science*, Fall 1994, p. 10: he calls *Hapgood* 'a play which derived from my belated recognition of the dual nature of light – particle and wave'

19. Bryson, Bill, *A Short History of Nearly Everything, op. cit.*, p. 9

20. Cf. Lord, M. G., *Astro Turf*, New York: Walker, 2005, p. 17

21. Hufbauer, Karl, *Exploring the Sun, op. cit.*, p. 123

22. Quoted by Hufbauer, Karl, *Exploring the Sun, op. cit.*, pp. 125–60

23. Cf. Neufeld, Michael J., *Von Braun, Dreamer of Space, Engineer of War*, New York: Knopf, 2007

24. 'The Space Age', *Science Times*, 25 September 2007, F1

25. Wilford, J. N., 'Remembering When the US Finally (and Really) Joined the Space Race', *The New York Times*, 29 January 2008, F3

26. Quoted in Sims, Michael, *Apollo's Fire, op. cit.*, p. 14

27. *Ibid.* The first American attempt, *Vanguard 1*, launched on 6 December 1957, was an embarrassing failure, shutting down a few feet off the launching pad, and being at once nicknamed 'Flopnik'

28. Cf. the Internet history of the Jet Propulsion Lab, editor and lead author Franklin O'Donnell, 2002

29. Porco, Carolyn, 'NASA Goes Deep', *The New York Times*, 20 February 2007, A19

30. Dupree, A. K., Beckers, J. M., and others, 1976. Report of the ad hoc committee on the interaction between solar physics and astrophysics (18 June), 1976. Cf. Hufbauer, Karl, *Exploring the Sun, op. cit.*, pp. 191–2

31. Cf. Hufbauer, Karl, *Exploring the Sun, op. cit.*, p. 245

32. Cf. *Solar System Plasma Physics*, Ed. Parker, E. N., Kennel, C. G., and Lanzerotti, L. J., Amsterdam: North-Holland, 1979, 1:3–49

33. Parker, E. N., 'Solar Physics in Broad Perspective', Boulder, Colorado: Westview Press, 1978, p. 3

34. Cf. Wilkinson, Alex, 'The Tenth Planet', *The New Yorker*, 24 July 2006, pp. 50–9

35. Cf. Frank, Adam, 'How Nature Builds a Planet', *Rochester Review*, summer 2006, pp. 15–21, and Overbye, Dennis, 'Dusty Planet-Forming Process May Be Playing Out in Miniature', *The New York Times*, 8 February 2005, F3. As recently as February 2008 the discovery was reported of a smaller version of our system 5,000 light years across the galaxy, with outer giant planets and room for smaller inner ones: two planets, one about two-thirds the mass of Jupiter, the other about ninety per cent the mass of Saturn, orbit a reddish star of about half the Sun's mass. Cf. *The New York Times*, 15 February 2008, A20. It did not even make the front page

36. *The Sunday Times*, 7 September 2003, News p. 9

37. *The New York Times*, 'The Space Age', date unknown, F8

38. Porco, Carolyn, 'NASA Goes Deep', *op. cit.*, A19

39. Neil Murphy, interview with the author, 6 December 2007

40. Marco Velli, interview with the author, 6 December 2007

41. Overbye, Dennis, *The New York Times*, 29 July 2003, F4

42. Kuiper, Gerard Peter, *The Solar System:* Vol. 1, *The Sun*, Chicago: University of Chicago Press, 1953

43. Prospectus of the Advanced Technology Solar Telescope, published by GONG

44. Cf. *The New York Times*, 25 September 2007, F3
45. *National Geographic*, date unknown, p. 10
46. Parker, E. N., 'Solar Physics in Broad Perspective', in *The New Solar Physics. Proceedings of an AAAS Selected Symposium 17*, ed. John Eddy, Boulder, Co: Westview Press, 1978, p.1

CHAPTER 30 UNDER THE WEATHER

1. Fitzgerald, Scott, *The Great Gatsby*, New York: Scribner, 1925, p. 124
2. Frost, Robert, *New Hampshire*, 1923; first published in *Harper's Magazine*, 1920, *The Poetry of Robert Frost*, New York: Holt, 2002
3. Asimov, Isaac, *A Choice of Catastrophes*, New York: Simon and Schuster, 1978
4. Sims, Michael, *Apollo's Fire, op. cit.*, p. 125. Further, 'it has been known for a decade or more that at levels routinely encountered in most American cities ozone burns through cell walls in lungs and airways. Tissues redden and swell.' Internet entry, Bates, David V., 'Smog: Nature's Most Powerful Purifying Agent' in *Health & Clean Air*, Fall 2002 newsletter, http://healthandcleanair.org/newsletters/issue.html
5. McNeill, J. R., *Something New Under the Sun: An Environmental History of the Twentieth-Century World*, New York: Norton, 2001, p. xxvi
6. Cf. Kolbert, Elizabeth, *Field Notes from a Catastrophe, Man, Nature, and Climate Change*, New York: Bloomsbury, 2006, p. 183
7. Schiff, Harold, and Dotto, Lydia, *The Ozone War*, Garden City, NY: Doubleday, 1978
8. Wildavsky, Aaron, *But Is It True? A Citizen's Guide to Environmental Health and Safety Issues*, Boston: Harvard University Press, 1995, p. 334
9. Cf. Blaustein, Andrew R., *Scientific American*, February 2003, Vol. 288, Issue No. 2, p. 60
10. Weiner, Jonathan, *The Next One Hundred Years, op. cit.*, p. 102
11. Cf. Houghton, Sir John, *Global Warming*, Cambridge: CUP, 1997, p. 12
12. Cf. *The New Yorker*, 20 November 2006, p. 69
13. Morton, Oliver, *Eating the Sun; How Plants Power the Planet*, London: Fourth Estate, 2007, p. 371
14. Kolbert, Elizabeth, 'The Climate of Man – II', *The New Yorker*, 2 May 2005, p. 70. In 2006 Kolbert published *Field Notes from a Catastrophe, op. cit.*, an eloquent extension of her *New Yorker* articles
15. Lynas, Mark, *Six Degrees: Our Future on a Hotter Planet*, London: Fourth Estate, 2007, p. xix
16. Cf. the *New Statesman* website for 19 December 2007 and 14 January 2008
17. Cf. Revkin, Andrew C., 'UN Report on Climate Details Risks of Inaction', *The New York Times*, 17 November 2007, A1. For several of the details of this history, cf. also Weart, Spenser R., *The Discovery of Global Warming*, http://www.aip.org/history/climate
18. *National Geographic*, September 2004, p. 10
19. Kolbert, Elizabeth, 'The Climate of Man – III', *The New Yorker*, 9 May 2005, p. 54

20. Cf. Kolbert, Elizabeth, 'The Catastrophist', *The New Yorker*, 29 June 2009, p. 42

21. Weiner, Jonathan, *The Next One Hundred Years, op. cit.*, p. 70

22. Rosen, Gary, 'More Heat Than Light', *The New York Times Magazine*, 8 July 2007, p. 20; cf. also McKibben, Bill, 'Carbon's New Math', *National Geographic*, October 2007, pp. 33–7

23. Vonnegut, Kurt, *Armageddon in Retrospect*, New York: Putnam, 2008, p. 26

24. Cf. Tierney, John, 'The Earth Is Warming? Adjust the Thermostat', *The New York Times*, 11 August 2009, D1

25. Cf. Weiner, Jonathan, *op. cit.*, p. 73

26. Cf. Chang, K., 'Globe Grows Darker as Sunshine Diminishes 10% to 37%', *The New York Times*, 13 May 2004

27. Cf. Adams, Cecil, 'The Straight Dope', *Washington City Paper*, 20 April 2007, p. 18

28. Lynas, Mark, *Six Degrees, op. cit.*, p. 6

29. Cf. Clark, Stuart, *The Sun Kings, op. cit.*, p. 180

30. 'The Sun Also Sets', *Investor's Business Daily*, 7 February 2008

31. Cf. Leake, Jonathan, 'So, Are We Going to Freeze or Fry?', *The Sunday Times*, December 2005, p. 16; cf. also Cooper, Richard N., *International Approaches to Global Climate Change*, Cambridge, Mass.: Harvard University Press, 2000

32. Cf. Revkin, A. C., 'Global Warming Is Delaying Ice Age, Study Finds', *The New York Times*, 4 September 2009

33. Kolbert, Elizabeth, *Field Notes from a Catastrophe, op. cit.*, p. 32

34. Wildavsky, Aaron, *But Is It True?, op. cit.*, pp. 352–3

35. Calder, Nigel, *The Weather Machine, op. cit.*, p. 76

36. Joyce, James, *Ulysses, op. cit.*, p. 75

37. Cf. *The Week*, 20 February 2010, p. 13

38. Cf. IPCC, *Summary for Policymakers, Climate Change 2007: The Physical Science Basis. Contribution of Working Group 1 to the Fourth Assessment Report of the Intergovernmental Panel on Climate Change*, Cambridge and New York: CUP, 2007

39. Cf. Kolbert, Elizabeth, 'The Copenhagen Diagnosis: Sobering Update on the Science', Yale Environment 360, 24 November 2009

40. 'The Sun Also Sets', *Investor's Business Daily, op. cit.*

41. Soon, Willie, letter to the author, 7 May 2009

42. Weart, Spencer R., *The Discovery of Global Warming, op. cit.*, p. 196

43. Kunzig, Robert, *Mapping the Deep, op. cit.*, pp. 318–19

CHAPTER 31 THE IMPOSSIBLE AND BEYOND

1. Clarke, Arthur C., *Astounding Days, A Science Fictional Autobiography*, London: Gollancz, 1989, p. 207

2. Wells, H. G., *The Time Machine*, New York: Airmont, 1964, p. 125

3. Cf. Needham, Joseph, *Science and Civilization in China, op. cit.*, p. 440

4. Cf. Nicolson, M. H., *Science and Imagination; Collected Essays on the Telescope and Imagination*, Ithaca: Cornell, 1965, p. 71

5. Swift, Jonathan, *Gulliver's Travels, op. cit.*, Pt. 3, Ch. 2

6. Maunder, E. Walter, *Are the Planets Inhabited?*, London: Harper & Bros, 1913, Ch. 3
7. Wells, H. G., *The Time Machine, op. cit.*, p. 111
8. For a full list of such stories, cf. the *Encyclopedia of Science Fiction*, Ed. Clute, J., and Nicholls, P. (published in the US as *The Science Fiction Encyclopedia*), New York: St Martin's Press, 2nd Edition, 1993, pp. 1177–8
9. Cf. Alkon, P. K., *Winston Churchill's Imagination*, Lewisburg: Bucknell University Press, 2006, p. 150
10. Quoted in Clarke, Arthur C., *Astounding Days, op. cit.*, p. 224
11. Heinlein, Robert, *The Man Who Sold the Moon*, New York: Signet, 1950
12. Overbye, Dennis, 'Setting Sail Into Space, Propelled by Sunshine', *The New York Times*, 10 November 2009, D1
13. Bradbury, Ray, 'The Golden Apples of the Sun', *Classic Stories 1*, London: Bantam, 1990, pp. 154–60
14. Pohl, Frederik, and Kornbluth, C. M., *Wolfbane*, New York: Ballantine, 1959, p. 28, p. 11
15. Asimov, Isaac, *In Memory Yet Green*, New York: Doubleday, 1979, pp. 295–6
16. Asimov, Isaac, *The Best of Isaac Asimov*, London: Sphere, 1973, 'Nightfall', pp. 37, 40 and 60
17. Asimov, Isaac, and Silverberg, Robert, *Nightfall*, London: Pan, 1990, p. 352
18. Cf. Liu, Charles, 'My Three Suns', *Natural History*, October 2006, pp. 70–1. On 19 July 2005 *The New York Times* reported that astronomers had discovered a planet with three suns
19. Clarke, Arthur C., *The Collected Stories of Arthur C. Clarke*, New York: Tor, 2000, pp. 517–21
20. Clarke, *Collected Stories*, 'A Slight Case of Sunstroke', pp. 687–92
21. Clarke, *Collected Stories*, 'Summertime on Icarus'
22. Clarke, *Collected Stories*, 'The Wind from the Sun', pp. 828–42
23. Cf. *Saga* magazine interview with Nina Myskow, quoted in *The Week* (UK edition), 7 July 2007, p. 10
24. Cf. Jonas, Gerald, obituary of Arthur C. Clarke, *The New York Times*, 19 March 2008, C12
25. Clarke, *Collected Stories*, 'Out of the Sun', pp. 652–7
26. Ballard, J. G., 1987, quoted by Strausbaugh, John, 'Aiming for Life's Jugular in Deadly Verbal Darts', *The New York Times*, 1 December 2004, E9

CHAPTER 32 THE DEATH OF THE SUN

1. Darwin, Erasmus, *The Botanic Garden*, Pt. 1, Canto IV (Air), ll. 380–3
2. Lewis, C. S., *The Last Battle*, New York: Macmillan, 1956, p. 148
3. Asimov, Isaac, *A Choice of Catastrophes, op. cit.*, pp. 92–130
4. Cf. Overbye, Dennis, 'Sun Might Have Exchanged Hangers-On With Rival Star', *The New York Times*, 12 February 2004, A25
5. *The Times*, 18 April 2005, p. 17
6. Chang, Kenneth, 'Prediction Proved: Light Speeds Up an Asteroid as it Spins', *The New York Times*, 13 March 2007

7. *The Sun*, 28 February 2005, pp. 8–10
8. Quoted by Sims, Michael, *Apollo's Fire, op. cit.*, p. 157
9. Overbye, Dennis, 'Kissing the Earth Goodbye in About 7.59 Billion Years', *The New York Times*, 11 March 2008
10. Allen, Woody, 'Strung Out', *op. cit.*, p. 96
11. Overbye, Dennis, 'Scientists' Good News: Earth May Survive Sun's Demise in 5 Billion Years', *The New York Times*, 13 September 2007
12. Cf. 'Scientists Spot Nearby "Super-Earth" Planet', Sphere News, 17 December 2009
13. Wolfe, Tom, 'One Giant Leap to Nowhere', *The New York Times*, 19 July 2009, p. 11
14. Cf. Austen, Ben, 'After Earth', *Popular Science*, March 2011, pp. 46–53, 98–102
15. Porco, Carolyn, 'NASA Goes Deep', *The New York Times*, 20 February 2007, A19
16. Kaku, Michio, *Physics of the Impossible, op. cit.*, pp. xvi–xvii
17. Cf. Clark, Stuart, 'The Death of the Sun', *Focus, op. cit.*, p. 34

"That can't be good."

SUNSET ON THE GANGES

1. Whitman, Walt, *Leaves of Grass*
2. Gayatri mantra, from Rig-Veda III, 62, 10
3. Ruffin, C. Bernard, *Last Words, A Dictionary of Deathbed Quotations*, London and North Carolina: McFarland, 1995
4. Hardy, Florence Emily, *The Early Life of Thomas Hardy*, New York: Macmillan, 1928, p. 19
5. Cf. Blehm, Eric, *The Last Season*, New York: HarperCollins, 2006
6. Van Gogh, Vincent, *Letters, op. cit.*, No. 617, p. 238. Others who have written supremely well about sunsets are Jules Verne, in *20,000 Leagues Under the Sea* (London: Penguin, 2004, p. 239); Mark Twain, in *Life on the Mississippi* (London: Penguin, p. 396); Nabokov, in *Speak, Memory* (*op. cit.*, p. 213); and T. E. Lawrence, *Seven Pillars of Wisdom, A Triumph* (London: Cape, 1973, p. 575)

7. Cf. Connolly, Michael, *The Closers*, New York: Little, Brown, 2005, p. 270

8. Cf. Adam, David, *Guardian*, 1 October 2007, and the AP report by Kate Schuman, 28 November 2007

9. Cf. Eck, Diana., *Banāras, City of Light*, New York: Columbia University Press, 1962, pp. 14–15

10. Caine, W. S., *Picturesque India, A Handbook for European Travellers*, London: Routledge, 1890, p. 302

11. Twain, Mark, *Following the Equator, A Journey Around the World*, New York: Dover, 1989, p. 496

12. Cf. Sengupta, Somini, 'Debate in India: Is Rule on Yoga Constitutional?', *The New York Times*, 26 January 2007

13. Cf. Vaudeville, Charlotte, *Kabīr*, Oxford: Clarendon Press, 1974, p. 267

14. Eck, Diana L., *Banāras, City of Light*, *op. cit.*, p. 182

15. Dyer, Geoff, *Jeff in Venice, Death in Varanasi*, New York: Pantheon, 2009, p. 173

AFTERWORD TO THE FIRST PAPERBACK EDITION

1. Cf. Chang, Kenneth, 'Twinkle, Twinkle, Perhaps Times Three', *The New York Times*, 2 December 2010, A17

2. MacKenzie, Debora, and Le Page, Michael, 'Light Diet: Animals That Eat Sunshine', *New Scientist*, 13 December 2010

3. *Science News*, 4 December 2010, p. 11

4. Cf. *The New York Times Book Review*, 5 December 2010, p. 58

LIST OF ILLUSTRATIONS AND
ILLUSTRATION CREDITS

p. xix "Solar Eclipse" Chiapas, Mexico, in 1991. Photo by Antonio Turok

p. 5 An 1857 painting of the sun goddess Amaterasu, with Mount Fuji in the background. Minneapolis Institute of Arts, Gift of Louis W. Hill Jr.

p. 9 The heroic archer Houyi, who shot down nine of the ten suns. Copyright © Mona Caron

p. 14 The sun chariot of Surya. Michaud Roland et Sabrina/Rapho/Eyedea Illustrations

p. 17 Max Müller. Walker & Cockerell, ph. Sc

p. 18 A map of the distribution of solar cultures round the world. *Graphis* magazine, 1962

p. 21 Audrey Hepburn and Gregory Peck in *Roman Holiday,* standing in front of the Bocca della Verità in the Church of Santa Maria, Rome. Paramount Pictures

p. 30 Celebrants jumping a bonfire during summer solstice festivities at St Jean in Alsace. Cabinet des Estampes et des Dessins de Strasbourg/Photo Musées de la Ville de Strasbourg/M. Bertola

p. 32 A Mexican sun- ball game. Reuters/Henry Romero

p. 37 The *Okipa*, the dance ritual of the Mandan Indians, as depicted by the artist George Catlin. Copyright © the Library, American Museum of Natural History

p. 41 The massive pillars of Stonehenge. Photo courtesy of Madanjeet Singh

p. 46 The passageway at Newgrange burial mound in Ireland. Courtesy of the Department of the Environment, Heritage and Local Government, Ireland

p. 48 The pyramid at Chichén Itzá. Damian Davies/Getty Images

p. 51 The Great Pyramid at Giza. Carolyn Brown/Photo Researchers, Inc.

p. 54 The sundial bridge at Turtle Bay in Redding, California, by Santiago Calatrava. Alan Karchmer/ESTO

p. 59 Aurora borealis, 1 March 1872. Étienne Léopold Trouvelot. Science, Industry & Business Library, The New York Public Library, Astor, Lenox and Tilden Foundations

p. 66 Eclipse at the battle of the Medes and Lydians, 28 May 585 BC. Camille Flammarion, *Astronomy for Amateurs,* D. Appleton and Company, New York and London, 1910. Painting by Rochegrosse

p. 73 Petroglyphs from the Camonica Valley in northern Italy. Sketch by Paul F. Jenkins

p. 78 Babylonian astronomers observing a meteor trail. From John F. Blake, *Astronomical Myths,* London: Macmillan and Co., 1877

p. 84 The family of Akhenaten offers sacrifice to Aton the sun god. Erich

Lessing/Art Resource, NY

p. 90 Pythagoras in Alexandria. Sheila Terry/Science Photo Library

p. 100 Hipparchus. SPL/Photo Researchers, Inc.

p. 108 Xi Chu with gnomon. HIP/Art Resource, NY

p. 114 The water- driven astronomical clock tower of Su Song. Drawing by John Christiansen, from Joseph Needham, *Science and Civilisation in China.* © Cambridge University Press. Reprinted by permission

p. 118 Eclipse in Siam, 1688. Bibliothèque nationale, Paris, France/Lauros/Giraudon/The Bridgeman Art Library

p. 119 The Imperial Observatory in Peking, in the late seventeenth century. SPL/Photo Researchers, Inc.

p. 128 Indian astronomers, seventeenth century. HIP/Art Resource, NY/British Library, London, Great Britain

p. 133 Indian folk drawing depicts Surya Mandala with a cobra. From Madanjeet Singh, *The Sun in Myth and Art,* London: Thames and Hudson, 1993

p. 135 The astrolabe of Lansberg. From Philips van Lansbergen, *Philippi Lansbergii Opera omnia,* Middleburgi Zelandi: Z. Roman, 1663

p. 148 The castle and observatory of Uraniborg (constructed by Tycho Brahe). Bibliothèque Mazarine, Paris, France/Archives Charmet/The Bridgeman Art Library

p. 150 The planetary spheres as depicted in Kepler's *Cosmographic Mystery* (1596). Mary Evans/Photo Researchers, Inc.

p. 154 Nicolaus Copernicus. Science Source/Photo Researchers, Inc.

p. 154 Tycho Brahe. Hulton Archive/Getty Images

p. 154 Johannes Kepler. Science Source/Photo Researchers, Inc.

p. 154 Galileo Galilei. SPL/Photo Researchers, Inc.

p. 156 Galileo's visit to the Vatican in 1611. Louvre, Paris, France/Peter Willi/The Bridgeman Art Library

p. 164 Sir Isaac Newton. Science Source/Photo Researchers, Inc.

p. 167 Newton using a prism to break white light into its spectrum. Image Select/Art Resource, NY

p. 173 *An Alchemist in His Workshop.* Courtesy of the Chemical Heritage Foundation Collections

p. 182 Dancers during eclipse in Antarctica, 2003. Photo by Woody Campbell

p. 184 A partial eclipse, seen against a crescent on Islamabad's grand Faisal mosque. AP Photo/B. K. Bangash

p. 189 Tintin uses an eclipse to outwit his Inca captors. Hergé/Moulinsart 2009

p. 192 Parisian women view eclipse, 1892. From *La Vie Illustrée: L'Eclipse Solaire du 30 Août, Rue de la Paix* (1905)

p. 195 Eclipse in Antarctica, 2003. Photo by Woody Campbell

p. 208 An advertisement for Tho-Radia cream. SPL/Photo Researchers, Inc.

p. 214 Albert Einstein and Robert Oppenheimer. USIA/AIP Photo Researchers, Inc.

p. 217 XX-28 George, a 225-kiloton atomic bomb, exploding over the Pacific Ocean on 8 May 1951. US Department of Energy/Photo Researchers, Inc.

p. 219 Sunbathing. Illustration by Lt. Col. Frank Wilson from Stephen Potter, *Lifemanship* (Henry Holt and Co., 1950)

p. 223   Galileo's sunspot drawings. Library of Congress

p. 227   Christoph Scheiner observing sunspots. SPL/Photo Researchers, Inc.

p. 231   Section of a Scotch pine showing the years of maximum sunspot activity. Courtesy of the Laboratory of Tree-Ring Research, University of Arizona

p. 241   Light streaming into the concourse of New York's Grand Central Station. Library of Congress

p. 251   A Cambodian farmer with a sunshade, 1952. Werner Bischof/Magnum Photos

p. 252   Horizontal light can be used to discover the site of lost villages. Ashmolean Museum, University of Oxford

p. 254   The great ball of Krishna, Mahabalipuram, on the southeast coast of India, 1971. Photo by Max Pam

p. 264   *Cleopatra Dissolving the Pearl*, 1759, by Joshua Reynolds.. Kitty Fisher: Kenwood House: The Iveagh Bequest/English Heritage Photo Library

p. 269   John F. Kennedy sunbathing, 1944. John F. Kennedy Library

p. 272   English children undergoing sun-lamp treatment. SVT-Bild/Das Fotoarchiv/Black Star

p. 278   'Heliotherapy'. Copyright © CAP/Roger-Viollet/The Image Works

p. 281   A mother and her albino child, Sudan, 1905. Michael Graham-Stewart/ The Bridgeman Art Library

p. 282   Freewheelin' Frank Reynolds, secretary of the San Francisco branch of the Hell's Angels, 1967. Photography by Larry Keenan.

p. 285   Protective clothing for the rare disease XP. Sarah Leen/National Geographic Image Collection

p. 287   Max Schreck in *Nosferatu: A Symphony of Horror*, 1922. Photofest

p. 304   Flower clock. Drawing by U. Schleicher-Benz

p. 310   The Texas blind salamander. Dante Fenolio/Photo Researchers, Inc.

p. 313   A digital recreation of a hydrothermal vent. David Batson/ DeepSeaPhotography.com

p. 319   The deep-sea angler fish. David Batson/DeepSeaPhotography.com

p. 327   *Father Time Carrying the Pope Back to Rome*, c. 1641. © The Trustees of the British Museum/Art Resource, NY

p. 340   Jonas Moore's *A New System of Mathematicks*, 1681. Lawrence H. Slaughter Collection, The Lionel Pincus and Princess Firyal Map Division, The New York Public Library, Astor, Lenox and Tilden Foundations

p. 342   Map of Commodore George Anson's attempt to circumnavigate the globe, 1740–44. Library of Congress, Rare books division

p. 356   William Hogarth, *The Election Dinner*, 1755. Private Collection/Ken Welsh/The Bridgeman Art Library

p. 360   Ethiopians celebrating the new millennium, 12 September 2007. Roberto Schmidt/AFP/Getty

p. 362   Borneo tribesmen measure the length of the Sun's shadow at summer solstice with a gnomon. From Charles Hose & William McDougall, The Pagan Tribes of Borneo, London: Macmillan & Co. 1912

p. 365   The Samrat Yantra, the giant sundial at the Jaipur Observatory. Science Museum/SSPL

p. 370   An illuminated manuscript page from *Les Très Riches Heures du Duc de Berry*,

1412–16. Réunion des Musées Nationauz/Art Resource, NY

p. 373 The calendar used by an Inuk hunter in the 1920s. Revillon Frères Museum, Moosonee, ON

p. 380 Atomic clock. National Physical Laboratory, Crown Copyright/SPL/Photo Researchers, Inc.

p. 387 Archimedes used mirrors to reflect the Sun's rays on to the Roman fleet. Archive Photos/Getty Images

p. 391 Augustine Mouchot's solar-powered printing press at the Paris World Exhibition of 1878. The Granger Collection, New York

p. 393 A revolving solarium at Aix-Les-Bains in September 1930. Fox Photos/Hulton/Getty Images

p. 409 Swastika sun symbol. Public domain

p. 414 Fireworks on the Grand Canal at Versailles, 1674. Réunion des Musées Nationaux/Art Resource, NY Chateaux de Versailles et de Trianon, Versailles, France

p. 415 Five young gypsies perform their salute to the Sun in the mid-1950s. Tziganes, Frans de Ville, 1956

p. 424 Indian woman stands naked in the water, facing the Sun. Michaud/Rapho/Eyedea

p. 428 St. Benedict during a solar eclipse. Courtesy of the Weltenberg Cathedral

p. 431 Turner, *The Fighting Téméraire*. National Gallery, London/Art Resource, NY

p. 434 Monet, *Impression, Sunrise*. Erich Lessing/Art Resource, NY

p. 442 Olafur Eliasson, *The Weather Project*. Tate, London 2010

p. 446 Pablo Picasso paints with light. Gjon Mili/Time Life Pictures/Getty Images

p. 452 *Man with Umbrella*, 1936. Brassaï. © Estate Brassaï/Agence photo RMN, Paris

p. 453 Gustave Le Gray, *The Great Wave, Sète* (1857). Image copyright © The Metropolitan Museum of Art/Art Resource, NY

p. 457 The Roman town of Bram, in France. Irène Alastruey/Agefotostock

p. 459 The Statue of Liberty. The Bridgeman Art Library

p. 460 The Walt Disney Concert Hall in downtown Los Angeles Hufton+Crow/View/Esto

p. 463 Judith Jamison as the Sun in the 1968 revival of Lucas Hoving's *Icarus*. Photo ©Jack Mitchell

p. 467 The Sun and Moon dance in traditional Japan. Courtesy of the author

p. 468 The set for the final scene of the second act of *The Magic Flute*, c. 1730. Bildarchiv Preussischer Kulturbesitz/Art Resource, NY

p. 477 Vladimir Nabokov. Philippe Halsman/Magnum Photos

p. 490 Aldous Huxley and D. H. Lawrence. Topham/The Image Works

p. 495 Norman Mailer with his sixth wife, Barbara. Photograph by Francis Delia

p. 499 No. 10 Downing Street. ImageState/age fotostock

p. 509 Charlie Chaplin in *The Great Dictator*. Photofest

p. 514 Mao's Little Red Book. Courtesy of the author.

p. 516 Joseph Ratzinger, now Pope Benedict XVI. Filippo Monteforte/AFP/Getty Images

p. 519 Sketch from the seventeenth-century manuscript *De thermis*. NOAA Library Collection

p. 526 Neutrino detector. Kamioka Observatory, ICRR (Institute for Cosmic Ray

Research), University of Tokyo

p. 530 'The Suicide Squad' tries to send a rocket into space. Courtesy of NASA

p. 546 Image of eruption in Eyjafjallajökull. Photo by Sigurdor Stefnisson

p. 553 Weather balloon. Novosti

p. 569 Still from *Sunshine*, 2007. Fox Searchlight/The Kobal Collection/The Picture Desk

p. 571 Sunset on Mars. JPL/NASA

p. 575 UFOs. From Harold T. Wilkins, *Flying Saucers Uncensored*, New York: Pyramid Books, 1955

p. 585 The Sun heats a dying Earth, five billion years from now. Detlev van Ravenswaay/Photo Researchers, Inc.

p. 588 Artist's impression of a manned base on Mars. Julian Baum/Science Photo Library

p. 596 Dasaswamedh Ghat, Varanasi, India. Courtesy of the Rare Book Division of the Library of Congress

p. 603 Sunset on the Ganges. Art Wolfe/Getty Images.

p. 612 'They're keeping me from practicing my religion' cartoon. © Frank Cotham/The New Yorker Collection/www.cartoonbank.com

p. 617 From *Peanuts*. Peanuts © United Features Syndicate, Inc.

p. 621 'Narcissism: Thinks World Revolves around It' cartoon. © Dolly Setton

p. 627 'Astronomy Beyond the Visible-light Spectrum' cartoon. © Dolly Setton

p. 629 From *Mutts*. © Patrick McDonnell. King Features Syndicate

p. 631 'Listen, pal, it's a free country' cartoon. © 2008 Leigh Rubin

p. 638 'Now that we can tell time' cartoon. © The New Yorker Collection. Tom Cheney, 2005, from cartoonbank.com. All rights reserved

p. 640 'Sunrise, Sunset, Noon' cartoon. © The New Yorker Collection. Jack Ziegler, 2007, from cartoonbank.com. All rights reserved

p. 644 From *Peanuts*. Peanuts © United Features Syndicate, Inc.

p. 649 'He's adjusting our solar panels' cartoon. © George Booth/The New Yorker Collection/www.cartoonbank.com

p. 653 'That can't be good' cartoon: © The New Yorker Collection. Danny Shanahan, 2006, from cartoonbank.com. All rights reserved

Plate section

The Sun viewed in extreme ultraviolet light by NASA's SOHO satellite. SOHO (ESA & NASA)

Diagram of parts of the Sun. SOHO (ESA & NASA)

A diamond ring during a total eclipse. SOHO (ESA & NASA)

Coronal mass ejections. SOHO (ESA & NASA)

The dark eye of a sunspot. SOHO (ESA & NASA)

The Sun's magnetic loops. SOHO (ESA & NASA)

An erupting prominence on the sun. SOHO (ESA & NASA)

Active regions on the Sun. SOHO (ESA & NASA)

Progress of a Coronal Mass Ejection. SOHO (ESA & NASA)

Aurora photos from top left, clockwise. Jan Curtis, Sigurdur Stefnisson, Jacki

Tolley, Sigurdur Stefnisson, Sigurdur Stefnisson

A time-lapse photograph of the Sun changing in colour and brightness as it sets. Pekka Parviainen/Photo Researchers, Inc.

Sunrise seen from Mount Fuji. Courtesy of the author

The seashore at Futamigaura, near the holy site of Ise. Courtesy of Madanjeet Singh

A pillar of light rises from the setting Sun. Courtesy of Michael Carlowicz

Louis XIV, aged fourteen, as Le Roi Soleil in Ballet de la Nuit, 1653. Photo: Bulloz Bibliothèque de l'Institut de France, Paris, France. Réunion des Musées Nationaux/Art Resource, NY

Inti mask. Museo Banco Central de Quito Ecuador/Gianni Dagli Orti/The Picture Desk

Three masks. Museum of the Sun in Riga, Latvia/Iveta Gražule; Photos courtesy of the author

Detail of ornamental sun. Richard I'Anson/Lonely Planet Images

The Sun (2002) by Yue Minjun

Girl with a sunflower (1962), a painting from the Croatian Naïve School by Ivan Rabuzin

Inti worshipping the Sun. By Théodore de Bry. Bridgeman Art Library/Service Historique de la Marine, Vincennes, France/Lauros/Giraudon

New Year celebrations at the Futamigaura seashore, Japan. From Madanjeet Singh, *The Sun in Myth and Art,* Thames and Hudson, 1993

Sunset at the Crimean Shores (1856) by Ivan Aivazovsky (1817–1900). Copyright © 2010 State Russian Museum, St Petersburg

Joseph Mallord William Turner's *Heidelberg Castle* (1840–45). Tate, London/ Art Resource, NY

The legend of the nine suns. British Museum, London, UK/The Bridgeman Art Library

Roy Lichtenstein's *Sinking Sun.* Estate of Roy Lichtenstein

Vincent Van Gogh, *The Sower.* Erich Lessing/Art Resource, NY

Edward Hopper, *Rooms by the Sea* (1951). Yale University Art Gallery/Art Resource, NY

David Hockney, *A Bigger Splash* (1967), Acrylic on Canvas, 96×96 © David Hockney/Art Resource, NY

INDEX

Abarihu, 18
Abdusamatov, Khabibullo, 559
Abstract Expressionism, 445
Abu Jirab, 52
Abul-Wafa, 122
Adams, Fred, 587
Adams, William Grylls, 390
adaptive optics, 543
Aeschylus, 321
Aesop's fables, 13
Afghanistan, 382
Age of Wonder (Holmes), xvii
Airy, Sir George, 378
Akhenaten, Pharaoh, 52
Akihito, Emperor, 512
Alaska purchase, 377n
Albers, Bill, 294, 299
albinism, 278–80
al-Biruni, 122, 126, 140
alchemy, 103, 172–4, 481
Alexander the Great, 67, 82, 87, 95, 97,
   99, 312, 336, 347
Alexandria, 123, 130, 132, 358
al-Fazari, Muhammad, 128
Alfonso X, King of Castile, 134
Alfred the Great, King, 67, 361
Algeria, 389, 404
al-Haytham, 387, 447
al-Khwarizmi, 122, 125
al-Kindi, 125
Allen, Paul G., 538
Allen, Woody, 382, 457, 587
Alloway, Lawrence, 445
al-Ma'mum, caliph, 125–6
al-Mansur, caliph, 125
al-Masudi, 122
Almería, 398–400
al-Razi, 122
Alter, David, 244
al-Tūsī, Nasīr al-Dīn, 129
Amaterasu, 2, 423, 510, 513
America (the name), 338n
American Civil War, 265, 375

American Declaration of Independence,
   185
American Revolution, 236n
Amundsen, Roald, 57
Anaxagoras, 68, 87, 91–2
Anaximander, 91, 361
Anaximenes, 91
'ancient lights' doctrine, 404–5n
Anderson, Poul, 571
Angkor Wat, 46
Anglo-Saxon Chronicle, 67, 224
animals, 306–9, 314–19
   albatrosses, 306
   albino king snake, 309
   and albinism, 279
   ants, 306, 307n, 308
   arrow worms, 316
   bats, 309
   bears, 309
   bees, 307
   birds, 12, 243n, 306, 308–9, 335
   blackbirds, 62
   blue tits, 306
   caribou, 308
   clams, 314
   cows, 61
   coyotes, 62
   crabs, 306, 314, 317n
   crickets, 536n
   dogs, 62, 297n
   eagles, 12
   eels, 308
   falcons, 12
   fish, 243n, 288n, 306, 317–19, 335, 549
   frogs, 13, 536n, 549
   hawks, 308
   hedgehogs, 34
   Heliconius butterflies, 306
   herring, 306
   horses, 11–12, 26–7, 83, 419, 454
   insects, 306–8, 548
   jaguars, 12, 62
   jellyfish, 315

animals – *continued*
  jumbo dumbo, 317n
  krill, 318
  lanternfish, 316
  Lightless Loosejaw (*Malacosteus niger*),
    318
  lions, 12
  lizards, 306
  'magnetic' termites, 309
  molluscs, 299–300
  mussels, 314
  octopuses, 314
  olm (*Proteus anguinus*), 309
  owls, 243n, 309
  pigeons, 235, 308
  reptiles, 309
  salmon, 308
  sea-cucumbers, 314, 317n
  Selevin's mouse, 309
  shrimps, 314, 316
  spiders, 307n
  squids, 316–18
  sunfish, 306
  tigers, 12
  toads, 308, 549
  tubeworms, 314, 317n
  turtles, 243n, 306, 308
  whalebone eater, 317n
  whales, 308, 315, 317
  Zambian mole rat, 308
Antioch earthquake, 132
Antiochus I, King, 82
Antonioni, Michelangelo, 457
Apollinaire, Guillaume, 488
Apollo, xv, 414–15, 488, 501n
Apollo programme, 422, 533n
Arabs, 109, 120–30, 133–4, 138, 251,
    263n, 331
Aramaic language, 77
Arawaks, 187
Arbela, battle of, 67–8
Archer, Mary, 394
Archimedes, 87, 122, 134, 386–8
architecture, 457–61
Aristarchus of Samos, 87, 98, 100, 140,
    143
Aristophanes, 67n, 92–3
Aristotle, 87, 90, 94–9, 103–4, 122, 134,
    146, 148, 157, 159n, 163, 166, 224,
    439 and celestial spheres, 95–6, 118,
    149, 482 and Christian Europe,
    137–8
Armitage, Simon, 494n

Arnold, Matthew, 16, 487, 494n
Arrhenius, Svante, 551
Aryabhatta, 125
Aryans, 16–17, 508–9
Asam, Aegid Quirin, 427
Ashton, Clare, 569
Ashurbanipal, King, 80n
Asim, Jabari, 270n
Asimov, Isaac, 174, 194, 545, 548–9, 572–
    4, 580–1, 586
Assyrians, 76–7, 80n
asteroids, 537n, 545, 582, 587
astroarchaeology, 49
astrology, 78–9, 99, 103–4, 136, 141
astronauts, 325n, 422, 593
astronomy, 75, 338, 345, 481
  Arab, 121–30
  and astrology, 78–9, 103–4
  Babylonian, 76–82
  and celestial spheres, 95–6, 118, 149–
    51, 482
  Chinese, 105–20
  and Christianity, 132–6
  Egyptian, 82–5
  Greek, 86–101
  and heliocentric universe, 98, 102,
    139–61
  Indian, 130–2
  Roman, 101–4
  post-war, 528–38
astrophysics, 242
Asuras, 18
Atahualpa, 421
Atget, Eugène, 451, 453
Athens, 91, 93–5, 98–9, 262n, 350–1
Atlas Mountains, 593
atomic bombs, 213–16
Attenborough, David, 301, 315
Atum, 85
aubades, 485
Auden, W. H., 49, 493
Augustus, Emperor, 104, 352
Aurillac, 34
Aurora, 422
auroras, 36, 56–9, 61, 228
Austen, Jane, 271, 275, 460n, 553n
Austerlitz, battle of, 246
Australia, 30, 259, 329, 331, 394, 402,
    412, 418, 424, 457, 544n
Australian Aboriginal chants, 536n
Autolycus, 136
Averroës, 134
Avicenna, 122

Aztecs, 12, 32, 47, 302, 345–6, 423

Babuyan volcano, 594
Babylonians
    astronomy, 76–82
    calendar, 81, 345, 347
    and eclipses, 63–4
    mathematics, 80–1
Bacon, Francis, 284n, 447
Bacon, Roger, 136–8, 153, 353, 387
Bahcall, John, 524, 527
Bailey, Anthony, 432
Bailey, Paul, 190–1
Bailey, Thomas, 140n
Baillie, Mike, 237–8
Baily, Francis, 186
Bakairi, 62
Baker, Josephine, 267
Ballard, Jim, 227
Ballard, J. G., 494n, 580
banking system, 125
Banville, John, 144n
Barbauld, Anna, 297n
Barbillius, 104
bare-knuckle boxing, 248
Barnes, William, 449n
Barrie, J. M., 249
Barrymore, Lionel, 455
Bartel, Julius, 535
Barthes, Roland, 450
Barton, Catherine, 166n
Bastiat, Frédéric, 390n
Bavaria, 13
Bayeux Tapestry, 171n
Beach Boys, 420, 472
Beardsley, Aubrey, 266
Beatles, 266n, 472
Beaumont and Fletcher, 253n
Beckett, Samuel
    Waiting for Godot, 379, 384
Becquerel, Alexandre-Edmond, 394
Becquerel, Henri, 207
Bede, Venerable, 140, 353
Beethoven, Ludwig van, 179, 469, 472,
    536n
Behte, Hans, 212, 216, 218
Belknap, Jeremy, 57
Bell Labs, 394–5
Bellarmine, Robert, 156n
Benedict XIV, Pope, 160
Benedict XVI, Pope (Joseph Ratzinger),
    160, 515–16
Benét, Stephen Vincent, 592–3

Bengalis, 14
Bennett, Arthur, 196n
Bergamo, 54
Bergerac, Cyrano de, 225, 488–9
Bergman, Ingrid, 266
Berlin, Irving, 472
Bermuda Triangle, 321
Bernard, Émile, 438
Bernhard, Oskar, 271–2
Bernini, Gian Lorenzo, 221
Berry, Chuck, 536n
Berry, Halle, 270n
Besançon, 396n
Bessel, Friedrich, 159n
Bethlehem, Star of, 145n
Betjeman, John, 494
Bezos, Jeffrey P., 538
Bible, 79, 80n, 135, 142, 156, 200, 245n,
    288, 426
    and age of the Earth, 203–5
    Apocryphal books, 57, 280n
    Genesis, 14, 191n, 203–4, 481
    Gospels, 68
    Isaiah, 482n
    New Testament, 27
    Old Testament, 203–4
    Psalms, 348, 486
    Revelation, 583
Bien, Reinhold, 357–8
Bierman, Ludwig, 534
Big Ben, 372n
Bigelow Aerospace, 579n
bioluminescence, 318
biomass energy, 299, 401
Black Death, 133
Black Drop effect, 176n
black dwarfs, 590
Black Elk, chief, 37
Black Hole of Calcutta, 297n
black holes, 200, 527, 545, 582
Black Sun, 492, 493n, 509–10
Blaise, Clark, 375–6
Blake, William, 434, 470, 486–7
Blanquilla Island, 185n
Blavatsky, Madame, 504–5, 506n, 509
blind people, 358n
Bliss, Sir Arthur, 445
blue flash, 60n
Boadicea, 43
boar's head, 31
Bohr, Niels, 211, 212n, 218, 522
bonfires, 30–1, 33
Bonnard, Pierre, 441

Book of Hours, 369
books, 102n
Boorstin, Daniel, 329, 332, 371
Born, Max, 522
Borodin, Alexander, 67n
Borodino, battle of, 246
Bouillaud, Ismaël, 165
Bowen, Elizabeth, 494n
Boy scouts, 362n
Boyd, Brian, 476
Bradbury, Ray, 572–3
Bradley, James, 241
Brahe, Tycho, 144–52, 159n, 371
Branson, Richard, 555n, 578
Braudel, Fernand, 335–6
Braun, Werner von, 530–1
Brecht, Bertolt, 152, 158n, 159
Breyer, Franziska, 396–8
Brin, David, 571
brise-soleils, 460n
Brockton Point, Vancouver, 360
Brodsky, Josef, 494
Broecker, Wallace S., 547n
Brooke, Rupert, 245, 274
Brougham, Henry Peter, 271
Brown, Joseph Epes, 37–8
Brown, Dan
    The Da Vinci Code, 54, 222, 280n, 331n
Browning, Robert, 16, 487, 592
Bruno, Giordano, xv, 158
Bryan, William Jennings, 191–2
Bryson, Bill, 200, 295, 528
Buddha, 109n, 597
Budyko, Mikhail, 557n
Buffon, Georges-Louis de, 204–5, 388
Buisson, Henri, 548
Bulldog survey ship, 316
Bunsen, Robert, 244
Bunyan, John
    The Pilgrim's Progress, 24
Burgess, Anthony, 368, 375
Bürgi, Jost, 371
Burnet, Thomas, 204
Burney, Fanny, 179
Burton, Robert, 157n
Bush, George W., 498, 578
Byron, Lord, 23, 205, 239, 487–8, 592

Calatrava, Santiago, 55
Calder, Nigel, 560
calendars, 221, 345–58
    Aztec, 345–6
    Babylonian, 81, 345, 347
    Chinese, 107, 113, 120, 345–7
    Egyptian, 345–6, 349, 352
    French Revolutionary, 352n, 354
    Greek, 345, 350–2
    Gregorian, 123, 345–6, 350, 351n, 353–7
    Inca, 345, 350
    Indian, 131
    Islamic, 127, 348–9, 357
    Japanese, 347, 355
    Jewish, 345, 348–9, 352–3
    Julian, 123, 145, 346, 351–5
    Mayan, 346, 349–50
    Persian, 123, 346, 348
    Roman, 351–2
    Russian, 355
caliphs, 124–5
Callendar, Guy Stewart, 551
Calvert, Edward, 434
Calvin, Melvin, 298
Cambridge University, 163, 165–6, 171, 204, 212
Cameron, Julia Margaret, 450
Camillo, Giulio, 484
Campanella, Tommaso, 484–5
Campbell, John W., 570, 574
Campbell, Naomi, 270n
Camus, Albert, 275, 496
Canaletto, Giovanni, 431
Candlemas bear, 34
carbon capture and storage (CCS), 555n
cardinal points, 81, 107
Carlos I, King of Spain, 419
Caroline, Queen, 189
Carrington, Richard, 227–8, 237
Carson, Rachel, 316
Carter, Jimmy, 403
Casanovas, Fr Juan, 222–4, 238
Castel Gandolfo, 221–2
Cathcart, Brian, 522
cathedrals, 53–4
Catullus, 592
Caux, Salomon de, 389
celestial equator, 77
celestial harmony, 151
Cervantes, Miguel de, 355n
    Don Quixote, 82n, 334–5
Cézanne, Paul, 436
ch'i, 111–12
Ch'un-ch'iu, 63
Chadwick, James, 212
Chaldeans, 77–80, 130
Chaliapin, Feodor, 471

Chamberlin, Thomas, 206
Chandigarh, 459
Chandler, Raymond, 418, 420
Chanel, Coco, 266
chaos theory, 238n
Chaplin, Charlie, xvii
Charlemagne, Emperor, 69, 331
Charles I, King, 263n, 364
Charles II, King, 71, 163, 263, 338
Chartres Cathedral, 49
Chatwin, Bruce, 44
Chaucer, Geoffrey, 323, 360–1, 481, 483
Chávez, Hugo, 382
Chen Jie, 535n
Chernobyl disaster, 62n
Chesterton, G. K., 283
Chichén Itzá, 32, 48
Chikyu, 314n
China
    astronomy, 105–20
    calendar, 107, 113, 120, 345–7
    cosmology, 31–2, 111–12
    eclipses, 63–5, 110, 114, 118, 120
    and mirrors, 423
    mythology, 12
    and navigation, 117, 329, 332
    philosophers, 87
    and sunspots, 222, 224
choreia, 30n
Christian IV, King of Denmark, 149
Christian penance, 37
Christianity, 27–9, 103–4, 123, 438, 500,
    501n, 505, 516–17
    and astronomy, 132–6
    and calendars, 352–5
Christie, Agatha, 275
Christmas, 27–9, 352, 354, 517
Chromo-Luminism, 437
chromosphere, 525, 543
Chu, Miyoko, 306
Chung-K'ang, Emperor, 64n
Church of England, 172n, 203
Churchill, Winston, 216, 569, 593
Cicero, 88, 143, 566
circadian rhythms, 382n
Clampitt, Amy, 494n
Clapton, Eric, 472
Clark, Stuart, 228, 244, 388
Clarke, Arthur C., 60n, 401, 573, 575–80,
    587
Clarke, Thurston, 257
Claude, 427–9, 594
Claudius, Emperor, 65, 363

Clavius, Christopher, 186, 353, 358
Clegg, Brian, 153, 525n
Clement IV, Pope, 137, 353
Clement VIII, Pope, 221
Clinton, Bill, 554, 563
clocks, 163, 361, 363–4, 369–72, 374
    astronomical, 115
    atomic, 378, 382–4
    caesium, 379n, 383
    chiming, 118, 363
    clepsydras, 359n, 361
    flower, 303–5
    French Revolutionary, 352n
    incense, 361
    'ion-trap', 384
    merkhet, 361
    and minute hands, 371
    and navigation, 341, 343
    notched-candle, 361
    pendulum, 364
    perpetual, 383–4
    sandglasses, 361
    water, 83, 361
    see also watches
clouds, 174, 254, 550
Cockcroft, John, 212
Codrington, Admiral, 372
Cohn, Roy, 236
Coleridge, Samuel Taylor, 179, 487
Colette, 494n
Collier, Richard, 259
Collins, Gail, 382
Colossus of Rhodes, 99n
colour, 166–70, 243–5
Columbus, Christopher, 140, 187–8, 331,
    338n, 421
Columella, 28
comets, 61, 107, 128, 148, 534, 537n,
    541, 545, 587
    Halley's Comet, 110, 171
computers, 81, 543
Comte, Auguste, 201, 244
Conan Doyle, Sir Arthur, xvii, 373
Congress of Vienna, 225n
Conquistadors, 39
Constable, John, 430, 437
Constantine, Emperor, 27, 28n
Constantinople, 132–3
constellations, 83–4, 110–11, 480
    Aquarius, 504
    Aries, 482
    Cassiopeia, 146

constellations – *continued*
  Gemini, 482
  Great Bear, 111
  Pegasus, 586
  Plough, 110
contact lenses, 283n
Cook, Captain James, 57, 176, 184–5, 296, 343
Copernicus, Nicolaus, 81, 91, 103, 123, 129–30, 141–5, 147, 151–2, 161, 174, 335
  and Catholic Church, 142–3, 157–8, 160
Copley, John Singleton, 594
Corbyn, Piers, 558, 561
Cordoba, 133–4
Coriolis Force, 201, 322
Cornelius, Pope, 45
Corot, Jean-Baptiste-Camille, 433
Cortés, Hernán, 47, 188, 422
Cosiguina volcano, 594
cosmic rays, 230, 545
Cottingham, E. T., 197–8
Council of Tours, 28
Coward, Noël, 258–9, 266
Cowley, Malcolm, 492
Cozens, Alexander, 594
Cradock, Admiral Christopher, 247
Crane, Hart, 493
Cromwell, Oliver, 71
Crosby, Harry, 489, 491–3
Crowley, Aleister, 504
Crump, Thomas, 72, 198
Crusaders, 129, 387
Crutzen, Paul, 549
Cullen, Dolores, 481
Cumont, Franz, 104
Curie, Irène, 212
Curie, Marie and Pierre, 207, 212
Curzon, Lord, 16
Cuzco, 20, 48, 350, 606
Cyrus II of Persia, 77, 116n

Daguerre, Louis, 448–9, 453
Dancing Stones of Namoratunga, 40–1
Danielson, Dennis, 144n
Dante, 160, 482
Darius, King, 67
dark matter, 200
Darrow, Clarence, 192n
Darwin, Charles, 89, 105, 142n, 144, 192n, 203, 205–6, 500, 584
Darwin, Sir George, 324–5, 336n

Davidar, David, 270n
Davis, Raymond, 527
Davis, Wade, 38
Davy, Humphry, 300
D-Day, 237n
De festo sancti Johannis Baptistae, 30
Deakin, Roger, 302n
Debussy, Claude, 470
Defence of the Realm Act (DORA), 368n
Defoe, Daniel, 566
  Robinson Crusoe, 252–3, 382
Degas, Edgar, 594
degrees, division of, 81
Delacroix, Eugène, 440
Delaney, Frank
  Ireland, 44–5
Delius, Frederick, 470
Democritus, 179
dendrochronology, 230–1, 233–4, 237–8
Deng Xiaoping, 515
Desanas tribe, 417
Descartes, René, 164, 168, 372
Desert Fathers, 359
Diamandis, Peter, 538
diamonds, 243, 300
Dickens, Charles, 439, 487
  Dombey and Son, 373
  Our Mutual Friend, 432n
Dickinson, Emily, 487–8
Dicks, D. R., 88
Dinesen, Thomas, 375n
Dingle peninsula, 185n
Diocletian, Emperor, 344
Diogenes Laertius, 88
Dionysius Exiguus, 344
dioramas, 448
diseases and conditions
  cardiovascular disease, 284
  cataracts, 260
  cholinergic urticaria, 260
  diabetes, 284
  HIV, 279
  homopolar ringworm, 261
  hypertension, 284
  influenza, 273, 284
  leprosy, 288
  miliaria, 160
  photoallergy, 261
  phototoxicity, 261
  porphyria, 286–8, 438
  premenstrual tension, 284
  psoriasis, 288–90, 476
  rheumatiod arthritis, 284

rickets, 285
seasonal affective disorder (SAD), 290–3
skin cancers, 277–81
smallpox, 601
solar eczema, 261
solar retinopathy, 283
tropical acne, 260
tuberculosis, 271–3, 548
xanthopsia, 438
see also skin cancers
Dixon, Jeremiah, 175
Dolphin, David, 287–8
Dondi, Gioivanni de', 370
Donne, John, 11, 358, 485–6
Donovan, 416
Douglass, Andrew, 230–1, 233
Dresel, Tom, 396–7
drugs, 416–17, 438
Dryden, John, 338
du Maurier, Daphne, 58n, 266
duels, 360n
Duke of York, 364
Dumas, Alexandre
   The Man in the Iron Mask, 45–6
Durant, Will, 89, 99
Durrell Lawrence, 275
dusk, 253
Dvořák, Antonín, 470
Dyer, Geoff, 605
Dylan, Bob, 247n
Dyson, Sir Frank, 197

Earth
   age, 132, 202–6, 210
   axial oscillations, 180, 234
   axial tilt, 26, 92, 234, 364, 561
   core, 208n, 325
   distance from Sun, xv–xvi, 102, 174–5, 177, 179
   elliptical motion, 364
   magnetic field, 164, 208n, 226–7, 233, 306, 417, 535, 561
   orbital variation, 26, 234
   precession, 49
   rotation, 321–2, 364, 378, 382–3, 567
   shape, 81, 90, 139–40
   temperature, 230, 235, 585
   and tidal friction, 61
earthquakes, 12, 47, 113, 132–3, 325n, 545
Easter Island, 44
Easter, 346, 352–3, 357, 517

Eastern Orthodox Church, 353, 355
Eberhart, Richard, 494n
Eck, Diana, 602
Eclipse Island, 185
eclipses, 56, 60–72, 79–80, 83, 88, 92, 107, 128, 137, 145, 181–98, 200
   Antarctic, 181–4
   and Baily's Beads, 186
   and Chinese, 63–5, 110, 114, 118, 120
   and eye damage, 283–4
   and Indians, 61–2, 131–2
   in literature, 187–90, 193–5
   in painting, 427
ecliptic, 60, 77, 92, 109, 111, 128
Eco, Umberto, 344
Eddington, Arthur, 196–8, 209–10
Eddy, John A., xix, 233–5, 237–8
Edison, Thomas, 408, 454, 456n
Edward the Confessor, King, 280n
Edward IV, King, 412–13
Edward VII, King, 343
Edward VIII, King, 274, 423n
Egyptians
   astronomy, 82–5
   calendar, 345–6, 349, 352
   measurement of time, 358, 361
   mythology, 12, 14, 18, 20
Eighth Army, 258
Einhard, 224
Einstein, Albert, 163, 196–8, 213, 216n, 218, 242–3, 374, 383
   and relativity, 521–2
Eisenhower, Dwight D., 533
El Dorado, 422
El Niño, 322
electric telegraphy, 237
electricity, and magnetism, 242
electromagnetic induction, 200
elements and compounds
   ammonia, 296, 391
   argon, 527
   arsenic, 262–3, 394
   barium, 244
   boron, 394, 590
   caesium-133, 379n, 383
   calcium, 244n, 583
   calcium hypochlorite, 263n
   carbon, 295, 584, 590
   carbon-14, 234
   carbon dioxide, 295–6, 298, 313, 315, 403, 551–3, 555–7, 559, 563, 585
   carbon monoxide, 296
   carbonic acid, 297n

elements and compounds – *continued*
  chlorine, 296, 527, 549
  chlorofluorocarbons (CFCs), 548–50,
    553, 556
  chromium, 244n
  copper, 244n, 313
  cordierite, 333
  deuterium, 200, 218
  gold, 420–3
  helium, 75, 196, 211–12, 406, 524,
    583–4, 586, 590
  hydrogen, xv, 211, 216, 228, 298n,
    299, 314–15, 406, 582–3, 589
  hydrogen chloride, 296
  hydrogen sulphide, 296, 314
  iodine, 319n
  iron, 208n, 244n, 313–14, 339n, 583
  lead, 263–4
  lithium deuteride, 218
  magnesium, 244n, 584
  magnetite, 314, 332
  mercury, 174, 263
  methane, 313, 556
  neon, 245n
  nickel, 208n, 244n
  nitrates, 313
  nitrogen, 58, 297, 590
  nitrous oxide, 296, 555
  oxygen, 58, 295–7, 298n, 314–15, 317,
    319, 403, 553n, 584, 590
  ozone, 530, 547–8, 556
  phosphates, 313
  pitchblende, 207
  radium, 207, 209
  selenium, 394
  silicon, 394, 584
  silicon tetrafluoride, 296
  silver, 424
  sodium, 243, 244n, 583
  strontium, 215, 244
  sulphates, 313
  sulphur, 314, 403n, 550
  sulphur dioxide, 296, 557n
  tetrachloroethylene, 528
  thorium, 208–9
  tritium, 218
  uranium, 207, 212–16
  xenon, 215
  zinc, 244n, 313
Elgar, Sir Edward, 470
Eliade, Mircea, 20, 417
Eliasson, Olafur, 443n
Eliot, T. S., 493

Elizabeth I, Queen, 164, 263, 354, 420
Elliot, Dr, 232n
Emerson, Ralph Waldo, 487, 574, 592
Emmett, Rowland, 386n
Empedocles, 240
Eneas, Aubrey, 391–2
energetic protons, 230
'ephemerides', 80
Epicurus, 87
equator, 25, 257, 331
equinoxes, 25–6, 34, 87, 330, 356, 567
  precession of, 166
  vernal equinox, 26, 54–5, 77, 111, 345,
    354
Ernst, Max, 440
'eros', 12
Eskimos, 13, 476
ether, 97, 242
Ethiopians, 345
Euclid, 87, 103, 122, 134, 136
Eudoxus of Cnidus, 87, 95–7
eunuchs, 119–20
Euripides, 480n
EuroDisney, 269
European Space Agency, 542n
evolution, 206, 310
Externsteine rocks, 507
eyes, 243n, 254, 443, 454
  of birds, 306–7
  and the Sun, 283–4
Eyjafjallajökull, 2, 238

Fabricius (Johann Goldsmid), 157, 222
Fabry, Charles, 548
'faint sun' paradox, 557
fairy stories, 29, 420
famines, 80, 132–3
Fatima, 416
Faulkner, William, 374n, 494n
Fawkes, Guy, 33n
Fechner, Gustav, 283
Fellini, Federico, 399
fencing, 248–9
Fenimore Cooper, James, 193, 568
Fenton, Roger, 450, 453
Ferdinand, Archduke, 150
Ferguson, Robert, 332
Fermi, Enrico, 212, 524, 528n
Ferris, Timothy, 101, 200, 206, 334
film-making, 453–7
films
  8½, 399
  Bananas, 382

Chitty Chitty Bang Bang, 386n
Cool Hand Luke, 457
Easy Rider, 457
The Falcon and the Snowman, 539
A Fistful of Dollars, 399
For a Few Dollars More, 399
From Here to Eternity, 270
The Good, the Bad and the Ugly, 399
The Great Dictator, xvii
Greed, 457
High Noon, 360n
Indiana Jones and the Last Crusade, 399
Lawrence of Arabia, 399
The Lives of Others, 498–9n
The Magnificent Seven, 399
A Man Called Horse, 37
The Man with the Golden Gun, 385
Manhattan, 457
Olympia, 457
On a Clear Day You Can See Forever, 305n
The Passenger, 457
Patton, 399
Ramona, 455–6
Rashômon, 457
South Pacific, 419
Star Wars, 570n
The Sun, 512n
Sunrise, 457
Sunshine, 570n
Walkabout, 457
The Wind and the Lion, 399
The Wizard of Oz, 455
Finsen, Niels, 272
Fisher, Kitty, 263–4
Fitzgerald, Edward, 122, 198
Fitzgerald, Scott, 266, 492
Fizeau, Armand-Hippolyte-Louis, 228n, 242
flags, 411–13
Flammarion, Camille, 569
Fleming, Ian, 236–7
   The Man with the Golden Gun, 385–6
Fleming, Sandford, 376
Fletcher, Valerie, 436
floods, 132–3
Follini, Stefania, 285–6n
Ford, Eileen, 281
Ford, Henry, 408
fossils, 204–5, 300
Foucault, Léon, 228n, 242
Fourier, Jean-Baptiste, 551
Fowles, John, 270–1, 274

Frank, Anne, 411
Frank, Robert, 451
Franklin, Benjamin, 379
Franz Ferdinand, Archduke, 392
Fraunhofer, Josef, 243
Frayn, Michael, 218
freckles, 265–6
Frederick the Great, 372
Frederick II, King of Denmark, 147, 149
'free planets', 582
Freiburg, 396–8, 400
Fremasonry, 469
French Revolution, 34–5, 81n, 205, 236, 289, 425, 500
   and calendar, 352n, 354
French Riviera, 266, 271
Frenk, Carlos, 544
Freud, Sigmund, 161, 422
Friedrich, Caspar David, 433, 594
Frisch, Karl von, 307
Frisch, Otto Robert, 213–14
Froehlich, Jack, 533n
Frost, Robert, 592
Frost, Sir Terry, 419
'fruit walls', 390
Fuh-hi, King, 106–7
Fussell, Paul, 265, 492

Gagarin, Yuri Alekseyevich, 532n
Gaia Theory, 310–11
Gainsborough, Thomas, 433, 594
galaxies
   Andromeda galaxy, 202
   formation of, 543
   Sombrero galaxy, 537
Galen, 134
Galileans, 155
Galileo, xvi, 70, 144, 152–3, 155–62, 164–5, 177, 324, 341, 486
   and speed of light, 240–1
   and sunspots, 155–6, 222, 224–5, 230
Galileo probe, 537
Gamow, George, 210–12, 393, 522
Gandhi, Indira, 596
Gandhi, Mohandas, 504
Garbo, Greta, 266
Garfunkel, Art, 289
Garland, Alex, 570n
Gautreau, Amélie, 265
Gauvain, Sir Henry, 273
Gehry, Frank, 460–1
Gell-Mann, Murray, 524n
Genghis Khan, 129

George II, King, 263
George III, King, 178, 288n
George V, King, 423n
geothermal energy, 314n, 386
Germanic mythology, 13
Germany
    nudism, 273–4, 505
    solar power, 396–8, 400, 404, 406
    Völkisch movement, 505–6, 508
    see also Nazis
Gibbon, Edward, 105, 123
Gide, André, 496
Gilbert, William, 164
Gilliss, Lieutenant, 190n
Gingerich, Owen, 152, 171n
Ginsberg, Allen, 592
Giordano, Umberto, 469
Gish, Lillian, 455
Giza, 40, 50
Gladstone, William Ewart, 16
Gleick, James, 174
Gleissberg cycle, 239
global warming, xix–xx, 235n, 239, 390,
        401, 403n, 547, 550–64
globes, 114, 128, 145, 337
'gnomonics', 361
gnomons, 54, 81, 83, 91, 110, 119–20,
        332, 567
    see also sundials
Goddard, Robert, 529
Godfrey, Thomas, 339
Goethe, Johann Wolfgang von, 170n,
        430, 469, 506–7
gold, 420–3
Goldberg, Leo, 531
Golding, William, 494n
Goldwyn, Sam, 366
Gorbachev, Mikhail, 498
Gore, Al, 305, 555, 557
Gorst, Martin, 203
Gould, Stephen Jay, 194
Gounod, Charles, 466n
Grafton, Augustus Henry Fitzroy, Duke
        of, 264
Grahame, Kenneth
    The Wind in the Willows, 253, 494n
Grann, David, 317
Grant, Cary, 267–8
gravity, 97, 140, 163, 165–6, 171, 200,
        323, 486
    and equivalence principle, 383
    and light, 196–8
Great Plains megaliths, 40

Greeks
    astronomy, 86–101
    calendar, 345, 350–2
    and hair colour, 420
    and measurement of time, 361, 364
    mythology and epics, 15, 245n, 261
    navigation, 329–31
    philosophers, 87, 163
    statuary, 262n
Green, Charles, 176
Green, Henry, 275
Green, Martin, 273–4
green flash, 59n
Gregory XIII, Pope, 221, 353–4
Gregory of Tours, 224
Griffith, D. W., 455–6
Grose, Francis, 31
Grotto del Cane, 297n
Groundhog Day, 34
Groves, General Leslie R., 215
Gu Yanwu, 109
Guesnerie, Roger, 238n
Gulf Stream, 560
Gunn, Thom, 494n
Gunpowder Plot, 33
Gunther, John, 510
Gurdjieff, George Ivanovich, 503
Gutenberg, Johannes, 141
Guy, Thomas, 203
Guzman, Abimael, 289
Gyldenstierne, Knud, 149n

Haddon, Mark, 201n
Hadley, James, 178
Hadley, John, 339
Haggard, H. Rider, 187–8
Hahn, Otto, 213
hair, human, 420–1
Hakluyt, Richard, 337
Hale, George Ellery, 228–9, 239
Hales, Stephen, 297
Hall, Conrad, 457
Halley, Edmond, 72, 165, 171, 175, 204
haloes, 28
Hamilton, James, 432, 433n
Hamilton-Paterson, James, 266n, 316
Hammurabi, King, 76
Hansen, James, 554, 556
Hansen, Truls Lynne, 291
Hanukkah lights, 31
Hanway, Jonas, 253n
Harakhty, 12
Harding, Warren G., 381

Hardy, G. H., 247–8
Hardy, Thomas, 33, 487, 593
Hari, Mata, 417
Harkhebi, 84
Harkness, William, 177
Harriot, Thomas, 157, 222, 283
Harris, Richard, 36
Harrison, George, 472
Harrison, John, 341, 343
Hartley, L. P., 494n
Hawkes, Jacquetta, 43
Hawking, Stephen, 200, 203n, 527
Hay, Harold R., 405
Haydn, Joseph, 471
Hazlitt, William, 248, 358, 365
Heaney, Seamus, 418
Heath, Edward, 499n
Hebrew language, 134
Hebrews, 130
Heezen, Bruce, 313
Hegel, Georg Wilhelm Friedrich, 253, 488
Heinlein, Robert, 570–1
Heisenberg, Werner, 218, 522
Heliodorus, 80
'heliography', 448
'heliopause', 536
Helios, 15, 488
helioseismology, 473
heliostats, 399
heliotherapy, 273, 459
Hellenistic philosophers, 99
Helmholtz, Hermann von, 206
henges, 40n
Henlein, Peter, 371
Henri IV, King, 225n
Henry I, King, 67
Henry II, King, 224
Henry VIII, King, 143, 230
Hepburn, Audrey, 266
Hepburn, Katharine, 266
Heracleides of Pontus, 98
Heraclitus, 87, 89
Herbert, Alan, 275
Hergé (Georges Prosper Remi), 189
Herod, King, 68
Herodotus, 25, 50, 64, 67, 77, 79, 82
Herrick, Robert, 483
Herschel, John, 180, 201–2, 210, 388–9
    and photography, 449–50
Herschel, William, xvi, 178–80, 202, 232, 243, 567
Hersey, John, 216n

Hesiod, 25, 86–7
Hesse, Hermann, 274
Higgins, Godfrey, 500
High, Philip E., 571
Hilton-Barber, Miles, 358n
Himmler, Heinrich, 508–9
Hindus, 37, 47, 61–2, 113, 130, 251, 345, 597, 600n
Hippalus, 331
Hipparchus of Nicaea, 64, 99–102, 104, 110
Hippocrates, 288, 292
Hirohito, Emperor, 510–12
Historia Anglorum, 43
Hitler, Adolf, xvii, 420, 507, 510
HMS Challenger, 316–17
HMS Resolution, 343
Hobbes, Thomas, 160
Hockney, David, 444–6, 452
Hodgson, William Hope, 569
Hofmann, Albert, 416
Hogarth, William, 594
Holbach, Paul Henri Dietrich d', 204
Holden, Anthony, 245n
Hölderlin, Friedrich, 494n
Holick, Michael, 284n
holidays, 265–8, 270
Holler, Clyde, 38
Holmes, Edmond, 487
Holmes, Richard, xvii
Holst, Gustav, 471
Homer (Iliad and Odyssey), 15, 25, 67n, 86, 111, 329, 368, 420, 479–81, 483, 566
Honecker, Erich, 498–9n
Hooke, Robert, 165, 170–2, 179
Hopman, Harry, 248
Hopper, Dennis, 457
Horrocks, Jeremiah, 165, 175
horse bones, 116n
horsepower, 390n
Horus, 466
hot climates, 256–61
House of Lords, 189
'House of the Rising Sun', 472n
Housman, A. E., 202, 488
Hsi and Ho, 64n
Huayna Capac, 15
Hubble, Edwin, 202
Hughes, Ted, 417, 484
Hugo, Victor
    The Hunchback of Notre Dame, 421
'Huitzilopochtli', 12

Hulāgū Khan, 129
human sacrifice, 32
Humboldt, Alexander von, 227
hurricanes, 25, 228
Hutton, Ronald, 28, 43
Huygens, Christopher, 171, 241
hydrogen bombs, 216, 218
hydrothermal vents, 313–14, 317n
Hypatia, Lady, 132
Hyperion, 488

I Ching, 224
Ibsen, Henrik, 501
Icarus, 577n
ice ages, 203, 226, 234–5, 553n, 559–60
Ilyas, Mohammad, 127
Impressionists, 435–6
Inca, 15, 18, 20–3, 47, 302, 345, 350, 421
India
    Anglo-Indians, 252, 258–60
    astronomy, 130–2
    dark laws, 250–1
    eclipses, 61–2, 131–2
    flag, 412
    partition, 460
    'Snow White Syndrome', 270n
    solar power, 390–1, 402
    solar temples, 46–7
Indian Mutiny, 259–60
Industrial Revolution, 200, 238n, 265,
        359, 388–90, 551
infrared radiation, 179
Ingenhousz, Jan, 298
Innocent IV, Pope, 115
instruments
    actinometer, 388
    armillary sphere, 114, 128
    astrolabe, 114, 119–20, 127–8, 334–5,
        481
    camera lucida, 177
    camera obscura, 137, 177, 447, 449
    cosmograph, 110–11
    gas analyser, 552n
    heliospectroscope, 200
    heliotropon, 92
    kamal, 331
    magnetic compass, 111, 117, 137, 236,
        331–3, 335, 340–1
    magnifying glass, 153n, 155
    orrery, 335
    quadrant, 128, 145, 335, 339
    seismoscope, 113
    sextant, 145, 174, 335, 339

spectroheliograph, 228, 533
spectroscope, 244
star spectroscope, 200
telespectroscope, 200
ultraviolet spectrograph, 530, 533
ultraviolet spectroheliometer, 533
see also telescopes
Intergovernmental Panel on Climate
        Change (IPCC), 554–6, 557n, 562–3
International Date Line, 345
Inti, 11, 20–3
Inuit, 13, 58
Ionesco, Eugene, 559n
ionosphere, 201, 531
Irving, Henry, 286
Isadore of Seville, 139
Islam, 19, 123–4, 126–7, 129–30
    and calendar, 127, 348–9, 357
Island of the Sun, 48
Isler, Martin, 52
Isocrates, 95

Jabir, 122
Jackson, Kevin, 369
Jackson, Michael, 270
Jainism, 131
James I, King, 70
James, Henry, 102n, 254, 568
James, William, 578
Janá ek, Leoš, 469
Janssen, Pierre-Jules-César, 196, 378
Jantar Mantar observatory, 49
Japan
    and atomic bombs, 216
    calendar, 347, 355
    and eclipses, 64, 66
    Edo Period, 395n
    flag, 412, 509
    and mirrors, 423–4
    Nippon (the name), 418, 424
    post-war, 510–13
    seasonal celebrations, 30–1
    solar power, 394–5, 404
    women's movement, 513n
Jardine, Lisa, 371
Jarnot, Lisa, 494n
Jefferson, Thomas, 364, 500
Jesuits, 37, 116–17, 143, 157, 486
Jesus Christ, 344–5, 508n, 516–17
Jet Propulsion Laboratory, 529, 531–2,
        533n, 536n, 538
Jews, 77, 122, 253, 263n, 416, 508n, 521
    and calendar, 345, 348–9, 352–3

Jhabvala, Ruth Prawer, 258
Jimmu, Emperor, 246, 512
John, Elton, 183, 289
John of Worcester, 224
John Paul II, Pope, 160
Johnson, Samuel, 7, 50, 120, 233n
Johnston, Edna, 480
Joliot, Frédéric, 212
Jones, James, 270
Jones, R. V., 213n
Josephus, 68
Joyce, James, 450, 493, 504
    Finnegans Wake, 524n
    Ulysses, 368, 421, 561
Julius Caesar, 28, 67n, 102n, 323, 344,
    351, 364, 556
Jung, Carl, 19–20, 255, 507–8, 592
Jungk, Robert, 209
Jupiter, 57, 141, 143, 145, 537, 569, 585,
    587, 588
    moons of, 155–6, 341, 537, 590
Justin Martyr, 28
Justinian, Emperor, 387, 404
Jyotisa Vedānga, 130

Kaaba, 126
Kabīr, 601
Kaku, Michio, 242, 589
kamikaze pilots, 510
Kan Tê, 110
Kandinsky, Wassily, 440
Kaner, H., 570
Kant, Immanuel, 180
Kao Tsu, Empress, 65
Kaplan, Raymond, 236
Kardashev, Nikolai, 407
Kármán, Theodore von, 529n
Katrina and the Waves, 472
Keats, John, 169, 186, 487
Keeling, Charles, 552
Keillor, Garrison, 494n
Kelvin, Lord (Sir William Thomson),
    324–5
Kemp, Clarence, 390
Kennedy, John F., 268–9
Kentridge, William, 468
Kepler, Johannes, 68, 144, 149–52, 155n,
    159, 174, 186, 324, 447, 566
Kerouac, Jack, 451
Keynes, John Maynard, 174n, 422
Khan, M. A. R., 124
Khayyám, Omar, 122–3, 198
Khepri, 12

Khrushchev, Nikita, 498
King, David, 126
King, Edward, 70
Kingsmill, Hugh, 572
Kinsey, Alfred, 504
Kipling, Rudyard, 258–9, 568
Kircher, Fr Athanasius, 338, 448
Kirchhoff, Gustav, 242, 244
Klee, Paul, 440
Klepper, Gernot, 555
Klimt, Gustav, 594
Knowles, Beyoncé, 270n
Ko Hung, 170n
Koch, Robert, 271
Koestler, Arthur, 121, 123, 138, 144
Kohl, Helmut, 397
Kojiki, 246
Kolbert, Elizabeth, 554–5, 560
Konso people, 358
Konungs Skuggsja, 332
Koontz, Dean R., 286n
Kooser, Ted, 177
Korycansky, Don, 587
Krakatoa, 2
Kryza, Frank, 387–8
Kublai Khan, 115
Kuhn, Thomas, 298n
Kuiper, G. P., 544
Kunzig, Robert, 320, 564
Kurosawa, Akira, 457

La Niña, 322
Lacock Abbey, 449
Lafayette, Marquis de, 364
Lalande, Joseph, 176, 201
Lamb, H. H., 553n
Land, Edwin, 451
Landig, Wilhelm, 509n
Laplace, Pierre-Simon, 180, 324
Larkin, Philip, 494n
Larsa, 76
Las Vegas 'Sky Beam', 245
Latham, Philip, 570
latitude, 83, 330, 332–3, 340
Laughlin, Gregory, 587
Lavoisier, Antoine, 298n
Lawrence, D. H., xix, 489–93, 496
Lawrence, T. E., 66, 374n
Le Corbusier, 273, 458–60
Le Gray, Gustave, 452
Le Roy Ladurie, Emmanuel, 33
Leadbetter, Charles, 184
Lee, Christopher, 385

Lee, Harper
  To Kill a Mockingbird, 280n
Leigh Hunt, James Henry, xvii
Lenin, Vladimir, 355
Leo the Great, Pope, 29, 53
Leo X, Pope, 142
Leo XIII, Pope, 222
Leo, Alan, 505
Leonardo da Vinci, 105, 144n, 153, 302, 388
Leone, Sergio, 399
Lessing, Doris, 494n
Letterman, David, 386n
Lévi-Strauss, Claude, 36
Lewis, C. S., 140n, 581
Lhuyd, Edward, 204
Li Ki (Book of Rites), 109
Li Youyuan, 514
Liberius, Bishop of Rome, 27
Liebig, Justus von, 424
Liewer, Paulett, 541
light, 166–70
  and gravity, 196–8
  speed of, 174, 240–3, 525n
  wave and particle nature, 171, 241
light years, 179n
lighthouses, 336
Lilius, Aloysius, 353
Linebarger, Paul, 571
Linnaeus, Carolus, 303–4
Lippershey, Hans, 153
Liu Shaoqi, 513
Livy, 102n
Lockyer, Sir Norman, 75, 85, 196
lodestones, 332
London Bridge, 323
Longfellow, Henry Wadsworth, 249, 592
longitude, 330, 338, 340–1
Loos, Anita, 266
Lopez, Barry, 24
Lopez, Jennifer, 270
Lorenz, Konrad, 307
Los Angeles, 391, 393, 444–5, 455–6, 460–1, 538
Louis the Pious, King, 69
Louis IX, King, 337
Louis XIV, King, 230, 371–2, 388, 413–15, 425, 499
Lovelock, James, 310–11
Lu Gwei-djen, 106
Lu Hao Dong, 515n
Lucian of Samosata, 565–6
Lumière, Louis and Auguste, 454

Luminists, 433–4, 436n
lunar bows, 168n
Luther, Martin, 143
Lydians, 67
Lyell, Charles, 205
Lyme Regis, 270–1, 274
Lynas, Mark, 555

Maalouf, Amin, 129n
Macao, 117
MacArthur, General Douglas, 257, 510–11, 512n
Macaulay, Lord, 16, 488
McClure, Ellen, 225n
MacDonald, John D., 572
MacNiece, Louis, 494n
McEvoy, James, 82
Machiavelli, Niccolò, 65
Machu Picchu, xvii, 20, 48
Mackerras, Charles, 466
McKinley, William, 191
MacNeice, 494n
McQueen, Steve, 248n
Macrobius, 27
Maddox, John, 528
magazine publishing, 270n
Magellan, Ferdinand, 333
magic lanterns, 448
magnetic storms, 532
magnetism, and electricity, 242
Mahabharata, 597
Mair, Victor H., 116n
Mairan, Jean-Jacques Dortous de, 205
Mallarmé, Stéphane, 488
Malory, Thomas, 483
Manchester, William, 140
Mandelstam, Osip, 493n
Manhattan Project, 215–16
Mann, William, 469
Mantel, Hilary, 412
Mao Zedong, 512–14
Maori, 132n
maps, 337–8
Marat, Jean-Paul, 289
Marcus Aurelius, Emperor, 28
Marey, Étienne-Jules, 454
Maria, Countess of Coventry, 263–4
Marley, Bob, 472
Marlowe, Christopher, 130
Mars, 97, 141, 143, 235, 299n, 310, 366, 481, 572, 578, 590
Marseilles Building, 459–60
'Martin Dessil', 53n

Martinez Soler, José, 400
Marvell, Andrew, 304, 483
Marx Brothers, 139
Mary Queen of Scots, 288n
Mascagni, Pietro, 469
Mason, Charles, 175
Massenet, Jules, 183, 469
Massey, Gerald, 505
Mastin, John, 570
mathematics, 80–1, 83, 89, 94, 113, 115,
    132, 165, 364
  algebra, 82n, 94, 122
  Arabic, 121–2
  calculus, 163
  cubic equations, 123, 125
  and decimal base, 132
  geometry, 82n, 83, 88–9, 92, 103
  sine function, 122, 125
  trigonometry, 125, 132, 481
  trilineal figures, 125
  and zero, 122, 132
Matisse, Henri, 440–4, 446
Maunder, Edward Walter, 229, 233,
    234n, 235, 567
Maurice of Nassau, Prince, 153
Maxwell, James Clerk, 242
Maya, 18–19, 32, 47, 346, 349–50
Mayakovsky, Vladimir, 471
Mayer, Julius Robert, 298
Medes, 67, 77
Medicine Mountain, 47
Mehta, Ved, 358n
Meitner, Lise, 213
melanin, 261, 265, 278–80
Melchior-Bonnet, Sabine, 425
Melville, Herman, 568
  Moby-Dick, 279–80, 329, 339–40, 494n
Mencken, H. L., 158n
Menelaus, King, 330
Ménestrier, Fr Claude-François, 414
menhirs, 40n
Mercator, Ferardus, 337
Mercury, 57, 98, 134, 141, 143, 196, 224,
    585
meridians, invention of, 330
Mesopotamia, 15, 76–9
Messier, Charles, 202n
meteors, 61, 210
meteorites, 92, 545
Meton of Athens, 92–3, 351
Meyer, Tobias, 185
Miao people, 285n
Michelangelo, 427

Michelson, Albert, 242–3
Midgley, Thomas, 548
Mid-Oceanic Ridge, 315
midsummer, 30–1, 42
Mighall, Robert, 438
Milankovi , Milutin, 234–5
Milky Way, 155, 178–9, 202, 245
millennium, 344–5
Milton, John, 70–2, 592
  Areopagitica, 160
  Lycidas, 70
  'On the Morning of Christ's
      Nativity', 483
  Paradise Lost, 71, 160, 173, 483–4
  Samson Agonistes, 71
Milvian Bridge, battle of, 27
Minoan Crete, 262
mirages, 79
Miró, Joán, 440
mirrors, 386–7, 423–5
Mishra, Sudharka, 62, 602–3
Mithraism, 500–1n, 508, 517
Miyagawa, Kazuo, 457
mock suns, 60n, 182
Modhera solar temple, 46–7, 596
Molière (Jean-Baptiste Poquelin), 413
Molina, Mario, 549
Moltke, Field Marshal Helmuth von, 374
Monastersky, Richard, 559n
Monet, Claude, 435–6, 443
Mongols, 417
monsoon winds, 331
Montalbini, Maurizio, 286n
Montezuma, 48
Montezuma II, 421
Monty Python, 140n
Moon
  brightness, 90
  manned landing, 49, 225, 588
  mountains, 164, 200
  in mythology, 13–14
  orbit, 165–6
  phases, 114
  research outpost, 538
  size relative to Sun, 60–1
  and solar power, 403
  and tides, 323
moonlight, 114, 254
Moorcock, Michael, 280n
Moore, Patrick, 143
More, Sir George, 485–6
Morgan, Dan, 305
Morgan, J. P., 492

Morley, Edward W., 243
Morton, Oliver, 301, 311, 407, 553n
Mosley, Nicholas, 198n
Mouchot, Augustin, 389
Mount Cayambe, 257
Mount Fuji, 1–8
Mount Sumeru, 131
Mount Tambora, 238–9, 594
moving pictures, 453–7
Moyle, Dorothy Mary, 106
Mozart, Wolfgang Amadeus, xvii, 183, 467–9, 473
Mu, Emperor, 65
Muhammad, Prophet, 123–4, 126, 348
Müller, Friedrich Max, 15–17, 19, 479, 508
Müller, Johann, 141
Munch, Edvard, 445, 501–5
Mundas, 14
Murneau, F. W., 457
Murphy, Neil, 539–42, 544
music and song, 462–74
musicals
    Guys and Dolls, 466n
    Hair, 416
    South Pacific, 419
Musk, Elon, 538
Muslims, 80n, 124–7, 130, 263n, 348, 422n, 600n
Muybridge, Eadweard, 454
Myers, Beverly, 119n

Nabokov, Vladimir, 253, 475–9, 496–7
Nanahuatzin, 12
Napoleon Bonaparte, 50, 180, 225n, 246, 258, 429, 551
Napoleon III, 389
NASA, 533–4, 538–9
NASCAR racing, 247
Nash, Paul, 440
navigation, 329–43
Nazis, 39, 411, 504, 506–8
Nebuchadnezzar, King, 77
nebulae, 179–80, 202
Needham, Joseph, 64n, 105–7, 109, 111–12, 114–15, 117, 361
Nefertiti, 52
Neo-Impressionists, 436
Neptune, 471, 588
Nergal, 15
Nero, Emperor, 104, 108
Nesbit, E., 246, 488
Nestorians, 122

Neugebauer, Otto, 87, 131, 144
Neuserre, 51
neutrinos, 526–8
Neville, Anne, 67
Newfoundland, 184–5, 382
Newgrange, 44–5, 52
Newton, Sir Isaac, 125, 143, 144n, 162–72, 174, 180, 204, 241–2, 283, 297, 521
    and gravity, 163, 165–6, 196–7, 200, 486
    and telescopes, 163, 167, 169, 178
    and tides, 163, 166, 324
Nicholas IV, Pope, 137
Nicholas of Cusa, 141
Nicholson, Jack, 457
Nielsen, Carl, 470, 473
Niépce, Joseph Nicéphore, 300, 448
Nietzsche, Friedrich, 38, 103, 488, 500–8
Nile, River, 19, 25, 52, 82, 111, 349
Nineveh, 80n
ninja fiction, 250
Nippur, 80n
Niven, Larry, 572
Nixon, Richard M., 269, 305, 533
noctilucence, 254
Noll, Richard, 505, 508
North American Indians
    astronomy, 233
    and medicine wheels, 47
    mythology and religion, 13, 19–20, 61–2, 168, 250
    sundances, 35–8
    and Wovoka movement, 190–1
North, John, 136
North Poles, 334n, 559
northern jet stream, 234n
Norwich Cathedral, 370
nuclear bombs, 545, 587
    see also atomic bombs; hydrogen bombs
nuclear fission, 213–16, 401
nuclear fusion, 211–12, 216, 218, 401, 406, 523–4, 583–4
nuclear reactors, 47, 213
nudism, 273–4, 276, 505
Nuremberg Chronicles, 371
nutation, 26

Obama, Barack, 257, 261, 407, 587
oceans, 312–25, 564
    currents, 320–2, 551
    monster waves, 322
    tides, 163, 166, 322–5, 336n

Odeillo Font-Romeu solar furnace, 402
Odenwald, Sten, 202n
Oenopides of Chios, 92
O'Hara, Frank, 494n
O'Hara, John, 494n
Olav Haraldson, King of Norway, 333
Olbers' Paradox, 201n
Olympia, 262n
On the Philosophy of Mythology (Müller), 16
opera, 462–70
opera glasses, 156n
Oppenheimer, J. Robert, 215–16, 218
optics, 165–70
oracle bones, 107
Oran, 196n
Oresme, Nicole, 138
Orloff, Valerii, 236n
Orwell, George (Eric Blair), 376n, 494n
Ott, John Nash, 305n
Overbye, Dennis, 584
Oviedo, Gonzalo Fernández de, 422
Owen, Wilfred, 494n
Oxford University, 136–7
ozone layer, 281, 547–50, 553

painting, 426–46, 450, 594
Palace of Westminster, 372n
Paley, William, 372
Palin, Sarah, 282
Palmer, Samuel, 434–5
Pan, 261
pandemics, 546
parallax, 148, 175n
parapegma, 92
Paris, City of Light, 450–1
Paris Commune, 236n
Paris Observatory, 241, 383
Parker, Dorothy, 266
Parker, Eugene N., 534–6, 544–5
Parmenides, 87, 90
Pascal, Blaise, 151
Passement, Claude-Simenon, 383
pastoral writing, 494n
Patmore, Coventry, 487
Patterson, Timothy, 562
Paul III, Pope, 143
Paul V, Pope, 156
Pauli, Wolfgang, 522
Payne, Cecilia, 210–12
Pearl Harbor, 345, 530
Péguy, Charles, 246
Peierls, Rudolf, 214
pendulums, 125

Pepys, Samuel, 42, 72, 177
Periander, 421
Pericles, 65, 91–2
perihelion, 481
Perry, Judith, 292
Perry, William James, and Grafton Elliot
    Smith
    Children of the Sun, 17–19
Persians, 77, 91, 95, 113, 122, 130, 417,
    500n
Peter of Maricourt, 333
Peter the Great, 355
peyote, 38n
phantasmagoria, 448
Pherekydes of Syros, 87
Philander, George, 555
Philip II of Macedon, 95
Philip II, King of Spain, 263n
Philolaus, 90
philosopher's stone, 172–3
'philosophy' (the word), 89
Phoebus, 488
phoenix, 12, 488–9
photography, 447–54
photosphere, 540, 542, 583
photosynthesis, 295–300, 305n, 314–15,
    319, 585
phototropism, 300–2
photoperiodism, 303
Piaf, Édith, 466n
Piazzi, Giuseppe, 201
Picasso, Pablo, 443, 446
Pickford, Mary, 455
Pied Piper of Hamelin, 345
Pink Floyd, 7
pirates, 254, 336
Pissarro, Camille, 436
Pitman, Joanna, 420
plagues, 58, 80, 93, 132–3, 165
Planck, Max, 522–3
planets, 29, 81, 83, 87, 89, 94–7, 151,
    155, 166
    distances from Sun, 143, 151, 174–5
    elliptical orbits, 141, 143, 151, 155n,
        171
    GJ 1214b, 588
    retrograde motion, 95–6
    V 391 Pegasi, 585–6
plants, 295–305
    albino cactus, 279
    algae, 299, 305, 315, 319, 403
    begonias, 301
    birch trees, 301

plants – *continued*
  calendulas, 303
  chrysanthemums, 303
  corals, 300
  daisies, 301, 303
  dandelions, 303
  day lilies, 304
  evening primroses, 304
  flowering tobacco, 304
  four o'clocks, 304
  garden lettuce, 303
  gentians, 303
  giant aroid, 301
  goat's beard, 304
  Lecidea cancriformis, 302
  marigolds, 303
  moonflowers, 304
  morning glory, 303
  night-blooming cereus, 304
  peanuts, 302n
  phytoplankton, 315
  poppies, 302–4
  Rhizanthella gardneri, 302n
  seaweed, 319n
  shepherd's sundial, 303
  snow buttercup, 302
  sunflowers, 301–2, 487
  truffles, 302n
  wild roses, 303
plasma, 535, 539–40, 542, 584
Plateau, Joseph, 454
Plato, 79, 87, 88n, 93–5, 97–8, 104, 122,
    134, 166, 240, 484
Plautus, 363
playing cards, 115
Pliny (Elder and Younger), 28, 64, 67,
    88, 139–40, 262n, 387
Plutarch, 65, 88, 143, 566
Poe, Edgar Allan, 487, 568
Pohl, Frederik, and C. M. Kornbluth
    Wolfbane, 573, 575
polarized light, 306, 307n, 318, 333, 451
Polaroid camera, 451
pollution, 545
Polo, Marco, 115
Polynesians, 18–19, 329
Pompey, 351
Pope, Alexander, 170
Porco, Carolyn, 589
Porter, Cole, 266
Portuguese, 331, 335
Potter, Dennis, 289
Potter, Stephen, 268

Pound, Ezra, 494n
Poussin, Nicolas, 427
precession, 49, 100, 126, 166
Priestley, Joseph, 296–8
printing press, 141
Prokofiev, Sergei, 471
Protestants, 39, 142
Proust, Marcel, 494n
proverbs, 419
Ptolemy, 87, 101–4, 129–32, 134, 138,
    141, 145, 147, 166, 482, 484, 521
  Almagest, 64, 102–3, 125, 131, 134, 136,
    140, 142, 144
  Geography, 140, 337
  and navigation, 334–5, 337
Puccini, Giacomo, 464
Pudor, Heinrich, 505
Pushkin, Alexander, 464, 493n
pyramids, 39, 40, 51n, 79
  Egyptian, 44, 49–53, 83
Pythagoras of Samos, 87, 89–90, 93, 104

Qasiyaun, 125
qibla, 126
Qizilchqa, 116n
quantum theory, 242, 522–4
Quintilian, 488
Qur'an, 62, 80n, 124, 127, 348

Ra, 12, 14, 18, 51–3, 423
rabbit skins, 231–2n
radar, 530
radio broadcasts, 235–7
radio waves, 530–1
radioactivity, 207–9
radiology, 137
Raeburn, Henry, 433
Ragnarok, 583
Rahu, 61
Raicho, Hiratsuka, 513n
railways, 374–6
Rainbow Gatherings, 168n
rainbows, 137, 168–9, 486
  in art, 442–3
Raleigh, Sir Walter, 157
Ramadan, 357
Ramses II, Pharaoh, 52
Ravel, Maurice, 470
Rawlins, Dennis, 101
Ray, John, 204
Ray, Man, 493
reading glasses, 137
Reagan, Ronald, 403, 549

red dwarves, 245
red giants, xvi, 586
refraction, 166–9
Regener, Erich, 531
Reifenstahl, Leni, 457
Rembrandt, 427, 594
Renoir, Pierre-Auguste, 435, 441, 443–4
Revelle, Roger, 552, 563
Rheticus, Georg, 142n, 144n
Ricci, Matteo, 117–20
Richard I, King, 357n
Richard III, King, 67
Richelieu, Cardinal, 371
Rimbaud, Arthur, 494–5
Rimsky-Korsakov, Nikolay, 463–4
rockets, 115, 529, 538
    V-2 rockets, 530–1
Roeg, Nicolas, 457
Roepke, Gabriela, 462–4, 466–7, 469–71
Roget, Peter Mark, 454
Rollier, Auguste, 273
Roman Catholic Church, 134, 481–2,
    508n, 516–17
    and calendars, 353–5
    and Copernicus, 142–3, 157–8, 160
    and Galileo, 156–61
Roman roads, 246
Romans, 26–7, 43, 82, 130, 323
    astronomy, 101–4
    calendar, 351–2
    court proceedings, 359n
    and hair colour, 420
    imperial religion, 517
    and measurement of time, 359, 361
    and Mithraism, 500–1n
Romans (France) carnival, 33–4
Romantic movement, 168–9, 270, 433,
    438, 478, 487, 500
Rome, fall of, 132
Rømer, Ole, 241
Ronan, Colin, 102–3
Rooney, David, 382, 384
Roosevelt, Franklin D., 214, 423
Rosenkrantz, Frederik, 149n
Rosenthal, Norman, 290–1
Ross, Alex, 470
Ross, Sir John, 316
Rossetti, Dante Gabriel, 122
Rotch, Josephine, 493
Rothschild, Lord, 521
Rouault, Georges, 440
Rousseau, Jean-Jacques, 484
Rowland, F. Sherwood, 549

Royal Society, 163, 165, 169–70, 172,
    176, 185, 204, 232, 296, 339
Rubens, Peter Paul, 594
Rubinstein, Artur, 471
Rubinstein, Helena, 266
rudder, invention of, 117
Ruders, Poul, 473
Rush, Norman, 494n
Ruskin, John, 16, 428, 431
Russell, Bertrand, 489
Russell, Henry Norris, 211
Russia, 31, 266n, 355, 374, 420, 493n
Russian Revolution, 236n, 493n
Rutherford, Ernest, 207–9, 211, 523
Ruysdael, Jacob van, 427–8
Rwala Bedouin, 13

Sabine, Edward, 226
Sabinianus, Pope, 369
Sachs, Julius von, 300
Sacsahuaman, 48
Sagan, Carl, 200, 227, 544
sailing ships, 117
St Augustine of Hippo, 29, 135, 332, 369
St Benedict, 369, 427
St Bridget's Day, 34
St Francis Xavier, 117
St Gregory of Nazianzen, 29
St Helena, 175
St John Gogarty, Oliver, 367–8
St Patrick, 29
St Paul, 330
St Thomas Aquinas, 103
St Vitus, 30n
S'akuntala, 251
Salisbury, Lord, 70
Samarkand observatory, 128–30
sandstorms, 79
Saqsaywaman, 21
Sargent, John Singer, 265
Sarton, George, 122–3, 125
Sartre, Jean-Paul, 496
Sassoon, Sir Philip, 593
satellites, 155n, 305, 395, 403, 405, 515,
    532, 542, 578
Saturn, 141, 143, 145, 588
Saturnalia, 26–7
Sawai Jai Singh II, Maharaja, 49
Sayers, Dorothy L., 193
Schama, Simon, 505
Scheele, Carl Wilhelm, 298n
Scheiner, Christopher, 157, 222
Schikaneder, Johann Josef, 468–9

Schliemann, Heinrich, 506n
Schopenhauer, Arthur, 466, 504, 510
Schrödinger, Erwin, 522
Schwabe, Samuel Heinrich, 225–7
Schweitzer, Albert, 257
science fiction, 564–79
'scientist' (the word), 324n
Scopes, John T., 192n
Scott, Captain Robert Falcon, 57
Scott, Walter, 58
Scriabin, Alexander, 445, 503–4
seasons, 345–6, 356
    celebrations, 24–39, 346
Sebastian I, King of Portugal, 257
Seinei, Emperor, 280n
Seleucus, 82, 347
Sendero Luminoso, 499
Seneca, 58n, 363
Sennacherib, King, 77
Seth, 466
Seurat, Georges, 436–7
Seuss cycle, 239
Sevilla Portillo, Alfonso, 400–1
shadows, 249–51
Shaffer, Peter, 494n
Shakespeare, William, 355n, 371, 485, 592
    Antony and Cleopatra, 70n
    Cymbeline, 363
    Hamlet, 70n, 142n, 149n
    Henry VI, 70n, 365, 413n
    Julius Caesar, 70n, 363
    King Lear, 69–70, 249
    Macbeth, 70n
    The Merry Wives of Windsor, 418
    A Midsummer Night's Dream, 31
    Much Ado about Nothing, 262
    Othello, 70n
    'The Phoenix and the Turtle', 488
    Richard II, 363
    Richard III, 413n
    Romeo and Juliet, 418, 485
    Sonnets, 70, 263
    The Taming of the Shrew, 354–5
    Twelfth Night, 262, 430
    Two Gentlemen of Verona, 262
    The Winter's Tale, 303
Shaw, George Bernard, 521
Shelburne, Lord, 296
Shelley, Percy Bysshe, 487
Shen Kua, 332
Shepard, Sam, 247n
Sherriff, R. C., 253
Shih Shen, 110

Shovell, Clowdesley, 341
Shugar, Daniel, 406
Shuman, Frank, 392–3
Sibelius, Jean, 470
Sidey, Hugh, 268n
Sidis, William James, 236n
Sidney, Philip, 484
silver, 424
Simak, Clifford D., 570
Simloki (Soldier Mountain), 249–50
Simpson, O. J., 270n
Sims, Michael, 456
Sing-Bonga, 14
Sippar, 76
Sironó tribe, 345
Sistine Chapel, 427
Skene, James, 430
skin
    cancers, 277–81
    colour, 261–70
    photoageing, 281, 300n
sky, colour of, 245
Smith, Adam, 238n
Smith, Cordwainer, 571
Smith, Mark, 377
Smith, Willoughby, 394
Smith, Zadie, 124n
sneezing, 284n
snipers, 254
Snow, C. P., 213
Sobel, Dava, 143, 144n, 200, 341, 362,
    364
Socrates, 91, 93–4
Soddy, Frederick, 207–9
Sokurov, Aleksandr, 512n
Solanki, Sami, 562
solar constant, 200
solar corona, 186–7, 200, 525, 534–5
    coronal ejections, xix, 239, 536, 543,
        561
solar cycles, 239, 539, 559n
solar flares, 56, 183, 228, 230, 237, 239,
    560
solar gales, 536
solar magnetic clusters, 525
solar maxima, 237
Solar Orbiter, 542n
solar particles, xix, 560–1
solar plexus, 419
solar power, 385–407
    photovoltaic cells, 394–5, 400
    solar initiatives, 396–400
    solar thermal energy, 400

space-based solar power (SSP), 403
and transport, 401–2, 593
Solar Probe Plus, 538–9, 541
solar prominences, 186, 196, 200, 230, 544
solar proximities, and cultures, 440–1
solar radii, 539
solar spectrum, 174, 179, 200, 244
solar storms, 237, 239, 537
solar system, xvi, 175, 180, 583
solar winds, xix, 230, 525, 534, 536–7, 540–2, 561–2, 571
solstices, 25, 84, 87–8, 356
    summer solstice, 25–6, 29–33, 42, 53
    winter solstice, 25–6, 28–33, 42, 44, 53, 354, 517
Soon, Willie, 563
Sorrell, Martin, 495
Sosigenes of Alexandria, 351
Soviet Union, 216, 305, 322, 532
Spain
    astronomy, 133–4
    solar power, 398–400, 404
Spanish Armada, 354
Spee, Admiral Graf Maximilian von, 247
Spence, Jonathan, 335
Spencer, Herbert, 89
Spender, Stephen, 274
Spenser, Edmund, 483
Speusippus, 95
spherical coordinates, 90
spinning-wheels, 29
Spinoza, 169
Spooner, Rev. William Archibald, 280n
sports and games, 247–9
    American football, 247
    baseball, 247
    cribbage, 484n
    cricket, 247–8
    Persian 'Sun game', 417
    polo, 417
    tennis, 248
    ullamaliztli, 32
sportswear, 283n
Spurling, Hilary, 441, 443
Sputnik, 532, 533n
Sriji (Arvind Singh Mewar), 594–6
Stalin, Josef, 216, 289, 420
star catalogues, 110, 114, 201, 202n
star maps, 123
star tables, 130
stars, 83, 86, 94, 96, 100, 143, 151, 166, 179, 201–2, 209

circumpolar stars, 111n, 331
decans, 349
explosions of, 544
fixed stars, 126, 200
in Homer, 480
planets of, 174
Pole Star, 111
Sirius, 82, 111, 349
temperature of, 209, 584
variable stars, 211
Steadman, Ralph, 499n
steamships, 137
Steel, Duncan, 354
Steiner, Rudolf, 504, 508n
Stetson, H. True, 231–2n
Stevens, Wallace, 415, 494n
Stevenson, Robert Louis, 253, 266
Stoke, Roger, 369–70
Stoker, Bram, 286
Stonehenge, 39, 40, 42–5, 49, 52, 55, 350, 556, 606
Stoppard, Tom, 149n, 156n, 162, 528
Strabo of Amasia, 78
Strassmann, Fritz, 213
Strauss, Richard, 470, 502
Stravinsky, Igor, 470–1
street slang, 418
Streisand, Barbra, 305n
Strindberg, August, 502, 504
Stroheim, Erich von, 457
Strout, Elizabeth, 497
Sturt, Charles, 259
Styron, William, 190
submarines, 305, 318, 531
Suckling, John, 484
suicide bombers, 275n
suicides, 235
Sumerians, 76, 107, 174n
'Sun devils', 31
Sun
    age, 200, 202, 210
    apparent velocity, 97
    contraction, 584
    core, 179, 200, 210, 542, 584
    definition, xv
    diameter, xv, 83, 88, 98, 100, 175n, 584
    distance from Earth, xv–xvi, 102, 174–5, 177, 179
    gravitational field, 584, 586
    luminosity, xv, 209
    M regions, 535
    magnetic fields, 239, 525, 542, 561–2

Sun – *continued*
    magnetic poles, 525, 542
    mass, xv, 209
    mechanism, 206–13, 526, 523–4, 526
    origin, 163
    relative density, 165
    relative mass, 165
    rotation, 156, 223
    size relative to Moon, 60–1
    temperature, xvi, 200, 209–11, 584
Sun Royal, 76
Sun Yaoting, 119n
suncream, 267
Sunday, 28
sundials, 101, 163, 359–6, 378, 381
sunglasses, 267
sunlight
    aphrodisiac qualities, 274–6
    and disease, 284–90
    and health, 270–4
    and vitamins, 284–5
sunsets, 283, 592–4
sunspots, xix, 98, 109–10, 131, 155–7,
        161, 221–39, 525, 534, 537
    cycles, 230, 238
    diameter, 229
    and magnetic fields, 226–30, 233, 239
    numbers, 540
    periodicity, 225–7
    photosphere, penumbra and umbra,
        229
    temperature, 229
    and tree rings, 230–1, 233–4, 237–8
sunstones, 333
suntans, 266–70, 281–3
supernovae, 146–7
Suryagrahana puja rite, 62
Suryavarman II, 46
Sutherland, Graham, 440
swastikas, 507–8n
Swerdlow, Noel, 82, 155
Swift, Jonathan, 166n, 569
    Gulliver's Travels, 341, 372, 565–6,
        569n
Sydney Opera House, 461
Sylvester II, Pope, 140
synaesthesia, 445, 476
Syracuse, 361, 386–7
Syrus, 27
Szilárd, Leó, 213–14

tachocline, 540
Tacitus, 505

Taj Mahal, 460n
Talbot, William Fox, 448–50
Tamur the Lame, King, 130
Tannenberg Memorial, 507
tanning salons, 282–3
Tarim mummies, 116n
Tate Modern, 443n
Tauranac, John, 457
Taverner, John, 470
Taylor, Stephen, 473
telephone, 237
telescopes, 137, 146, 152–3, 155, 156n,
        161, 163, 167, 174, 177–8, 202
    advanced, 532–3, 542–3, 582, 587
    and navigation, 335, 339, 341
    reflecting, 169, 178, 180
    and sunspots, 155, 224–5
television, 137
Teller, Edward, 218
Tellez, Balthazar, 278
Tellier, Charles, 391–2
Tennyson, Alfred Lord, 16, 485, 487
'terraforming', 588
Tertullian the Convert, 104
Tesibius, 361
Teuchitlán pyramids, 47–8
'Tezcatlipoca', 12, 423
Thacker, Jeremy, 341
Thackeray, William Makepeace, 16
Thales of Miletus, 67–8, 88, 91
Thatcher, Margaret, 558
Theodosius, 136
Theophrastus, 98, 222
Thomas, Dylan, 494n
Thomas, Theodore L., 571
Thomson, David, 456
Tiberius, Emperor, 58n, 104, 108, 592
tidal energy, 386
time, 344–84
    daylight saving, 379–82
    days, 325, 359
    hours, 358
    Internet Time, 384
    Islamic year, 348, 357
    leap seconds, 382–3
    measurement of, 358–66
    minutes, 371
    months, 345, 347
    seconds, 382–3
    sidereal months, 110
    solar year, 88, 99, 107, 123
    standardization of, 373–9
    'time immemorial', 357n

tropical year, 126
UTC, 378–9
weeks, 358
see also clocks; sundials
time balls, 360
timekeeping, 131
Timurids, 18
Todd, Mrs D. P., 187
Toffana, Giulia, 262
Tolkien, J. R. R., 140n, 250, 286n, 356n
Tolstoy, Leo
War and Peace, 258
Tower of Babel, 79
Tower of the Four Winds, 353
Trogloditae, 285n
Trojan War, 111, 130n, 330
Tromsø, 291–3, 445
Troy, excavation of, 507n
Truman, Harry S., 216
tsunamis, 238
Turner, Joseph Mallord William, 428–34, 438, 442, 450, 594
Turner, Nat, 189–90
Turner, Tina, 270n
Twain, Mark, 188, 190, 205, 232–3, 431, 568, 600
Huckleberry Finn, 24
twilight, 254–5
Tynan, Kenneth, 474
Tyndall, John, 170n, 244, 551
Tyson, Neil deGrasse, 55, 406–7

ultraviolet light, 170, 234n, 261, 273, 280–1, 284, 289, 300n, 306–7, 316, 319n, 525, 530–1, 533, 548–9
Ulugh Beg, 130
umbrellas, 251–3
universe, 201n, 545
age of, 202, 581
universities, 136–7
University of East Anglia, 563n
Updike, John, 194, 289–90, 494n, 524, 527
Uranus, 178, 471, 588
Urban VIII, Pope, 158–9
Ussher, James, 203–5

Valentino, Rudolph, 266
Valkyries, 58
vampire myth, 286–8
Van Gogh, Vincent, 437–40, 444–5, 593
Varanasi, 61–2, 450, 594–606
Varro, 102n

Vedas, xv, 16, 62, 130–1, 245n, 471
Vega, Garcilaso de la, 15
Velli, Marco, 539, 541–2
Venice, 152–3, 226, 421, 425, 431, 597
Venus, 98, 134, 141, 143, 146, 155, 235, 350, 585, 587
transits, 174–7, 224
Verdi, Giuseppe, 463
Verlaine, Paul, 494–5
Verne, Jules, 59n, 378, 567
Versailles, 414–15, 425
Vespasian, Emperor, 18
Vespucci, Amerigo, 338n
Victoria, Queen, 16, 288n
Vidal, Gore, 269
Vigilius, Pope, 53
Vikings, 332–3
Virgil, 224
Virgin Mary, 28n
visions, 416
vitamins, 284–5
Vitruvius, 364
volcanoes, 238–9, 545, 550–51, 559, 594
Volney, Comte de, 500, 504n, 517
Voltaire (François-Marie Arouet), 166n, 354
Vonnegut, Kurt, 476, 556

Wagner, Richard, 463, 465–7, 503–5
Wald, George, 299n
Wallis, George C., 569
Walton, Ernest, 212
Wang Mang, 113
Wanika people, 249
warfare, 246–7, 258
and solar energy, 386–7
Warhol, Andy, 593
Washington, George, 257, 364
Watchers of the Dawn, 52
watches, 343, 371–2, 375
water desalination, 390
water vapour, 550, 556, 559
Watson, William, 343
Watt, James, 390n
Waugh, Evelyn, 279n, 494n
wave power, 401
Wedgwood, Josiah, 262
Weiner, Jonathan, 261
Wells, H. G., 213, 299n, 588
The Time Machine, 279n, 568–9
Wheeler, Sir Hugh, 260
wheels, 29–30, 76, 417
Whewell, William, 324n

whirlwinds, 228
White, Patrick, 494n
white dwarfs, xvi, 198, 590
Whitehead, Alfred North, 160, 563
Whitehouse, David, 555n
Whitman, Walt, 274, 494n
'widdershins', 53n
Wilde, Oscar, 368, 374, 376, 496
Willett, William, 379, 381
Williams, Tennessee, 494n
Willsie, Henry E., 392
Wilson, E. O., 308
Wilson, Harold, 499n
Wilson, Tom, 284n
wind power, 401
windmills, 125
Winfrey, Oprah, 270n
Wöhrbach, Otto, 396
Wolf, Rudolf, 226–7
Wolfe, Gene, 570
Wolfe, Tom, 588
Wolff, Geoffrey, 492
Wollaston, William, 243
women, and colour, 262–4, 267, 420
Wood, John, the Younger, 460n
Woodhenge, 252n
Woodward, John, 204–5
Woolf, Virginia, 193, 245, 372
Wordsworth, William, 179

Wovoka (Jack Wilson), 190–1
Wren, Sir Christopher, 165, 170–1
Wright, Thomas, 178
Wrigley, E. A., 238n
writing, 76, 83, 346
Wu Hsien, 110

Xerxes, 67, 258
Xihi, 12
Xuan Yuan, the Yellow Emperor,
    106–7
Xue Shenwei, 116n

Yates, Frances, 484
Yeats, W. B., 488, 494, 504
yin and yang, 32, 111–12, 115, 515n
yoga, 417
Yule, 29, 31

Zalda festival, 33
Zeno of Elea, 87
Zhāng Hêng, 113–15, 566
ziggurats, 79
Zinneman, Fred, 360n
zodiac, 77, 82, 84, 109n, 480
Zola, Émile
    Germinal, 35
Zoroastrianism, 33, 508
Zulus, 249

NOTE ON THE AUTHOR

Richard Cohen is a former publishing director of Hutchinson and Hodder & Stoughton and the founder of Richard Cohen Books. The acclaimed author of *By the Sword: A History of Gladiators, Musketeers, Samurai, Swashbucklers and Olympic Champions*, he has written for *The Times, Guardian, Independent* and most other leading London newspapers, and has appeared on BBC radio and television. He lives in New York City.